Chemical Biology of Neurodegeneration

Chemical Biology of Neurodegeneration

A Molecular Approach

Pedro Merino

WILEY-VCH

Author

Pedro Merino
Universidad de Zaragoza
Departamento de Qumica Orgánica
Campus San Francisco
Zaragoza
E-50009 Aragón
Spain

Cover Images Images were kindly provided by Pedro Merino (University of Zaragoza, Spain)

Library of Congress Card No.: applied for

British Library Cataloguing-in-Publication Data
A catalogue record for this book is available from the British Library.

Bibliographic information published by the Deutsche Nationalbibliothek
The Deutsche Nationalbibliothek lists this publication in the Deutsche Nationalbibliografie; detailed bibliographic data are available on the Internet at <http://dnb.d-nb.de>.

Print ISBN: 978-3-527-34424-6
ePDF ISBN: 978-3-527-81343-8
ePub ISBN: 978-3-527-81345-2
oBook ISBN: 978-3-527-81342-1

Typesetting SPi Global, Chennai, India
Printing and Binding Markono Print Media Pte Ltd, Singapore
Printed on acid-free paper

10 9 8 7 6 5 4 3 2 1

Contents

Preface

Neurodegeneration is the progressive loss of structure or function of neurons and causes many neurodegenerative disorders. Diseases affecting the central nervous system (CNS) that implicate directly not neurons (e.g. multiple sclerosis) or those in which neurons die because of acute episodes (infections, hypoxia, edema, trauma, etc.) are not considered neurodegenerative. Several types of neurodegenerative diseases have been reported, and many of them have a clinic and pathologic overlap. In general, they are differentiated by the predominant region affected in the brain although a more detailed classification is made on the basis of symptoms and predominant clinical features [1].

The main neurodegenerative disorders are Alzheimer's disease (AD), Parkinson's disease (PD), Huntington's disease (HD), Lewis body dementia (LBD), and amyotrophic lateral sclerosis (ALS). Almost all neurodegenerative disorders have an origin unknown, and only devastating effects are observed. Neurodegenerative disorders are age associated, and the etiology of neurodegenerative diseases regarding the role of the different risk factors, especially positive family history, unhealthy lifestyle, and toxic environment, is still under debate [2]. With few exceptions, a combination of both causes (family story and toxic environment) should be considered. Recent advances on mechanisms underlying neurodegenerative diseases focus on neurotoxicity, neurovascular dysfunction, accumulation of misfolded proteins, and selective vulnerability of certain neurons and other brain cells such as microglia [3].

Currently, treatment of symptoms is the only therapeutic approach to neurodegenerative diseases. Prior to search a cure, it is necessary to understand the chemical processes involved in the diseases, and for that goal, the molecular structures involved in the affected brain areas also need to be identified. This book focuses on the main actors of CNS involved in neurodegenerative disorders at a molecular level with great emphasis on structural aspects and modes of action.

The first part introduces the neurotransmitter systems. Clinical aspects are reduced to a minimum since there are many exclusive books and treatises on this topic. In this respect, this part contains an introduction on the neurotransmitter system including a general explanation of the synapse and a description of the main structures involved in neurotransmission that can be considered therapeutic targets for neurodisorders, including receptors and transporters. Pharmacological features (of special interest in characterization of receptors) are not considered in detail because books and encyclopedia collecting all details for each receptor are available. Neurotransmitter systems are deeply discussed to give the reader a general overview. All aspects are not treated in detail, because several volumes need to be devoted for those topics alone.

The second part presents molecular and chemical aspects directly involved in neurodegeneration including metabolism of neurotransmitters, enzymes processing neurotransmitters, protein misfolding, and therapeutic agents.

References

1 Przedborski, S., Vila, M., and Jackson-Lewis, V. (2003). Neurodegeneration: what is it and where are we? *J. Clin. Invest.* 111 (1): 3–10.
2 Farooqui, T. and Farooqui, A.A. (2009). Aging: an important factor for the pathogenesis of neurodegenerative diseases. *Mech. Ageing Dev.* 130 (4): 203–215.
3 Issue, S. (2018). Focus on neurodegenerative disease. *Nat. Neurosci.* 21 (10): 21.

Part I

Neurotransmitter Systems

1

The Dopaminergic System

1.1 Introduction

The dopaminergic system can be defined as a group of nerve cells (dopaminergic neurons) mainly located in the midbrain that extend their axons to different sections of the forebrain. Five pathways of dopaminergic system are identified [1] (Figure 1.1): (i) *Nigrostriatal*, going from the substantia nigra to dorsal striatum, involved in facilitating movements, (ii) *Mesocortical*, going to the frontal lobes, particularly the prefrontal cortex, involved in modulating cognitive function, (iii) *Mesolimbic*, innervating the ventral striatum (nucleus accumbens), involved in emotions and reward, (iv) *Tuberoinfundibular*, located in the arcuate nucleus of the hypothalamus and controlling prolactin secretion from the anterior pituitary gland, and (v) *Thalamic*, described in primates. It innervates the thalamus, and its function is not well known although it is involved in sleep and arousal mechanisms.

The central dopaminergic neurons can be divided into ascending, descending, and local neuron systems [2] and are directly implicated in Parkinson's disease (PD) [3] and related complications such as depression, [4] schizophrenia [5], and several neuroendocrine disorders. Depending on the affected pathway, a different neurological disorder can arise. The most known, affecting 2% of population, is PD resulting from the death of neurons in the substantia nigra (nigrostriatal pathway) [6, 7] and, consequently, causing movement disorders such as shaking, rigidity, slowness, and difficulty with walking. Chorea and addiction are also related to nigrostriatal pathway. Schizophrenia is related to mesocorticolimbic projection; in particular, mesocortical and mesolimbic pathways are related to positive and negative symptoms of the pathology, respectively. The mesocorticolimbic pathways are also linked to addiction and attention-deficit hyperactivity disorder (ADHD). Finally, the tuberoinfundibular pathway is affected in hypoprolactinemia.

At a molecular level, the main actors of the dopaminergic system are the neurotransmitter dopamine (4-(2-aminoethyl)benzene-1,2-diol; DA) and dopamine receptors (DRs). DA is biosynthesized from the amino acid tyrosine by the action of tyrosine hydroxylase, which forms dihydroxyphenylalanine (DOPA) that is decarboxylated by DOPA decarboxylase to yield DA (Figure 1.1).

The DRs play a crucial role in DA signaling since they are responsible for the signal transduction. In fact, their drug-mediated stimulation can compensate the

Chemical Biology of Neurodegeneration: A Molecular Approach, First Edition. Pedro Merino.

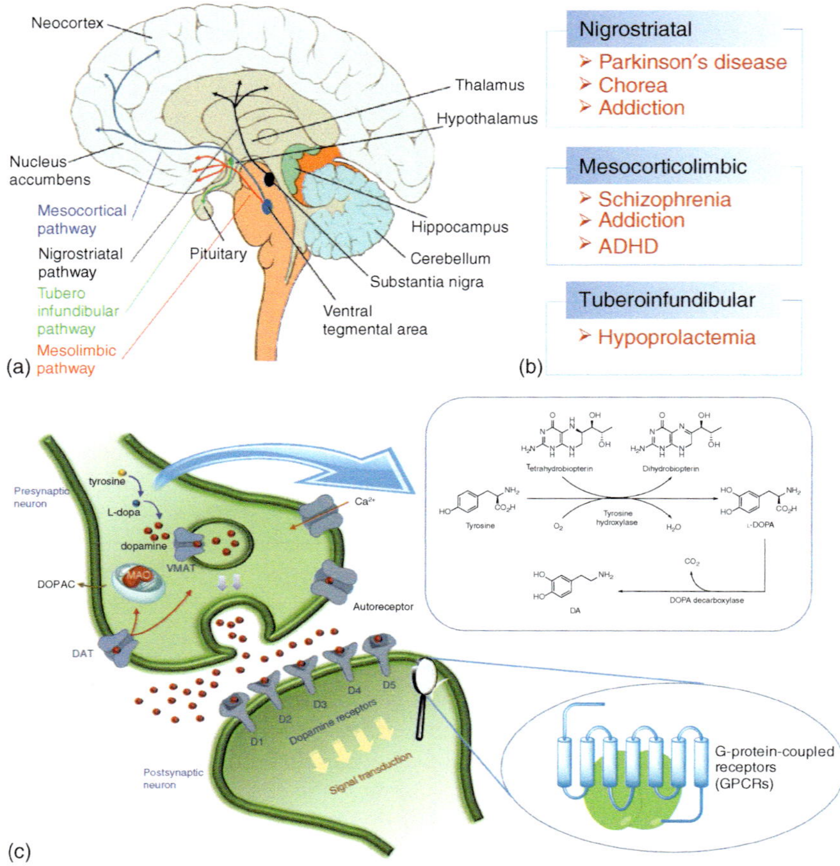

Figure 1.1 The dopaminergic system. (a) Pathways of dopaminergic system (only the four identified in humans are shown). (b) Neurological disorders associated with the pathways. (c) Dopaminergic synapse. Dopamine (DA) is synthesized from tyrosine inside the dopaminergic neuron. After packaging in vesicles by VMAT, DA is released into the synapse. DA is recognized by dopamine receptors that produce the signal. The excess of DA is removed from the synaptic area by DAT and reintroduced in the neuron for being reused or catabolized by mitochondrial MAO. VMAT: vesicular monoamine transporter; DAT: dopamine transporter; D_1–D_5: dopamine receptors; MAO: monoamine oxidase; DOPAC: 3,4-dihydroxyphenylacetic acid.

lack of DA in the first stages of neurodisorders such as PD [8, 9]. DRs are part of the catecholamine-binding family of G-protein-coupled receptors [10]. The superfamily of receptors are membrane proteins for which several computational predictions of their structures have been reported [11]. Recently, some X-ray structures have been resolved, but none are related to dopamine receptors [12, 13]. Biological aspects of DRs related to selective drugs, modulation of signaling, and trafficking have been collected in a volume edited by Neve in 2010 [14]. The treatise also includes links between DRs and neurodisorders such as schizophrenia, PD, and other neuropsychiatric disorders.

Once DA is synthesized inside dopaminergic neurons, it is packed into synaptic vesicles by a vesicular monoamine transporter (VMAT) and then released at nerve terminals into the synapse (Figure 1.1). The released DA binds to DRs to produce a signal in the postsynaptic neuron. However, DRs are widely expressed away from dopamine synapses; hence it is not evident how dopamine synapses are involved in dopaminergic transmission. Recent studies have demonstrated that dopamine synapse is a contact formed between dopaminergic presynaptic and GABAergic postsynaptic structures [15]. After the signal transduction, to stop signaling, dopamine is removed from the synaptic cleft by the dopamine transporter (DAT) that reenters DA to the presynaptic neuron for repackaging into vesicles. DA accumulated in the cytosol is recycled to form 3,4-dihydroxyphenylacetic acid, DOPAC, by the action of monoamine oxidase, MAO.

1.2 Dopamine Receptors

1.2.1 Classification

Dopamine receptors are classified into two groups according to initial pharmacological and biochemical studies [16]. These studies suggested two types of dopamine receptors: (D_1) activating adenylyl cyclase (AC) (the enzyme that converts adenosine triphosphate, ATP, into cyclic adenosine monophosphate, cAMP) and the other inhibiting the enzyme (D_2) [17]. Further analysis of cloned full-length cDNAs by several groups [18–21] clearly identified the D_1 receptor. Further cloning revealed two additional receptors named D_3 [22] and D_4 [23]. According to their pharmacology and structure, the two new receptors were included in the initially assigned D_2-type. At the same time, a novel D_1-like receptor (based on its sequence and pharmacological profile) was cloned and identified as D_5 [24, 25]. So, two groups are defined on the basis of their ability to activate/inhibit adenylyl cyclase: D_1-like receptors and D_2-like receptors [26]. The D_1-like group includes D_1 and D_5 receptors, which in some texts are referred to as D_{1A} and D_{1B}, respectively. These receptors present very high homology in their transmembrane domains and ligand-binding features. The D_2-like group comprising D_2, D_3, and D_4 receptors also shares a similar homology between them. Genetics also serves to identify the two groups of dopamine receptors. While D_1-like receptor genes are intronless [27], the genes encoding D_2-like receptors are interrupted by introns [28]. The presence of introns leads to the generation of splice variants and thus there are two D_2 receptor isoforms: D_{2L} and D_{2S} [29], different nonfunctional isoforms of D_3 [30] and polymorphic forms within the coding sequence of D_4 receptor [31].

1.2.2 Location

Dopamine receptors are localized in different parts of the brain. D_1 receptors are the most widespread and expressed than any other receptor in the central nervous system (CNS) [32]. Associated with chromosome 5, they are highly expressed in the caudate putamen, the nucleus accumbens, the zona reticulata of substantia nigra, and the olfactory tubercle. Lower expression of D_1 receptors has been found in the dorsolateral prefrontal cortex, cingulate cortex, habenular, and hippocampus [33]. D_5 receptors are associated with chromosome 4 and expressed in the hippocampus, cerebral cortex, and hypothalamus, so they are related to pain stimuli. Actually, D_1 and D_5 receptors cannot be distinguished by radioligand binding, the usual methodology for locating dopamine receptors; hence, their distribution has been determined using receptor-selective antibodies [34]. D_5 receptors are predominant in pyramidal cells of the hippocampus and frontal cortex, and they are also found in the cerebellum [35]. D_2 receptors, associated with chromosomes 6 and 7, are preferentially located in the zona compacta of the substantia nigra. Other locations include glomerular layer of the olfactory bulb, bed nucleus of the stria terminalis, hypothalamus, habenula, hippocampus, and the intermediate lobe of pituitary [36]. D_3 receptors, associated with chromosome 3, are mainly expressed in two regions of the limbic system called the islands of Calleja and nucleus accumbens as well as the olfactory tubercle [37, 38]. D_4 receptors are associated with chromosome 11, and they are mainly located in the hippocampus and caudate putamen [39]; less predominant but in similar amounts they are also located at olfactory tubercle and substantia nigra. So, D_4 receptors are associated with both limbic structures and motor areas [40].

1.2.3 General Structural Considerations

Dopamine receptors such as other G-protein-coupled receptors (GPCRs) are characterized by seven transmembrane domains with three intracellular loops (il_{1-3}) and three extracellular loops (ol_{1-3}), the amino and carbon termini being extracellular and cytoplasmic, respectively [41]. Despite the similarity in the transmembrane regions, there are several structural differences between D_1-like and D_2-like receptors. In D_1-like receptors, the C-terminal tail is about seven times longer and the third intracellular loop is shorter, compared to D_2-like receptors (Figure 1.2) [42, 43]. In both D_1-like and D_2-like receptors, the C-terminal tail is rich in threonine and serine residues, containing a cysteine residue. Figure 1.2 also collects the amino acid sequence for the five dopamine receptors. Dopamine receptors are subject to posttranslational modifications including glycosylation in Asn residues, palmitoylation in Cys residues, and phosphorylation in Ser or Thr residues (Figure 1.2).

The presence of isoforms for dopamine receptors furnishes a wider range of choices to couple to G-proteins. Due to this diversity along with the heterotrimeric nature of the G-proteins themselves (Figure 1.2, a,b), Dopamine receptors can couple different families of G-proteins [44]. Indeed, both types of dopamine receptors form dimers and oligomers. Examples of homo- and hetero-oligomerization are reported in several studies. Homo-oligomerization of D_1 [45] and D_3 [46] dopamine receptors has been demonstrated. The formation of homo-oligomers for D_2 receptors has also been evidenced by ligand binding studies [47], cysteine cross-linking [48], and visualization of oligomeric complexes by Western blotting [49]. Hetero-oligomers of D_2 receptors with D_1 [50] and D_3 [51] have been demonstrated *in vitro*. Heteromers D_1–D_3 have been studied in detail as a target for the treatment of schizophrenia and motor dysfunctions

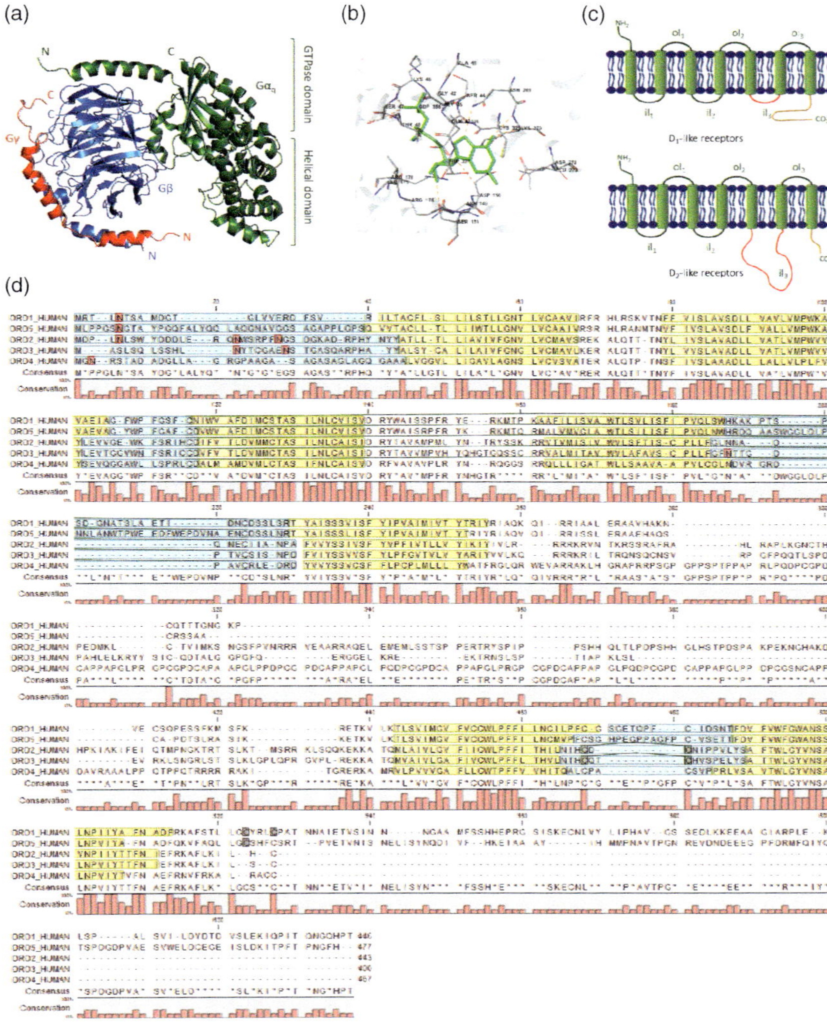

Figure 1.2 (a) Overall structure of the $G\alpha_{i/q}\beta\gamma$ heterotrimer (PDB ID: 3AH8). The heterotrimer is viewed with the expected orientation at the plasma membrane. $G\alpha_{i/q}$ (green) consists of the GTPase and the helical domains. Gβ and Gγ are blue and red, respectively. (b) Details on the interactions between GDP (green) and residues of $G\alpha_{i/q}$ (PDB ID: 1GG2). (c) General schematic structures of D_1- and D_2-like dopamine receptors. Differences are highlighted for the third intracellular loop (red) and the C-terminal tail (brown). (d) Sequence alignment of the five dopamine receptors. Conservation graphic and consensus sequence are shown. Disulfide bonds (green), N-glycosylation sites (red), palmitoylation sites (brown), extracellular domains (light blue), intracellular domains (white), and transmembrane domains (yellow) are indicated on sequences.

[52]. D_2 receptors also form hetero-oligomers with other receptors such as CB_1 cannabinoid receptor [53], A_{2A} adenosine receptor [54], and somatostatin receptor subtype SSTR5 [55]. D_1 receptors also form hetero-oligomers with A_1 adenosine [56] and glutamate *N*-methyl-D-aspartate (NMDA) receptors [57].

1.2.4 Effector Mechanisms

Dopamine receptors activate multiple and diverse signaling pathways through selective coupling to different G-proteins. Indeed, both D_1-like and D_2-like dopamine receptors have a "typical" GPCR structure, but differ in that the D_1 receptor has a smaller third intracellular loop (il_3) but a longer intracellular C-terminal tail compared to the D_2 receptor. GPCRs are a superfamily of membrane proteins, which sense diverse molecules outside of the cell and transfer the information inside the cell where they regulate a number of signaling pathways [58]. GPCRs have been extensively studied, and several comprehensive dedicated treatises have been published elsewhere [59–61]. A number of various volumes published are dedicated to particular aspects such as ligand design [62], pharmaceutical targets [63] and drug discovery [64, 65], pharmacology [66, 67], modeling [68], methods [69], structure and function [70], and their relationship with neurobiology [71], and related diseases [72]. The reader is directed to these volumes for detailed information on GPCRs.

Regarding effector mechanisms, D_1-type receptors couple to G_s (and G_{olf} in the striatum), which causes activation of adenylyl cyclase and consequently, increasing cAMP formation, which activates protein kinase A (PKA) (Figure 1.3). On the contrary, D_2-type receptors couple to $G_{i/o}$ guanosine triphosphate (GTP)-binding proteins inhibiting cAMP formation and dissociating $\beta\gamma$ subunits from $G_{i/o}$. The dissociated $\beta\gamma$ subunits inhibit voltage-dependent calcium channels (VDCCs) and activate voltage-sensitive potassium channels G-protein-coupled inwardly rectifying potassium (GIRKs) as well as a phospholipase C isozyme, resulting in an increase in intracellular calcium, which leads to activation of kinases and phosphatases including mitogen-activated protein kinase (MAPK), protein kinase C (PKC), and calmodulin-stimulated enzymes such as calcium/

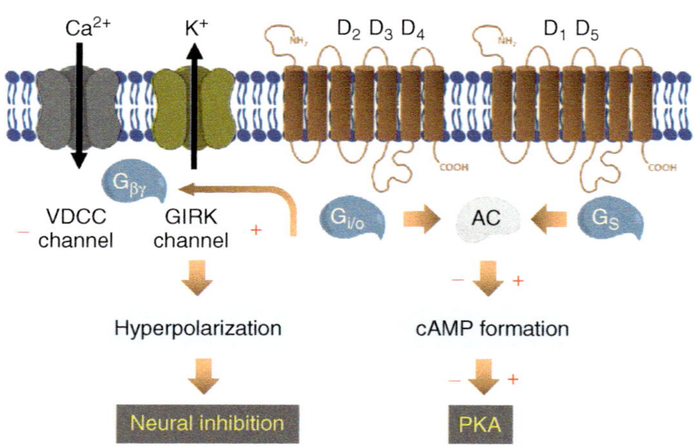

Figure 1.3 Signal transduction of dopaminergic receptors via coupling to individual G-proteins (e.g. $G_{i/o}$, G_s, and G_q). AC: adenylyl cyclase; GIRK: G-protein-coupled inwardly rectifying potassium; VDCC: voltage-dependent calcium channel; PLC: phospholipase C; PKA: protein kinase A.

calmodulin-stimulated protein kinases (CaMK), as well as protein phosphatase-2B (PP-2B).

1.2.5 Implication in Neurodisorders

The complexity of dopamine signaling – largely conserved in both invertebrates and vertebrates – is increased as it can be affected by minor alterations in receptor/G-protein coupling. Such alterations may be responsible for imbalanced signaling detected in dopamine-related disorders such as schizophrenia, PD, and ADHD. Even though both D_1-like and D_2-like dopamine receptors are important for dopamine function in basal ganglia and frontal cortical areas, D_2-like receptors are the ones targeted therapeutically by antipsychotic and anti-Parkinson's disease drugs. In fact, D_2-like receptors are involved in several neurodisorders including PD, Tourette's syndrome, schizophrenia, and hyperprolactinemia. D_2-like receptors are blocked in varying degrees by antipsychotic drugs. D_1, D_2, and D_3 receptors are directly involved in motor activity. D_1 and D_2 receptors are also involved in reward and reinforcing properties of different drugs of abuse as well as in learning and memory [26]. D_2 receptor is directly involved in psychiatric disorders, especially schizophrenia [73], and D_3 receptor has been suggested as target for the treatment of depression [74]. In general, D_2, D_3, and D_4 receptors have similar pharmacology, although subtle differences are emerging with the development of more specific ligands; for instance, D_4 receptors have a particularly high affinity for the antipsychotic drug clozapine. All three subtypes have postsynaptic locations in dopamine terminal areas.

1.2.6 D_1-like Receptors

D_1-like receptors couple to G_S and α-units of G_{olf} in the striatum and stimulate adenylate cyclase to increase the concentration of intracellular cAMP [75]. This coupling has been demonstrated in neostriatum, where $G\alpha_{olf}$ is abundant and expression of D_1 is high, whereas expression of $G\alpha_s$ is low [76]. Similarly, in basal ganglia nuclei, the $G\alpha_{olf}$ is the primary G-protein coupled to D_1Rs [75]. Additional *in vitro* studies indicated that D_1 receptors can also couple to other proteins such as $G\alpha_z$, $G\alpha_o$, $G\alpha_{i1}$, and $G\alpha_{i2}$ [77]. Details of the interactions between D_1 receptors and G-proteins have been obtained using several fluorescence techniques. In the absence of receptor activation, no preassociated complexes were detected, which suggests that D_1 receptors alter the distribution of $G\alpha_s$ and $G\alpha_{i3}$ subunits inside the membrane [78]. In PD patients, $G\alpha_{olf}$ is highly expressed in striatum, where dopamine is deficient; on the contrary, $G\alpha_s$ is less abundant. D_1 receptors are also involved in the regulation of calcium, sodium, and potassium channels in striatal and prefrontal cortical neurons. The intracellular Ca^{2+} concentration is a key parameter for neurotransmitter release and synaptic plasticity [79]. However, the mechanism through which D_1-mediated Ca^{2+} current regulates neuronal activity is still unknown.

The structure of D_1 receptors is common for all the mammals having 446 residues. The general structure for D_1-like receptors is shown in Figure 1.3. Potential sites for N-glycosylation (Asn5 at the amino terminus and Asn175 in the third extracellular domain) have been identified, but no effect of glycosylation on the function (or expression) has been revealed [80]. Phosphorylation by cyclic-adenosine monophosphate (c-AMP)-dependent kinase (PKA) takes place at (i) Thr136 in the second cytoplasmic loop, within the sequence R-K-X-T, (ii) Thr268, at the C-terminal end of the third cytoplasmic loop, within the less commonly phosphorylated sequence R-X-T, and (iii) Ser380 (R-X-S) [81]. Most isoforms, but not all, of PKC can phosphorylate D_1 receptors, regulating signaling [82]. Phosphorylation by the GPCR kinase family is also possible at multiple Ser and Thr residues present in the long cytoplasmic tail in proximity to acidic residues. At Cys347 and Cys351, the polypeptide is palmitoylated [83]. Notably, when palmitoylation is blocked by mutation of Cys347 and Cys351, D_1 receptor activation of G-proteins is not affected [84]. On the other hand, replacement of Leu344 and Leu345 at C-terminus by alanines decreased glycosylation, suggesting an important role in plasma membrane trafficking for such di-leucine motif [85]. The N-terminus plays a crucial role in defining the specific ligand (D_1 or D_5) affinity and ligand-induced adenylate cyclase activation [86].

The difficulty in obtaining suitable crystals for GPCRs, for which the only available high-resolution crystal structure is that of bovine rhodopsin [87], has limited the availability of three-dimensional structures of D_1-like receptors. To gain insights into dopamine receptors function and structure, homology-modeling techniques based on the crystal structure of bovine rhodopsin have been used [88–91]. However, further studies demonstrated that due to the low sequence identity of GPCRs with rhodopsin (less than 20%), homology-based approaches provide unreliable three-dimensional structures to be used for making predictions [92]. Actually, homology models based on bovine rhodopsin provide good correlation for antagonists but not for agonists [93]. Other authors have used MembStruk [94, 95] and HierDock [96] methods for prediction of structure and ligand-binding sites of several dopamine receptors [97–99]. In particular, the structure of D_1 dopamine receptors has been predicted based on the modeled D_2 structure using the MembStruk procedure. Seven transmembrane helices of 19–29 residues per helix were predicted (Figure 1.4) [101].

The binding site of dopamine was determined using the HierDock procedure [97], and selective pharmacophore models have been developed [102]. Several interactions between dopamine and D_1 receptors have been reported. At physiological pH condition, the amine functionality is protonated, and it is known it interacts with receptors [103]. In particular, dopamine forms a salt bridge with the highly conserved Asp103 located in transmembrane segment 3 (TM3) [104]. The catechol moiety of dopamine forms H-bonds with conserved Ser198, Ser199, and Ser202 located in TM5 [105, 106]. In fact, any mutation from serine to alanine in that position reduces activity of dopamine and catechol-agonists of D_1 receptors [107]. The conserved aromatic residues Trp285, Phe289, Ph296, and Phe297 have been suggested to participate in π-interactions playing an important role in D_1 activation [103]; however, it has not been tested with D_1 receptors yet. The interaction of D_1 receptors with non-catechol derivatives has been studied with ergoline-based ligands [108], which are known to have antipsychotic properties [109]. Conserved aromatic residues at TM6 are crucial for binding affinity of ergolines [110].

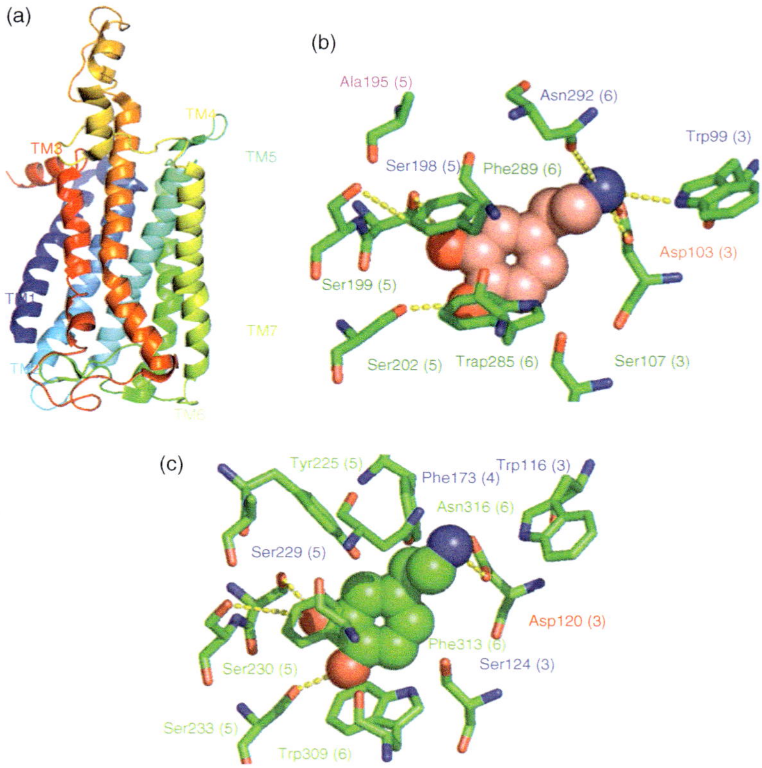

Figure 1.4 (a) Predicted structure for D_1 dopamine receptor. The seven transmembrane helices are given in different colors. (b, c) Predicted binding sites and interactions for dopamine in D_1 (b) and D_5 (c) receptors (numbers in the brackets indicate the TM helices to which the residues belong to). Source: (b, c) Reproduced with permission from Kalani 2004 [100]. Copyright 2004, California Institute of Technology.

D_5 receptors have 477 residues, and although they are genetically different from D_1 receptors, both receptors share c. 60% amino acid identity overall and 82% identity in the transmembrane-spanning regions [111]. D_5 receptors couple to $G\alpha_s$ [112] and in a minor extent to $G\alpha_{olf}$ [113, 114]. Binding of dopamine or an agonist induces a conformational change that promotes interaction and, thus, activation of $G\alpha_s$. Through this interaction, exchange of guanosine diphosphate (GDP) for GTP is facilitated, leading to dissociation of the α and β-γ subunits of G-proteins and stimulation of adenylate cyclase with subsequent production of c-AMP and activation of PKA. There is also further evidence for coupling of D_5 with $G\alpha_z$ [114], suggesting various signaling responses upon simultaneous stimulation of both D_1 and D_5 receptors [113].

Despite the structural and pharmacological similarity between D_1 and D_5 receptors, there are functional differences in the cellular expression and distribution, indicating that they cannot reconstitute each other when a pathology is present [115]. However, until now there are no highly selective D_1 or D_5 receptor ligands, dopamine being the most discriminating (D_5 receptors have a higher affinity for DA than D_1 receptors); consequently, their design constitutes a current challenge [116]. Recent studies have demonstrated that D_5 stimulation may enable cells to better support adjacent neurons [117]. The structure of D_5 dopamine receptors has been modeled based on the template of high-resolution crystal structure of human β_2 adrenergic receptor (PDB ID: 2RH1A) [118]. The predicted binding sites included all the seven transmembrane helices corresponding to the general structure outlined in Figure 1.4. Sites for N-glycosylation are found in first and third extracellular (ol_3) domains. While glycosylation at Asn7 and Asn198 is required for receptor trafficking [119], it is not necessary for ligand binding [80]. The palmitoylation sites of D_5 are conserved from D_1 receptors. Phosphorylation by PKA and G-protein receptor kinases takes place at the third cytoplasmic loop (il_3) and carboxy-terminal tail, respectively. The D_5 dopamine receptors form a complex with the GABA-A ligand-gated ion channel, leading to reciprocal inhibition of both inhibitors [120]. Deficiencies in D_5 receptors are associated with learning problems directly connected to ADHD [121]. Polymorphism of D_5 receptors is associated with schizophrenia [122] but not with PD [123].

As in the case of D_1 receptors, the structure of D_5 receptors has been predicted using the MembStruk procedure. The binding site of dopamine has been determined using the HierDock procedure and a library of 11 ligands has been studied [124]. Seven helical motifs were identified and very similar to those of D_1 receptors. From the predicted structure, docked conformation of dopamine showed a stabilizing salt bridge with Asp120 located at TM3. In addition, a network of H-bonds formed by Ser229, Ser230, and Ser233 located at TM5 was found to interact with catechol moiety. Up to eight residues (Trp116, Ser124, Phe173, Tyr225, Ala226, Trp309, Phe313, and Asn316) were identified in the binding site as candidates to constitute a hydrophobic pocket. A model based on the X-ray structure of human β_2 adrenergic receptor led to very similar results in predicting binding sites [118].

1.2.7 D_2-like Receptors

D_2-like receptors couple to heterotrimeric $G\alpha_i$ or $G\alpha_o$ GTP-binding proteins to inhibit adenylate cyclase activity, reducing the concentration of cAMP, and modulate potassium and calcium ion channel activities. The large third intracellular loop il_3 – essential for G-coupling [125–127] – is a characteristic of all D_2-like receptors. Depending on the receptor subtype, different subunits are coupled. Several studies on coupling between D_{2S} and D_{2L} receptors and G-proteins have been carried out. Depending on the studies, evidences on preferential coupling with $G\alpha_{i2}$, $G\alpha_{i2}$, $G\alpha_{i3}$, or $G\alpha_{o1}$ subunits have been observed [128–130], the latter being the most strongly activated subtype [131]. Initially, D_3 receptor was reported not to couple G-proteins [132]; however, it has been demonstrated in some cell lines that it can inhibit adenylate cyclase [133], suggesting coupling with $G\alpha_q$ or $G\alpha_{11}$ subunits [134]. The human D_4 receptor activates $G\alpha_{i2}$, $G\alpha_{i3}$, $G\alpha_{o1}$ subunits in a similar way to D_2 [135]. The rat D_4 receptor preferentially couples to $G\alpha_z$ and the transducin subtype $G\alpha_{t2}$ [136].

As in D_1-like receptors, in D_2-like receptors there are potential sites for N-glycosylation and phosphorylation although the regulation of the latter is not as well understood as that of the former. Nevertheless, it is known that phosphorylation is exerted by second-messenger activated and GPCR-specific protein kinases, and it has been demonstrated for D_2 and D_3 subtypes. In D_2-like receptors, the short C-terminus ends with a region containing a cysteine residue, conserved in many GPCRs, and identified as the palmitoylation site.

The two isoforms of D_2 receptor, D_{2S} and D_{2L}, differ in the insertion of a stretch of 29 amino acids in the third intracellular loop il_3. In general, the D_{2L} isoform is predominant over the D_{2S} isoform although the latter is the sole autoreceptor on dopaminergic neurons [137]. A third isoform (longer) has been identified in which Val270 is replaced by the sequence Val-Val-Gln.

Initial studies using predicted structures suggested that highly conserved residues within the hydrophobic TM domains define a binding pocket corresponding to the agonist-binding site [97]. In particular, the carboxyl group of Asp114 (a residue located in TM3) forms a salt bridge with the amino group of dopamine, as demonstrated by mutagenesis [138]. Hydroxyl groups at meta- and para-positions in dopamine form hydrogen bonds with Ser193 and Ser197 (located at TM5), respectively. It has been hypothesized that Ser194 might be an alternative to Ser193 for a slightly different conformation resulting from activation. Residues Asp114, Ser193, Ser194, and Ser197 are conserved in all five human dopamine receptors. A hydrophobic pocket for dopamine is formed in D_{2L} receptor by residues Phe110, Met117, Cys118, Phe164, Phe189, Val190, Trp386, Phe390, and His394 [97]. Transmembrane proline residues are highly conserved in all GPCRs, particularly Pro201, Pro388, and Pro423, the most conserved residues in class A GPCRs. Those three prolines as well as Pro89 and Pro169 have been located in the transmembrane region of the D_2 receptor, suggesting an important role in its function by eliminating a backbone hydrogen donor when necessary [139].

The structure of D_2 receptors has been predicted from primary sequence using the MembStruk procedure [97]. The predicted structure has been validated against experimental evidences on affinities of dopamine, three agonists (antiparkinsonian) and seven antagonists (antipsychotic), and correctly predicts the critical residues for binding dopamine (Figure 1.5a).

Differences between the binding sites of agonists and antagonists have also been identified. The different interactions for class I and II antagonists, represented by clozapine and haloperidol, respectively, are illustrated in Figure 1.5b,c.

Binding interactions of dopamine and the agonist apomorphine with high and low states of human D_2 receptor have been determined using computational and experimental techniques. Monomeric and dimeric forms of D_2 were considered. The observed results showed that dimerization of D_2 receptors have negative cooperative effect on agonist ligand binding [140].

The only X-ray structure resolved in which D_2 receptor is involved corresponds to the neuronal calcium sensor-1 (NCS-1) in complex with two peptides derived from the receptor [141]. Although structural information on the receptor cannot be obtained, the structure provides very important information about

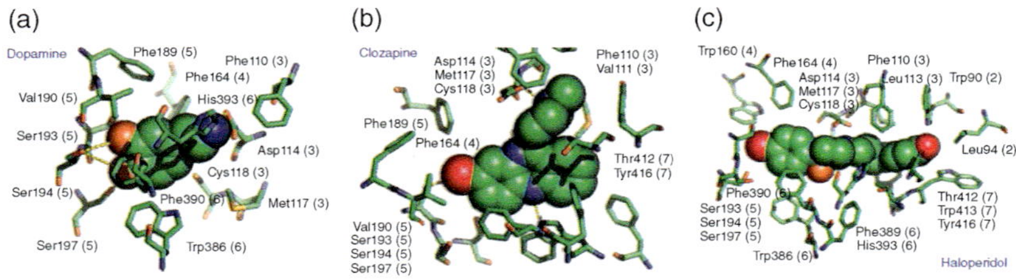

Figure 1.5 Residues within 5.5 Å of dopamine (a), clozapine (b), and haloperidol (c) bound to human D_2 receptor. Source: Reproduced with permission from Kalani et al. 2004 [97]. Copyright 2004, National Academy of Sciences, USA.

binding of D_2 with NCS-1. Two peptides (*a* and *b*) are bound as amphipathic helices with their C-termini pointing toward the center of NCS-1 in which a conformational change is induced (Figure 1.6a). The same residues are involved in both peptides in hydrophobic interactions with NCS-1. These residues are Ile431, Phe433, Ala436, Phe437, Ile440, and Leu441. In peptide *b* Leu438 is also involved (Figure 1.6b). The two copies of D2R peptides interact between them through hydrophobic interactions (*a*Leu438 and *b*Leu441) and hydrogen bonds (*a*Glu431 and *b*Ser433; *a*Lys435 and *b*His442; *a*His442 and *b*Arg434, and *a*Ser433 and *b*Arg434).

The murine D_3 receptor also presents two functional isoforms, short and long, similarly to the D_2 receptor [30]. Although such splice variation has not been found in humans, nonfunctional splice variants resulting in truncated chains have been elucidated. The variant $D3_{nf}$ has been suggested to be involved in schizophrenia [142]. Polymorphism has also been identified for D_3 receptor but no alterations in the function are detected. Some studies, however, suggested a relationship between schizophrenia and a polymorphism consisting of a Ser to Gly change in the ninth residue of the amino terminus [143].

The only full X-ray structure of a human dopamine receptor has been resolved for D_3 receptor in complex with the antagonist eticlopride [144]. The main fold of D_3 receptor consists of the seven-TM bundle of α-helices characteristic of GPCRs (Figure 1.6c). The extracellular region has high sequence diversity among the GPCR family, and two different conformations of intracellular loop il_2 are observed. Starting from the crystal structure of D_3 receptor, a homology model of D_2 receptor was created. A comparison of both models revealed subtle differences that might be the origin of the different selectivity between the receptors.

The utility and reliability of homology models has been further demonstrated through studies carried out with D_3 receptor in which the docking screen against the homology model was no less effective than that against the crystal structure [145]. Additional molecular dynamics refinement contributes to a better approach to the three-dimensional structure of the receptor [146] as it has been demonstrated for both D_1 and D_2 receptors [147] and GPCRs in general [148, 149]. These studies concluded that the homology models suit at least as well as the X-ray structure, thus being excellent surrogates to aid in drug discovery, [150] particularly for GPCRs [99, 151]. The resolution of the crystal structure of

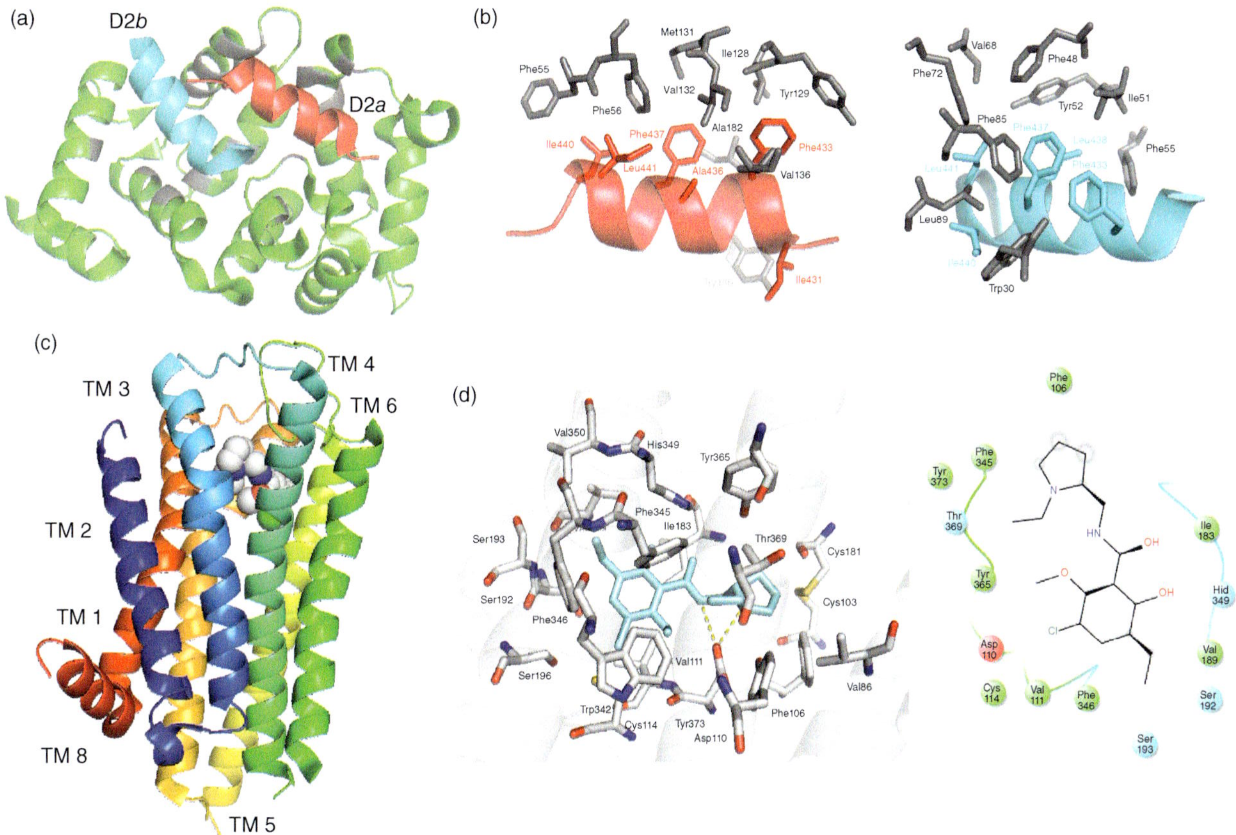

Figure 1.6 (a) Cartoon representation of NCS-1 in complex with two D_2 receptor chains bound independently. (b) Details on hydrophobic interactions between NCS-1 and D_2 receptor chains a (red) and b (cyan). (c) Crystal structure of D_3 receptor in complex with eticlopride (space-filling representation). (d) Details on interactions of eticlopride (cyan) at the binding site.

D_3 receptor has promoted further advances with D_2 receptor using reliable homology models [152]. These models served for determining key receptor–ligand interactions of several well-known dopaminergic compounds such as piperazines [153] and have been used as templates for antipsychotic virtual screening [154].

The functional role of various residues located at TM3 has been studied with homology models derived from D_3 X-ray structure. It was concluded that a microdomain packed by aromatic–aromatic interactions and connecting helices 3, 5, and 6 forms a barrier that prevents dopamine from binding further toward the intracellular surface [155]. At a molecular level, a sulfur–π interaction is responsible for those interactions. Homology models have been extended to D_2-Gα_i complex in the presence of dopamine and compared with aripiprazole-type partial agonists. It has been found that agonists induced different conformations of important structural motifs, including the extracellular loop regions, the binding pocket and, more importantly, intracellular G-protein-binding domains. Consequently, partial agonist efficacy is caused by an impaired coupling between the receptor and G-proteins [156]. Molecular dynamics of D_2 receptor interaction with the GIPC1-PDZ domain suggested a mechanistic role for helix 8 (located immediately after the end of TM7) in GPCRs in the course of signaling [157].

The selectivity of ligands for D_3 receptor over D_2 has also been studied by multiple fragment docking [158] and in combination with evaluations against chimeric and mutant receptors, suggesting that the selectivity of ligands is due to interactions at the extracellular end of TM7 [159]. The binding affinity of D_2 [160] and D_3 [161] receptors for a series of [4-(4-carboxamidobutyl)]-1-arylpiperazines has been evaluated at density functional theory (DFT) level. From these results, partial bidimensional pharmacophores have been proposed (Figure 1.7).

Key specific residues involved in activation of D_3 receptor signaling have been identified in TM6 and ol_3 [162]. The quaternary structure of the D_3 receptor has been investigated, and the formation of not only homodimers but also higher order oligomers has been confirmed. Those oligomers are formed by four protomers organized in a rhombic arrangement (Figure 1.8) [163].

D_4 receptor presents the structural features characteristic of D_2-like receptors including a carboxyl-terminal cysteine, which is believed to be a palmitoylation site. It has been demonstrated that glycosylation can take place on a single conserved Asn3 [164]. This glycosylation is not involved in ligand binding or receptor trafficking. D_4 receptor shares 53% identity with D_2 receptor in the transmembrane domains [43]. The human D_4 receptor is one of the most known polymorphic proteins [165]. Most of the diversity is found at the third intracellular loop il_3 – unusually proline-rich [166] for which 20 different variants of the receptor have been identified [167]. The $D_{4.4}$ variant is predominant followed by $D_{4.7}$ and $D_{4.2}$ [168]. Polymorphic variations encoding amino terminus are also known [165] although no major differences have been observed in G-protein coupling or signaling properties [169]. In this context, dimerization of D4 receptor is a key factor for the correct folding of the GPCR [170]. Several studies have investigated links between polymorphisms of D_4 receptors and disorders such as ADHD [171, 172] and schizophrenia [173, 174].

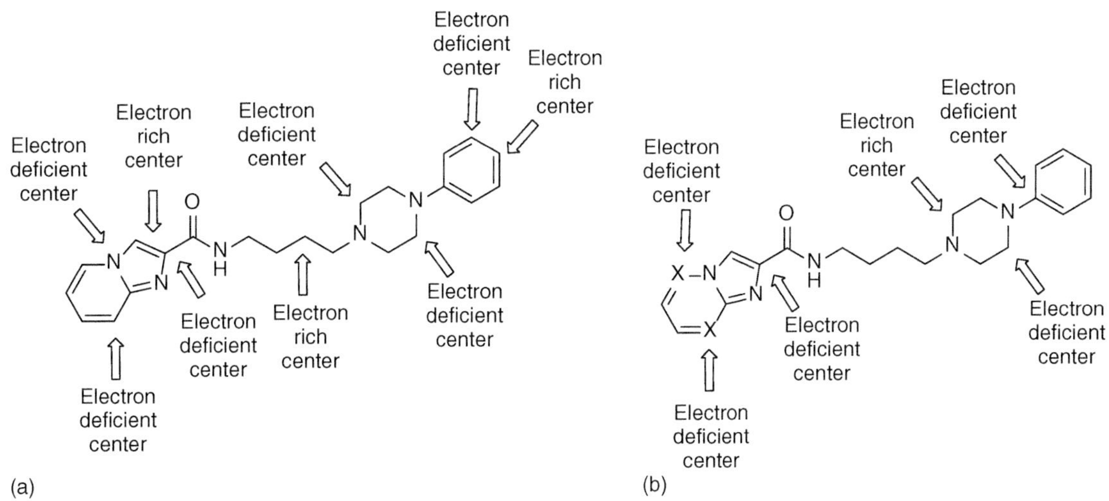

Figure 1.7 Examples of 2D pharmacophores calculated at DFT level for D_2 (a) and D_3 (b) receptors.

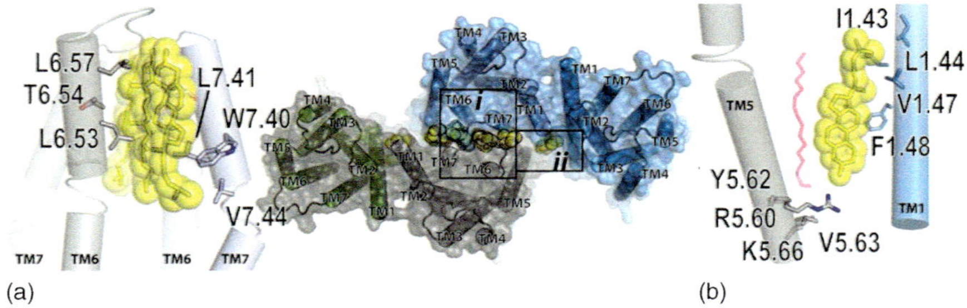

Figure 1.8 Molecular modeling of human D_3 receptor in tetrameric arrangements. Each dimer is shown as a semitransparent surface, whereas predicted cholesterols are shown as yellow spheres forming a buffer between the two dimers. Inset (a) shows details of the TM6 and TM7 interface, and the residues shown in sticks (gray and light blue) were found experimentally to affect hD3 quaternary structure. Yellow sticks and spheres depict predicted cholesterol molecules in positions as observed in adenosine A2A receptor, mu-opioid receptor, and the P2Y12 receptor structures. Inset (b) shows details of the predicted interaction between (transmembrane domain) TMD V Arg-5.60 and Tyr-5.63 (in gray sticks) of one dimer and the TMD I cholesterol (in yellow sticks and spheres). A predicted palmitoyl moiety, bound to Cys-8.60, is also shown in magenta semitransparent sticks. Source: Reproduced with permission from Marsango et al. 2015 [163]. Copyright 2015, JBC.

1.3 Vesicular Monoamine Transporter (VMAT)

The VMAT is a membrane protein that introduces monoamine neurotransmitters into storage vesicles to allow subsequent release into the synapse [175]. VMAT accumulates both newly synthesized and reuptaken from the synapse neurotransmitter molecules into the intraneuronal vesicles [176]. The VMAT exists in two distinct forms with different pharmacologies and tissues distribution: VMAT1 and VMAT2 [177]. These transporters are relatively nonspecific and transport the monoamine neurotransmitters serotonin, ephedrine, and norephedrine in addition to dopamine [178]. The neuronal selectivity for a given monoamine is determined by the particular biosynthetic enzymes expressed in the corresponding neuron. On the other hand, there are other plasma membrane neurotransmitter transporters that are selective for each monoamine, i.e. DAT for dopamine, SERT for serotonin, and NET for norepinephrine. In this section, an overview of VMAT2, present in the CNS, and DAT is presented, and NERT and SERT are discussed in Chapters 2 and 3, respectively.

VMAT1, previously known as the chromaffin granule amine transporter, is located in extraneural tissues although there is some evidence that it may also be present in the human brain, perhaps concentrated in the substantia nigra. Some studies suggest a relationship between VMAT1 gene and neuropsychiatric disorders [179] such as schizophrenia [180] and bipolar disorder [181]. VMAT2, previously known as the synaptic VMAT, is primarily located in neuronal cells of the central, peripheral, and enteric nervous systems [182]. VMAT2 selectively recognizes and transports all of the amine neurotransmitters across biomembranes, that is, dopamine, serotonin, histamine, norepinephrine, and epinephrine. VMAT2 is not selective, but the quantitative assessment of its density by positron emission tomography (PET) scanning has been clinically useful for early diagnosis and monitoring of the progression of Parkinson's and Alzheimer's diseases and drug addiction [183, 184]. Genetic studies connected alterations of VMAT2 with depression [185]. The immunochemical analysis of VMAT2 has also been suggested to be used in the diagnosis of PD [186].

VMAT2 (as VMAT1) is a transmembrane glycoprotein with 12 transmembrane domains with both N- and C-termini in the cytoplasm and a large, hydrophobic, N-glycosylated loop between TM1 and TM2 facing the vesicle lumen. The structure has been predicted (Figure 1.9), but no crystal structures have been resolved. VMAT2 has 742 residues. It can be N-glycosylated by Glc-NAc at Asn84 and Asn91. A posttranslational phosphorylation takes place at Ser511 and Ser513.

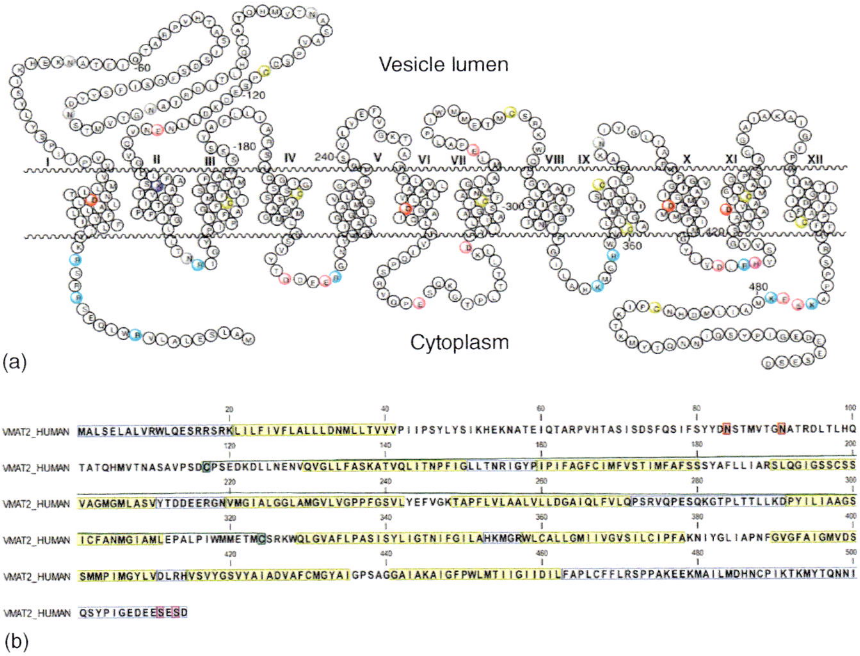

Figure 1.9 (a) Predicted secondary structure of the human vesicular monoamine transporter 2 (VMAT2) Conserved (in hVMAT1, hVMAT2, rVMAT1, rVMAT2, and bVMAT2) transmembrane Asp33, 262, 399, 426 (red), and Lys138 (blue); proposed vesicle lumen disulfide bridged Cys117, and 324 (yellow); and glycosylation sites Asn (gray) are shown. The conserved Cys (117, 302, 324, 360, 374, 430, 467, and 488) (yellow); cytoplasmic domain charged residues Lys (354, 476, and 479) (cyan), Arg (10, 16, 19, 155, 217, 357, and 413) (cyan), Glu (216, 278, 477, and 478) (pink), Asp (213, 291, and 411) (pink), His414 (red), and vesicle lumen Glu (127 and 312) (pink) are also shown (amino acid numbering is based on the sequence of hVMAT2). Source: Reproduced with permission from Van Liefferinge et al. 2013 [187]. Copyright subjected to CC-BY 4.0 License terms. http://journal.frontiersin.org/article/10.3389/fncel.2013.00139/full. (b) Amino acids sequence of human VMAT2. Disulfide bonds (green), N-glycosylation sites (red), extracellular domains (white), intracellular domains (light blue), and transmembrane domains (yellow) are indicated on sequences.

Studies with mammalian neurons suggested that VMAT2-mediated transport of monoamines is subject to regulation by two calcium-dependent activator proteins of secretion (CAPS1 and 2) linked to the heterotrimeric GTPase G_o [188]. The established proposed model for introduction of monoamines into vesicles is also based on efflux of two protons coupled with influx of the monoamine (Figure 1.10a) [175].

An intravesicular proton induces a conformational change, increasing amine-binding affinity outside the vesicle; the second proton induces a second and large conformational change that introduces the monoamine in the vesicle, reducing amine-binding affinity inside the vesicle. Finally, the monoamine is released inside the vesicle and two protons are liberated to cytosolic phase. The first proton-induced conformational change is assisted by His419. Mutagenesis experiments identified the residues involved in substrate recognition [190] and suggested that Lys139 and Asp427 interact to form an ion pair in the native protein, promoting high-affinity interaction with the substrate [191]. However, recent studies revealed that, during the transport cycle, VMAT2 is expected to occupy at least three different conformations: cytoplasm-facing, occluded, and lumen-facing (Figure 1.9b) [189]. The lumen to cytoplasm-facing transition, facilitated by protonation of at least one of the essential membrane-embedded carboxyls, generates a binding site for reserpine, a competitive inhibitor of both VMAT1 and VMAT2, that locks the

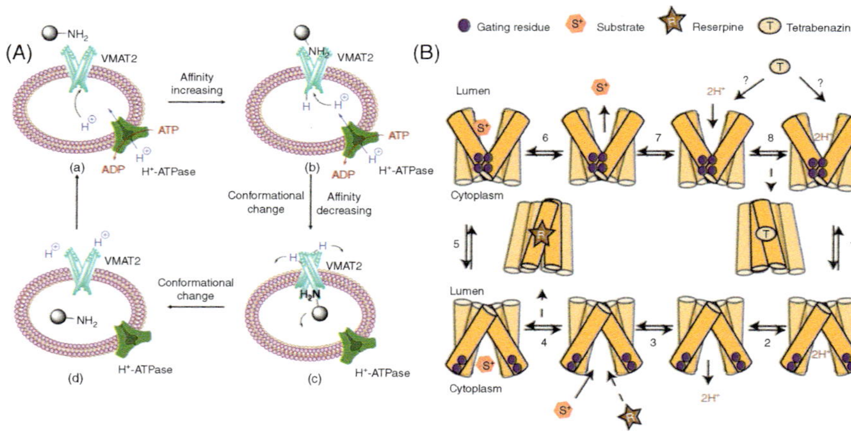

Figure 1.10 Models for introduction of monoamine neurotransmitters into intraneuronal vesicles mediated by VMAT2. (A) Simplified model. A conformational change induced by a proton (a) increases affinity by the amine (b), which is introduced into the vesicle through a second conformational change induced by a second proton (c). The amine is released and the two protons initially introduced by H^+-ATPase are liberated to cytosol (d). (B) Illustration of a third conformation. For simplicity, only six TMs are shown. In the absence of a proton gradient, the dominant population is of the lumen-facing conformation, as indicated by the transporter's ability to bind tetrabenazine but not reserpine. Binding of protons enables the conformational switch to the cytoplasm-facing conformation (step 1), whereas binding of substrate enables the change to the lumen-facing conformation (step 5). Source: Reproduced with permission from Yaffe et al. 2016 [189]. Copyright 2016.

transporter in a dead-end conformation. However, reserpine binding competes with substrate binding, suggesting that the reserpine-bound conformation is cytoplasm-facing. The VMAT2-selective inhibitor, tetrabenazine, locks the transporter in a conformation that appears incompatible with substrate binding and is therefore presumably not cytoplasm-facing.

1.4 The Dopamine Transporter (DAT)

The DAT is a transmembrane protein belonging to the family of sodium/chloride-dependent neurotransmitters, which removes the excess of dopamine from synaptic and presynaptic spaces and reuptake it into dopaminergic neurons through a process driven by the symport of sodium and chloride ions along their concentration gradients [192]. In the brain, DAT is expressed exclusively in dopaminergic neurons within the CNS, especially in midbrain neurons originating in the substantia nigra and ventral tegmental area. Dysfunction of DAT has been associated with several neurodisorders and psychiatric disorders including ADHD, bipolar disorder, clinical depression, drug addiction, and alcoholism [193]. Abnormal DAT densities are found in PD [194], Tourette's syndrome [195], dementia with Lewy bodies (DLB) [196], depression [197], and Lesch–Nyhan disease [198]. The quantitative assessment of DAT using PET and single photon emission computed tomography (SPECT) (DaTscan) imaging is found to be a powerful technique [199] for the diagnosis of PD [200] and other neurodisorders [201–203] as well as for the quantification of DAT in heroin- [204] and cocaine-dependent [205] subjects.

1.4.1 Structural Aspects

The human DAT has 620 residues with 12-transmembrane rings with a glycosylated large second extracellular loop and intracellular amino- and carboxy-termini (Figure 1.11a). N-glycosylation can take place at Asn181, Asn188, and Asn205. The transporter interacts with two sodium ions through residues Gly75, Val78, Leu418, Asp421, and Ser422 for one ion and Ala77, Asn82, Ser321, and Asn353 for the second one. Natural variants include replacement of Gly121 by Ser, Arg237 by Gln, Leu368 by Gln, Pro395 by Leu, Val471 by Ile, and Arg544 by Ser. Although the crystal structure of human dopamine transporter (hDAT) has not been determined, the X-ray structure of the *drosophila* dopamine transporter (dDAT) has been resolved (Figure 1.12). Figure 1.11b illustrates aligned sequences of both receptors showing a high degree of conservation. Hence, homology models of hDAT can be constructed from dDAT structural data.

The structure of a transport-inactive dDAT was resolved in complex with nortriptyline (PDB ID: 4M48) (Figure 1.12) and served for elucidating antidepressant mechanism [206] consisting of stabilizing the open conformation by sterically preventing closure of the extracellular gate and thus blocking the transporter from binding substrate and isomerizing to the inward-facing conformation. The same authors reported, two years later, X-ray structures of dDAT in complex with antidepressants nisoxetine (PDB ID: 4XNU) and reboxetine

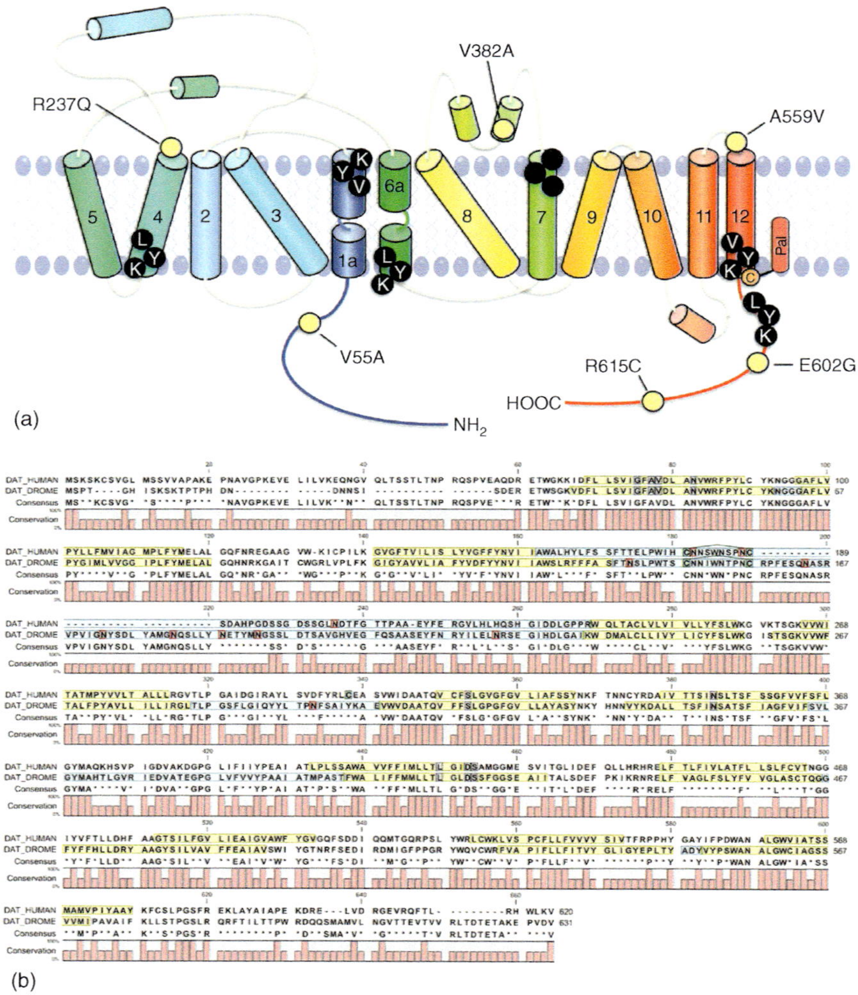

Figure 1.11 (a) Coding variants known to alter dopamine transporter (DAT) function (numbered yellow circles) and helical topological 2D architecture of DAT depicting essential cholesterol-interacting residues in putative Cholesterol Recognition Amino Acid Consensus (CRAC) motifs (black circles with white letters). Source: Reprinted with permission from Vaughan and Foster 2013 [193]. Copyright 2013, Elsevier. (b) Sequence alignment of hDAT and dDAT. Conservation graphic and consensus sequence are shown. Disulfide bonds (green), N-glycosylation sites (red), sodium interaction residues (gray), extracellular domains (light blue), intracellular domains (white), and transmembrane domains (yellow) are indicated on sequences.

(PDB ID: 4XNX) (Figure 1.12) [207]. The ligands occupy the substrate-binding site, contributing to the same stabilizing effect of the outward-open conformation observed with nortriptyline. These structures established that aromatic moieties and extended amine groups of antidepressants are important for generating high-affinity inhibition. The structural study further revealed that

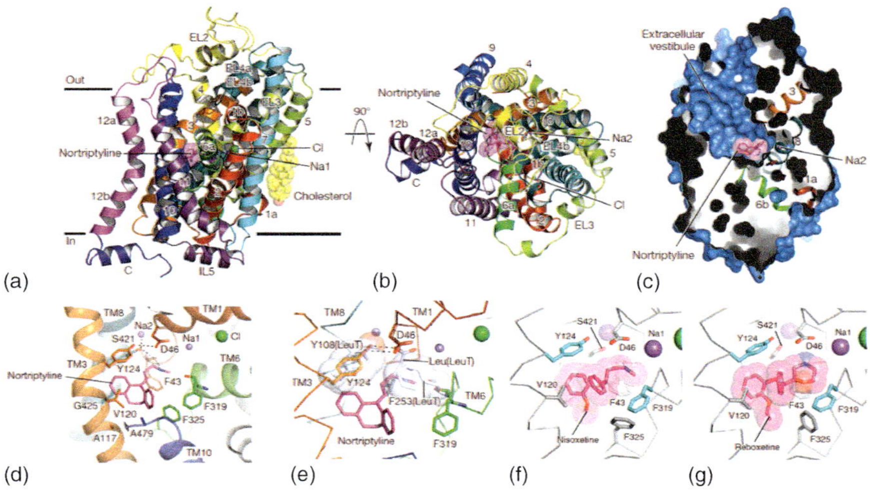

Figure 1.12 (a–c) Architecture of Drosophila DATcryst. (a) Structure of DATcryst viewed parallel to membrane. Nortriptyline, sodium ions, a chloride ion, and a cholesterol molecule are shown in sphere representation in magenta, purple, green, and yellow, respectively. (b) View of DATcryst from the extracellular face. (c) Surface representation showing that ligand- and ion-binding sites are accessible from the extracellular vestibule. Nortriptyline and TMs 1, 3, 6, and 8 are colored as in (a). (d) Close-up view of the drug-binding pocket. Sodium and chloride ions are shown as spheres. Nortriptyline is represented as sticks (magenta). The amino group of nortriptyline is 2.7 Å from the carbonyl oxygen of Phe43 (TM1a), and the *N*-methyl group of nortriptyline is 3.1 Å from the carbonyl oxygen of Phe319. Residues lining the drug-binding pocket with interfacial areas greater than 10 $Å^2$ are represented as sticks. (e) Comparison of the drug- or substrate-binding pocket of DATcryst with that of LeuT (PDB ID: 2A65). The distance between the carboxylate group of leucine and Tyr108 (spheres) is 2.7 Å in the occluded state (PDB ID: 2A65) and 5.1 Å in the inhibitor-bound state (PDB ID: 3F3A) of LeuT, whereas the equivalent interaction in DATcryst between Asp46 and Tyr124 is 3.1 Å. Source: (a–e) Reprinted with permission from Penmatsa et al. 2015 [206]. Copyright 2013, Macmillan Publishers Ltd. (f, g) Close-up views of the binding pocket for nisoxetine-bound (f) and reboxetine-bound (g) dDAT. Cyan, Tyr124 and Phe319; purple spheres, sodium ions; green spheres, chloride ions. Source: Reprinted with permission from Penmatsa et al. 2017 [207]. Copyright 2015, Macmillan Publishers Ltd.

chemical modifications on the aromatic rings of antidepressants encode selectivity between catecholamine transporters opening a door for the structure-directed design of more-selective biogenic amine transport inhibitors.

Several crystal structures of dDAT with cocaine (PDB ID: 4XP4), cocaine analogues (PDB ID: 4XP5), amphetamine (PDB ID: 4XP9), methamphetamine (PDB ID: 4XP6), 3,4-dichlorophenetylamine (PDB ID: 4XPA), and the natural substrate dopamine (PDB ID: 4XP1) were also reported [208]. The same outward-face orientation was found in all cases. Dopamine is located in the central binding site of the outward-face conformation defined by TMs 1, 3, 6, and 8. The primary amino group interacts with Asp46, and the catechol moiety is placed in a pocket formed by Ala117, Val120, Asp121, Tyr124, Ser422, and Phe325. All the resolved structures of dDAT have an outward-face orientation and similar interactions between the ligands and the residues forming the binding pocket. Amphetamines lacking the catechol moiety present less affinity by the transporter although they are also placed in the binding site. However, amphetamines have a considerable weaker affinity by dDAT than by hDAT, probably due to differences in residues in the subpart of the binding pocket where the catechol unit is recognized. In fact, in mammalian DATs, amphetamines are closely as effective as cocaine at inhibition of dopamine uptake [209], suggesting that hDAT is more complementary to those drugs of abuse than DATs from invertebrates.

The resolution of dDAT served for the construction of reliable homology models of hDAT, which were the basis for unraveling the molecular mechanism of dopamine transport by hDAT [210]. By using homology models in combination with accelerated molecular dynamics calculations, it has been established that during dopamine transport, four conformational states are differentiated corresponding to global helical motions and local conformational switches. Initially, dopamine binding induces a spontaneous closure of extracellular gates (Arg85-Asp476 and Tyr156-Phe320) accompanied by 10–15° inward tilting of TM1 and TM6 that rapidly evolves to a more stable conformation. Disruption of Asn82-Asn353 hydrogen bond promoted by Na_2 dislocation induces permeation of intracellular water molecules while TM1 outward tilting breaks salt bridges Lys66-Asp345 and Arg60-Asp436. These transformations complete transition to inward-face conformation and release of chloride and one sodium cation. Finally, disruption of the H-bond between dopamine and Asp79 releases dopamine inside the neuron (Figure 1.9).

Recent studies revealed that the large N- and C-termini extended into the cytoplasm, synergistically contribute to dopamine affinity [211]. Molecular dynamics simulations combined with mutagenesis studies suggested that initial binding to DAT is favored by Val152 that leaves enough space for the substrate [212]. Regulation of domains responsible for binding is exerted through post-translational modifications such as phosphorylation [213] and palmitoylation [214] that control DAT kinetics [215]. The human DAT has a tetrahedral zinc(II) allosteric binding site that might be of interest for activating dopamine transport [216].

1.4.2 Molecular Mechanism

The DAT belongs to the SLC6 family of sodium- and chloride-dependent transporters. The cotransport of sodium and chloride ions provides the energy for the transporter to pump dopamine into the terminals against its concentration gradient. The sodium gradient in the membrane is maintained by the sodium–potassium pump Na^+, K^--ATPase.

The general reuptaking process of dopamine by DAT starts by binding one molecule of dopamine. Further binding of two sodium cations and one chloride induces a conformational change to an inward-facing conformation that releases both ions and dopamine into the cytoplasm. Actually, this process involves complex interactions between ions and DAT, initially foreseen by previous investigations [217] and ultimately identified by structural and computational studies (Figure 1.13). The use of homology models of human DAT, based on X-ray, determined dDAT and was complemented by full-atomic microseconds

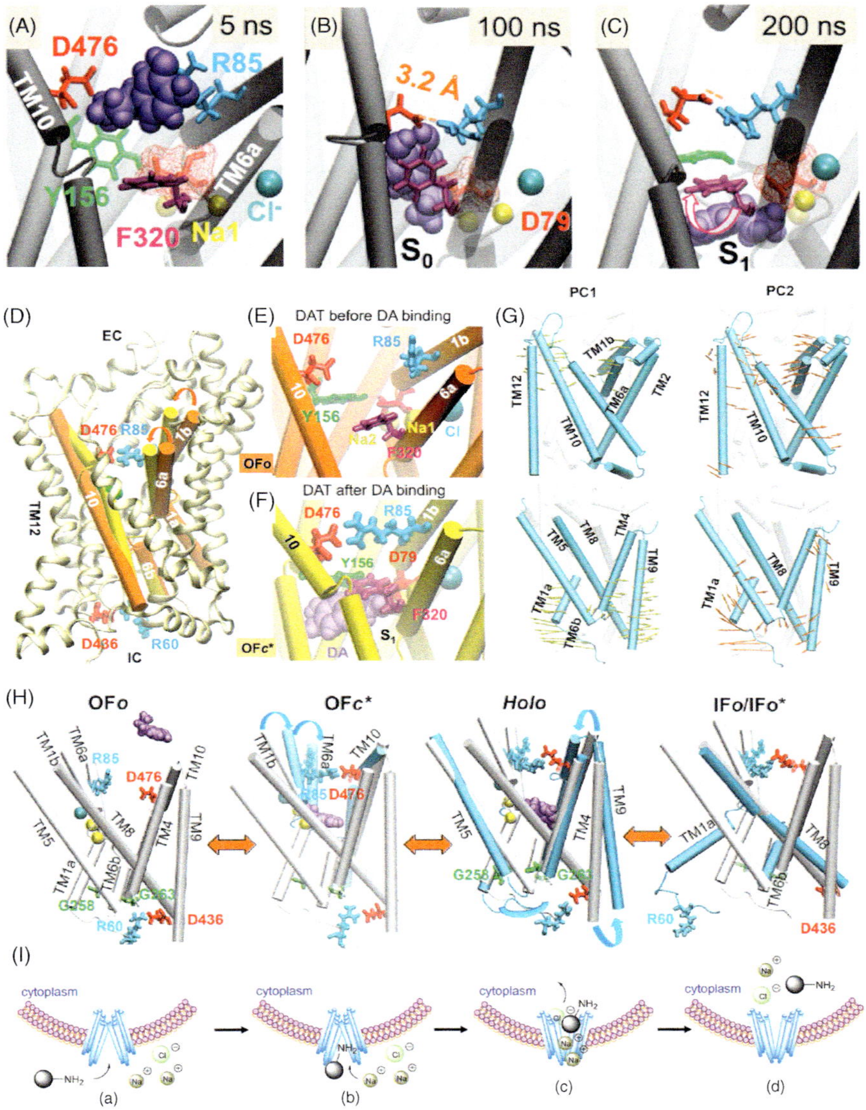

Figure 1.13 (A–C) Positions of DA (purple van der Waals spheres) with respect to Arg85, Asp476, Tyr156, Phe320, and Asp79 (red stick with mesh), at 5, 100, and 200 ns. (D) MD-resolved hDAT in the outward-facing closed (OFc*) state (cream ribbons). Alignment of a representative OFc* conformation (yellow) sampled in MD simulations with the initial hDAT in the OFo state (orange). (E) Substrate-binding pocket before DA binding, in the OFo state. (F) Same pocket after binding. Isomerization of F320 brings its aromatic side chain on top of DA. Tyr156-Phe320 and salt bridge by Arg85-Asp476 serve as EC gates. (G, H) hDAT Trajectory from outward-facing open to inward-facing open state, via holo-occluded state. (G) Movements along PC1 (yellow arrows) and PC2 (brown arrows). (H) Conformation states visited by hDAT along with substrate translocation. Four states are distinguished by global helical motions (cyan) together with local conformational switches. (I) Schematic mechanism of dopamine transport. Source: (A–H) Reprinted with permission from Cheng and Bahar 2015 [210]. Copyright 2015, Elsevier.

accelerated simulations, allowed determining the sequence of events during dopamine reuptake into dopaminergic neurons.

Dopamine binding to DAT induces closure of extracellular gates and a holo-occluded intermediate is stabilized. Then concerted motions of TM1 and TM7, accompanied by the TM1a–TM1b loop reordering, break the hydrogen bond between Asn82 and Asn353, causing exposure to intracellular water triggered by Na_2 dislocation. This process facilitates the release of sodium and chloride ions through a redistribution of a network of salt bridges at the intracellular surface in the inward-facing state and concerted tilting of intracellular helices. Finally, dopamine is released after protonation of Asp79.

The role of DAT in combination with VMAT2 is crucial for the subtle balance of neuronal dopamine exerted by dopaminergic neurons [218]. Increasing dopamine concentration in cytoplasm can lead to oxidative stress and neurotoxicity. On the other hand, enough but not excessive dopamine must be present in the synaptic and presynaptic environments. Thus, a correct compartmentalization of dopamine is critical for neuron function, and any misbalance results in severe neurodisorders. DAT and VMAT2 are responsible for a correct function and, consequently, therapeutic targets in diseases.

References

1 Stahl, S.M. (2013). *Stahl's Essential Psychopharmacology, Neuroscientific Basis and Practical Applications*, 4e, 86–96. Cambridge: Cambridge University Press.

2 Fuxe, K., Agnati, L.F., Kalia, M. et al. (1985). Dopaminergic systems in the brain and pituitary. In: *The Dopaminergic System* (ed. E. FLuckiger, E.E. Müller and M.O. Thorner), 11–26. Heidelberg: Springer-Verlag.

3 Agid, Y. (1989). Parkinson's disease: some physiopathologic aspects and clinical practice. *Rev. Prat.* 39 (8): 641–646.

4 Winter, C., von Rumohr, A., Mundt, A. et al. (2007). Lesions of dopaminergic neurons in the substantia nigra pars compacta and in the ventral tegmental area enhance depressive-like behavior in rats. *Behav. Brain Res.* 184 (2): 133–141.

5 Howes, O., McCutcheon, R., and Stone, J. (2015). Glutamate and dopamine in schizophrenia: an update for the 21st century. *J. Psychopharmacol.* 29 (2): 97–115.

6 Giza, E., Gotzamani-Psarrakou, A., and Bostantjopoulou, S. (2012). Imaging beyond the striatonigral dopaminergic system in Parkinson's disease. *Hell. J. Nucl. Med.* 15 (3): 224–232.

7 Janezic, S., Threlfell, S., Dodson Paul, D. et al. (2013). Deficits in dopaminergic transmission precede neuron loss and dysfunction in a new Parkinson model. *Proc. Natl. Acad. Sci. U.S.A.* 110 (42): E4016–E4025.

8 Ansari, J.A., Siraj, A., and Inamdar, N.N. (2010). Pharmacotherapeutic approaches of Parkinson's disease. *Int. J. Pharmacol.* 6 (5): 584–590.

9 Das, B., Modi, G., and Dutta, A. (2015). Dopamine D_3 agonists in the treatment of Parkinson's disease. *Curr. Top. Med. Chem.* 15 (10): 908–926.

10 Xhaard, H., Rantanen, V.-V., Nyrönen, T., and Johnson, M.S. (2006). Molecular evolution of adrenoceptors and dopamine receptors: implications for the binding of catecholamines. *J. Med. Chem.* 49 (5): 1706–1719.

11 Rossetti, G., Dibenedetto, D., Calandrini, V. et al. (2015). Structural predictions of neurobiologically relevant G-protein coupled receptors and intrinsically disordered proteins. *Arch. Biochem. Biophys.* 582: 91–100.

12 Shonberg, J., Kling, R.C., Gmeiner, P., and Löber, S. (2015). GPCR crystal structures: medicinal chemistry in the pocket. *Bioorg. Med. Chem.* 23 (14): 3880–3906.

13 Coudrat, T., Christopoulos, A., Sexton, P.M., and Wootten, D. (2017). Structural features embedded in G protein-coupled receptor co-crystal structures are key to their success in virtual screening. *PLoS One* 12 (4): e0174719.

14 Neve, K.A. (2010). *The Dopamine Receptors*, 2e, 650. New York, NY: Humana Press.

15 Uchigashima, M., Ohtsuka, T., Kobayashi, K., and Watanabe, M. (2016). Dopamine synapse is a neuroligin-2-mediated contact between dopaminergic presynaptic and GABAergic postsynaptic structures. *Proc. Natl. Acad. Sci. U.S.A.* 113 (15): 4206–4211.

16 Spano, P.F., Govoni, S., and Trabucchi, M. (1978). Studies on the pharmacological properties of dopamine receptors in various areas of the central nervous system. *Adv. Biochem. Psychopharmacol.* 19: 155–165.

17 Kebabian, J.W. and Calne, D.B. (1979). Multiple receptors for dopamine. *Nature* 277 (5692): 93–96.

18 Dearry, A., Gingrich, J.A., Falardeau, P. et al. (1990). Molecular cloning and expression of the gene for a human D_1 dopamine receptor. *Nature* 347 (6288): 72–76.

19 Monsma, F.J., Mahan, L.C., McVittie, L.D. et al. (1990). Molecular cloning and expression of a D_1 dopamine receptor linked to adenylyl cyclase activation. *Proc. Natl. Acad. Sci. U.S.A.* 87 (17): 6723–6727.

20 Zhou, Q.Y., Grandy, D.K., Thambi, L. et al. (1990). Cloning and expression of human and rat D_1 dopamine receptors. *Nature* 347 (6288): 76–80.

21 Gingrich, J.A., Dearry, A., Falardeau, P. et al. (1991). Molecular characterization of G-protein coupled receptors: isolation and cloning of a D_1 dopamine receptor. *J. Recept. Res.* 11 (1–4): 521–534.

22 Sokoloff, P., Giros, B., Martres, M.P. et al. (1990). Molecular cloning and characterization of a novel dopamine receptor (D_3) as a target for neuroleptics. *Nature* 347 (6289): 146–151.

23 Van Tol, H.H., Bunzow, J.R., Guan, H.C. et al. (1991). Cloning of the gene for a human dopamine D_4 receptor with high affinity for the antipsychotic clozapine. *Nature* 350 (6319): 610–614.

24 Grandy, D.K., Zhang, Y.A., Bouvier, C. et al. (1991). Multiple human D_5 dopamine receptor genes: a functional receptor and two pseudogenes. *Proc. Natl. Acad. Sci. U.S.A.* 88 (20): 9175–9179.

25 Jarviet, K.R., Tiberit, M., Silvia, C. et al. (1993). Molecular cloning, stable expression and desensitization of the human dopamine D1B/D5 receptor. *J. Recept. Res.* 13 (1–4): 573–590.

26 Missale, C., Nash, S.R., Robinson, S.W. et al. (1998). Dopamine receptors: from structure to function. *Physiol. Rev.* 78 (1): 189–225.

27 Gentles, A.J. and Karlin, S. (1999). Why are human G-protein-coupled receptors predominantly intronless? *Trends Genet.* 15 (2): 47–49.
28 Niznik, H.B. and Van Tol, H.H. (1992). Dopamine receptor genes: new tools for molecular psychiatry. *J. Psychiatry Neurosci.* 17 (4): 158–180.
29 Wang, C., Liu, Y., Wang, H. et al. (2014). Molecular characterization and differential expression of multiple goose dopamine D_2 receptors. *Gene* 535 (2): 177–183.
30 Fishburn, C.S., Belleli, D., David, C. et al. (1993). A novel short isoform of the D_3 dopamine receptor generated by alternative splicing in the third cytoplasmic loop. *J. Biol. Chem.* 268 (8): 5872–5878.
31 Shimada, M.K., Inoue-Murayama, M., Ueda, Y. et al. (2004). Polymorphism in the second intron of dopamine receptor D_4 gene in humans and apes. *Biochem. Biophys. Res. Commun.* 316 (4): 1186–1190.
32 Cadet, J.L., Jayanthi, S., McCoy, M.T. et al. (2010). Dopamine D_1 receptors, regulation of gene expression in the brain, and neurodegeneration. *CNS Neurol. Disord. Drug Targets* 9 (5): 526–538.
33 Savasta, M., Dubois, A., and Scatton, B. (1986). Autoradiographic localization of D_1 dopamine receptors in the rat brain with [^{3}H]SCH 23390. *Brain Res.* 375 (2): 291–301.
34 Ciliax, B.J., Nash, N., Heilman, C. et al. (2000). Dopamine D_5 receptor immunolocalization in rat and monkey brain. *Synapse* 37 (2): 125–145.
35 Khan, Z.U., Gutierrez, A., Martin, R. et al. (2000). Dopamine D_5 receptors of rat and human brain. *Neuroscience* 100 (4): 689–699.
36 Wamsley, J.K., Gehlert, D.R., Filloux, F.M., and Dawson, T.M. (1989). Comparison of the distribution of D-1 and D-2 dopamine receptors in the rat brain. *J. Chem. Neuroanat.* 2 (3): 119–137.
37 Levant, B. (1997). The D_3 dopamine receptor: neurobiology and potential clinical relevance. *Pharmacol. Rev.* 49 (3): 231–252.
38 Joyce, J.N. (2001). Dopamine D_3 receptor as a therapeutic target for antipsychotic and antiparkinsonian drugs. *Pharmacol. Ther.* 90 (2–3): 231–259.
39 Lahti, R.A., Roberts, R.C., Cochrane, E.V. et al. (1998). Direct determination of dopamine D_4 receptors in normal and schizophrenic postmortem brain tissue: a [^{3}H]NGD-94-1 study. *Mol. Psychiatry* 3 (6): 528–533.
40 Defagot, M.C. and Antonelli, M.C. (1997). Autoradiographic localization of the putative D_4 dopamine receptor in rat brain. *Neurochem. Res.* 22 (4): 401–407.
41 Strader, C.D., Fong, T.M., Tota, M.R. et al. (1994). Structure and function of G protein-coupled receptors. *Annu. Rev. Biochem.* 63: 101–132.
42 Probst, W.C., Snyder, L.A., Schuster, D.I. et al. (1992). Sequence alignment of the G-protein coupled receptor superfamily. *DNA Cell. Biol.* 11 (1): 1–20.
43 Civelli, O., Bunzow, J.R., and Grandy, D.K. (1993). Molecular diversity of the dopamine receptors. *Annu. Rev. Pharmacol. Toxicol.* 33: 281–307.
44 Pandey, P., Mersha, M.D., and Dhillon, H.S. (2013). A synergistic approach towards understanding the functional significance of dopamine receptor interactions. *J. Mol. Signaling* 8 (1): 13.
45 George, S.R., Lee, S.P., Varghese, G. et al. (1998). A transmembrane domain-derived peptide inhibits D_1 dopamine receptor function without affecting receptor oligomerization. *J. Biol. Chem.* 273 (46): 30244–30248.

46 Elmhurst, J.L., Xie, Z., O'Dowd, B.F., and George, S.R. (2000). The splice variant $D3_{nf}$ reduces ligand binding to the D_3 dopamine receptor: evidence for heterooligomerization. *Mol. Brain Res.* 80 (1): 63–74.

47 Vivo, M., Lin, H., and Strange, P.G. (2006). Investigation of cooperativity in the binding of ligands to the D_2 dopamine receptor. *Mol. Pharmacol.* 69 (1): 226–235.

48 Guo, W., Shi, L., Filizola, M. et al. (2005). Crosstalk in G protein-coupled receptors: changes at the transmembrane homodimer interface determine activation. *Proc. Natl. Acad. Sci. U.S.A.* 102 (48): 17495–17500.

49 Lee, S.P., O'Dowd, B.F., Rajaram, R.D. et al. (2003). D_2 dopamine receptor homodimerization is mediated by multiple sites of interaction, including an intermolecular interaction involving transmembrane domain 4. *Biochemistry* 42 (37): 11023–11031.

50 Dziedzicka-Wasylewska, M., Faron-Gorecka, A., Andrecka, J. et al. (2006). Fluorescence studies reveal heterodimerization of dopamine D_1 and D_2 receptors in the plasma membrane. *Biochemistry* 45 (29): 8751–8759.

51 Scarselli, M., Novi, F., Schallmach, E. et al. (2001). D_2/D_3 dopamine receptor heterodimers exhibit unique functional properties. *J. Biol. Chem.* 276 (32): 30308–30314.

52 Fiorentini, C., Savoia, P., Bono, F. et al. (2015). The D_3 dopamine receptor: from structural interactions to function. *Eur. Neuropsychopharmacol.* 25 (9): 1462–1469.

53 Kearn, C.S., Blake-Palmer, K., Daniel, E. et al. (2005). Concurrent stimulation of cannabinoid CB_1 and dopamine D_2 receptors enhances heterodimer formation: a mechanism for receptor cross-talk? *Mol. Pharmacol.* 67 (5): 1697–1704.

54 Ferré, S., Ciruela, F., Canals, M. et al. Adenosine A_{2A}–dopamine D_2 receptor–receptor heteromers. Targets for neuro-psychiatric disorders. *Parkinsonism Relat. Disord.* 10 (5): 265–271.

55 Rocheville, M., Lange, D.C., Kumar, U. et al. (2000). Receptors for dopamine and somatostatin: formation of hetero-oligomers with enhanced functional activity. *Science* 288 (5463): 154–157.

56 Gines, S., Hillion, J., Torvinen, M. et al. (2000). Dopamine D_1 and adenosine A_1 receptors form functionally interacting heteromeric complexes. *Proc. Natl. Acad. Sci. U.S.A.* 97 (15): 8606–8611.

57 Missale, C., Fiorentini, C., Collo, G., and Spano, P. (2010). The neurobiology of dopamine receptors: evolution from the dual concept to heterodimer complexes. *J. Recept. Signal Transduction* 30 (5): 347–354.

58 Rosenbaum, D.M., Rasmussen, S.G.F., and Kobilka, B.K. (2009). The structure and function of G-protein-coupled receptors. *Nature* 459 (7245): 356–363.

59 Haga, T. and Berstein, G. (1999). *G Protein-Coupled Receptors*. Boca Raton, FL: CRC Press.

60 Devi, L.A. (2005). *The G Protein-Coupled Receptors Handbook*, 413. Totowa, NJ: Humana Press.

61 Shukla, A.K. (2016). *G Protein-Coupled Receptors: Signaling, Trafficking and Regulation*, Methods in Cell Biology Series, vol. 132, 132, 475. Amsterdam: Elsevier-Academic Press.

62 Rognan, D. (2006). *Ligand Design for G Protein-Coupled Receptors*, 265. Weinheim: Wiley-VCH.
63 Seifert, R. and Wieland, T. (2005). *G Protein-Coupled Receptors as Drug Targets*, 295. Weinheim: Wiley-VCH.
64 Neve, K.A. (2009). *Functional Selectivity of G-Protein-Coupled Receptor Ligands. New Opportunities for Drug Discovery*, 287. New York, NY: Humana Press and Springer.
65 Filizola, M. (2015). *Methods in Molecular Biology*, 2e, 327. New York, NY: Humana Press and Springer.
66 Vauquelin, G. and Von Mentzer, B. (2007). *G Protein-Coupled Receptors: Molecular Pharmacology*, 264. Wiley.
67 Neubig, R.R. (2011). *Pharmacology of G Protein-Coupled Receptors*, 408. Amsterdam: Academic Press.
68 Filizola, M. (2014). *Advances in Experimental Medicine and Biology*, 228. Heidelberg: Springer-Verlag.
69 Poyner, D.R. and Wheatley, M. (2010). *G Protein-Coupled Receptors: Essential Methods*, 289. Hoboken, NJ: Wiley-Blackwell.
70 Giraldo, J. and Pin, J.-P. (2011). *RSC Drug Discovery*, 548. Oxford: RSC Publishing.
71 Pangalos, M.N. and Davies, C.H. (2002). *Molecular and Cellular Neurobiology*, 653. Oxford: Oxford University Press.
72 Tao, Y.-X. (2009). *Progress in Molecular Biology and Translational Science*, vol. 88, 215. London: Academic Press.
73 Ye, N., Song, Z., and Zhang, A. (2014). Dual ligands targeting dopamine D_2 and serotonin 5-HT1A receptors as new antipsychotical or anti-Parkinsonian agents. *Curr. Med. Chem.* 21 (4): 437–457.
74 Leggio, G.M., Salomone, S., Bucolo, C. et al. (2013). Dopamine D_3 receptor as a new pharmacological target for the treatment of depression. *Eur. J. Pharmacol.* 719 (1–3): 25–33.
75 Corvol, J.C., Studler, J.M., Schonn, J.S. et al. (2001). Galpha(olf) is necessary for coupling D_1 and A_{2A} receptors to adenylyl cyclase in the striatum. *J. Neurochem.* 76 (5): 1585–1588.
76 Watson, J.B., Coulter, P.M. II, Margulies, J.E. et al. (1994). G-protein gamma 7 subunit is selectively expressed in medium-sized neurons and dendrites of the rat neostriatum. *J. Neurosci. Res.* 39 (1): 108–116.
77 Sidhu, A. and Niznik, H.B. (2000). Coupling of dopamine receptor subtypes to multiple and diverse G proteins. *Int. J. Dev. Neurosci.* 18 (7): 669–677.
78 Mystek, P., Tworzydlo, M., Dziedzicka-Wasylewska, M., and Polit, A. (2015). New insights into the model of dopamine D_1 receptor and G-proteins interactions. *Biochim. Biophys. Acta, Mol. Cell. Res.* 1853 (3): 594–603.
79 Cull-Candy, S., Kelly, L., and Farrant, M. (2006). Regulation of Ca^{2+}–permeable AMPA receptors: synaptic plasticity and beyond. *Curr. Opin. Neurobiol.* 16 (3): 288–297.
80 Karpa, K.D., Lidow, M.S., Pickering, M.T. et al. (1999). N-linked glycosylation is required for plasma membrane localization of D_5, but not D_1, dopamine receptors in transfected mammalian cells. *Mol. Pharmacol.* 56 (5): 1071–1078.

81 Mason, J.N., Kozell, L.B., and Neve, K.A. (2002). Regulation of dopamine D_1 receptor trafficking by protein kinase A-dependent phosphorylation. *Mol. Pharmacol.* 61 (4): 806–816.

82 Rankin, M.L. and Sibley, D.R. (2010). Constitutive phosphorylation by protein kinase C regulates D_1 dopamine receptor signaling. *J. Neurochem.* 115 (6): 1655–1667.

83 Jin, H., Xie, Z., George, S.R., and O'Dowd, B.F. (1999). Palmitoylation occurs at cysteine 347 and cysteine 351 of the dopamine D_1 receptor. *Eur. J. Pharmacol.* 386 (2–3): 305–312.

84 Jin, H., Zastawny, R., George, S.R., and O'Dowd, B.F. (1997). Elimination of palmitoylation sites in the human dopamine D_1 receptor does not affect receptor-G protein interaction. *Eur. J. Pharmacol.* 324 (1): 109–116.

85 Guo, Y. and Jose, P.A. (2011). C-terminal di-leucine motif of dopamine D_1 receptor plays an important role in its plasma membrane trafficking. *PLoS One* 6 (12): e29204.

86 D'Aoust, J.-P. and Tiberi, M. (2010). Role of the extracellular amino terminus and first membrane-spanning helix of dopamine D_1 and D_5 receptors in shaping ligand selectivity and efficacy. *Cell. Signalling* 22 (1): 106–116.

87 Palczewski, K., Kumasaka, T., Hori, T. et al. (2000). Crystal structure of rhodopsin: a G protein-coupled receptor. *Science* 289 (5480): 739–745.

88 Trumpp-Kallmeyer, S., Hoflack, J., Bruinvels, A., and Hibert, M. (1992). Modeling of G-protein-coupled receptors: application to dopamine, adrenaline, serotonin, acetylcholine, and mammalian opsin receptors. *J. Med. Chem.* 35 (19): 3448–3462.

89 Teeter, M.M., Froimowitz, M., Stec, B., and DuRand, C.J. (1994). Homology modeling of the dopamine D_2 receptor and its testing by docking of agonists and tricyclic antagonists. *J. Med. Chem.* 37 (18): 2874–2888.

90 Neve, K.A., Cumbay, M.G., Thompson, K.R. et al. (2001). Modeling and mutational analysis of a putative sodium-binding pocket on the dopamine D_2 receptor. *Mol. Pharmacol.* 60 (2): 373–381.

91 Varady, J., Wu, X., Fang, X. et al. (2003). Molecular modeling of the three-dimensional structure of dopamine 3 (D_3) subtype receptor: discovery of novel and potent D_3 ligands through a hybrid pharmacophore- and structure based database searching approach. *J. Med. Chem.* 46 (21): 4377–4392.

92 Baker, D. and Sali, A. (2001). Protein structure prediction and structural genomics. *Science* 294 (5540): 93–96.

93 Furse, K.E. and Lybrand, T.P. (2003). Three-dimensional models for beta-adrenergic receptor complexes with agonists and antagonists. *J. Med. Chem.* 46 (21): 4450–4462.

94 Trabanino, R.J., Hall, S.E., Vaidehi, N. et al. (2004). First principles predictions of the structure and function of G-protein-coupled receptors: validation for bovine rhodopsin. *Biophys. J.* 86 (4): 1904–1921.

95 Michino, M., Chen, J., Stevens, R.C., and Brooks, C.L. (2010). Fold GPCR: structure prediction protocol for the transmembrane domain of G protein-coupled receptors from class A. *Proteins* 78 (10): 2189–2201.

96 Vaidehi, N., Floriano, W.B., Trabanino, R. et al. (2002). Prediction of structure and function of G protein-coupled receptors. *Proc. Natl. Acad. Sci. U.S.A.* 99 (20): 12622–12627.
97 Kalani, M.Y.S., Vaidehi, N., Hall, S.E. et al. (2004). The predicted 3D structure of the human D_2 dopamine receptor and the binding site and binding affinities for agonists and antagonists. *Proc. Natl. Acad. Sci. U.S.A.* 101 (11): 3815–3820.
98 Li, Y., Zhu, F., Vaidehi, N. et al. (2007). Prediction of the 3D structure and dynamics of human DP G-protein coupled receptor bound to an agonist and an antagonist. *J. Am. Chem. Soc.* 129 (35): 10720–10731.
99 Abrol, R., Griffith, A.R., Bray, J.K., and Goddard, W.A. III (2012). Structure prediction of G protein-coupled receptors and their ensemble of functionally important conformations. In: *Membrane Protein Structure and Dynamics*, Methods in Molecular Biology (Methods and Protocols), vol. 914 (ed. N. Vaidehi and J. Klein-Seetharaman), 237–254. Totowa, NJ: Humana Press.
100 Kalani, Y. (2004). Structure and function studies of the human dopamine receptors. PhD dissertation. California Institute of Technology.
101 Sudandiradoss, C., Priya Doss, C.G., Rajasekaran, R. et al. (2008). Investigations on the interactions of scorpion neurotoxins with the predicted structure of D_1 dopamine receptor by protein–protein docking method. A bioinformatics approach. *C.R. Biol.* 331 (7): 489–499.
102 Malo, M., Brive, L., Luthman, K., and Svensson, P. (2010). Selective pharmacophore models of dopamine D_1 and D_2 full agonists based on extended pharmacophore features. *ChemMedChem* 5 (2): 232–246.
103 Floresca, C.Z. and Schetz, J.A. (2004). Dopamine receptor microdomains involved in molecular recognition and the regulation of drug affinity and function. *J. Recept. Signal Transduction* 24 (3): 207–239.
104 Ballesteros, J.A., Shi, L., and Javitch, J.A. (2001). Structural mimicry in G protein-coupled receptors: implications of the high-resolution structure of rhodopsin for structure-function analysis of rhodopsin-like receptors. *Mol. Pharmacol.* 60 (1): 1–19.
105 Pollock, N.J., Manelli, A.M., Hutchins, C.W. et al. (1992). Serine mutations in transmembrane V of the dopamine D_1 receptor affect ligand interactions and receptor activation. *J. Biol. Chem.* 267 (25): 17780–17786.
106 Bonner, L.A., Laban, U., Chemel, B.R. et al. (2011). Mapping the catechol binding site in dopamine D_1 receptors: synthesis and evaluation of two parallel series of bicyclic dopamine analogues. *ChemMedChem* 6 (6): 1024–1040.
107 Chemel, B.R., Bonner, L.A., Watts, V.J., and Nichols, D.E. (2012). Ligand-specific roles for transmembrane 5 serine residues in the binding and efficacy of dopamine D_1 receptor catechol agonists. *Mol. Pharmacol.* 81 (5): 729–738.
108 Lee, S.-M., Bhattacharya, S., Nagarajan, V., and Mailman, R. (2013). Conserved serines of transmembrane 5 in the dopamine D_1 receptor interact with non-catechol D_1 ligands. *FASEB J.* 27 (1 Suppl): 883–884.
109 Zivin, M., Sprah, L., and Sket, D. (1996). The D_1 receptor-mediated effects of the ergoline derivative LEK-8829 in rats with unilateral 6-hydroxydopamine lesions. *Br. J. Pharmacol.* 119 (6): 1187–1196.

110 Lee, S.-M. and Mailman, R. (2014). Critical interaction of the dopamine D_1 receptor with ergolines: the roles of conserved aromatic residues in transmembrane six segment (656.5). *FASEB J.* 28 (1 Suppl).

111 Jarvie, K.R. and Caron, M.G. (1993). Heterogeneity of dopamine receptors. *Adv. Neurol.* 60: 325–333.

112 Kimura, K., Sela, S., Bouvier, C. et al. (1995). Differential coupling of D_1 and D_5 dopamine receptors to guanine nucleotide binding proteins in transfected GH_4C_1 rat somatomammotrophic cells. *J. Neurochem.* 64 (5): 2118–2124.

113 Sidhu, A. (1998). Coupling of D_1 and D_5 dopamine receptors to multiple G proteins. *Mol. Neurobiol.* 16 (2): 125–134.

114 Sidhu, A., Kimura, K., Uh, M. et al. (1998). Multiple coupling of human D_5 dopamine receptors to guanine nucleotide binding proteins Gs and Gz. *J. Neurochem.* 70 (6): 2459–2467.

115 Demchyshyn, L.L., O'Dowd, B.F., and George, S.R. (2003). Structure of mammalian D_1 and D_5 dopamine receptors and their function and regulation in cells. *Neurol. Dis. Ther.* 56: 45–76.

116 Giorgioni, G., Piergentili, A., Ruggieri, S., and Quaglia, W. (2008). Dopamine D_5 receptors: a challenge to medicinal chemists. *Mini Rev. Med. Chem.* 8 (10): 976–995.

117 Guha, S., Baltazar, G.C., Tu, L.-A. et al. (2012). Stimulation of the D_5 dopamine receptor acidifies the lysosomal pH of retinal pigmented epithelial cells and decreases accumulation of autofluorescent photoreceptor debris. *J. Neurochem.* 122 (3–4): 823–833.

118 Chakraborty, R., Ganguli, S., Chakraborty, H.J., and Datta, A. (2010). Structural analysis and molecular modeling of human dopamine-receptor 5 (DRD5). *Int. J. Bioinf. Res.* 2 (2): 96–102.

119 Thompson, D. and Whistler, J.L. (2011). Trafficking properties of the D_5 dopamine receptor. *Traffic* 12 (5): 644–656.

120 Liu, F., Wan, Q., Pristupa, Z.B. et al. (2000). Direct protein-protein coupling enables cross-talk between dopamine D_5 and gamma-aminobutyric acid A receptors. *Nature* 403 (6767): 274–280.

121 Medin, T., Rinholm, J.E., Owe, S.G. et al. (2013). Low dopamine D_5 receptor density in hippocampus in an animal model of attention-deficit/hyperactivity disorder (ADHD). *Neuroscience* 242: 11–20.

122 Muir, W.J., Thomson, M.L., McKeon, P. et al. (2001). Markers close to the dopamine D_5 receptor gene (DRD5) show significant association with schizophrenia but not bipolar disorder. *Am. J. Med. Genet.* 105 (2): 152–158.

123 Wang, J., Liu, Z.L., and Chen, B. (2001). Dopamine D_5 receptor gene polymorphism and the risk of levodopa-induced motor fluctuations in patients with Parkinson's disease. *Neurosci. Lett.* 308 (1): 21–24.

124 Kalani, M.Y.S. (2004). *Structure and Function Studies Of The Human Dopamine Receptors*. Pasadena, CA: California Institute of Technology.

125 Simonovic, M., Soskic, V., and Joksimovic, J. (1998). Quantification of human dopamine D_{2S} receptor interactions with G alpha(i,1,2)- and G alpha(o)-proteins. *Neurochem. Int.* 33 (3): 271–275.

126 Filteau, F., Veilleux, F., and Levesque, D. (1999). Effects of reciprocal chimeras between the C-terminal portion of third intracellular loops of the human dopamine D_2 and D_3 receptors. *FEBS Lett.* 447 (2–3): 251–256.
127 Senogles, S.E., Heimert, T.L., Odife, E.R., and Quasney, M.W. (2004). A region of the third intracellular loop of the short form of the D_2 dopamine receptor dictates Gi coupling specificity. *J. Biol. Chem.* 279 (3): 1601–1606.
128 Grunewald, S., Reilander, H., and Michel, H. (1996). In vivo reconstitution of dopamine D_{2S} receptor-mediated G protein activation in baculovirus-infected insect cells: preferred coupling to G_{i1} versus G_{i2}. *Biochemistry* 35 (48): 15162–15173.
129 O'Hara, C.M., Tang, L., Taussig, R. et al. (1996). Dopamine D_{2L} receptor couples to G alpha i2 and G alpha i3 but not G alpha i1, leading to the inhibition of adenylate cyclase in transfected cell lines. *J. Pharmacol. Exp. Ther.* 278 (1): 354–360.
130 Wolfe, S.E. and Morris, S.J. (1999). Dopamine D_2 receptor isoforms expressed in AtT20 cells differentially couple to G proteins to acutely inhibit high voltage-activated calcium channels. *J. Neurochem.* 73 (6): 2375–2382.
131 Jiang, M., Spicher, K., Boulay, G. et al. (2001). Most central nervous system D_2 dopamine receptors are coupled to their effectors by Go. *Proc. Natl. Acad. Sci. U.S.A.* 98 (6): 3577–3582.
132 Tang, L., Todd, R.D., Heller, A., and O'Malley, K.L. (1994). Pharmacological and functional characterization of D_2, D_3, and D_4 dopamine receptors in fibroblast and dopaminergic cell lines. *J. Pharmacol. Exp. Ther.* 268 (1): 495–502.
133 Potenza, M.N., Graminski, G.F., Schmauss, C., and Lerner, M.R. (1994). Functional expression and characterization of human D_2 and D_3 dopamine receptors. *J. Neurosci.* 14 (3 Pt. 2): 1463–1476.
134 Newman-Tancredi, A., Cussac, D., Marini, L., and Millan, M.J. (2002). Antibody capture assay reveals bell-shaped concentration-response isotherms for h5-HT_{1A} receptor-mediated Galpha(i3) activation: conformational selection by high-efficacy agonists, and relationship to trafficking of receptor signaling. *Mol. Pharmacol.* 62 (3): 590–601.
135 Liu, L.X., Burgess, L.H., Gonzalez, A.M. et al. (1999). D_{2S}, D_{2L}, D_3, and D_4 dopamine receptors couple to a voltage-dependent potassium current in N18TG2 x mesencephalon hybrid cell (MES-23.5) via distinct G proteins. *Synapse* 31 (2): 108–118.
136 Yamaguchi, I., Harmon, S.K., Todd, R.D., and O'Malley, K.L. (1997). The rat D_4 dopamine receptor couples to cone transducin (Galphat2) to inhibit forskolin-stimulated cAMP accumulation. *J. Biol. Chem.* 272 (26): 16599–16602.
137 Khan, Z.U., Mrzljak, L., Gutierrez, A. et al. (1998). Prominence of the dopamine D_2 short isoform in dopaminergic pathways. *Proc. Natl. Acad. Sci. U.S.A.* 95 (13): 7731–7736.
138 Shi, L. and Javitch, J.A. (2002). The binding site of aminergic G protein-coupled receptors: the transmembrane segments and second extracellular loop. *Annu. Rev. Pharmacol. Toxicol.* 42: 437–467.
139 Van Arnam, E.B., Lester, H.A., and Dougherty, D.A. (2011). Dissecting the functions of conserved prolines within transmembrane helices of the D_2 dopamine receptor. *ACS Chem. Biol.* 6 (10): 1063–1068.

140 Durdagi, S., Salmas, R.E., Stein, M. et al. (2016). Binding interactions of dopamine and apomorphine in D_2 high and D_2 low states of human dopamine D_2 receptor using computational and experimental techniques. *ACS Chem. Neurosci.* 7 (2): 185–195.

141 Pandalaneni, S., Karuppiah, V., Saleem, M. et al. (2015). Neuronal calcium sensor-1 binds the D_2 dopamine receptor and G-protein-coupled receptor Kinase 1 (GRK1) peptides using different modes of interactions. *J. Biol. Chem.* 290 (30): 18744–18756.

142 Schmauss, C. (1996). Enhanced cleavage of an atypical intron of dopamine D_3-receptor pre-mRNA in chronic schizophrenia. *J. Neurosci.* 16 (24): 7902–7909.

143 Schwartz, J.C., Diaz, J., Pilon, C., and Sokoloff, P. (2000). Possible implications of the dopamine D_3 receptor in schizophrenia and in antipsychotic drug actions. *Brain Res. Rev.* 31 (2–3): 277–287.

144 Chien, E.Y., Liu, W., Zhao, Q. et al. (2010). Structure of the human dopamine D_3 receptor in complex with a D_2/D_3 selective antagonist. *Science* 330 (6007): 1091–1095.

145 Carlsson, J., Coleman, R.G., Setola, V. et al. (2011). Ligand discovery from a dopamine D_3 receptor homology model and crystal structure. *Nat. Chem. Biol.* 7 (11): 769–778.

146 Platania, C.B.M., Salomone, S., Leggio, G.M. et al. (2012). Homology modeling of dopamine D_2 and D_3 receptors: molecular dynamics refinement and docking evaluation. *PLoS One* 7 (9): e44316.

147 Yap, B.K., Buckle, M.J.C., and Doughty, S.W. (2012). Homology modeling of the human 5-HT_{1A}, 5-HT_{2A}, D_1, and D_2 receptors: model refinement with molecular dynamics simulations and docking evaluation. *J. Mol. Model.* 18 (8): 3639–3655.

148 Tarcsay, A., Paragi, G., Vass, M. et al. (2013). The impact of molecular dynamics sampling on the performance of virtual screening against GPCRs. *J. Chem. Inf. Model.* 53 (11): 2990–2999.

149 Vaidehi, N., Bhattacharya, S., and Larsen, A.B. (2014). Structure and dynamics of G-protein coupled receptors. In: *G Protein-Coupled Receptors--Modeling and Simulation*, Advances in Experimental Medicine and Biology, vol. 796 (ed. M. Filizola), 37–54. Dordrecht: Springer.

150 Levoin, N., Calmels, T., Krief, S. et al. (2011). Homology model versus X-ray structure in receptor-based drug design: a retrospective analysis with the dopamine D_3 receptor. *ACS Med. Chem. Lett.* 2 (4): 293–297.

151 Beuming, T. and Sherman, W. (2012). Current assessment of docking into GPCR crystal structures and homology models: successes, challenges, and guidelines. *J. Chem. Inf. Model.* 52 (12): 3263–3277.

152 Ostopovici-Halip, L. and Rad-Curpan, R. (2014). Modeling of ligand binding to the dopamine D_2 receptor. *J. Serb. Chem. Soc.* 79 (2): 175–183.

153 Sukalovic, V., Soskic, V., Sencanski, M. et al. (2013). Determination of key receptor-ligand interactions of dopaminergic arylpiperazines and the dopamine D_2 receptor homology model. *J. Mol. Model.* 19 (4): 1751–1762.

154 Duan, X., Zhang, M., Zhang, X. et al. (2015). Molecular modeling and docking study on dopamine D_2-like and serotonin 5-HT_{2A} receptors. *J. Mol. Graphics Model.* 57: 143–155.

155 Daeffler, K.N.M., Lester, H.A., and Dougherty, D.A. (2012). Functionally important aromatic-aromatic and sulfur-π interactions in the D_2 dopamine receptor. *J. Am. Chem. Soc.* 134 (36): 14890–14896.

156 Kling, R.C., Tschammer, N., Lanig, H. et al. (2014). Active-state model of a dopamine D_2 receptor – Gαi complex stabilized by aripiprazole-type partial agonists. *PLoS One* 9 (6): e100069/100061–e100069/100010, 100010 pp.

157 Sensoy, O. and Weinstein, H. (2015). A mechanistic role of helix 8 in GPCRs: computational modeling of the dopamine D_2 receptor interaction with the GIPC1-PDZ-domain. *Biochim. Biophys. Acta, Biomembr.* 1848 (4): 976–983.

158 Vass, M., Agai-Csongor, E., Horti, F., and Keseru, G.M. (2014). Multiple fragment docking and linking in primary and secondary pockets of dopamine receptors. *ACS Med. Chem. Lett.* 5 (9): 1010–1014.

159 Ananthan, S., Saini, S.K., Zhou, G. et al. (2014). Design, synthesis, and structure-activity relationship studies of a series of [4-(4-carboxamidobutyl)]-1-arylpiperazines: insights into structural features contributing to dopamine D_3 versus D_2 receptor subtype selectivity. *J. Med. Chem.* 57 (16): 7042–7060.

160 Gomez-Jeria, J.S. and Valdebenito-Gamboa, J. (2015). A density functional study of the relationships between electronic structure and dopamine D_2 receptor binding affinity of a series of [4-(4-carboxamidobutyl)]-1-arylpiperazines. *Res. J. Pharm., Biol. Chem. Sci.* 6 (6): 203–218.

161 Gomez-Jeria, J.S. and Valdebenito-Gamboa, J. (2015). Electronic structure and docking studies of the dopamine D_3 receptor binding affinity of a series of [4-(4-carboxamidobutyl)]-1-arylpiperazines. *Pharma Chem.* 7 (7): 323–347.

162 Kota, K., Kuzhikandathil, E.V., Afrasiabi, M. et al. (2015). Identification of key residues involved in the activation and signaling properties of dopamine D_3 receptor. *Pharmacol. Res.* 99: 174–184.

163 Marsango, S., Caltabiano, G., Pou, C. et al. (2015). Analysis of human dopamine D_3 receptor quaternary structure. *J. Biol. Chem.* 290 (24): 15146–15162.

164 Lanau, F., Brockhaus, M., Pink, J.R. et al. (1997). Development and characterization of antibodies against the N terminus of the human dopamine D_4 receptor. *J. Neurochem.* 69 (5): 2169–2178.

165 Oak, J.N., Oldenhof, J., and Van Tol, H.H.M. (2000). The dopamine D_4 receptor: one decade of research. *Eur. J. Pharmacol.* 405 (1–3): 303–327.

166 Woods, A.S. (2010). The dopamine D_4 receptor, the ultimate disordered protein. *J. Recept. Signal Transduction* 30 (5): 331–336.

167 Rondou, P., Haegeman, G., and Van Craenenbroeck, K. (2010). The dopamine D_4 receptor: biochemical and signalling properties. *Cell. Mol. Life Sci.* 67 (12): 1971–1986.

168 Ding, Y.C., Chi, H.C., Grady, D.L. et al. (2002). Evidence of positive selection acting at the human dopamine receptor D_4 gene locus. *Proc. Natl. Acad. Sci. U.S.A.* 99 (1): 309–314.

169 Jovanovic, V., Guan, H.C., and Van Tol, H.H. (1999). Comparative pharmacological and functional analysis of the human dopamine D4.2 and D4.10 receptor variants. *Pharmacogenetics* 9 (5): 561–568.

170 Van Craenenbroeck, K., Borroto-Escuela, D.O., Skieterska, K. et al. (2014). Role of dimerization in dopamine D_4 receptor biogenesis. *Curr. Protein Pept. Sci.* 15 (7): 659–665.
171 LaHoste, G.J., Swanson, J.M., Wigal, S.B. et al. (1996). Dopamine D_4 receptor gene polymorphism is associated with attention deficit hyperactivity disorder. *Mol. Psychiatry* 1 (2): 121–124.
172 DiMaio, S., Grizenko, N., and Joober, R. (2003). Dopamine genes and attention-deficit hyperactivity disorder: a review. *J. Psychiatry Neurosci* 28 (1): 27–38.
173 Nanko, S., Hattori, M., Ikeda, K. et al. (1993). Dopamine D_4 receptor polymorphism and schizophrenia. *Lancet* 341 (8846): 689–690.
174 Semba, J. (2004). Functional role of dopamine D_4 receptor in schizophrenia. *Japan. J. Physchopharm.* 24 (1): 13–20.
175 Wimalasena, K. (2011). Vesicular monoamine transporters: structure-function, pharmacology, and medicinal chemistry. *Med. Res. Rev.* 31 (4): 483–519.
176 Henry, J.P., Sagné, C., Bedet, C., and Gasnier, B. (1998). The vesicular monoamine transporter: from chromaffin granule to brain. *Neurochem. Int.* 32 (3): 227–246.
177 Weihe, E., Schäfer, M.K.H., Erickson, J.D., and Eiden, L.E. (1994). Localization of vesicular monoamine transporter isoforms (VMAT1 and VMAT2) to endocrine cells and neurons in rat. *J. Mol. Neurosci.* 5 (3): 149–164.
178 Eiden, L.E., Schäfer, M.K.H., Weihe, E., and Schütz, B. (2004). The vesicular amine transporter family (SLC18): amine/proton antiporters required for vesicular accumulation and regulated exocytotic secretion of monoamines and acetylcholine. *Pflugers Archiv. Eur. J. Physiol.* 447 (5): 636–640.
179 Lohoff, F.W. (2010). Genetic variants in the vesicular monoamine transporter 1 (VMAT1/SLC18A1) and neuropsychiatric disorders. *Methods Mol. Biol.* 637: 165–180.
180 Richards, M., Iijima, Y., Kondo, H. et al. (2006). Association study of the vesicular monoamine transporter 1 (VMAT1) gene with schizophrenia in a Japanese population. *Behav. Brain Funct.* 2: 39.
181 Lohoff, F.W., Dahl, J.P., Ferraro, T.N. et al. (2006). Variations in the vesicular monoamine transporter 1 gene (VMAT1/SLC18A1) are associated with bipolar I disorder. *Neuropsychopharmacology* 31 (12): 2739–2747.
182 Nirenberg, M.J., Chan, J., Liu, Y. et al. (1997). *Advances in Pharmacology*, vol. 42, 240–243. Elsevier B.V.
183 Frey, K.A., Koeppe, R.A., and Kilbourn, M.R. (2001). Imaging the vesicular monoamine transporter. *Adv. Neurol.* 86: 237–247.
184 Kilbourn, M. (2013). Imaging the vesicular monoamine transporter (VMAT2) in neurodegenerative diseases. In: *Catecholamine Research in the 21st Century: Abstracts and Graphical Abstracts, 10th International Catecholamine Symposium, 2012*, 20. Elsevier B.V.
185 Schwartz, K., Yadid, G., Weizman, A., and Rehavi, M. (2003). Decreased limbic vesicular monoamine transporter 2 in a genetic rat model of depression. *Brain Res.* 965 (1–2): 174–179.
186 Miller, G.W., Erickson, J.D., Perez, J.T. et al. (1999). Immunochemical analysis of vesicular monoamine transporter (VMAT2) protein in Parkinson's disease. *Exp. Neurol.* 156 (1): 138–148.

187 Van Liefferinge, J., Massie, A., Portelli, J. et al. (2013). Are vesicular neurotransmitter transporters potential treatment targets for temporal lobe epilepsy? *Front. Cell. Neurosci.* 7 (August): 139/131–139/124.

188 Brunk, I., Blex, C., Speidel, D. et al. (2009). Ca^{2+}–dependent activator proteins of secretion promote vesicular monoamine uptake. *J. Biol. Chem.* 284 (2): 1050–1056.

189 Yaffe, D., Vergara-Jaque, A., Forrest, L.R., and Schuldiner, S. (2016). Emulating proton-induced conformational changes in the vesicular monoamine transporter VMAT2 by mutagenesis. *Proc. Natl. Acad. Sci. U.S.A.* 113 (47): E7390–E7398.

190 Merickel, A., Rosandich, P., Peter, D., and Edwards, R.H. (1995). Identification of residues involved in substrate recognition by a vesicular monoamine transporter. *J. Biol. Chem.* 270 (43): 25798–25804.

191 Merickel, A., Kaback, H.R., and Edwards, R.H. (1997). Charged residues in transmembrane domains II and XI of a vesicular monoamine transporter form a charge pair that promotes high affinity substrate recognition. *J. Biol. Chem.* 272 (9): 5403–5408.

192 Shan, J., Javitch, J.A., Shi, L., and Weinstein, H. (2011). The substrate-driven transition to an inward-facing conformation in the functional mechanism of the dopamine transporter. *PLoS One* 6 (1): e16350.

193 Vaughan, R.A. and Foster, J.D. (2013). Mechanisms of dopamine transporter regulation in normal and disease states. *Trends Pharmacol. Sci.* 34 (9): 489–496.

194 Frost, J.J., Rosier, A.J., Reich, S.G. et al. (1993). Positron emission tomographic imaging of the dopamine transporter with ^{11}C-WIN 35,428 reveals marked declines in mild Parkinson's disease. *Ann. Neurol.* 34 (3): 423–431.

195 Malison, R.T., McDougle, C.J., van Dyck, C.H. et al. (1995). [123I]beta-CIT SPECT imaging of striatal dopamine transporter binding in Tourette's disorder. *Am. J. Psychiatry* 152 (9): 1359–1361.

196 Siepel, F.J., Dalen, I., Grüner, R. et al. (2016). Loss of dopamine transporter binding and clinical symptoms in dementia with lewy bodies. *Mov. Disord.* 31 (1): 118–125.

197 Laasonen-Balk, T., Kuikka, J., Viinamaki, H. et al. (1999). Striatal dopamine transporter density in major depression. *Psychopharmacology* 144 (3): 282–285.

198 Wong, D.F., Harris, J.C., Naidu, S. et al. (1996). Dopamine transporters are markedly reduced in Lesch–Nyhan disease in vivo. *Proc. Natl. Acad. Sci. U.S.A.* 93 (11): 5539–5543.

199 Brooks, D.J. (2016). Molecular imaging of dopamine transporters. *Ageing Res. Rev.* 30: 114–121.

200 Yoshii, F., Ryo, M., Baba, Y. et al. (2017). Combined use of dopamine transporter imaging (DAT-SPECT) and ^{123}I-metaiodobenzylguanidine (MIBG) myocardial scintigraphy for diagnosing Parkinson's disease. *J. Neurol. Sci.* 375: 80–85.

201 Terbeek, J., Latour, P., Van Laere, K., and Vandenberghe, W. (2017). Abnormal dopamine transporter imaging in adult-onset Niemann–Pick disease type C. *Parkinsonism Relat. Disord.* 36: 107–108.

202 Washida, K., Kowa, H., Yamamoto, Y. et al. (2017). Dopamine transporter imaging as a diagnostic modality for atypical Alzheimer's disease mimicking corticobasal degeneration. *Psychogeriatrics* 17 (1): 73–75.

203 Yildiz, Y., Pektas, E., Tokatli, A., and Haliloglu, G. (2017). Hereditary dopamine transporter deficiency syndrome: challenges in diagnosis and treatment. *Neuropediatrics* 48 (1): 49–52.

204 Xu, S., Liu, Y., Li, Y. et al. (2017). Availability of dopamine transporters in heroin-dependent subjects: a ^{18}F-FECNT PET imaging study. *Psychiatry Res. Neuroimaging* 263: 121–126.

205 Karila, L., Leroy, C., Dubol, M. et al. (2016). Dopamine transporter correlates and occupancy by modafinil in cocaine-dependent patients: a controlled study with high-resolution PET and [^{11}C]-PE2I. *Neuropsychopharmacology* 41 (9): 2294–2302.

206 Penmatsa, A., Wang, K.H., and Gouaux, E. (2013). X-ray structure of dopamine transporter elucidates antidepressant mechanism. *Nature* 503 (7474): 85–90.

207 Penmatsa, A., Wang, K.H., and Gouaux, E. (2015). X-ray structures of *Drosophila* dopamine transporter in complex with nisoxetine and reboxetine. *Nat. Struct. Mol. Biol.* 22 (6): 506–508.

208 Wang, K.H., Penmatsa, A., and Gouaux, E. (2015). Neurotransmitter and psychostimulant recognition by the dopamine transporter. *Nature* 521 (7552): 322–327.

209 Han, D.D. and Gu, H.H. (2006). Comparison of the monoamine transporters from human and mouse in their sensitivities to psychostimulant drugs. *BMC Pharmacol.* 6: 6.

210 Cheng, M.H. and Bahar, I. (2015). Molecular mechanism of dopamine transport by human dopamine transporter. *Structure* 23 (11): 2171–2181.

211 Sweeney, C.G., Tremblay, B.P., Stockner, T. et al. (2017). Dopamine transporter amino and carboxyl termini synergistically contribute to substrate and inhibitor affinities. *J. Biol. Chem.* 292 (4): 1302–1309.

212 Seddik, A., Geerke, D.P., Stockner, T. et al. (2016). Combined simulation and mutation studies to elucidate selectivity of unsubstituted amphetamine-like cathinones at the dopamine transporter. *Mol. Inf.* 36 (5–6): 1600094.

213 Foster, J.D. and Vaughan, R.A. (2016). Phosphorylation mechanisms in dopamine transporter regulation. *J. Chem. Neuroanat.* 83–84: 10–18.

214 Rastedt, D.E., Vaughan, R.A., and Foster, J.D. (2016). Palmitoylation mechanisms in dopamine transporter regulation. *J. Chem. Neuroanat.* 83–84: 3–9.

215 Moritz, A.E., Rastedt, D.E., Stanislowski, D.J. et al. (2015). Reciprocal phosphorylation and palmitoylation control dopamine transporter kinetics. *J. Biol. Chem.* 290 (48): 29095–29105.

216 Li, Y., Mayer, F.P., Hasenhuetl, P.S. et al. (2017). Occupancy of the zinc-binding site by transition metals decreases the substrate affinity of the human dopamine transporter by an allosteric mechanism. *J. Biol. Chem.* 292 (10): 4235–4243.

217 Chen, N. and Reith, M.E. (2000). Structure and function of the dopamine transporter. *Eur. J. Pharmacol.* 405 (1–3): 329–339.

218 Lohr, K.M., Masoud, S.T., Salahpour, A., and Miller, G.W. (2017). Membrane transporters as mediators of synaptic dopamine dynamics: implications for disease. *Eur. J. Neurosci.* 45 (1): 20–33.

2

The Noradrenergic System

2.1 Introduction

The noradrenergic system is a neuronal organization (noradrenergic neurons) located in both the central and peripheral nervous systems and responsible for the synthesis, storage, and release of norepinephrine (NE), also known as noradrenaline [1]. Although the central and peripheral noradrenergic systems are generally considered as two separate systems sharing a common neurotransmitter, there are some evidences of a direct link between them [2]. Herein, we focus our attention on the central noradrenergic system (CNA), which is closely associated with various cerebral functions such as sleeping, memory, learning, and emotions. Several neurodisorders course with an abnormal noradrenergic activity. The highest activity has been observed in all forms of stress [3], including panic disorders [4], posttraumatic stress disorder [5, 6], fear [7], and anxiety [8, 9]. In fact, enhanced noradrenergic activity in the amygdala has been related to posttraumatic stress disorder [10]. Other disorders showing abnormalities in the noradrenergic system include attention-deficit hyperactivity disorder (ADHD) [11], Tourette's syndrome [12], reward-dependence related disorders [13], and Alzheimer's disease [14].

There are two major projections of the CNA (Figure 2.1): (i) originated in the locus coeruleus (LC), responsible for almost the entire output to the cerebral cortex [15] and (ii) originated in the lateral tegmental area, responsible for the output to the spinal cord. Ascending and descending fibers from the two projections innervate the cerebral cortex, brainstem, thalamus, hypothalamus, limbic forebrain structures, cerebellum, and spinal cord. Neurons located at the LC degenerate early and progressively in Parkinson's disease and Alzheimer's disease [16].

The noradrenergic system originated in the LC uses norepinephrine as neurotransmitter. Norepinephrine is biosynthesized from the amino acid tyrosine via dopamine. Actually, norepinephrine is a degradation product of dopamine. Noradrenergic neurons may contain and release neurotransmitters other than norepinephrine through a process called "cotransmission" [17] that allows modulating the action of norepinephrine both at the presynaptic and postsynaptic sites. The main cotransmitters present in central noradrenergic neurons are adenosine triphosphate (ATP) [18], glutamate [19], and neuropeptides Y (NPY, 97 residues), enkephalins (in both variants, i.e. pentapeptides ended with Leu or

Chemical Biology of Neurodegeneration: A Molecular Approach, First Edition. Pedro Merino.

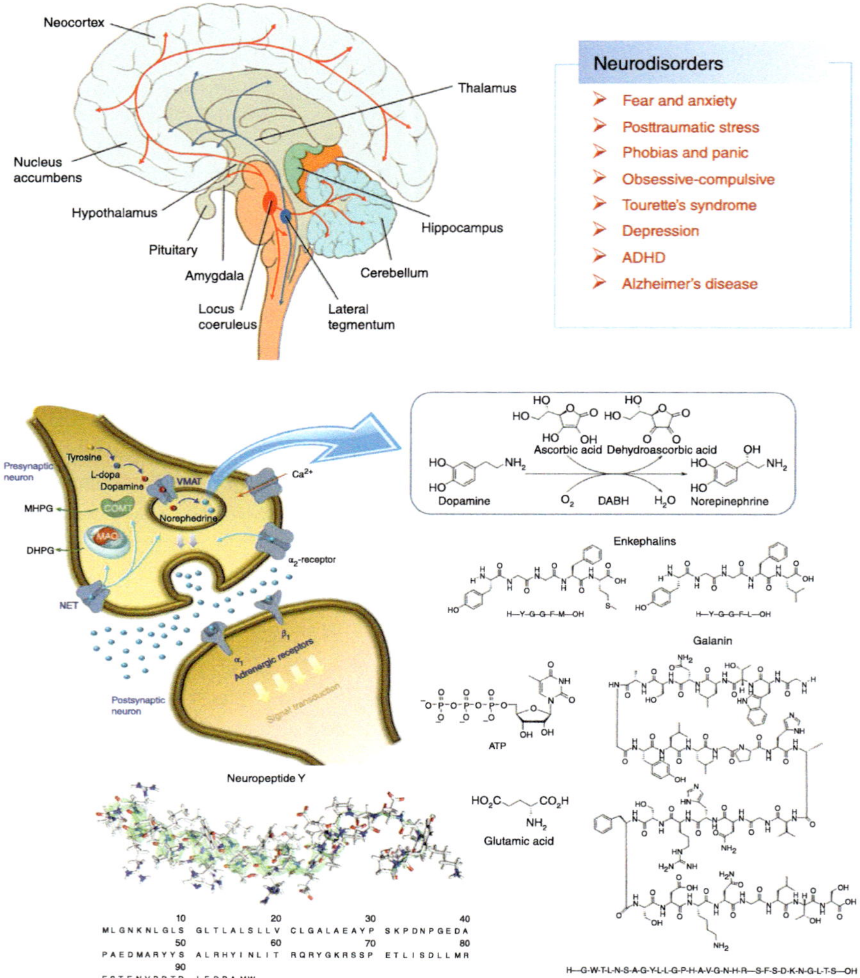

Figure 2.1 The noradrenergic system. (a) Pathways of the noradrenergic system. (b) Neurological disorders associated with noradrenergic system. (c) Details of the noradrenergic synapse and neurotransmitters involved in the noradrenergic system. Norepinephrine (NE) is synthesized from dopamine inside the noradrenergic neuron. After packaging in vesicles by VMAT, NE is released into the synapse. NE is recognized by adrenoreceptors α_1 and β_1 that produce the signal. The excess of NE is removed from the synaptic area by DAT and adrenoreceptors α_2 and reintroduced in the neuron for being reutilized or catabolized by MAO. VMAT: vesicular monoamine transporter; NET: norepinephrine transporter; α_1, α_2, and β_1, adrenoreceptors; MAO: monoamine oxidase; COMT: catechol-*O*-methyl transferase; DHPG: 3,4-dihydroxyphenylglycol; MHPG: 3-methoxy-4-hydroxyphenylglycol.

Cys) [20], and galanin (30 residues) [21, 22] (Figure 2.1). Indeed, neuropeptides are the most common signaling molecules in the central nervous system (CNS). It has been suggested that dysregulation of these neuropeptides may be of importance for neurodisorders associated with cognitive impairments [23] due to their direct implication in learning and memory [24].

Autoregulation of the noradrenergic neuronal activity is facilitated by different transporters and receptors located in the vesicular and cell membranes. Adrenoreceptors, belonging to the family of G-protein-coupled receptors (GPCRs), are stimulated by norepinephrine and epinephrine [25]. They can either activate or inhibit release of neurotransmitters; there are three classes: α_1-adrenoreceptors (excitatory), α_2-adrenoreceptors (inhibitory), and β-adrenoreceptors (excitatory or inhibitory). We focus our attention on those located in the brain and their relationship with neurodisorders. Reports from the last 10 years will only be considered. For detailed aspects of adrenergic receptors reported earlier, the reader is directed to a book entitled "The adrenergic receptors in the 21st century" and published in 2005 [26]. Several aspects of adrenoreceptors related to structure–function, imaging, genetically altered models, and pharmacogenomics are collected in a treatise edited by Perez in 2006 [27]. The cotransmitters have also specific receptors whose mode of action is coupled to adrenoreceptors. Receptors for neuropeptides NPY [28], galanin [29], and enkephalins (opiate receptors) [30] are also GPCRs. As neuropeptides are involved in neurodisorders associated with cognitive impairments, neuropeptide receptors are therapeutic targets for a number of disorders involving cognitive deficiencies [31].

Norepinephrine (NE) is synthesized from dopamine by dopamine-β-hydroxylase (DBH) within storage vesicles and then released into the synaptic area to bind reversibly to adrenoreceptors of postsynaptic neuron. The α_2-adrenoreceptors located at the presynaptic membrane inhibit the release of additional NE when it is detected in the synapse. At that time, removal of NE from the synaptic space terminates the adrenergic impulse. Reuptake of norepinephrine is carried out by the norepinephrine transporter (NET). NET is responsible for the termination of synaptic activity, and it can be inhibited by cocaine and tricyclic antidepressants, resulting in an increase in NE activity in the synaptic cleft. As in the case of dopamine transporter (DAT), NET cotransports the neurotransmitter with Na^+ and Cl^-. Mitochondrial monoamine oxidase (MAO) and catechol-*O*-methyl transferase (COMT) are responsible for the catabolism of norepinephrine, which yields 3,4-dihydroxyphenylglycol(DHPG) and 3-methoxy-4-hydroxyphenylglycol (MHPG), respectively [32, 33].

2.2 Adrenoreceptors

2.2.1 Classification

Initially, adrenoreceptors were divided into two major types [34]: α and β and further subdivided into α_1/α_2 [35] and β_1/β_2 [36]. The currently accepted classification based on both structural and pharmacological evidences identifies three major types: α_1, α_2, and β [37]. In addition, several subtypes of adrenoreceptors have been characterized pharmacologically based on their affinities for selective ligands. The α_1-adrenoceptor subclass includes the subtypes α_{1A}, α_{1B}, and α_{1D}. Historically, after identification of α_{1A}-adrenoceptor, another one was cloned from bovine brain and identified as α_{1C}-subtype [38]. However, it was demonstrated that, pharmacologically, it corresponds to the α_{1A}-subtype [39].

To avoid confusion, α_{1C}-subtype was no longer considered and naming was continued with letter D [40]. The α_2-adrenoceptor subclass comprises the subtypes $\alpha_{2A/D}$, α_{2B}, and α_{2C} [41]. The β-adrenoceptor subclass is constituted by three subtypes: β_1, β_2, and β_3 [42]. Most of the receptor subtypes have been genetically defined and cloned, and all of them have been identified in the CNS: α_1-adrenoceptor subtypes [43–46], α_2-adrenoceptor subtypes [47, 48], and β-adrenoceptor subtypes [49–52]. Presynaptic adrenoceptors, both auto- and heteroreceptors, belong to the $\alpha_{2A/D}$ subtype and have been specifically identified [53, 54].

2.2.2 Location

Adrenoreceptors are located in various tissues. We focus our attention on those located in the brain. The α_1-adrenoceptors are located in the brain, in the cerebral cortex (α_{1A} and α_{1D} subtypes), and fetal brain (α_{1B} subtype). α_2-Adrenoceptors are located in several parts of the brain. In particular, α_{2A}-subtype is located in the LC, projecting noradrenergic signal to the hypothalamus, especially the paraventricular nucleus (PVN) [55, 56]. In the hypothalamus, adrenergic receptors are important components of the ascending arousal pathway [57]. α_{2A}-Receptors are also located in the hippocampus. The α_{2B}-subtype is located in the thalamus and the α_{2C}-subtype in the olfactory bulb and hippocampus. Initial experiments with quantitative autoradiography showed that β-receptors in human brain was predominant in all subfields of the hippocampus, followed by cerebellum, and then thalamic nuclei, basal ganglia, midbrain, and cerebral cortex. They were also found in white matter and hypothalamus but in lower concentration [58]. In particular, β_1-subtype is located in the cerebral cortex and hypothalamus, whereas β_2-subtype is located in the cerebellum, hippocampus, cerebral cortex, and olfactory bulb [59]. The location of β_3-receptors is not clear. In the synapse, all adrenoreceptors are in postsynaptic locations, but there are evidences of presynaptic locations of α_2-adrenoceptors to regulate the release of norepinephrine acting as an autoreceptor.

2.2.3 General Structure Considerations

Adrenoreceptors (adrenergic receptors) are GPCRs, thus exhibiting the common motif of a single-peptide chain looped back and forth through the cell membrane seven times (seven transmembrane domains identified from TM1 to TM7) with the extracellular N-terminus and the intracellular C-terminus. Aligned sequences of all subtypes of adrenoreceptors indicating extracellular, intracellular, and transmembrane residues, as well as positions for N-glycosylation, O-phosphorylation, and S-palmitoylation will be given in Sections 2.2, 2.2.6, and 2.2.8 for each subtype. Phosphorylation by protein kinase A (PKA) and G-protein receptor kinase (GRK) is of special importance because it regulates GPCR activity and, in particular, can downregulate β-adrenoreceptors.

2.2.4 Effector Mechanisms

Each receptor type couples to a different G protein; consequently, the three groups present different signal transduction pathways (Figure 2.2). The α_1-receptors

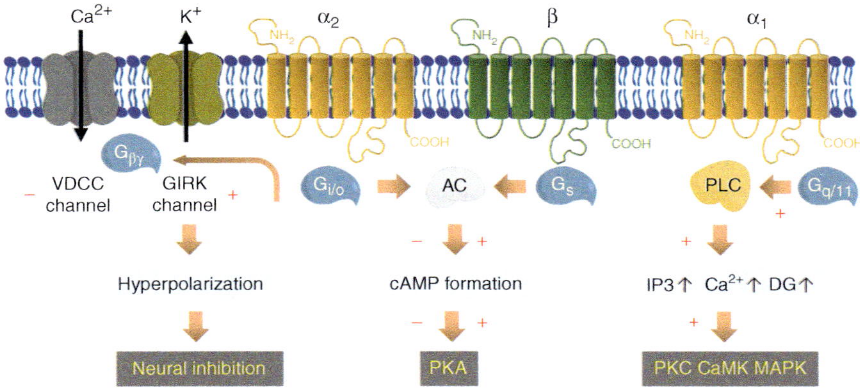

Figure 2.2 Signal transduction of adrenergic receptors via coupling to individual G proteins (e.g. $G_{i/o}$, G_s, and $G_{q/11}$). AC, adenylate cyclase; CaMK: calcium/calmodulin-dependent protein kinase; DG: diacylglycerol; GIRK: G-protein-coupled inwardly rectifying potassium; IP3: inositol triphosphate; PLC: phospholipase C; VDCC: voltage-dependent calcium channel; MAPK: mitogen-activated protein kinase; PKA: protein kinase A; PKC: protein kinase C.

couple to $G_q/G_{11,}$ increasing calcium ion release by activating phospholipase C (PLC) and protein kinase C (PKC) through increasing the concentration of inositol triphosphate and diacylglycerol. These receptors also activate mitogen-activated protein kinases (MAPK) and phospholipase D. The α_2-receptors couple to G_i/G_o proteins and inhibit adenylyl cyclase, decreasing the formation of cyclic adenosine monophosphate (cAMP), but stimulate phospholipase A_2. Activation of α_2-receptors leads to the release of $G_{\beta\gamma}$ resulting in activation of potassium ion channels G-protein-coupled inwardly rectifying potassium (GIRK) and inhibition of calcium channels. Activation of GIRK channels results in membrane hyperpolarization, leading to neural inhibition. The β-receptors couple to G_S thereby activating adenylyl cyclase and increasing the release of cAMP. Some studies suggest additional activation of MAPK in a similar way to α_1-receptors. α_2- and β-adrenergic receptors couple to G proteins such as D_2-type and D_1-type dopaminergic receptors, respectively (see Section 1.2.4).

2.2.5 Implication in Neurodisorders

Adrenoreceptors are directly involved in basic functions such as memory and alertness that diminish with aging. In fact, age-related changes in arousal, mood, and memory have been correlated with the loss of adrenergic receptors [60]. At a neuronal level, stimulation of α_1-adrenoceptors in combination with D_1 dopamine receptor produces anorexia and is related to the efficacy of appetite suppressants in the treatment of obesity [61]. Activation of α_{1A}-adrenoceptor regulates inhibition in the olfactory bulb [62] and leads to wakefulness, while blockade of these receptors elicits sedation. Activation of α_1- and β-adrenergic receptors enhances long-term memory consolidation in the amygdala and hippocampus. Antipsychotic drugs block α_1-adrenergic receptors, and antidepressant drugs increase norepinephrine levels.

Dysfunction of the noradrenergic system has been demonstrated in schizophrenia [63], major depression [64], Parkinson's disease, [65], posttraumatic stress disorder [6], ADHD [66], and autism-related disorders [67]. In particular, adrenergic receptors have been related to schizophrenia and mood disorders [68].Dysregulation of α_2-receptors is a characteristic of panic disorder [69].

Binding studies with α_2-, β_1-, and β_2-receptors of the prefrontal cortex and/or hippocampus demonstrated the key role of the adrenergic system in schizophrenia [70]. Changes in the concentration of the same receptors have been found in patients with Alzheimer's disease. No differences were found in concentrations within the frontal cortex or hypothalamus, whereas significant variations were found in the cerebellum [71]. In particular, increased levels of α_2-receptors were found in aggressive Alzheimer's disease (AD) patients, whereas the concentration of β_1- and β_2-receptors were up to 25% lower in those patients compared to nonaggressive subjects. The number of β_2-receptors remains unchanged in patients with Parkinson's disease, whereas, α_1- and β_1-receptors increase, particularly in demented subjects [72].

2.2.6 α_1-Adrenoceptors

The three α_1-adrenoceptor subtypes have been identified by cloning, i.e. α_{1A}-adrenoceptor was cloned from rat cortex [73]; α_{1B}-adrenoceptor from hamster smooth muscle [74]; and α_{1D}-adrenoceptor from rat cortex [75]. Four isoforms have been identified for α_{1A}-adrenoceptors by alterative splicing [76, 77], but no functional differences were found among them. Although a fourth α_1-adrenoceptor subtype, denominated as $\alpha_{1L,}$ has been proposed [78], it is, actually, a conformational state of the α_{1A}-adrenoceptor [79].

The human α_{1A}-adrenoceptor encoded by a gene on chromosome 8 has 466 residues (Figure 2.3). The receptor can be N-glycosylated with GlcNAc at Asn7, Asn13, and Asn22. Palmitoylation takes place at Cys345. Homology models have been elaborated, based on the crystal structure of β_2-adrenoceptors (Figure 2.4) [81].

The human α_{1B}-adrenoceptor is encoded by a gene on chromosome 5. The receptor has 520 residues that can be N-glycosylated with GlcNAc at Asn10, Asn24, Asn29, and Asn34. Palmitoylation takes place at Cys365. Only homology models of limited residue ranges have been approached. The extracellular surface residues of the α_{1B}-adrenoceptor have been shown to be critical for G-protein coupling [82]. A homology model of this receptor subtype identified Asp191, in the second extracellular loop, as a potential regulator of agonist binding. Further mutagenesis studies confirmed that Asp191 acts as a "point of first contact" for endogenous agonists [83].

The human α_{1D}-adrenoceptor, encoded by a gene on chromosome 20, has 520 residues that can be N-glycosylated with GlcNAc at Asn65 and Asn82. Palmitoylation takes place at Cys419. It has been suggested that processing of the N-terminal domain of α_{1D}-adrenoceptor is a physiological mechanism for modulating pharmacodynamic properties [84].

The structure and function of α_1-adrenoceptor subtypes has been studied in detail [85]. Since there are no solved structures for α_1-adrenoceptors, structural

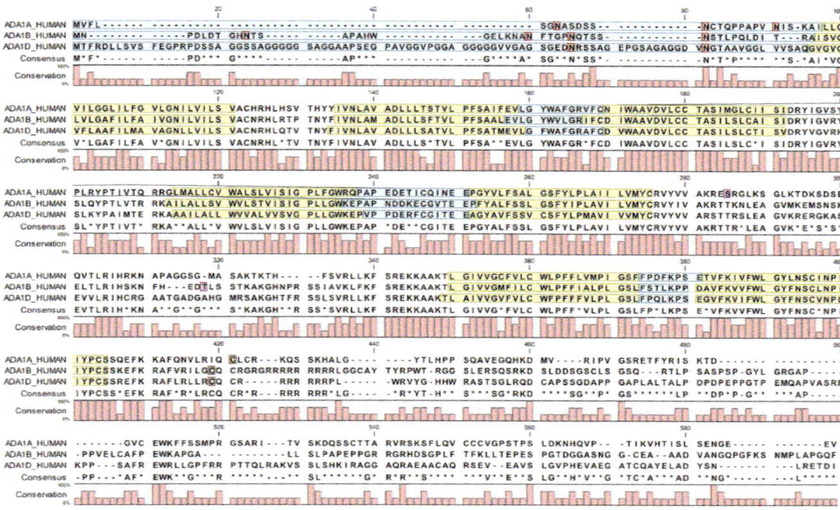

Figure 2.3 Sequence alignment of the human α_1-receptors. Conservation graphic and consensus sequence are shown. Disulfide bonds (green), N-glycosylation sites (red), palmitoylation sites (brown), phosphorylation sites (magenta), extracellular domains (light blue), intracellular domains (white), and transmembrane domains (yellow) are indicated on sequences.

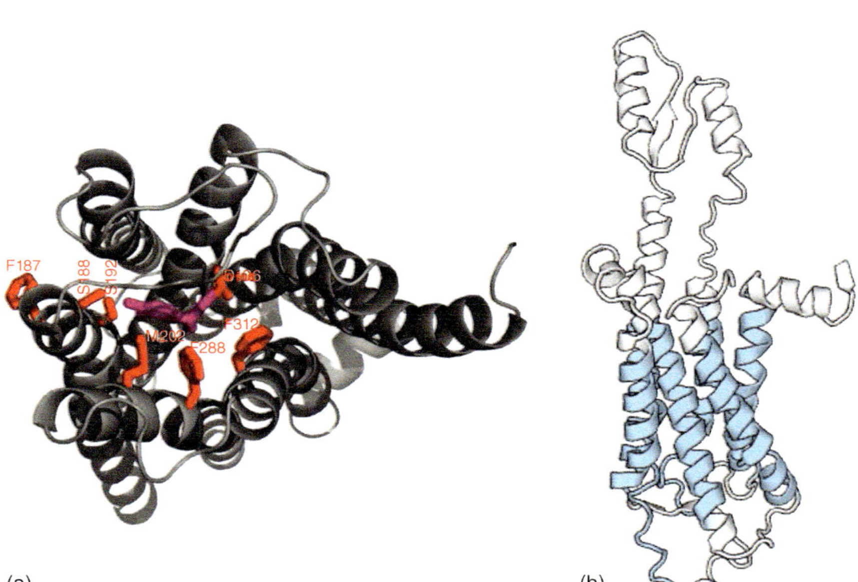

Figure 2.4 (a) Homology modeling of epinephrine binding site in the α_{1A}-adrenoceptor. Residues involved in binding are indicated in red. Epinephrine is indicated in violet. Source: Maiga et al. 2014 [80]. Reproduced with permission of AGRIS. (b) Homology model of α_{1A}-adrenoceptor showing transmembrane domains in light blue.

and binding information is obtained from mutagenesis studies and homology models (Figure 2.4) [86]. Asp125 in TM3 interacts with the protonated amine of (nor)epinephrine. Hydrophobic interactions with the catechol ring are established by Phe163 in TM4, Phe187 in TM5, and Phe310 in TM6. Ser 188 in TM5 binds to the meta-hydroxyl of (nor)epinephrine.

The residues involved in agonist binding of α_1-adrenoceptors have been identified by mutagenesis. Cys 128 in TM3 is involved in agonist-selective signaling. Ala 204 and Val185 (TM 5) and Leu314 and Met 293 (TM6) are involved in conferring agonist selectivity between the α_1-adrenoceptor subtypes. It has also been found that α_1-adrenoceptors can accommodate bulky substituents in the para-position of catechol ring structures facilitated by Leu 290 and Phe 360 in TM6. Lys331 in TM7 is involved in α_1-adrenoceptors agonism by forming a salt-bridge to Asp125 in TM3. Similarly, residues involved in antagonist binding have also been identified [87, 88]. Phe86 in TM2, and Gln196, Ile197, and Asn 198, which are in the second extracellular loop are responsible for the several discriminations observed among selective antagonists. In particular, Asp125 and Thr130 in TM3 are major sites of α_{1B}-adrenoceptor antagonist binding [89]. A combined study of mutagenesis and molecular dynamics with homology models based on the crystal structure of rhodopsin provided direct evidence that Asp106 and Gln167 are responsible for important interactions with antagonists [90]. Phe308 and Phe312 in TM7 involved in aromatic contacts with imidazoline-type agonists also present contacts with most α_1-adrenoceptor antagonists. Mutagenesis studies demonstrated the interaction of Phe312 with epinephrine. A moderate interaction with Asp106 was also found [80]. Ligand-based pharmacophores have been developed for the three subtypes of α_1-adrenoceptors [91].

2.2.7 α_2-Adrenoceptors

The different subtypes of α_2-adrenoceptors are identified pharmacologically. Prazosin has relatively low affinity by α_{2A}-adrenoceptor but high affinity by α_{2B}- and α_{2C}-adrenoceptors. The opposite is observed for oxymetazoline. The three subtypes α_{2A}-adrenoceptor [92], α_{2B}-adrenoceptor [93], and α_{2C}-adrenoceptor [94] have been cloned from humans. A fourth subtype cloned from the rat, identified as α_{2D}-adrenoceptor, has been suggested [95]. However, considering the amino acid sequence, it is a species orthologue of the human α_{2A}-adrenoceptor; in consequence, it is not considered a different subtype. Activation of α_2-adrenoceptors has been suggested as an effective treatment for ADHD and related disorders [66].

The human α_{2A}-adrenoceptor (encoded by a gene on chromosome 10) has 450 residues that can be N-glycosylated with GlcNAc at Asn10 and Asn14 (Figure 2.5). Palmitoylation takes place at Cys442. Some polymorphisms consisting of replacement of Asn251 by a Lys residue [96] or Cys401 by a serine have been reported [97]. The structure of the second intracellular loop (positions 118–149) of the human α_{2A}-adrenoceptor has been determined by NMR spectroscopy evidencing a cytoplasmic helix [98]. Mutagenesis studies confirmed involvement of Asp130, Ser200, and Ser204 in agonist recognition [99]. Those studies also demonstrated the role of Phe412 in antagonist binding [100].

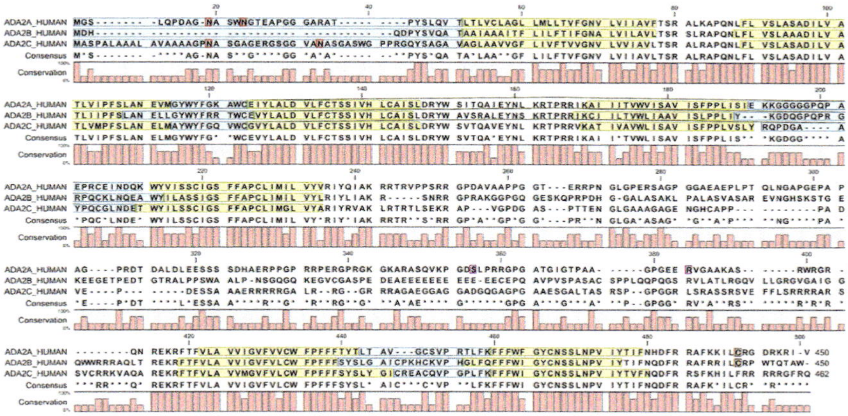

Figure 2.5 Sequence alignment of the human α_2-receptors. Conservation graphic and consensus sequence are shown. Disulfide bonds (green), N-glycosylation sites (red), palmitoylation sites (brown), phosphorylation sites (magenta), extracellular domains (light blue), intracellular domains (white), and transmembrane domains (yellow) are indicated on sequences.

The human α_{2B}-adrenoceptor (encoded by a gene on chromosome 2) has also 450 residues that can be palmitoylated at Cys442 as the α_{1B}-adrenoceptor. A rare polymorphic frameshift at position 451 produces a protein of 545 residues. Natural variants include replacement of Gly211 by Ala, Val376 by Ile, and Val379 by Gly or Ile. The fragment His225-Gly226-Gly227-Ala228-Leu229 can be replaced by Gln-Phe-Gly-Arg a mutation that has been directly related to adult familial myoclonic epilepsy [101]. A missing polymorphism of residues Glu301–Glu303 has been reported to induce enhanced emotional memory, greater amygdala responses to emotional stimuli, and greater intrusiveness of traumatic memories [102]. Furthermore, the deletion seems to selectively predict stress-induced enhancement of long-term memory in females [103].

The human α_{2C}-adrenoceptor (encoded by a gene on chromosome 4) has also 462 residues that can be N-glycosylated with GlcNAc at Asn19 and Asn33. Natural variants include missing residues 322–325 in the third intracellular loop [104, 105]. The structures of α_2-adrenoceptor subtypes have been studied and compared using homology modeling (Figure 2.6) [106, 107]. Binding sites of the three α_2-adrenoceptor subtypes are very similar in the key residues, and the observed small differences are being responsible for the particular selectivity of α_2-adrenoreceptor ligands. The computational predicted results are in agreement with previous mutagenesis-based observations [99] and literature data regarding GPCRs [108].

2.2.8 β-Adrenoceptors

An early classification comparing rank orders of potency of agonists identified two subtypes of β-adrenoceptors [36]: β_1-adrenoceptor, equally sensitive to norepinephrine and epinephrine and β_2-adrenoceptor, much less sensitive to

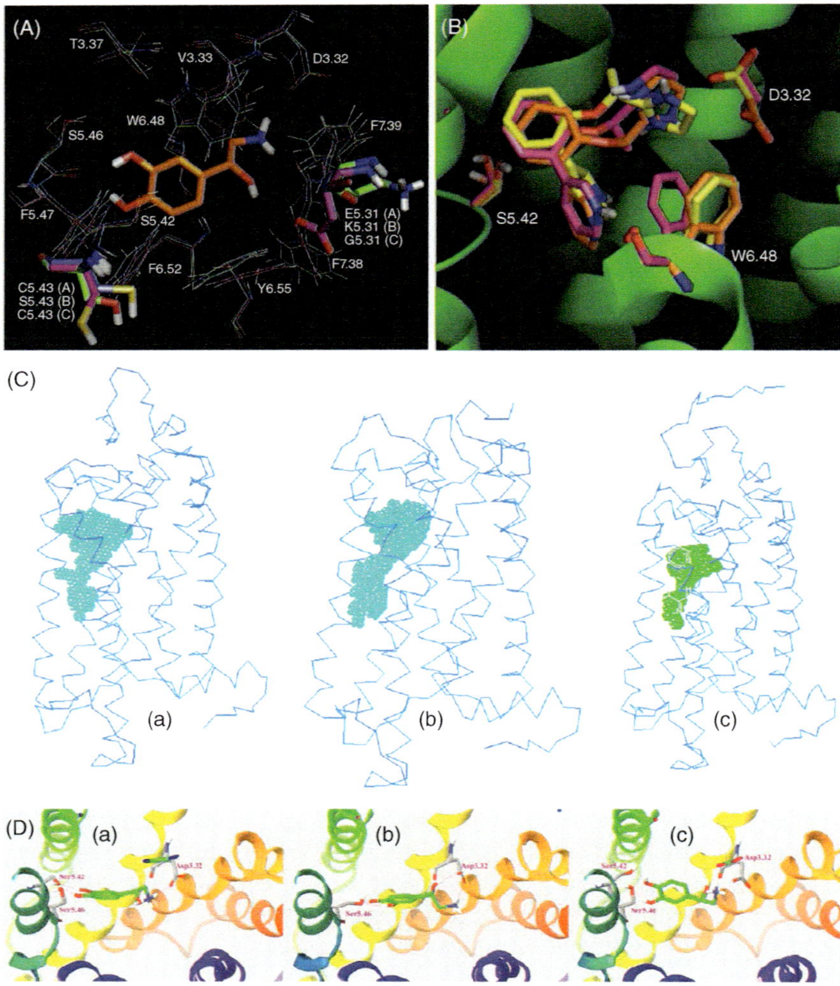

Figure 2.6 (A) Key binding residues of adrenoceptor subtypes α_{2A} (red), α_{2B} (green), and α_{2C} (blue). Norepinephrine (orange) according to subtype α_{2A} docking is also shown. (B) The best docking conformations of an imidazoline-containing ligand to subtypes α_{2A} (yellow), α_{2B} (pink), and α_{2C} (orange). The most significant interaction between imidazoline-N and Asp3.32 (Asp 113) is well recognized, but there is no significant difference in the orientation of side chain of Trp6.48 (Trp 258) in cases of subtypes which may lead to subtype-selective binding. (C) Predicted binding sites for subtypes α_{2A} (a), α_{2B} (b), and α_{2C} (c). (D) Docked norepinephrine in subtypes α_{2A} (a), α_{2B} (b), and α_{2C} (c). Source: (A, B, C) Reprinted with permission from Balogh et al. 2009 [106]. Copyright 2009, Elsevier. Source: (D) Reprinted with permission from Ostopovici-Halip 2011 [107]. Copyright 2011, Elsevier.

norepinephrine. A third subtype, β_3-adrenoceptor, was further identified [109]. The existence of a fourth subtype is controversial. There are pharmacological evidences of its existence [110], but some of them overlap with β_3-adrenoceptor and no definitive evidences have been reported [111]. Moreover, additional evidences present the new proposed subtype as a "state" of the β_1-adrenoceptor

[112, 113]. The β_1-adrenoceptor proved difficult to be cloned although both β_2-[114, 115] and β_3-adrenoceptors [116] have been cloned from humans. β-adrenoceptors play crucial roles in brain processes. As an example, during enhanced arousal and learning, β-adrenoceptors are stimulated by the release of norepinephrine from the locus coeruleus [117]. The excessive activation of β-adrenoceptors by catecholamines may induce hypermetabolism and increase both cardiac and cerebral oxygen demands, resulting in traumatic brain injury (TBI). In fact, the use of β-blockers is recommended for patients with TBI [118]. β-Adrenoceptors are predominant in peripheral tissues. In particular, β_1-adrenoceptors are predominant in the heart, β_2-adrenoceptors in the respiratory system, and β_3-adrenoceptors in the adipose tissues [119]. They are also present in the brain [120]. Thus, RT-PCR analyses revealed that microglia expressed β_1- and β_2-adrenoreceptors but not β_3-adrenoreceptor [121]. The same technique allowed detecting β_3-adrenoreceptor mRNA in the human brain [122].

β-Adrenoceptors share some structural features. Human β_2-adrenoceptor shares 54% identity with β_1-adrenoceptor and human β_3-adrenoceptor shares 51% and 46% identity with β_1- and β_2-adrenoceptors amino acid sequences, respectively (Figure 2.6). The N-terminus of β-adrenoceptors is extracellular and glycosylated and the C-terminus is intracellular. While β_1- and β_2-adrenoceptors are phosphorylated in specific sites by PKA and β-adrenoceptor kinase (βARK), β_3-adrenoceptor has no site for phosphorylation. All the three β-adrenoceptors are N-glycosylated with GlcNAc at specific Arg residues located in the extracellular domain. Also, palmitoylation takes place in Cys residues located in the fourth intracellular domain. It has been suggested that palmitoylation promotes the insertion of several adjacent residues in the membrane, forming an additional intracellular loop that results in an active conformation for G-protein coupling [123]. This process mediates adenylyl cyclase stimulation by agonist-bound receptors.

The human β_1-adrenoceptor (encoded by a gene on chromosome 10) has 477 residues that can be N-glycosylated with GlcNAc at Asn15 (Figure 2.7). Palmitoylation takes place at Cys392. Several polymorphisms have been described for the β_1-adrenoceptor: Ala26 replaced by Val, Ala29 by Thr, Arg31 by Gln, Ser49 by Gly [124], Arg389 by Gly or Leu [125], Arg399 by His, and His405 by Tyr. Phosphorylation may take place at Ser312, Ser412, and Ser428.

In contrast to α_1- and α_2-adrenoceptors, several structures of β-adrenoceptors have been elucidated. The structure of an analogue of the fragment 197–221 of the human β_1-adrenoceptor has been elucidated by NMR spectroscopy. The first complete structure of the thermostabilized turkey (*Meleagris gallopavo*) β_1-adrenoceptor was solved in complex with the partial agonist cyanopindolol (Figure 2.8, PDB ID: 2VT4) [126]. The same authors reported the crystal structures of the same β_1-adrenoceptor in complex with full agonists carmoterol (PDB ID: 2Y02) and isoprenaline (PDB ID: 2Y03), and the partial agonists salbutamol (PDB ID: 2Y04) and dobutamine (PDB ID: 2Y01) [127], as well as with antagonists bucindolol (PDB ID: 4AMI) and carvedilol (PDB ID: 4AMJ) [128]. Similar interactions were observed for the different agonists studied, which induced a 1 Å contraction of the catecholamine-binding pocket relative to antagonist-bound structures [127].

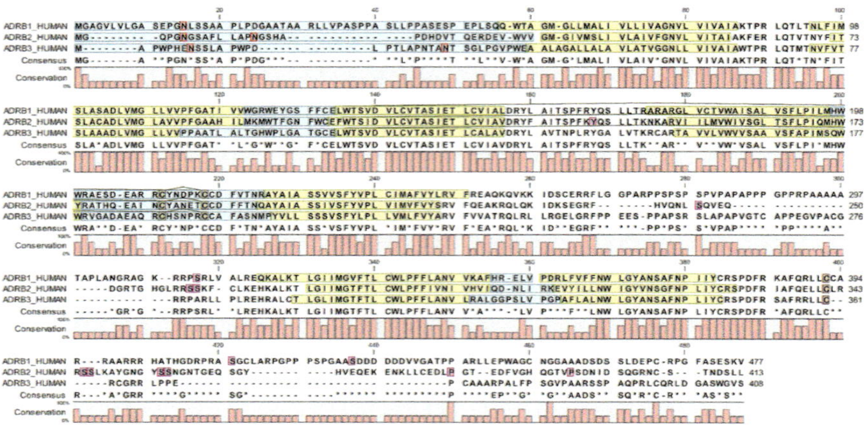

Figure 2.7 Sequence alignment of the human β-receptors. Conservation graphic and consensus sequence are shown. Disulfide bonds (green), N-glycosylation sites (red), palmitoylation sites (brown), phosphorylation sites (magenta), extracellular domains (light blue), intracellular domains (white), and transmembrane domains (yellow) are indicated on sequences.

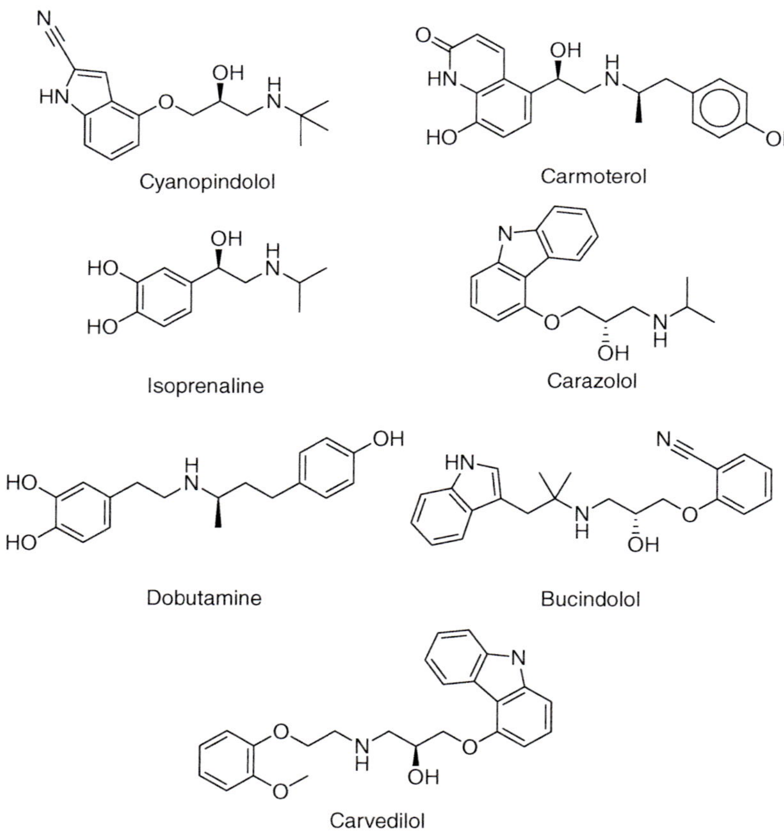

Figure 2.8 β-Adrenoceptor ligands.

Antagonist-bound structures showed two different conformations of helix TM6 [129]. The structures with bucindolol and carvedilol showed additional interactions in extracellular loop 2 and TM7, which could promote G-protein-independent signaling and offer an explanation to the observed pharmacological differences arising from Gly389/Arg polymorphism [128]. Structures in complex with antagonists carazolol (PDB ID: 2YCW), cyanopindolol (PDB IDs: 2YCX, 2YCY), and iodocyanopindolol (PDB ID: 2YCZ) have been reported by Schertler and coworkers [129].

Structures of an ultra-thermostable turkey β_1-adrenoceptor with cyanopindolol (PDB ID: 4BVN) and 7-methylcyanopindolol (PDB ID: 5A8E) bound have also been reported [130]. The structure with cyanopindolol bound served to identify an intramembrane sodium ion coordinated to Asp87, Ser128, and three water molecules that stabilizes the ligand-free receptor (Figure 2.9). The water

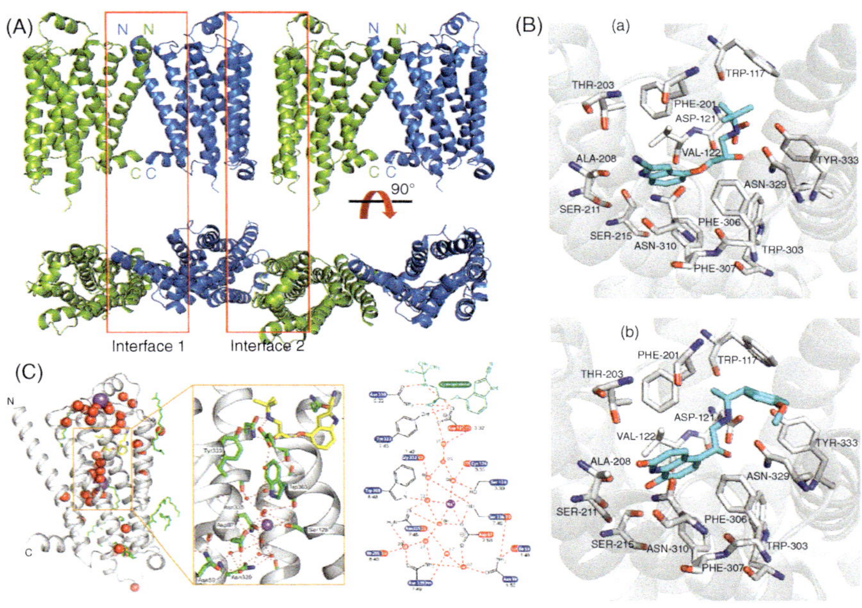

Figure 2.9 X-ray structures of β_1-adrenoceptor. (A) Representation of oligomers of β_1-adrenoceptor (PDB ID: 4GPO). Within the same layer, β_1-adrenoceptor forms oligomers with two dimer interfaces. The N and C termini are indicated, and representation of the top view (from the extracellular surface). (B) Structures of the ligand-binding pocket for antagonist cyanopindolol ((a) PDB ID: 2VT4) and agonist carmoterol ((b), PDB ID: 2Y02) and the interacting amino acid residues. (C) Structure of the intramembrane Na^+ ion binding site (PDB ID: 4BVN). (Left) Cartoon of β_1-adrenoceptor (gray) depicting the positions of ordered water molecules (red spheres), Na^+ ions (purple spheres), lipids (green sticks), and cyanopindolol (stick representation: carbon, yellow; nitrogen, blue; oxygen, red). The extracellular surface is at the top of the figure with the N-terminus (N) and C-terminus (C) labeled appropriately. (Center) Details of the Na^+ binding site with portions of H2 and H3 removed for clarity: red spheres, water molecules; purple sphere, Na^+ ion; green sticks, amino acid side chains; yellow sticks, cyanopindolol; red dashed lines, polar contacts, as defined by PyMOL. (Right). Diagrammatic representation of the hydrogen bond network. Source: Miller-Gallacher et al. 2014 [130]. https://creativecommons.org/licenses/by/4.0/.

molecules are part of a more extensive network of water molecules within a cavity formed by helices TM1, TM2, TM3, TM6, and TM7.

The crystal structure of oligomeric turkey β_1-adrenoceptor in ligand-free basal state (PDB ID: 4GPO) has been reported in 2013 (Figure 2.5), which was found to be in an inactive conformation [131]. It displays two dimer interfaces, one of which involves TM1, TM2, the C-terminal His8, and the extracellular loop 1. The other interface comprises residues from TM4, TM5, intracellular loop 2, and extracellular loop 2.

The human β_2-adrenoceptor (encoded by a gene on chromosome 5) has 413 residues that can be N-glycosylated with GlcNAc at Asn6 and Asn15. Palmitoylation takes place at Cys341 [132]. Phosphorylation can take place at several positions including Tyr141 [133], Ser246, Ser261, Ser262, Ser345, Ser346 [134], Ser355, and Ser356 [135]. 4-Hydroxyprolines can be found at positions 382 and 395 [135]. A variety of polymorphisms have also been described for the β_2-adrenoceptor: Asn15 replaced by Ser, Gly16 replaced by Arg [136], Glu27 replaced by Gln [114], Val34 replaced by Met [137], Ile159 replaced by Phe or Leu [138], Thr164 replaced by Ileu [137], Ser220 replaced by Cys, and Lys 375 replaced by Arg [138].

A number of X-ray structures have been reported for the human β_2-adrenoceptor. The structure of a human β_2-adrenoceptor-T4 lysozyme fusion protein (Figure 2.10a) bound to the partial agonist carazolol (PDB ID: 2RH1) showed a location of the ligand (Figure 2.10b) very similar to that of retinal in rhodopsin [139]. The same authors reported the structure of the human β_2-adrenoceptor in a lipid environment bound to an inverse agonist and in complex with Fab (PDB ID: 2R4R, 2R4S) [140]. The structure of the receptor bound to cholesterol and the partial inverse agonist timolol (PDB ID: 3D4S) showed a binding site for cholesterol between TM1, TM2, TM3, and TM4, similar to that previously found and illustrated in Figure 2.5a [141]. Cholesterol plays an important role in stability and structure of the β_2-adrenoceptor (Figure 2.10c).

Studies of the extracellular surface of the β_2-adrenoceptor have been carried by X-ray crystallography of a methylated derivative in complex with Fab (PDB ID: 3KKJ6) and NMR spectroscopy [142]. A comparison of the structure of the β_2-adrenoceptor in complex with epinephrine (PDB ID: 4LDO) and with agonists such as BI167107 (PDB ID: 4LDE) and hydroxybenzyl isoproterenol (HBI) (PDB ID: 4LDL) revealed a highly conserved overall ligand recognition. While the couples BI167107/HBI (Figure 2.11A(a) and epinephrine (or adrenaline)/HBI (Figure 2.11A(b)) showed a highly conserved agonist-binding mode, a comparison between epinephrine and BI167107 showed similar polar networks but a clear difference in the H-bonding of Ann293 (TM6) to the amide group of BI167107 and the *meta*-OH of epinephrine (Figure 2.11A(c)). Due to this difference, Asn 293 and TM6 shift inward in the epinephrine-bound structure, leading to a cascade of changes culminating in a rearrangement of ECL3 (Figure 2.11A(d)) [143].

Several structures in complex with inverse agonists such as ICI 118,5551 (PDB ID: 3NY8) or a novel derivative (PDB ID: 3NY9) and antagonists such as alprenolol (PDB ID: 3NYA) have been reported [144]. The structure of the human β_2-adrenoceptor with a specifically designed agonist covalently bound through a disulfide bond (PDB ID: 3PDS) served to confirm long-timescale molecular

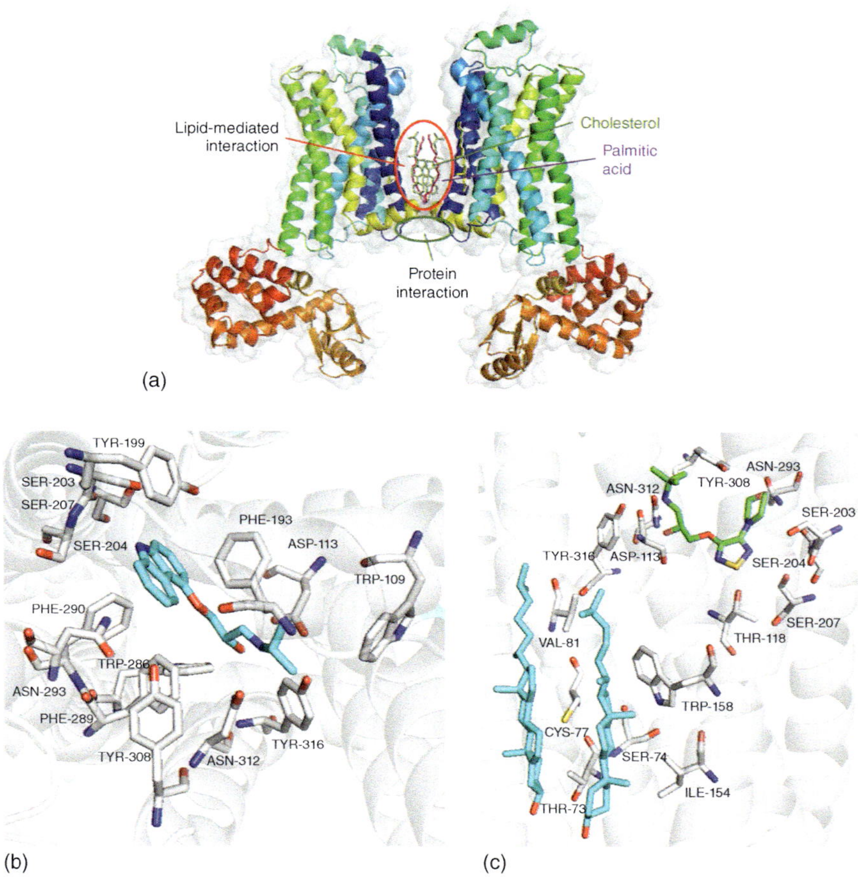

Figure 2.10 (a) β_2-Adrenoceptor crystal-packing interface (PDB ID: 2RH1) with two cholesterol molecules (green) and two palmitic acid (magenta) molecules. (b) Details of the carazolol (cyan) binding site (PDB ID: 2RH1). (c) Details of timolol (green) and cholesterol (blue) binding sites (PDB ID: 3D4S).

dynamics simulations pointing out that an agonist-bound active conformation spontaneously relaxes to an inactive-like conformation in the absence of a G protein [145]. These studies also suggest that binding energy from a G protein or nanobody interaction is required to stabilize conformational changes. Indeed, stabilization using an antibody fragment (nanobody) revealed subtle changes in the binding site but induced an 11 Å outward movement of the cytoplasmic end of TM6 and rearrangements of TM5 and TM7 (PDB ID: 3P0G) [146]. The dynamic behavior of GPCRs activation has been confirmed by stabilizing two distinct β_2-adrenoceptor conformations by using conformationally sensitive nanobodies (PDB ID: 5JQH) [147]. Further studies with covalent β_2-adrenoceptor agonist FAUC50 (PDB ID: 4QKX) demonstrated an easier crystallogenesis when conformationally selective antibody (nanobody) is also added [148]. A new method has been recently implemented for obtaining high-resolution crystal

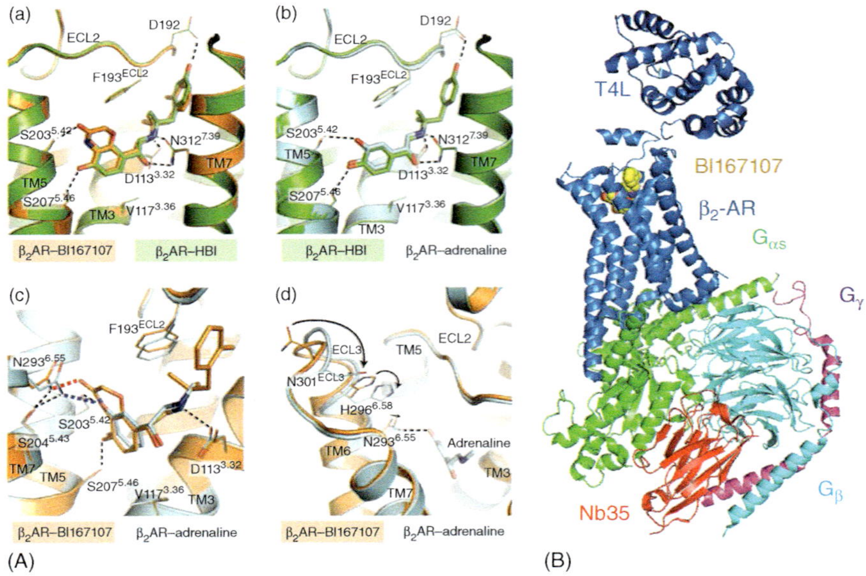

Figure 2.11 (A) Comparison of agonist-binding modes. (a) BI167107 (orange) with HB (green). (b) Epinephrine (or adrenaline) (cyan) and HBI. (c) BI167107 (orange) with adrenaline (cyan). (d) Conformational changes of Asn293 and TM6 in the adrenaline-bound structure. Source: Reprinted with permission from Ring et al. 2013. [143]. Copyright (2013). Macmillan Publishers Ltd. (B) Overall structure of the asymmetric unit contents (PDB ID: 3SN6) showing the β_2-adrenoceptor (blue) bound to an agonist (yellow spheres) and interacting with $G\alpha s$ (green).

structures of membrane proteins in a lipid phase at 100 K [149]. The methodology has been successfully applied to the β_2-adrenoceptor (PDB IDs: 5D5A and 5D5B). A ternary complex formed by the β_2-adrenoceptor bound to agonist BI167107 and nucleotide-free G_s heterotrimer has been crystallized and characterized (PDB ID: 3SN6) (Figure 2.12b) [150].

Important interactions between the β_2-adrenoceptor and G protein involve opening of TM5 and TM6 in the β_2-adrenoceptor, thus forming a cavity on the intracellular side into which the α5-helix of $G_{\alpha s}$ docks (Figure 2.12). The T4 lysozyme fusion facilitates crystallization (PDB ID: 4GBR) [151].

The human β_3-adrenoceptor (encoded by a gene on chromosome 8) has 408 residues that can be N-glycosylated with GlcNAc at Asn8 and Asn26. Palmitoylation takes place at Cys361. Splice variants have been reported [152] and, consequently, polymorphisms can be found: Trp64 replaced by Arg [153, 154], Glu249 replaced by Lys, Tyr265 replaced by Met [155], and Arg353 replaced by Cys. Unlike β_1- and β_2-subtypes, β_3-adrenoceptor has no sites for phosphorylation. The disulfide bond between Cys110 and Cys189 is essential for the activity of the receptor and ligand binding. For β_3-adrenoceptor, there is a higher homology between human and other species such as bovine, guinea pig, monkey, hamster, rat, and mouse than between different β-adrenoceptor subtypes [156]. The β_3-adrenoceptor, closely related to cardiovascular diseases [157], has a limited presence in the brain [158] although it is closely related to

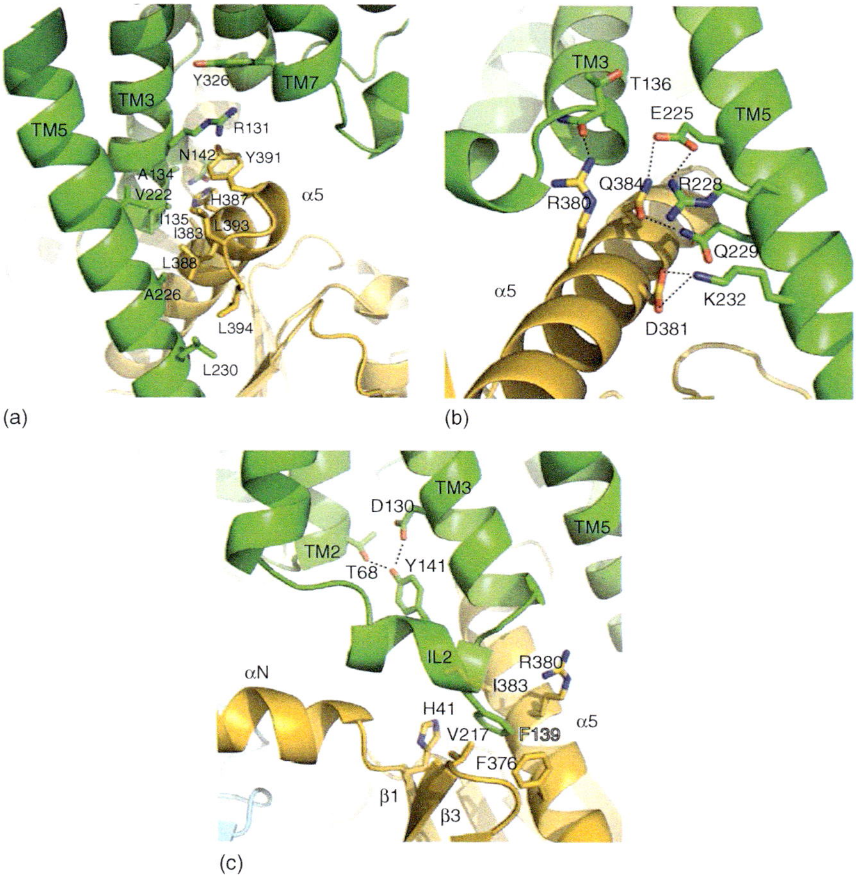

Figure 2.12 (a) Polar interactions illustrated by packing of Tyr391 of the α_5-helix against Arg131 of the conserved DRY sequence in TM3. Arg131 also packs against Tyr326 of the conserved NPxxY sequence in TM7. (b) Polar interactions with TM5 and TM3. (c) Receptor residues Thr68 and Asp130 interact with the ICL2 helix of the β_2-adrenoceptor via Tyr141, positioning the helix so that Phe139 of the receptor docks into a hydrophobic pocket on the G-protein surface. Source: Reprinted with permission from Rasmussen et al. 2011 [150]. Copyright 2011, Macmillan Publishers Ltd.

chronic stress [159], depression and schizophrenia [70], and memory consolidation [160]. Indeed, norepinephrine activates adult hippocampal precursors via the β_3-adrenoceptor [161]. It has also been reported that β_3-adrenoceptor activation produces an anxiolytic response that correlates with GABAergic synaptic responses [162], and there are evidences that β_3-adrenoceptor agonists have antidepressant/anxiolytic-like profiles [163–165]. Activation of both β_2- and β_3-adrenoreceptors resulted in an increase in brain tryptophan [166].

The structure of the β_3-adrenoceptor has only been predicted using homology models based on bacteriorhodopsin. The predicted model includes 3D structures of G_s protein, the β_3-adrenoceptor, and their complex and is in agreement with experimental observations [167]. The model predicts that various residues of N- and C-termini of G_s protein and the third intracellular loop of the β_3-adrenoceptor participate in binding. More precise results have been achieved by developing pharmacophore models based on 144 β_3-adrenoceptor agonists as well as homology models based on the well-established structure of the β_2-adrenoceptor (Figure 2.13) [168]. Indeed, docking models are in agreement with experimental evidences. It has been determined that agonists formed H-bond interactions with Asp117, Arg315, and Asn332, π–π stacking interactions with Phe308, and hydrophobic interactions with Val118, Val121, Ala197, Phe198, Ala199, Phe309, and Phe328. The major difference between the homology models of the β_3-adrenoceptor with the crystal structures of β_1- and β_2-adrenoreceptors is the H-bond interaction with Arg315, which corresponds to Asn313 of β_2-adrenoreceptor and His296 of β_2-adrenoreceptor, respectively.

2.3 Neuropeptides Receptors

2.3.1 Neuropeptide Y Receptors

NPY (Figure 2.1) is not only a noradrenergic cotransmitter but also a neurohormone directly involved in basic processes of learning and memory [169]. NPY is one of the most abundant peptides in the CNS and its name is due to the presence of a Tyr (Y) residue at the N-terminus and a Tyr-amide at the C-terminus. It is mainly synthesized in the hypothalamus and in smaller amounts in the hippocampus, too [170, 171]. Nevertheless, NPY crosses the blood–brain barrier [172], so the origin of brain NPY might also be peripheral. In addition to depression and addictions to alcohol and food intake, NPY has been related to anxiety, schizophrenia, and posttraumatic stress disorder [173].

NPY is recognized by five types of receptors, namely, Y1, Y2, Y4, Y5, and Y6 [174], diversely distributed in the brain [175]. In particular, Y1 and Y5, related to feeding behavior [176, 177] as well as Y2 have been detected in different parts of the hippocampus [178]. mRNA expression of Y1, but not Y2 or Y5, has been observed in the subventricular zone [179, 180] On the other hand, Y5 mRNA expression [181] and immunoreactivity [182] are also present in the CA1–CA3 regions and in the dentate gyrus. Y1 receptor immunoreactivity has been detected in the molecular layer of the dentate gyrus (at very high levels) and in the CA regions (at moderate levels) [183]. Y2 receptor immunoreactivity has also been detected at moderate levels in the CA subfields after colchicine treatment [184]. NPY stimulates neurogenesis in the hippocampus and the olfactory epithelium [185] without affecting self-renewal of subventricular zone stem cells [186]. The neuroproliferative effect is mediated by receptor Y1, and it has been suggested that stimulation of endogenous subventricular zone neural stem cells by NPY could be a therapeutic approach against neurodisorders affecting the striatum as Huntington's disease [187].

NPY receptors are GPCRs having the typical seven transmembrane helical domains. The sequences for human Y1 (384 residues), Y2 (381 residues), Y4 (375 residues), and Y5 (445 residues) and mouse Y6 (371 residues) receptors are given

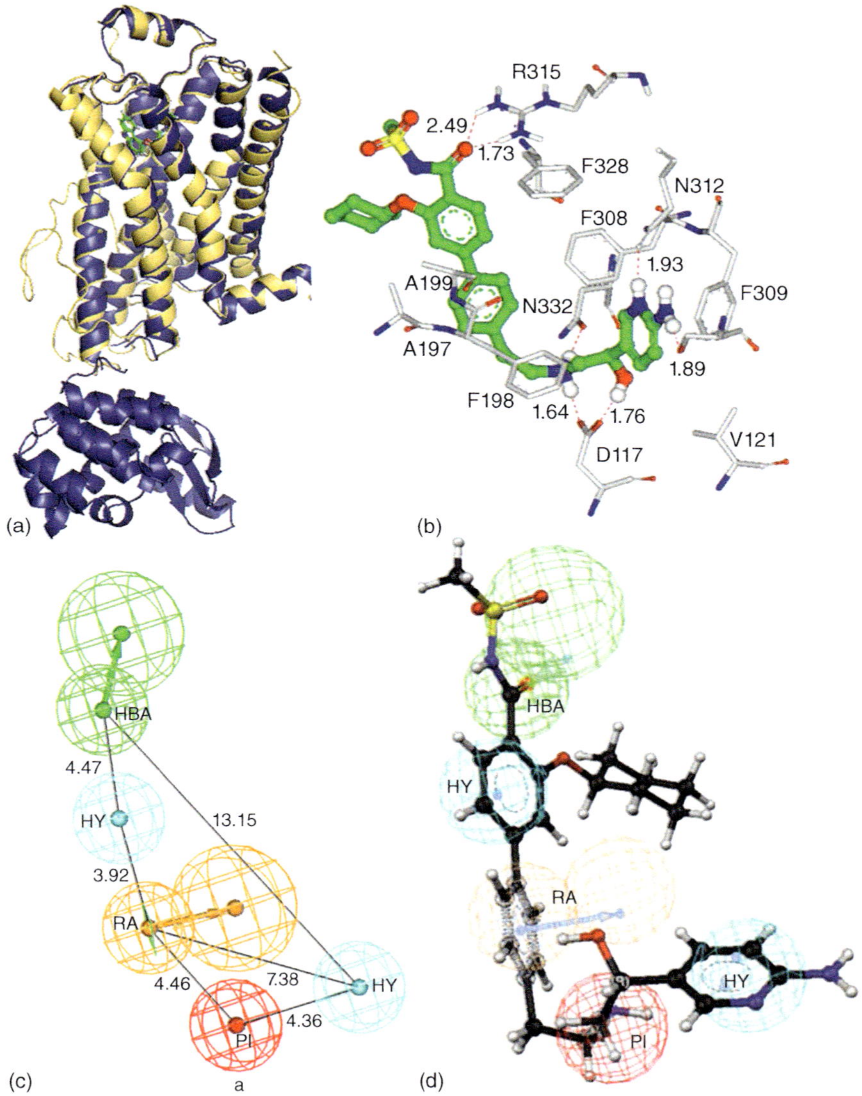

Figure 2.13 (a) The superimposition of the 3D structure of the template and of the homology model for β_3-adrenoceptor. Template and homology model are colored deep blue and pale yellow, respectively. (b) Docking model of the best agonist with the β_3-adrenoceptor homology model. (c) The best pharmacophore for β_3-adrenoreceptor model with distance constraints (in angstrom, Å). Pharmacophore features are color-coded with blue for hydrophobic groups (HY), red for positively ionizable feature (PI), green for H-bond acceptor (HBA), and orange for ring aromatic (RA). (d) Most active agonist (left bottom) mapped onto the pharmacophore model of β_3-adrenoreceptors. Source: Reprinted with permission from Jin et al. 2011 [168]. Copyright 2011, Springer Science+Business Media B.V.

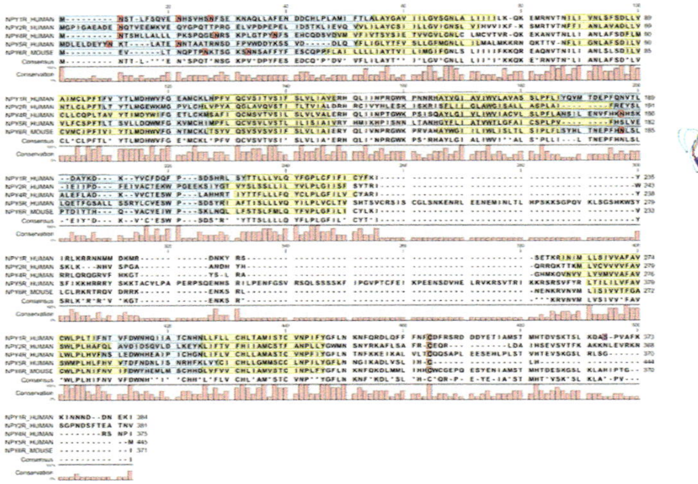

(a) (b)

Figure 2.14 (a) Sequence alignment of the human NPY receptors Y1, Y2, Y4, and Y5 and rat NPY receptor Y6. Conservation graphic and consensus sequence are shown. Disulfide bonds (green), N-glycosylation sites (red), palmitoylation sites (brown), phosphorylation sites (magenta), extracellular domains (light blue), intracellular domains (white), and transmembrane domains (yellow) are indicated on sequences. (b) Superimposition of four homology models of human NPY receptors: NPYR1 (green), NPYR2 (red), NPYR4 (blue), and NPYR5 (cyan).

in Figure 2.14. Currently, it is accepted for humans that the so-called Y-receptor family is formed by Y1, Y2, Y4, and Y5 receptors since the suggested Y6 receptor is actually another member of the Y1-receptor subfamily [188]. NPY receptors, as other GPCRs, can be N-glycosylated with GlcNAc in specific positions (NPYR1 in Asn2, Asn11, Asn17; NPYR2 in Asn11; NPYR4 in Asn2, Asn19, Asn29, and Asn187; NPYR5 in Asn10 and Asn17; NPYR6 in Asn11, Asn18, and Asn182) and palmitoylated in Cys residues of the last intracellular loop (NPR1 in Cys338; NPR2 in Cys342; NPR4 in Cys340; NPR5 in Cys442; NPR6 in Cys336). Only Y1 receptor is phosphorylated at Ser368. For Y1 a natural variant in which Phe96 is replaced by Lys has been found [189]. Y4 receptor can exchange Ala99 by Ser [190] and Arg240 by Cys.

No X-ray structures of NPY receptors have been resolved but homology models have been calculated for Y1, Y2, Y4, and Y5 human receptors. By combining several techniques including peptide synthesis and receptor mutagenesis, binding and signaling has been established at a molecular level. A number of studies have been carried out for the various NPY receptors, and a revision has been reported in 2008 [28]. More recently, a computational model of Y2 receptor based on mutagenesis and interactions with native peptide agonists PYY (peptide tyrosine tyrosine) and PP (pancreatic polypeptide) suggested the presence of an H-bond network formed by the highly conserved residues Thr2.61, Gln3.32, and His7.39. Also, Tyr5.38 and Leu 6.51 might make contacts with the C-terminus of NPY [191].

2.3.2 Galanin Receptors

Galanin is noradrenergic cotransmitter formed by 30 residues (Figure 2.1) that is recognized by three receptors denominated GALR1–3 [192]. Both the peptide and receptors are expressed in the CNS, [193] particularly on regions responsible for the memory processes [194] including hippocampal neurons and fibers [194]. Moreover, it has been found a hyperinnervation of galanin-containing fibers in patients with Alzheimer's disease and in patients with Parkinson's disease [195] as well as increased galanin autoantibody levels in the cerebrospinal fluid of patients with Alzheimer's disease correlating negatively with cognitive impairment, neurodegenerative signs, and age [196]. These studies support the neuroprotective role of galanin [192, 197]. It has been documented that galanin affects norepinephrine release in the locus coeruleus [198], suggesting a close relationship with mood disorders. The ability of galanin for modulating hippocampal excitability by inhibiting glutamate release relates the neuropeptide with epilepsy.

The three receptors have been cloned [199], but only GALR1 [200] and GALR2 [201] are present in relative abundance and with a specific spatial distribution in the hippocampus. GALR1 expression has been identified in the central and peripheral nervous systems [202], where it is detected in the hypothalamus, amygdala, ventral hippocampus, thalamus, brainstem (medulla oblongata, locus coeruleus, and lateral parabrachial nucleus), and dorsal horn of the spinal cord [203]; broader central and peripheral tissue distribution has also been reported [204]. The GALR2 receptor is abundant in the dorsal and ventral parts of the dentate gyrus, but it is also found in the hypothalamus, olfactory and cortical regions, cerebellum and in the brainstem and spinal cord [205, 206]. GALR3 is present in the hypothalamus, and the expression is confined to the paraventricular, ventromedial, and dorsomedial hypothalamic nuclei [207].

Activation of GALR1 results in pertussis toxin-sensitive inhibition of adenylate cyclase (AC) through interaction with $G\alpha_i/G\alpha_0$ G proteins [208] and MAPK activity [209]. GALR2 stimulates multiple intracellular pathways through multiple classes of G proteins. In particular, it activates the stimulatory pathway of $G\alpha_{q/11}$ G proteins. As a consequence, PLC activity and intracellular phosphatidylinositol turnover are generated, increasing intracellular Ca^{2+} and opening Ca^{2+}-dependent channels. GALR2 also activates MAPK [209] affecting phosphoinositide-dependent protein kinase I (PDK-1) activity but only weakly coupled to AC inhibition [197, 210]. The activity of GAL3 receptor is mediated by $G\alpha_i$/$G\alpha_o$, which are strongly coupled to the inhibition of AC [210]. Both GALR1 and GALR2 are involved in the anticonvulsant effect of galanin, with a direct implication in epilepsy. While the former affects the initiation phase, the latter affects the maintenance phase of status epilepticus [211].

Unraveling the molecular mechanisms of galanin receptors is crucial to understand the neurological effects exerted by galanin, and thus, galanin receptors are therapeutic targets for neurological diseases [212]. From the three receptors of galanin, GALR1 and GALR2 are involved in the inhibition of long-term potentiation (LTP), the basis of memory consolidation [213], in the hippocampus, although some aspects remain unclear [169]. GALR2 is responsible for protective features of galanin against β-amyloid neurotoxicity [214]. Further studies with selective receptor agonists in a rat model of AD demonstrated that the memory improvement and inhibiting effects of the hippocampal β-amyloid deposition are mediated by the GALR2 receptors [215]. The role of the galanin receptor subtypes and receptor subtype-specific agonists and antagonists as well as their involvement in different biological and pathological functions has been reviewed [216].

Galanin receptors are GPCRs presenting seven transmembrane helical domains. The sequences for human GALR1 (349 residues), GALR2 (387 residues), and GALR3 (368 residues) receptors are given in Figure 2.15. Glycosylation of galanin receptors with GlcNAc takes place in Asn7, Asn12, and Asn183 for GALR1; in Asn2 and Asn11 for GALR2, and in Asn6 for GALR3. Palmitoylation is found in Cys320 for GALR1 and Cys308 for GALR3, without evidences of palmitoylation of GALR2. Replacement of Trp15 by Cys [218], Ser3334 by Asn [219], or Pro342 by Leu can account in GALR1. In GALR3, replacement of Arg342 by Gly and Gln349 by Arg can be found.

The ligand-binding site of human GALR2 has been characterized and subtype-selective interactions have been identified [220]. In particular, His252 and His253 in TM6 and Phe 264 and Tyr271 in extracellular loop 3 play crucial roles in binding [221]. A 3D model has been generated for each galanin receptor, and substantial differences have been found among the three types [222], accounting for the ligand specificity observed experimentally. Docking studies with a number of ligands allowed studying the involvement of transmembrane and loop residues in ligand interactions (Figure 2.16). In the absence of X-ray data, homology models of GALR2 in combination with docking and molecular dynamics calculations [217] allowed identifying His252, His253, Ile256, Phe264, and Tyr271 as important residues involved in binding. A potential binding pocket can be designed (Figure 2.15) in which the following residues are involved: Ile75$^{2.54}$, Gln82$^{2.61}$, Ile85$^{2.64}$, Tyr86$^{2.65}$, Trp91^{ECL1}, Val101$^{3.28}$, Ile105$^{3.32}$, Met109$^{3.36}$, Tyr160$^{4.60}$, Tyr163^{ECL2}, Val174^{ECL2}, Cys175^{ECL2}, Arg184^{ECL2}, Asp188$^{5.40}$, Arg274$^{7.35}$, His 2787.39, and Tyr282$^{7.43}$.

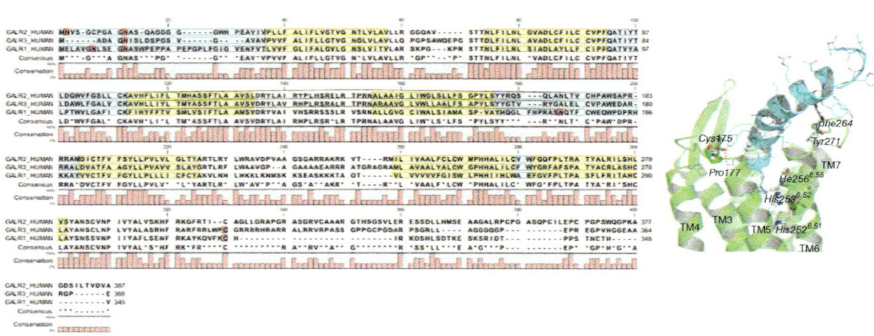

(a) (b)

Figure 2.15 (a) Sequence alignment of the human galanin receptors GALR1, GALR2, and GALR3. Conservation graphic and consensus sequence are shown. Disulfide bonds (green), N-glycosylation sites (red), palmitoylation sites (brown), extracellular domains (light blue), intracellular domains (white), and transmembrane domains (yellow) are indicated on sequences. (b) Molecular docking results for GALR2 with galanin. Source: Reprinted with permission from Hui et al. [217]. Copyright 2016, Springer Science+Business Media B.V.

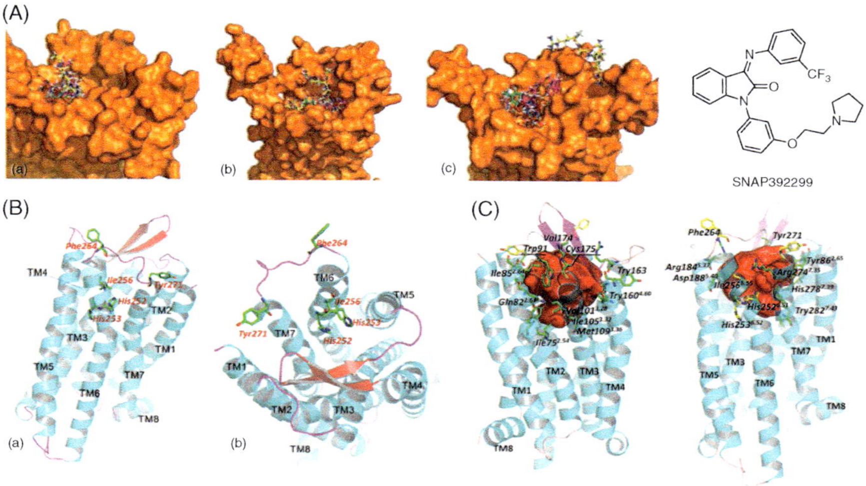

Figure 2.16 (A) Binding sites of galanin receptor subtypes with best binding ligands: (a) GALR1 and (b) GALR2 docked to galanin (2–6) and SNAP398299; GALR3 in complex with (c) two poses of galanin (2–6) and two poses of SNAP398299. Source: (A) Reprinted with permission from Jurkowski et al. 2013 [222]. Copyright 2013, Taylor & Francis. (B) 3D structure of GALR2 based on δ-opioid receptor after energy minimization (sequence identity: 29.1 ± 2%, PDB ID: 4N6H, resolution: 1.80 Å). Some important residues are highlighted using green sticks: (a) side view from the membrane; (b) top view from the periplasmic side. (C) Potential binding pocket of GALR2, and the residues that comprise it. The potential binding pocket is highlighted in red. Residues that are involved in the binding pocket are highlighted as yellow sticks; other important residues involved in the binding pocket are highlighted as green sticks. Source: (B, C) Reprinted with permission from Lundström et al. 2005 [217]. Copyright 2005, Springer Science+Business Media B.V.

The model constructed from the δ-opioid receptor was selected as the best one providing the basis for docking studies with galanin, which showed high consistency with experimental data. Potential allosteric binding was also explored, and three potential allosteric binding pockets were found. Based on the docking results and literature experimental data, the authors identified the pocket formed from TM2, TM3, ECL1, and ECL2 as the most favorable allosteric binding site. A possible mechanism for the allosteric modulation is suggested to be disruption of the Na^+ ion binding site and/or the Na^+ ion pathway caused by the binding of a ligand at the pocket.

2.4 Norepinephrine Transporter (NET)

The NET regulates the action of norephedrine through reuptake of released neurotransmitter [223–225]. It is located in the plasma membrane of noradrenergic neurons and, as DAT, belongs to the family of Na^+/Cl^-- dependent transporters [226]. There is a 66% homology in the amino acid sequence between human DAT and NET. Several studies directed to understand the regulation of NET have been carried out and demonstrated that NET is not a static protein but it is regulated by phosphorylation, which controls its cell surface expression. A variety of intra- and extracellular signaling molecules, including cyclic adenosine monophosphate (AMP) and other neurotransmitters different from norepinephrine are also responsible for regulation of NET [227].

In the CNS, NET is associated with mood regulation, sleep regulation, expression of behavior, and general alertness and arousal. NET is the target of psychomotor stimulants such as cocaine and amphetamine as well as for therapeutic antidepressants.

The canonical isoform of human NET, encoded by a gene on chromosome 12, has 617 residues (Figure 2.17) that can be N-glycosylated with GlNAc in Asn184, Asn192, and Asn198. A polymorphism by exchanging Ala457 by Phe has been

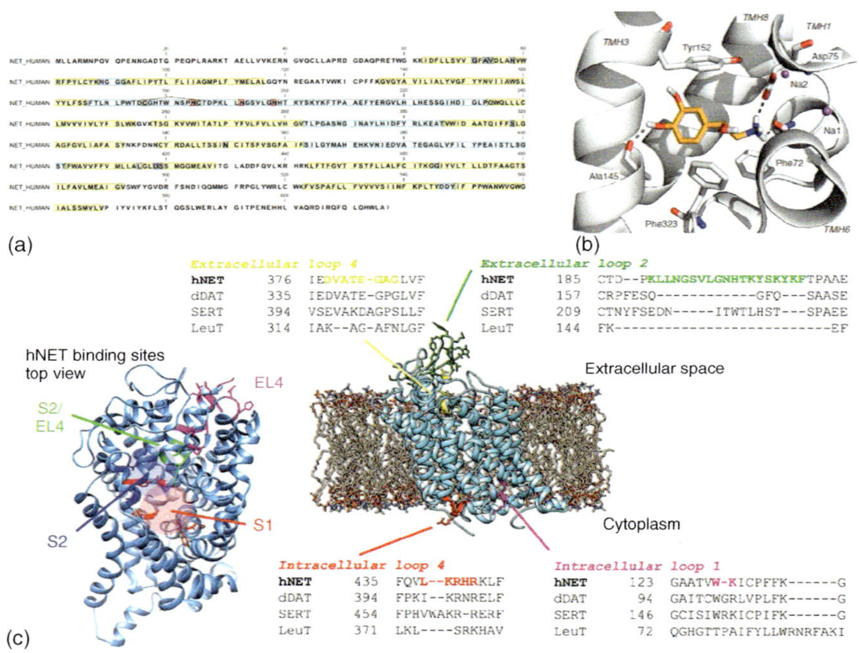

Figure 2.17 (a) Sequence of the human norepinephrine transporter (NET). Disulfide bonds (green), N-glycosylation sites (red), palmitoylation sites (brown), extracellular domains (light blue), intracellular domains (white), and transmembrane domains (yellow) are indicated on sequences. (b) Predicted structure of the NET–norepinephrine complex. Norepinephrine is colored in orange, with oxygen, nitrogen, and hydrogen atoms in red, blue, and white, respectively. Sodium ions are visualized as purple spheres. NET's transmembrane helices are depicted as white ribbons. Key residues are displayed as sticks; the three hydrogen bonds between norepinephrine and NET (involving residues Ala145, Phe72, and Asp75) are shown as dotted gray lines. Source: Reproduced with permission from Schlessinger et al. 2011 [228]. Copyright 2011, National Academy of Sciences, USA. (c) Two divergent extracellular loops and two intracellular loops are shown. A top view of hNET is shown on the left. Low-conserved residues in binding sites as described [229] are highlighted in colors: central binding site (S1), secondary binding site (S2), and extracellular loop 4 (EL4). Red and blue circles highlight S1 and S2 sites, respectively. Source: Reproduced with permission from Haddad et al. 2017 [230]. https://creativecommons.org/licenses/by/4.0/.

reported and associated with orthostatic intolerance and tachycardia [231]. Other natural variants include replacement of Asn7 by Lys; Val69, Thr99, Val245, and Val449 by Ileu; Thr283 and Lys463 by Arg; Asn292 and Ileu549 by Thr; Val365 by Leu; Ala369 and Ala457 by Pro; Asn375 and Gly478 by Ser; Phe528 by Cys; and Tyr 548 by His. Two additional isoforms have been reported for human NET, one having an alternative sequence for residues 1–135 and the other one for residues 611–617. Mutations in the gene encoding NET have been identified as potential causes for psychiatric illness. Thirteen NET DNA sequence variants were identified in patients suffering from schizophrenia and bipolar affective disorders [232]. An overabundance of NET is associated with ADHD [233, 234].

No X-ray data are reported; consequently, a homology model based on prokaryotic leucine transporter (LeuT) [235] was calculated [228]. The modeled binding site was validated by known NET ligands and subsequent docking studies suggested that norepinephrine mainly interacts with Asp75, Ala145, Phe72, and Tyr152. More accurate results were obtained by application of new experimental models from drosophila dopamine transporter (dDAT) and human serotonin transporter (SERT) (Figure 2.8) [236]. A comparison between the sequences of the known dDAT, SERT, and LeuT allowed identifying four major alignment gaps between them: the glycosylated extracellular loop 2 (EL2), extracellular loop 4 (EL4), and two intracellular loops [230]. The homology model of human NET identified residues in disallowed (Phe133(IL1), Phe134(IL1), Val379(EL4), and Ala384(EL4)) and allowed (Trp173(EL2), Asp175(EL2), Thr200(EL2), His441(IL4), Tyr467(TM10), Phe540(EL6), and Asn555(TM12)) regions [237]. This binding site strongly resembles the position of dopamine (DA) within DAT [238]. NET exists as a functional oligomer although as individual homomer is still functional as transporter.

References

1 Szabadi, E. (2013). Functional neuroanatomy of the central noradrenergic system. *J. Psychopharmacol. (London, UK)* 27 (8): 659–693, 635 pp.

2 Viljoen, M. and Panzer, A. (2007). The central noradrenergic system: an overview. *Af. J. Psychiatry (South Africa)* 10 (3): 135–141.

3 McCall, J.G., Al-Hasani, R., Siuda, E.R. et al. (2015). CRH engagement of the locus coeruleus noradrenergic system mediates stress-induced anxiety. *Neuron* 87 (3): 606–621.

4 Sullivan, G.M., Coplan, J.D., Kent, J.M., and Gorman, J.M. (1999). The noradrenergic system in pathological anxiety: a focus on panic with relevance to generalized anxiety and phobias. *Biol. Psychiatry* 46 (9): 1205–1218.

5 Berridge, C.W. (2009). The locus coeruleus-noradrenergic system and stress: implications for post-traumatic stress disorder. In: *Post-Traumatic Stress Disorder: Basic Science and Clinical Practice* (ed. P.J. Shiromani, T.M. Keane and J.E. LeDoux), 213–230. New York, NY: Humana Press.

6 de Berardis, D., Marini, S., Serroni, N. et al. (2015). Targeting the noradrenergic system in posttraumatic stress disorder: a systematic review and meta-analysis of prazosin trials. *Curr. Drug Targets* 16 (10): 1094–1106.

7 Davies, M.F., Tsui, J.Y., Flannery, J.A. et al. (2003). Augmentation of the noradrenergic system in alpha-2 adrenergic receptor deficient mice: anatomical changes associated with enhanced fear memory. *Brain Res.* 986 (1–2): 157–165.
8 Charney, D.S. and Redmond, D.E. Jr. (1983). Neurobiological mechanisms in human anxiety evidence supporting central noradrenergic hyperactivity. *Neuropharmacology* 22 (12): 1531–1536.
9 Itoi, K. and Sugimoto, N. (2010). The brainstem noradrenergic systems in stress, anxiety and depression. *J. Neuroendocrinol.* 22 (5): 355–361.
10 Ronzoni, G., Del Arco, A., Mora, F., and Segovia, G. (2016). Enhanced noradrenergic activity in the amygdala contributes to hyperarousal in an animal model of PTSD. *Psychoneuroendocrinology* 70: 1–9.
11 Engert, V. and Pruessner, J.C. (2008). Dopaminergic and noradrenergic contributions to functionality in ADHD: the role of methylphenidate. *Curr. Neuropharmacol.* 6 (4): 322–328.
12 Sandyk, R. (1986). Tourette syndrome: a case for noradrenergic and opiatergic mechanisms. *Int. J. Neurosci.* 30 (4): 303–309.
13 Kalpachidou, T., Raftogianni, A., Melissa, P. et al. (2016). Effects of a neonatal experience involving reward through maternal contact on the noradrenergic system of the rat prefrontal cortex. *Cereb. Cortex* 26 (9): 3866–3877.
14 Phillips, C., Fahimi, A., Das, D. et al. (2016). Noradrenergic system in down syndrome and Alzheimer's disease a target for therapy. *Curr. Alzheimers Res.* 13 (1): 68–83.
15 Berridge, C.W. and Waterhouse, B.D. (2003). The locus coeruleus-noradrenergic system: modulation of behavioral state and state-dependent cognitive processes. *Brain Res. Rev.* 42 (1): 33–84.
16 Szot, P. (2012). Common factors among Alzheimer's disease, Parkinson's disease, and epilepsy: possible role of the noradrenergic nervous system. *Epilepsia* 53 (Suppl. 1): 61–66.
17 Burnstock, G. (2004). Cotransmission. *Curr. Opin. Pharmacol.* 4 (1): 47–52.
18 Burnstock, G. (2009). Purinergic cotransmission. *Exp. Physiol.* 94 (1): 20–24.
19 Fung, S.I., Chan, J.Y., Manzoni, D. et al. (1994). Cotransmitter-mediated locus coeruleus action on motoneurons. *Brain Res. Bull.* 35 (5–6): 423–432.
20 Fung, S.J., Reddy, V.K., Zhuo, H. et al. (1994). Anatomical evidence for the presence of glutamate or enkephalin in noradrenergic projection neurons of the locus coeruleus. *Microsc. Res. Tech.* 29 (3): 219–225.
21 Xu, Z.Q., Shi, T.J., and Hokfelt, T. (1998). Galanin/GMAP- and NPY-like immunoreactivities in locus coeruleus and noradrenergic nerve terminals in the hippocampal formation and cortex with notes on the galanin-R_1 and -R_2 receptors. *J. Comp. Neurol.* 392 (2): 227–251.
22 Le Maitre, E., Barde, S.S., Palkovits, M. et al. (2013). Distinct features of neurotransmitter systems in the human brain with focus on the galanin system in locus coeruleus and dorsal raphe. *Proc. Natl. Acad. Sci. U.S.A.* 110 (6): E536–E545.
23 Ögren, S.O., Kuteeva, E., Elvander-Tottie, E., and Hökfelt, T. (2010). Neuropeptides in learning and memory processes with focus on galanin. *Eur. J. Pharmacol.* 626 (1): 9–17.

24 Borbely, E., Scheich, B., and Helyes, Z. (2013). Neuropeptides in learning and memory. *Neuropeptides (Oxford, UK)* 47 (6): 439–450.

25 Strosberg, A.D. (1993). Structure, function, and regulation of adrenergic receptors. *Protein Sci.* 2 (8): 1198–1209.

26 Perez, D.M. (2005). *The Receptors* (ed. K.A. Neve), 402. Totowa, NJ: Humana Press.

27 Perez, D.M. (2006). *The Receptors* (ed. K.A. Neve), 402. Totowa, NJ: Humana Press.

28 Lindner, D., Stichel, J., and Beck-Sickinger, A.G. (2008). Molecular recognition of the NPY hormone family by their receptors. *Nutrition* 24 (9): 907–917.

29 Lang, R., Gundlach, A.L., and Kofler, B. (2007). The galanin peptide family: receptor pharmacology, pleiotropic biological actions, and implications in health and disease. *Pharmacol. Ther.* 115 (2): 177–207.

30 Grossman, A. and Clement-Jones, V. (1983). Opiate receptors: enkephalins and endorphins. *Clin. Endocrinol. Metab.* 12 (1): 31–56.

31 Hoyer, D. and Bartfai, T. (2012). Neuropeptides and neuropeptide receptors: drug targets, and peptide and non-peptide ligands: a tribute to Prof. Dieter Seebach. *Chem. Biodivers.* 9 (11): 2367–2387.

32 Glowinski, J. and Baldessarini, R.J. (1966). Metabolism of norepinephrine in the central nervous system. *Pharmacol. Rev.* 18 (4): 1201–1238.

33 Gitlow, S.E., Mendlowitz, M., Bertani, L.M. et al. (1971). Human norepinephrine metabolism: its evaluation by administration of tritiated norepinephrine. *J. Clin. Invest.* 50 (4): 859–865.

34 Ahlquist, R.P. (1948). A study of the adrenotropic receptors. *Am. J. Physiol.* 153 (3): 586–600.

35 Berthelsen, S. and Pettinger, W.A. (1977). A functional basis for classification of alpha-adrenergic receptors. *Life Sci.* 21 (5): 595–606.

36 Lands, A.M., Arnold, A., McAuliff, J.P. et al. (1967). Differentiation of receptor systems activated by sympathomimetic amines. *Nature* 214 (5088): 597–598.

37 Bylund, D.B., Eikenberg, D.C., Hieble, J.P. et al. (1994). International union of pharmacology nomenclature of adrenoceptors. *Pharmacol. Rev.* 46 (2): 121–136.

38 Schwinn, D.A., Lomasney, J.W., Lorenz, W. et al. (1990). Molecular cloning and expression of the cDNA for a novel alpha 1-adrenergic receptor subtype. *J. Biol. Chem.* 265 (14): 8183–8189.

39 Ford, A.P., Williams, T.J., Blue, D.R., and Clarke, D.E. (1994). Alpha 1-adrenoceptor classification: sharpening Occam's razor. *Trends Pharmacol. Sci.* 15 (6): 167–170.

40 Hieble, J.P., Bylund, D.B., Clarke, D.E. et al. (1995). International union of pharmacology. X. Recommendation for nomenclature of alpha 1-adrenoceptors: consensus update. *Pharmacol. Rev.* 47 (2): 267–270.

41 Civantos Calzada, B. and Aleixandre de Artinano, A. (2001). Alpha-adrenoceptor subtypes. *Pharmacol. Res.* 44 (3): 195–208.

42 Hieble, J.P. (2007). Subclassification and nomenclature of alpha- and beta-adrenoceptors. *Curr. Top. Med. Chem.* 7 (2): 129–134.

43 Day, H.E.W., Campeau, S., Watson, S.J. Jr., and Akil, H. (1997). Distribution of α1a-, α1b- and α1d-adrenergic receptor mRNA in the rat brain and spinal cord. *J. Chem. Neuroanat.* 13 (2): 115–139.

44 Domyancic, A.V. and Morilak, D.A. (1997). Distribution of α1A adrenergic receptor m RNA in the rat brain visualized by in situ hybridization. *J. Comp. Neurol.* 386 (3): 358–378.

45 Shen, H., Peri, K.G., Deng, X.F. et al. (2000). Distribution of alpha1-adrenoceptor subtype proteins in different tissues of neonatal and adult rats. *Can. J. Physiol. Pharmacol.* 78 (3): 237–243.

46 Szot, P., White, S.S., Greenup, J.L. et al. (2005). Alpha1-adrenoreceptor in human hippocampus: binding and receptor subtype mRNA expression. *Brain Res. Mol. Brain Res.* 139 (2): 367–371.

47 Flugge, G., van Kampen, M., Meyer, H., and Fuchs, E. (2003). Alpha2A and alpha2C-adrenoceptor regulation in the brain: alpha2A changes persist after chronic stress. *Eur. J. Neurosci.* 17 (5): 917–928.

48 Owesson, C.A., Seif, I., McLaughlin, D.P., and Stamford, J.A. (2003). Different alpha(2) adrenoceptor subtypes control noradrenaline release and cell firing in the locus coeruleus of wildtype and monoamine oxidase-A knockout mice. *Eur. J. Neurosci.* 18 (1): 34–42.

49 Joyce, J.N., Lexow, N., Kim, S.J. et al. (1992). Distribution of beta-adrenergic receptor subtypes in human post-mortem brain: alterations in limbic regions of schizophrenics. *Synapse* 10 (3): 228–246.

50 Hillman, K.L., Doze, V.A., and Porter, J.E. (2005). Functional characterization of the beta-adrenergic receptor subtypes expressed by CA1 pyramidal cells in the rat hippocampus. *J. Pharmacol. Exp. Ther.* 314 (2): 561–567.

51 Paschalis, A., Churchill, L., Marina, N. et al. (2009). beta1-Adrenoceptor distribution in the rat brain: an immunohistochemical study. *Neurosci. Lett.* 458 (2): 84–88.

52 Ursino, M.G., Vasina, V., Raschi, E. et al. (2009). The beta3-adrenoceptor as a therapeutic target: current perspectives. *Pharmacol. Res.* 59 (4): 221–234.

53 Starke, K. (2001). Presynaptic autoreceptors in the third decade: focus on alpha2-adrenoceptors. *J. Neurochem.* 78 (4): 685–693.

54 Gilsbach, R. and Hein, L. (2012). Are the pharmacology and physiology of alpha(2) adrenoceptors determined by alpha(2)-heteroreceptors and autoreceptors respectively? *Br. J. Pharmacol.* 165 (1): 90–102.

55 Hwang, K.R., Chan, S.H., and Chan, J.Y. (1998). Noradrenergic neurotransmission at PVN in locus ceruleus-induced baroreflex suppression in rats. *Am. J. Physiol.* 274 (4 Pt. 2): H1284–H1292.

56 Wang, B., Wang, Y., Wu, Q. et al. (2017). Effects of alpha2A adrenoceptors on norepinephrine secretion from the locus coeruleus during chronic stress-induced depression. *Front. Neurosci.* 11: https://doi.org/10.3389/fnins.2017.00243.

57 Mitchell, H.A. and Weinshenker, D. (2010). Good night and good luck: norepinephrine in sleep pharmacology. *Biochem. Pharmacol.* 79 (6): 801–809.

58 Reznikoff, G.A., Manaker, S., Rhodes, C.H. et al. (1986). Localization and quantification of beta-adrenergic receptors in human brain. *Neurology* 36 (8): 1067–1073.

59 Nicholas, A.P., Pieribone, V.A., and Hokfelt, T. (1993). Cellular localization of messenger RNA for beta-1 and beta-2 adrenergic receptors in rat brain: an in situ hybridization study. *Neuroscience* 56 (4): 1023–1039.

60 Sscarpace, P.J. and Abrass, I.B. (1988). Alpha- and beta-adrenergic receptor function in the brain during senescence. *Neurobiol. Aging* 9: 53–58.

61 Cheng, J.T. and Kuo, D.Y. (2003). Both alpha1-adrenergic and D_1-dopaminergic neurotransmissions are involved in phenylpropanolamine-mediated feeding suppression in mice. *Neurosci. Lett.* 347 (2): 136–138.

62 Zimnik, N.C., Treadway, T., Smith, R.S., and Araneda, R.C. (2013). α(1A)-Adrenergic regulation of inhibition in the olfactory bulb. *J. Physiol.* 591 (Pt. 7): 1631–1643.

63 Fitzgerald, P.J. (2014). Is elevated norepinephrine an etiological factor in some cases of Parkinson's disease? *Med. Hypotheses* 82 (4): 462–469.

64 Maletic, V., Eramo, A., Gwin, K. et al. (2017). The role of norepinephrine and its α-adrenergic receptors in the pathophysiology and treatment of major depressive disorder and schizophrenia: a systematic review. *Front. Psychiatry* 8: https://doi.org/10.3389/fpsyt.2017.00042.

65 Espay, A.J., Le Witt, P.A., and Kaufmann, H. (2014). Norepinephrine deficiency in Parkinson's disease: the case for noradrenergic enhancement. *Mov. Disord.* 29 (14): 1710–1719.

66 Bidwell, L.C., Dew, R.E., and Kollins, S.H. (2010). Alpha-2 adrenergic receptors and attention-deficit/hyperactivity disorder. *Curr. Psychiatry Rep.* 12 (5): 366–373.

67 Beversdorf, D.Q. (2010). The role of the noradrenergic system in autism spectrum disorders. In: *The Neurochemical Basis of Autism: From Molecules to Minicolumns* (ed. G.J. Blatt), 175–184. New York, NY: Springer US.

68 Catapano, L.A. and Manji, H.K. (2007). G Protein-coupled receptors in major psychiatric disorders. *Biochim. Biophys. Acta* 1768 (4): 976–993.

69 Gurguis, G.N.M., Antai-Otong, D., Vo, S.P. et al. (1999). Adrenergic receptor function in panic disorder: I. Platelet [alpha]2 receptors Gi protein coupling, effects of imipramine, and relationship to treatment outcome. *Neuropsychopharmacology* 20 (2): 162–176.

70 Klimek, V., Rajkowska, G., Luker, S.N. et al. (1999). Brain noradrenergic receptors in major depression and schizophrenia. *Neuropsychopharmacology* 21 (1): 69–81.

71 Russo-Neustadt, A. and Cotman, C.W. (1997). Adrenergic receptors in Alzheimer's disease brain: selective increases in the cerebella of aggressive patients. *J. Neurosci.* 17 (14): 5573–5580.

72 Cash, R., Ruberg, M., Raisman, R., and Agid, Y. (1984). Adrenergic receptors in Parkinson's disease. *Brain Res.* 322 (2): 269–275.

73 Lomasney, J.W., Cotecchia, S., Lorenz, W. et al. (1991). Molecular cloning and expression of the cDNA for the alpha 1A-adrenergic receptor. The gene for which is located on human chromosome 5. *J. Biol. Chem.* 266 (10): 6365–6369.

74 Cotecchia, S., Schwinn, D.A., Randall, R.R. et al. (1988). Molecular cloning and expression of the cDNA for the hamster alpha 1-adrenergic receptor. *Proc. Natl. Acad. Sci. U.S.A.* 85 (19): 7159–7163.

75 Perez, D.M., Piascik, M.T., and Graham, R.M. (1991). Solution-phase library screening for the identification of rare clones: isolation of an alpha 1D-adrenergic receptor cDNA. *Mol. Pharmacol.* 40 (6): 876–883.

76 Hirasawa, A., Shibata, K., Horie, K. et al. (1995). Cloning, functional expression and tissue distribution of human alpha 1c-adrenoceptor splice variants. *FEBS Lett.* 363 (3): 256–260.

77 Chang, D.J., Chang, T.K., Yamanishi, S.S. et al. (1998). Molecular cloning, genomic characterization and expression of novel human alpha1A-adrenoceptor isoforms. *FEBS Lett.* 422 (2): 279–283.

78 Oshita, M., Kigoshi, S., and Muramatsu, I. (1991). Three distinct binding sites for [3H]-prazosin in the rat cerebral cortex. *Br. J. Pharmacol.* 104 (4): 961–965.

79 Ford, A.P., Daniels, D.V., Chang, D.J. et al. (1997). Pharmacological pleiotropism of the human recombinant alpha1A-adrenoceptor: implications for alpha1-adrenoceptor classification. *Br. J. Pharmacol.* 121 (6): 1127–1135.

80 Maiga, A., Dupont, M., Blanchet, G. et al. (2014). Molecular exploration of the alpha(1A)-adrenoceptor orthosteric site: binding site definition for epinephrine, HEAT and prazosin. *FEBS Lett.* 588 (24): 4613–4619.

81 Li, M., Fang, H., Du, L. et al. (2008). Computational studies of the binding site of alpha1A-adrenoceptor antagonists. *J. Mol. Model.* 14 (10): 957–966.

82 Ragnarsson, L., Andersson, A., Thomas, W.G., and Lewis, R.J. (2015). Extracellular surface residues of the alpha1B-adrenoceptor critical for G protein-coupled receptor function. *Mol. Pharmacol.* 87 (1): 121–129.

83 Campbell, A.P., MacDougall, I.J., Griffith, R., and Finch, A.M. (2014). An aspartate in the second extracellular loop of the alpha(1B) adrenoceptor regulates agonist binding. *Eur. J. Pharmacol.* 733: 90–96.

84 Kountz, T.S., Lee, K.S., Aggarwal-Howarth, S. et al. (2016). Endogenous N-terminal domain cleavage modulates alpha1D-adrenergic receptor pharmacodynamics. *J. Biol. Chem.* 291 (35): 18210–18221.

85 Perez, D.M. (2007). Structure-function of alpha1-adrenergic receptors. *Biochem. Pharmacol.* 73 (8): 1051–1062.

86 Bremner, J.B., Griffith, R., and Coban, B. (2001). Ligand design for alpha(1) adrenoceptors. *Curr. Med. Chem.* 8 (6): 607–620.

87 Nagaoka, Y., Ahmed, M., Hossain, M. et al. (2008). Amino acids of the human alpha1d-adrenergic receptor involved in antagonist binding. *J. Pharm. Sci.* 106 (1): 114–120.

88 Du, L. and Li, M. (2010). Modeling the interactions between alpha(1)-adrenergic receptors and their antagonists. *Curr. Comput. Aided Drug Des.* 6 (3): 165–178.

89 Takahashi, K., Hossain, M., Ahmed, M. et al. (2007). Asp125 and Thr130 in transmembrane domain 3 are major sites of alpha1b-adrenergic receptor antagonist binding. *Biol. Pharm. Bull.* 30 (10): 1891–1894.

90 Ahmed, M., Hossain, M., Bhuiyan, M.A. et al. (2008). Mutational analysis of the alpha 1a-adrenergic receptor binding pocket of antagonists by radioligand binding assay. *Biol. Pharm. Bull.* 31 (4): 598–601.

91 MacDougall, I.J.A. and Griffith, R. (2006). Selective pharmacophore design for α1-adrenoceptor subtypes. *J. Mol. Graphics Modell.* 25 (1): 146–157.

92 Kobilka, B.K., Matsui, H., Kobilka, T.S. et al. (1987). Cloning, sequencing, and expression of the gene coding for the human platelet alpha 2-adrenergic receptor. *Science* 238 (4827): 650–656.

93 Weinshank, R.L., Zgombick, J.M., Macchi, M. et al. (1990). Cloning, expression, and pharmacological characterization of a human alpha 2B-adrenergic receptor. *Mol. Pharmacol.* 38 (5): 681–688.

94 Regan, J.W., Kobilka, T.S., Yang-Feng, T.L. et al. (1988). Cloning and expression of a human kidney cDNA for an alpha 2-adrenergic receptor subtype. *Proc. Natl. Acad. Sci. U.S.A.* 85 (17): 6301–6305.

95 Lanier, S.M., Downing, S., Duzic, E., and Homcy, C.J. (1991). Isolation of rat genomic clones encoding subtypes of the alpha 2-adrenergic receptor. Identification of a unique receptor subtype. *J. Biol. Chem.* 266 (16): 10470–10478.

96 Small, K.M., Forbes, S.L., Brown, K.M., and Liggett, S.B. (2000). An asn to lys polymorphism in the third intracellular loop of the human alpha 2A-adrenergic receptor imparts enhanced agonist-promoted Gi coupling. *J. Biol. Chem.* 275 (49): 38518–38523.

97 Small, K.M., Brown, K.M., Seman, C.A. et al. (2006). Complex haplotypes derived from noncoding polymorphisms of the intronless alpha2A-adrenergic gene diversify receptor expression. *Proc. Natl. Acad. Sci. U.S.A.* 103 (14): 5472–5477.

98 Chung, D.A., Zuiderweg, E.R., Fowler, C.B. et al. (2002). NMR structure of the second intracellular loop of the alpha 2A adrenergic receptor: evidence for a novel cytoplasmic helix. *Biochemistry* 41 (11): 3596–3604.

99 Wang, C.D., Buck, M.A., and Fraser, C.M. (1991). Site-directed mutagenesis of alpha 2A-adrenergic receptors: identification of amino acids involved in ligand binding and receptor activation by agonists. *Mol. Pharmacol.* 40 (2): 168–179.

100 Suryanarayana, S., Daunt, D.A., Von Zastrow, M., and Kobilka, B.K. (1991). A point mutation in the seventh hydrophobic domain of the alpha 2 adrenergic receptor increases its affinity for a family of beta receptor antagonists. *J. Biol. Chem.* 266 (23): 15488–15492.

101 De Fusco, M., Vago, R., Striano, P. et al. (2014). The alpha2B-adrenergic receptor is mutant in cortical myoclonus and epilepsy. *Ann. Neurol.* 75 (1): 77–87.

102 Small, K.M., Brown, K.M., Forbes, S.L., and Liggett, S.B. (2001). Polymorphic deletion of three intracellular acidic residues of the alpha 2B-adrenergic receptor decreases G protein-coupled receptor kinase-mediated phosphorylation and desensitization. *J. Biol. Chem.* 276 (7): 4917–4922.

103 Zoladz, P.R., Kalchik, A.E., Hoffman, M.M. et al. (2014). ADRA2B deletion variant selectively predicts stress-induced enhancement of long-term memory in females. *Psychoneuroendocrinology* 48: 111–122.

104 Small, K.M., Forbes, S.L., Rahman, F.F. et al. (2000). A four amino acid deletion polymorphism in the third intracellular loop of the human alpha 2C-adrenergic receptor confers impaired coupling to multiple effectors. *J. Biol. Chem.* 275 (30): 23059–23064.

105 Small, K.M., Mialet-Perez, J., Seman, C.A. et al. (2004). Polymorphisms of cardiac presynaptic alpha2C adrenergic receptors: diverse intragenic variability with haplotype-specific functional effects. *Proc. Natl. Acad. Sci. U.S.A.* 101 (35): 13020–13025.

106 Balogh, B., Szilagyi, A., Gyires, K. et al. (2009). Molecular modelling of subtypes (alpha(2A), alpha(2B) and alpha(2C)) of alpha(2)-adrenoceptors: a comparative study. *Neurochem. Int.* 55 (6): 355–361.

107 Ostopovici-Halip, L., Curpan, R., Mracec, M., and Bologa, C.G. (2011). Structural determinants of the alpha2 adrenoceptor subtype selectivity. *J. Mol. Graphics Modell.* 29 (8): 1030–1038.
108 Rosenbaum, D.M., Rasmussen, S.G.F., and Kobilka, B.K. (2009). The structure and function of G-protein-coupled receptors. *Nature (London, UK)* 459 (7245): 356–363.
109 Bond, R.A. and Clarke, D.E. (1988). Agonist and antagonist characterization of a putative adrenoceptor with distinct pharmacological properties from the alpha- and beta-subtypes. *Br. J. Pharmacol.* 95 (3): 723–734.
110 Kaumann, A.J. (1997). Four beta-adrenoceptor subtypes in the mammalian heart. *Trends Pharmacol. Sci.* 18 (3): 70–76.
111 Strosberg, A.D., Gerhardt, C.C., Gros, J. et al. (1998). On the putative existence of a fourth α-adrenoceptor: proof is still missing. *Trends Pharmacol. Sci.* 19: 165–166.
112 Kompa, A.R. and Summers, R.J. (1999). Desensitization and resensitization of beta 1- and putative beta 4-adrenoceptor mediated responses occur in parallel in a rat model of cardiac failure. *Br. J. Pharmacol.* 128 (7): 1399–1406.
113 Konkar, A.A., Zhai, Y., and Granneman, J.G. (2000). beta1-adrenergic receptors mediate beta3-adrenergic-independent effects of CGP 12177 in brown adipose tissue. *Mol. Pharmacol.* 57 (2): 252–258.
114 Chung, F.Z., Lentes, K.U., Gocayne, J. et al. (1987). Cloning and sequence analysis of the human brain beta-adrenergic receptor. Evolutionary relationship to rodent and avian beta-receptors and porcine muscarinic receptors. *FEBS Lett.* 211 (2): 200–206.
115 Kobilka, B.K., Dixon, R.A., Frielle, T. et al. (1987). cDNA for the human beta 2-adrenergic receptor: a protein with multiple membrane-spanning domains and encoded by a gene whose chromosomal location is shared with that of the receptor for platelet-derived growth factor. *Proc. Natl. Acad. Sci. U.S.A.* 84 (1): 46–50.
116 Emorine, L.J., Marullo, S., Briend-Sutren, M.M. et al. (1989). Molecular characterization of the human beta 3-adrenergic receptor. *Science* 245 (4922): 1118–1121.
117 Hansen, N. and Manahan-Vaughan, D. (2015). Hippocampal long-term potentiation that is elicited by perforant path stimulation or that occurs in conjunction with spatial learning is tightly controlled by beta-adrenoreceptors and the locus coeruleus. *Hippocampus* 25 (11): 1285–1298.
118 Alali, A.S., McCredie, V.A., Golan, E. et al. (2014). Beta blockers for acute traumatic brain injury: a systematic review and meta-analysis. *Neurocritical Care* 20 (3): 514–523.
119 Skeberdis, V.A. (2004). Structure and function of beta3-adrenergic receptors. *Medicina (Kaunas, Lithuania)* 40 (5): 407–413.
120 Levin, B.E. (1982). Presynaptic location and axonal transport of beta 1-adrenoreceptors in the rat brain. *Science* 217 (4559): 555–557.
121 Tanaka, K.F., Kashima, H., Suzuki, H. et al. (2002). Existence of functional beta1- and beta2-adrenergic receptors on microglia. *J. Neurosci. Res.* 70 (2): 232–237.
122 Rodriguez, M., Carillon, C., Coquerel, A. et al. (1995). Evidence for the presence of beta 3-adrenergic receptor mRNA in the human brain. *Brain Res. Mol. Brain Res.* 29 (2): 369–375.

123 Moffett, S., Mouillac, B., Bonin, H., and Bouvier, M. (1993). Altered phosphorylation and desensitization patterns of a human beta 2-adrenergic receptor lacking the palmitoylated Cys341. *EMBO J.* 12 (1): 349–356.

124 Moore, J.D., Mason, D.A., Green, S.A. et al. (1999). Racial differences in the frequencies of cardiac beta(1)-adrenergic receptor polymorphisms: analysis of c145A>G and c1165G>C. *Hum. Mutat.* 14 (3): 271.

125 Mason, D.A., Moore, J.D., Green, S.A., and Liggett, S.B. (1999). A gain-of-function polymorphism in a G-protein coupling domain of the human beta1-adrenergic receptor. *J. Biol. Chem.* 274 (18): 12670–12674.

126 Warne, T., Serrano-Vega, M.J., Baker, J.G. et al. (2008). Structure of a beta1-adrenergic G-protein-coupled receptor. *Nature* 454 (7203): 486–491.

127 Warne, T., Moukhametzianov, R., Baker, J.G. et al. (2011). The structural basis for agonist and partial agonist action on a β1-adrenergic receptor. *Nature (London, UK)* 469 (7329): 241–244.

128 Warne, T., Edwards, P.C., Leslie, A.G., and Tate, C.G. (2012). Crystal structures of a stabilized beta1-adrenoceptor bound to the biased agonists bucindolol and carvedilol. *Structure* 20 (5): 841–849.

129 Moukhametzianov, R., Warne, T., Edwards, P.C. et al. (2011). Two distinct conformations of helix 6 observed in antagonist-bound structures of a beta1-adrenergic receptor. *Proc. Natl. Acad. Sci. U.S.A.* 108 (20): 8228–8232.

130 Miller-Gallacher, J.L., Nehme, R., Warne, T. et al. (2014). The 2.1 A resolution structure of cyanopindolol-bound beta1-adrenoceptor identifies an intramembrane Na^+ ion that stabilises the ligand-free receptor. *PLoS One* 9 (3): e92727.

131 Huang, J., Chen, S., Zhang, J.J., and Huang, X.Y. (2013). Crystal structure of oligomeric beta1-adrenergic G protein-coupled receptors in ligand-free basal state. *Nat. Struct. Mol. Biol.* 20 (4): 419–425.

132 O'Dowd, B.F., Hnatowich, M., Caron, M.G. et al. (1989). Palmitoylation of the human beta 2-adrenergic receptor. Mutation of Cys341 in the carboxyl tail leads to an uncoupled nonpalmitoylated form of the receptor. *J. Biol. Chem.* 264 (13): 7564–7569.

133 Valiquette, M., Parent, S., Loisel, T.P., and Bouvier, M. (1995). Mutation of tyrosine-141 inhibits insulin-promoted tyrosine phosphorylation and increased responsiveness of the human beta 2-adrenergic receptor. *EMBO J.* 14 (22): 5542–5549.

134 Moffett, S., Rousseau, G., Lagace, M., and Bouvier, M. (2001). The palmitoylation state of the beta(2)-adrenergic receptor regulates the synergistic action of cyclic AMP-dependent protein kinase and beta-adrenergic receptor kinase involved in its phosphorylation and desensitization. *J. Neurochem.* 76 (1): 269–279.

135 Xie, L., Xiao, K., Whalen, E.J. et al. (2009). Oxygen-regulated beta(2)-adrenergic receptor hydroxylation by EGLN3 and ubiquitylation by pVHL. *Sci. Signal.* 2 (78): ra33.

136 Kobilka, B.K., Frielle, T., Dohlman, H.G. et al. (1987). Delineation of the intronless nature of the genes for the human and hamster beta 2-adrenergic receptor and their putative promoter regions. *J. Biol. Chem.* 262 (15): 7321–7327.

137 Reihsaus, E., Innis, M., MacIntyre, N., and Liggett, S.B. (1993). Mutations in the gene encoding for the beta 2-adrenergic receptor in normal and asthmatic subjects. *Am. J. Respir. Cell Mol. Biol.* 8 (3): 334–339.
138 Rupert, J.L., Monsalve, M.V., Devine, D.V., and Hochachka, P.W. (2000). Beta2-adrenergic receptor allele frequencies in the Quechua, a high altitude native population. *Ann. Hum. Genet.* 64 (Pt. 2): 135–143.
139 Cherezov, V., Rosenbaum, D.M., Hanson, M.A. et al. (2007). High-resolution crystal structure of an engineered human beta2-adrenergic G protein-coupled receptor. *Science* 318 (5854): 1258–1265.
140 Rasmussen, S.G., Choi, H.J., Rosenbaum, D.M. et al. (2007). Crystal structure of the human beta2 adrenergic G-protein-coupled receptor. *Nature* 450 (7168): 383–387.
141 Hanson, M.A., Cherezov, V., Griffith, M.T. et al. (2008). A specific cholesterol binding site is established by the 2.8 A structure of the human beta2-adrenergic receptor. *Structure* 16 (6): 897–905.
142 Bokoch, M.P., Zou, Y., Rasmussen, S.G. et al. (2010). Ligand-specific regulation of the extracellular surface of a G-protein-coupled receptor. *Nature* 463 (7277): 108–112.
143 Ring, A.M., Manglik, A., Kruse, A.C. et al. (2013). Adrenaline-activated structure of beta2-adrenoceptor stabilized by an engineered nanobody. *Nature* 502 (7472): 575–579.
144 Wacker, D., Fenalti, G., Brown, M.A. et al. (2010). Conserved binding mode of human beta2 adrenergic receptor inverse agonists and antagonist revealed by X-ray crystallography. *J. Am. Chem. Soc.* 132 (33): 11443–11445.
145 Rosenbaum, D.M., Zhang, C., Lyons, J.A. et al. (2011). Structure and function of an irreversible agonist-beta(2) adrenoceptor complex. *Nature* 469 (7329): 236–240.
146 Rasmussen, S.G., Choi, H.J., Fung, J.J. et al. (2011). Structure of a nanobody-stabilized active state of the beta(2) adrenoceptor. *Nature* 469 (7329): 175–180.
147 Staus, D.P., Strachan, R.T., Manglik, A. et al. (2016). Allosteric nanobodies reveal the dynamic range and diverse mechanisms of G-protein-coupled receptor activation. *Nature* 535 (7612): 448–452.
148 Weichert, D., Kruse, A.C., Manglik, A. et al. (2014). Covalent agonists for studying G protein-coupled receptor activation. *Proc. Natl. Acad. Sci. U.S.A.* 111 (29): 10744–10748.
149 Huang, C.Y., Olieric, V., Ma, P. et al. (2016). In meso in situ serial X-ray crystallography of soluble and membrane proteins at cryogenic temperatures. *Acta Crystallogr., Sect. D: Struct. Biol.* 72 (Pt. 1): 93–112.
150 Rasmussen, S.G., DeVree, B.T., Zou, Y. et al. (2011). Crystal structure of the beta2 adrenergic receptor-Gs protein complex. *Nature* 477 (7366): 549–555.
151 Zou, Y., Weis, W.I., and Kobilka, B.K. (2012). N-terminal T4 lysozyme fusion facilitates crystallization of a G protein coupled receptor. *PLoS One* 7 (10): e46039.
152 Evans, B.A., Papaioannou, M., Hamilton, S., and Summers, R.J. (1999). Alternative splicing generates two isoforms of the beta3-adrenoceptor which are differentially expressed in mouse tissues. *Br. J. Pharmacol.* 127 (6): 1525–1531.

153 Clement, K., Vaisse, C., Manning, B.S. et al. (1995). Genetic variation in the beta 3-adrenergic receptor and an increased capacity to gain weight in patients with morbid obesity. *N. Engl. J. Med.* 333 (6): 352–354.

154 Candelore, M.R., Deng, L., Tota, L.M. et al. (1996). Pharmacological characterization of a recently described human beta 3-adrenergic receptor mutant. *Endocrinology* 137 (6): 2638–2641.

155 Halushka, M.K., Fan, J.B., Bentley, K. et al. (1999). Patterns of single-nucleotide polymorphisms in candidate genes for blood-pressure homeostasis. *Nat. Genet.* 22 (3): 239–247.

156 Walston, J., Lowe, A., Silver, K. et al. (1997). The beta3-adrenergic receptor in the obesity and diabetes prone rhesus monkey is very similar to human and contains arginine at codon 64. *Gene* 188 (2): 207–213.

157 Balligand, J.L. (2013). Beta3-adrenoreceptors in cardiovasular diseases: New roles for an "old" receptor. *Curr. Drug Delivery* 10 (1): 64–66.

158 Zelaszczyk, D., Zakrzeska, A., Kwolek, G. et al. (2005). A search for presynaptic beta3-adrenoceptors in the rat. *Fundam. Clin. Pharmacol.* 19 (2): 147–153.

159 Chuang, J.C., Krishnan, V., Yu, H.G. et al. (2010). A beta3-adrenergic-leptin-melanocortin circuit regulates behavioral and metabolic changes induced by chronic stress. *Biol. Psychiatry* 67 (11): 1075–1082.

160 Gibbs, M.E., Hutchinson, D.S., and Summers, R.J. (2008). Role of beta-adrenoceptors in memory consolidation: beta3-adrenoceptors act on glucose uptake and beta2-adrenoceptors on glycogenolysis. *Neuropsychopharmacology* 33 (10): 2384–2397.

161 Jhaveri, D.J., Mackay, E.W., Hamlin, A.S. et al. (2010). Norepinephrine directly activates adult hippocampal precursors via beta3-adrenergic receptors. *J. Neurosci.* 30 (7): 2795–2806.

162 Greene, R.W. (2010). An anxiolytic response exerted by beta 3-adrenoreceptor activation: correlation with an enhanced subset of GABAergic synaptic responses. *Neuropsychopharmacology* 35 (9): 1839–1840.

163 Claustre, Y., Leonetti, M., Santucci, V. et al. (2008). Effects of the beta3-adrenoceptor (Adrb3) agonist SR58611A (amibegron) on serotonergic and noradrenergic transmission in the rodent: relevance to its antidepressant/anxiolytic-like profile. *Neuroscience* 156 (2): 353–364.

164 Overstreet, D.H., Stemmelin, J., and Griebel, G. (2008). Confirmation of antidepressant potential of the selective beta3 adrenoceptor agonist amibegron in an animal model of depression. *Pharmacol. Biochem. Behav.* 89 (4): 623–626.

165 Stemmelin, J., Cohen, C., Terranova, J.P. et al. (2008). Stimulation of the beta3-Adrenoceptor as a novel treatment strategy for anxiety and depressive disorders. *Neuropsychopharmacology* 33 (3): 574–587.

166 Lenard, N.R., Gettys, T.W., and Dunn, A.J. (2003). Activation of beta2- and beta3-adrenergic receptors increases brain tryptophan. *J. Pharmacol. Exp. Ther.* 305 (2): 653–659.

167 Mahmoudian, M. (1994). The complex of human Gs protein with the beta 3 adrenergic receptor: a computer-aided molecular modeling study. *J. Mol. Graphics* 12 (1): 22–28, 34.

168 Jin, F., Lu, C., Sun, X. et al. (2011). Insights into the binding modes of human beta(3)-adrenergic receptor agonists with ligand-based and receptor-based methods. *Mol. Diversity* 15 (4): 817–831.
169 Beck, B. and Pourié, G. (2013). Ghrelin, neuropeptide Y, and other feeding-regulatory peptides active in the hippocampus: role in learning and memory. *Nutr. Rev.* 71 (8): 541–561.
170 Chronwall, B.M., DiMaggio, D.A., Massari, V.J. et al. (1985). The anatomy of neuropeptide-Y-containing neurons in rat brain. *Neuroscience* 15 (4): 1159–1181.
171 Caberlotto, L., Fuxe, K., and Hurd, Y.L. (2000). Characterization of NPY mRNA-expressing cells in the human brain: co-localization with Y2 but not Y1 mRNA in the cerebral cortex, hippocampus, amygdala, and striatum. *J. Chem. Neuroanat.* 20 (3–4): 327–337.
172 Kastin, A.J. and Akerstrom, V. (1999). Nonsaturable entry of neuropeptide Y into brain. *Am. J. Physiol.* 276 (3 Pt. 1): E479–E482.
173 Eaton, K., Sallee, F.R., and Sah, R. (2007). Relevance of neuropeptide Y (NPY) in psychiatry. *Curr. Top. Med. Chem.* 7 (17): 1645–1659.
174 Blomqvist, A.G. and Herzog, H. (1997). Y-receptor subtypes--how many more? *Trends Neurosci.* 20 (7): 294–298.
175 Widdowson, P.S. (1993). Quantitative receptor autoradiography demonstrates a differential distribution of neuropeptide-Y Y1 and Y2 receptor subtypes in human and rat brain. *Brain Res.* 631 (1): 27–38.
176 Beck, B. (2001). KO's and organisation of peptidergic feeding behavior mechanisms. *Neurosci. Biobehav. Rev.* 25 (2): 143–158.
177 Beck, B. (2006). Neuropeptide Y in normal eating and in genetic and dietary-induced obesity. *Philos. Trans. R. Soc. London, Ser. B: Biol. Sci.* 361 (1471): 1159–1185.
178 Redrobe, J.P., Dumont, Y., St-Pierre, J.A., and Quirion, R. (1999). Multiple receptors for neuropeptide Y in the hippocampus: putative roles in seizures and cognition. *Brain Res.* 848 (1–2): 153–166.
179 Parker, R.M. and Herzog, H. (1999). Regional distribution of Y-receptor subtype mRNAs in rat brain. *Eur. J. Neurosci.* 11 (4): 1431–1448.
180 Decressac, M., Prestoz, L., Veran, J. et al. (2009). Neuropeptide Y stimulates proliferation, migration and differentiation of neural precursors from the subventricular zone in adult mice. *Neurobiol. Dis.* 34 (3): 441–449.
181 Guo, H., Castro, P.A., Palmiter, R.D., and Baraban, S.C. (2002). Y5 receptors mediate neuropeptide Y actions at excitatory synapses in area CA3 of the mouse hippocampus. *J. Neurophysiol.* 87 (1): 558–566.
182 Morin, S.M. and Gehlert, D.R. (2006). Distribution of NPY Y5-like immunoreactivity in the rat brain. *J. Mol. Neurosci.* 29 (2): 109–114.
183 Kopp, J., Xu, Z.Q., Zhang, X. et al. (2002). Expression of the neuropeptide Y Y1 receptor in the CNS of rat and of wild-type and Y1 receptor knock-out mice. Focus on immunohistochemical localization. *Neuroscience* 111 (3): 443–532.
184 Stanic, D., Brumovsky, P., Fetissov, S. et al. (2006). Characterization of neuropeptide Y2 receptor protein expression in the mouse brain. I. Distribution in cell bodies and nerve terminals. *J. Comp. Neurol.* 499 (3): 357–390.

185 Decressac, M., Wright, B., David, B. et al. (2011). Exogenous neuropeptide Y promotes in vivo hippocampal neurogenesis. *Hippocampus* 21 (3): 233–238.
186 Thiriet, N., Agasse, F., Nicoleau, C. et al. (2011). NPY promotes chemokinesis and neurogenesis in the rat subventricular zone. *J. Neurochem.* 116 (6): 1018–1027.
187 Vähätalo, L.H., Ruohonen, S.T., Mäkelä, S. et al. (2015). Role of the endocannabinoid system in obesity induced by neuropeptide y overexpression in noradrenergic neurons. *Nutr. Diabetes* 5 (4): e151.
188 Pedragosa-Badia, X., Stichel, J., and Beck-Sickinger, A.G. (2013). Neuropeptide Y receptors: how to get subtype selectivity. *Front. Endocrinol.* 4: https://doi.org/10.3389/fendo.2013.00005.
189 Herzog, H., Hort, Y.J., Ball, H.J. et al. (1992). Cloned human neuropeptide Y receptor couples to two different second messenger systems. *Proc. Natl. Acad. Sci. U.S.A.* 89 (13): 5794–5798.
190 Yan, H., Yang, J., Marasco, J. et al. (1996). Cloning and functional expression of cDNAs encoding human and rat pancreatic polypeptide receptors. *Proc. Natl. Acad. Sci. U.S.A.* 93 (10): 4661–4665.
191 Xu, B., Fallmar, H., Boukharta, L. et al. (2013). Mutagenesis and computational modeling of human G-protein-coupled receptor Y2 for neuropeptide Y and peptide YY. *Biochemistry* 52 (45): 7987–7998.
192 Counts, S.E., Perez, S.E., and Mufson, E.J. (2008). Galanin – 25 years with a multitalented neuropeptide. *Cell. Mol. Life Sci.* 65 (12): 1842–1853.
193 Lu, X., Mazarati, A., Sanna, P. et al. (2005). Distribution and differential regulation of galanin receptor subtypes in rat brain: effects of seizure activity. *Neuropeptides* 39 (3): 147–152.
194 Skofitsch, G. and Jacobowitz, D.M. (1985). Immunohistochemical mapping of galanin-like neurons in the rat central nervous system. *Peptides* 6 (3): 509–546.
195 Bowser, R., Kordower, J.H., and Mufson, E.J. (1997). A confocal microscopic analysis of galaninergic hyperinnervation of cholinergic basal forebrain neurons in Alzheimer's disease. *Brain Pathol.* 7 (2): 723–730.
196 Costa, A., Bini, P., Hamze-Sinno, M. et al. (2011). Galanin and alpha-MSH autoantibodies in cerebrospinal fluid of patients with Alzheimer's disease. *J. Neuroimmunol.* 240–241: 114–120.
197 Ding, X., MacTavish, D., Kar, S., and Jhamandas, J.H. (2006). Galanin attenuates beta-amyloid (Abeta) toxicity in rat cholinergic basal forebrain neurons. *Neurobiol. Dis.* 21 (2): 413–420.
198 Kehr, J., Yoshitake, T., Wang, F.H. et al. (2002). Galanin is a potent in vivo modulator of mesencephalic serotonergic neurotransmission. *Neuropsychopharmacology* 27 (3): 341–356.
199 Branchek, T.A., Smith, K.E., Gerald, C., and Walker, M.W. (2000). Galanin receptor subtypes. *Trends Pharmacol. Sci.* 21 (3): 109–117.
200 Burgevin, M.C., Loquet, I., Quarteronet, D., and Habert-Ortoli, E. (1995). Cloning, pharmacological characterization, and anatomical distribution of a rat cDNA encoding for a galanin receptor. *J. Mol. Neurosci.* 6 (1): 33–41.
201 O'Donnell, D., Ahmad, S., Wahlestedt, C., and Walker, P. (1999). Expression of the novel galanin receptor subtype GALR2 in the adult rat CNS: distinct distribution from GALR1. *J. Comp. Neurol.* 409 (3): 469–481.

202 Waters, S.M. and Krause, J.E. (2000). Distribution of galanin-1, -2 and -3 receptor messenger RNAs in central and peripheral rat tissues. *Neuroscience* 95 (1): 265–271.

203 Gustafson, E.L., Smith, K.E., Durkin, M.M. et al. (1996). Distribution of a rat galanin receptor mRNA in rat brain. *Neuroreport* 7 (4): 953–957.

204 Sullivan, K.A., Shiao, L.L., and Cascieri, M.A. (1997). Pharmacological characterization and tissue distribution of the human and rat GALR1 receptors. *Biochem. Biophys. Res. Commun.* 233 (3): 823–828.

205 Ahmad, S., O'Donnell, D., Payza, K. et al. (1998). Cloning and evaluation of the role of rat GALR-2, a novel subtype of galanin receptor, in the control of pain perception. *Ann. N.Y. Acad. Sci.* 863: 108–119.

206 Xia, C.Y., Qi, Y.T., Zhang, Y., and Yuan, C.G. (2005). The expression of galanin and galanin receptors in olfactory ensheathing cells and effects of galanin on cells proliferation. *Shi Yan Sheng Wu Xue Bao* 38 (2): 98–104.

207 Mennicken, F., Hoffert, C., Pelletier, M. et al. (2002). Restricted distribution of galanin receptor 3 (GalR3) mRNA in the adult rat central nervous system. *J. Chem. Neuroanat.* 24 (4): 257–268.

208 Wang, S., Hashemi, T., He, C. et al. (1997). Molecular cloning and pharmacological characterization of a new galanin receptor subtype. *Mol. Pharmacol.* 52 (3): 337–343.

209 Wang, S., Hashemi, T., Fried, S. et al. (1998). Differential intracellular signaling of the GalR1 and GalR2 galanin receptor subtypes. *Biochemistry* 37 (19): 6711–6717.

210 Kolakowski, L.F. Jr., O'Neill, G.P., Howard, A.D. et al. (1998). Molecular characterization and expression of cloned human galanin receptors GALR2 and GALR3. *J. Neurochem.* 71 (6): 2239–2251.

211 Mazarati, A., Lundstrom, L., Sollenberg, U. et al. (2006). Regulation of kindling epileptogenesis by hippocampal galanin type 1 and type 2 receptors: the effects of subtype-selective agonists and the role of G-protein-mediated signaling. *J. Pharmacol. Exp. Ther.* 318 (2): 700–708.

212 Freimann, K., Kurrikoff, K., and Langel, U. (2015). Galanin receptors as a potential target for neurological disease. *Expert Opin. Ther. Targets* 19 (12): 1665–1676.

213 Bliss, T.V. and Collingridge, G.L. (1993). A synaptic model of memory: long-term potentiation in the hippocampus. *Nature* 361 (6407): 31–39.

214 Elliott-Hunt, C.R., Holmes, F.E., Hartley, D.M. et al. (2011). Endogenous galanin protects mouse hippocampal neurons against amyloid toxicity in vitro via activation of galanin receptor-2. *J. Alzheimers Dis.* 25 (3): 455–462.

215 Li, L., Yu, L., and Kong, Q. (2013). Exogenous galanin attenuates spatial memory impairment and decreases hippocampal beta-amyloid levels in rat model of Alzheimer's disease. *Int. J. Neurosci.* 123 (11): 759–765.

216 Webling, K., Runesson, J., Bartfai, T., and Langel, Ü. (2012). Galanin receptors and ligands. *Front. Endocrinol.* 3: https://doi.org/10.3389/fendo.2012.00146.

217 Hui, W.Q., Cheng, Q., Liu, T.Y., and Ouyang, Q. (2016). Homology modeling, docking, and molecular dynamics simulation of the receptor GALR2 and its interactions with galanin and a positive allosteric modulator. *J. Mol. Model.* 22 (4): 90.

218 Lorimer, D.D., Matkowskj, K., and Benya, R.V. (1997). Cloning, chromosomal location, and transcriptional regulation of the human galanin-1 receptor gene (GALN1R). *Biochem. Biophys. Res. Commun.* 241 (2): 558–564.
219 Jacoby, A.S., Webb, G.C., Liu, M.L. et al. (1997). Structural organization of the mouse and human GALR1 galanin receptor genes (Galnr and GALNR) and chromosomal localization of the mouse gene. *Genomics* 45 (3): 496–508.
220 Lundström, L., Lu, X., Langel, Ü., and Bartfai, T. (2005). Important pharmacophores for binding to galanin receptor 2. *Neuropeptides* 39 (3): 169–171.
221 Lundström, L., Sollenberg, U.E., Bartfai, T., and Langel, Ü. (2007). Molecular characterization of the ligand binding site of the human galanin receptor type 2, identifying subtype selective interactions. *J. Neurochem.* 103 (5): 1774–1784.
222 Jurkowski, W., Yazdi, S., and Elofsson, A. (2013). Ligand binding properties of human galanin receptors. *Mol. Membr. Biol.* 30 (2): 206–216.
223 Pacholczyk, T., Blakely, R.D., and Amara, S.G. (1991). Expression cloning of a cocaine- and antidepressant-sensitive human noradrenaline transporter. *Nature* 350 (6316): 350–354.
224 Amara, S.G. and Kuhar, M.J. (1993). Neurotransmitter transporters: recent progress. *Annu. Rev. Neurosci.* 16: 73–93.
225 Blakely, R.D., De Felice, L.J., and Hartzell, H.C. (1994). Molecular physiology of norepinephrine and serotonin transporters. *J. Exp. Biol.* 196: 263–281.
226 Nelson, N. (1998). The family of Na^+/Cl^- neurotransmitter transporters. *J. Neurochem.* 71 (5): 1785–1803.
227 Mandela, P. and Ordway, G.A. (2006). The norepinephrine transporter and its regulation. *J. Neurochem.* 97 (2): 310–333.
228 Schlessinger, A., Geier, E., Fan, H. et al. (2011). Structure-based discovery of prescription drugs that interact with the norepinephrine transporter, NET. *Proc. Natl. Acad. Sci. U.S.A.* 108 (38): 15810–15815.
229 Andersen, J., Ringsted, K.B., Bang-Andersen, B. et al. (2015). Binding site residues control inhibitor selectivity in the human norepinephrine transporter but not in the human dopamine transporter. *Sci. Rep.* 5 (1): https://doi.org/10.1038/srep15650.
230 Haddad, Y., Heger, Z., and Adam, V. (2017). Targeting neuroblastoma cell surface proteins: recommendations for homology modeling of hNET, ALK, and TrkB. *Front. Mol. Neurosci.* 10: https://doi.org/10.3389/fnmol.2017.00007.
231 Shannon, J.R., Flattem, N.L., Jordan, J. et al. (2000). Orthostatic intolerance and tachycardia associated with norepinephrine-transporter deficiency. *N. Engl. J. Med.* 342 (8): 541–549.
232 Stober, G., Nothen, M.M., Porzgen, P. et al. (1996). Systematic search for variation in the human norepinephrine transporter gene: identification of five naturally occurring missense mutations and study of association with major psychiatric disorders. *Am. J. Med. Genet.* 67 (6): 523–532.
233 Schroeter, S., Apparsundaram, S., Wiley, R.G. et al. (2000). Immunolocalization of the cocaine- and antidepressant-sensitive l-norepinephrine transporter. *J. Comp. Neurol.* 420 (2): 211–232.
234 Tellioglu, T. and Robertson, D. (2001). Genetic or acquired deficits in the norepinephrine transporter: current understanding of clinical implications. *Expert Rev. Mol. Med.* 3 (29): https://doi.org/10.1017/s1462399401003878.

235 Yamashita, A., Singh, S.K., Kawate, T. et al. (2005). Crystal structure of a bacterial homologue of Na^+/Cl^--dependent neurotransmitter transporters. *Nature* 437 (7056): 215–223.
236 Haddad, Y., Heger, Z., and Adam, V. (2016). Guidelines for homology modeling of dopamine, norepinephrine, and serotonin transporters. *ACS Chem. Neurosci.* 7 (11): 1607–1613.
237 Koldso, H., Christiansen, A.B., Sinning, S., and Schiott, B. (2013). Comparative modeling of the human monoamine transporters: similarities in substrate binding. *ACS Chem. Neurosci.* 4 (2): 295–309.
238 Koldsø, H., Grouleff, J., and Schiøtt, B. (2015). Insights to ligand binding to the monoamine transporters—from homology modeling to LeuBAT and dDAT. *Front. Pharmacol.* 6: https://doi.org/10.3389/fphar.2015.00208.

3

The Serotonergic System

3.1 Introduction

The serotonergic system is one of the oldest amine-based transmitter systems in the brain consisting of a diverse group of neurons innervating virtually all regions of the CNS but concentrated in the cerebral cortex, limbic structures, basal ganglia, several regions of the brainstem, and the gray matter of the spinal cord. The complex anatomy of the serotonergic system has been largely studied using histofluoresecence and immunohistochemical methods [1] as well as autoradiographic techniques [2], which has allowed a complete characterization of the topography and morphology of serotonergic cell bodies in the brain. The serotonergic system, involved in almost every physiological function, interacts with other neurotransmitter systems at different molecular levels [3]. It also modulates neural activities in the entorhinal cortex, considered as the gate to control the flow of information into and out of the hippocampus and crucial for functions such as emotional control, learning, and memory [4]. Consequently, dysfunction of the serotonergic system at the CNS level is believed to be related to a variety of neurodisorders, the more representative being autism [5, 6], anxiety [7], depression [8], anorexia [9], restless legs syndrome [10], schizophrenia [11], and Alzheimer's disease [12]. Of particular relevance is the relationship between Parkinson's disease and the serotonergic system [13, 14], which can affect motor and nonmotor manifestations of the disease [15], including depression [16, 17], levodopa-induced dyskinesias [18], and other therapy side effects [19]. These connections emphasize the therapeutic roles of the serotonergic system in Parkinson's disease [20, 21].

The serotonergic pathways are schematically illustrated in Figure 3.1. For a more detailed representation, the reader is referred to a review on brain serotonergic circuitries [3]. The main serotonergic pathway (rostral group) is located in the brainstem area denominated the Raphe nucleus. The median raphe and dorsal raphe nuclei provide parallel and overlapping projections to many forebrain structures including the neocortex and the limbic system. The dorsal raphe nucleus projects to the cortex, neostriatum, amygdala, and substantia nigra. The median raphe nucleus projects into the hippocampus, hypothalamus, septal nuclei, ventral tegmental area, and cortex. A second pathway (caudal group) descends down the spinal cord.

Chemical Biology of Neurodegeneration: A Molecular Approach, First Edition. Pedro Merino.

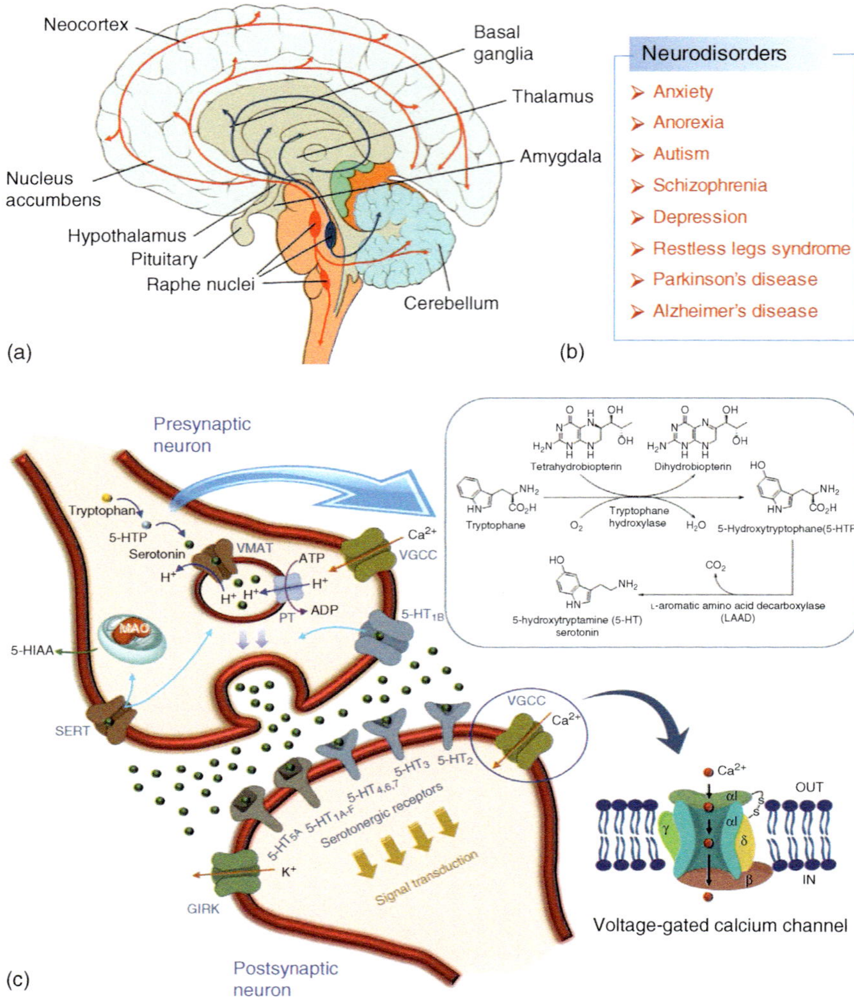

Figure 3.1 The serotonergic system. (a) Pathways of the serotonergic system. (b) Neurological disorders associated with the serotonergic system. (c) Details of the noradrenergic synapse and neurotransmitters involved in noradrenergic system. Serotonin (5-HT) is synthesized from tryptophan inside the serotonergic neuron. After packaging in vesicles by VMAT, 5-HT is released into the synapse. 5-HT is recognized by serotonin receptors that produce the signal. The excess of 5-HT is removed from the synaptic area by 5-HT_{1b} and SERT and reintroduced in the neuron for being reutilized or catabolized by MAO. VMAT: vesicular monoamine transporter; SERT: serotonin transporter; 5-HT_x: serotonin receptors, MAO: monoamine oxidase; 5-HTP: 5-hydroxytryptophan; 5-HIAA: 5-hydroxyindolacetic acid; VGCC: voltage-gated calcium channel; GIRK: G-protein-coupled inwardly rectifying potassium channel; PT: proton transporter.

The serotonergic system uses serotonin (or 5-hydroxytryptamine, 5-HT) as neurotransmitter. Serotonin is biosynthesized from L-tryptophan, uptaken into the serotonergic neuron by tryptophan hydroxylase (L-tryptophan-5-monooxygenase, LTH) that converts the amino acid into 5-hydroxytryptophan (5-HTP). Further decarboxylation of 5-HTP by amino acid decarboxylase (AAD), the enzyme responsible for forming dopamine from L-DOPA (see Chapter 1), yields serotonin (5-HT). The synthesized 5-HT is stored in synaptic vesicles by the monoamine transporter (VMAT). The transport is driven by the electrochemical gradient generated by a proton-pumping ATPase in such a way the uptake of serotonin in the vesicle is coupled with the efflux of a proton. The vesicle, under appropriate conditions, migrates to and fuse with the cell membrane, releasing serotonin into the synapse (Figure 3.1). In the synaptic cleft, 5-HT can act on both postsynaptic and presynaptic receptors. Released serotonin interacts with the pre- and post-synaptic G-protein coupled receptors (GPCRs), which activate second messenger systems that convey information to other neurons. Interaction with presynaptic receptors regulates 5-HT synthesis and release. The action of serotonin is terminated either by diffusion away from the synaptic area and subsequent catabolism or by reuptake into the serotonergic neuron by the action of 5-HT transporter (SERT, serotonin reuptake transporter) and 5-HT_{1B} autoreceptors that suppress excess serotonin release (see Section 3.3). Once inside the presynaptic neuron, 5-HT can be re-stored in vesicles or metabolized by mitochondrial MAO, resulting in the metabolite 5-hydroxyindolacetic acid (5-HIAA) and/or 5-HT *N*-acetyltransferase. As other neurotransmitters, release of serotonin depends on calcium. When calcium channels are blocked, neurotransmitter release is inhibited. When the action potential reaches the nerve terminal, voltage-gated calcium channels (VGCCs) open and calcium rushes into the neuron terminal due to a greater extracellular concentration. Once in the presynaptic neuron, calcium triggers a complex series of events, common to all neurotransmitters, that regulates the synaptic vesicle life cycle [22].

3.2 Serotonin Receptors

3.2.1 Classification

The classification of serotonin receptors, as in the case of other receptors, was initially based on pharmacological and physiological studies. By using pharmacological criteria and functional responses, three families 5-HT_1, 5-HT_2, and 5-HT_3 were identified. However, application of molecular biology techniques identified additional families and subtypes of receptors. The current classification system identifies seven types of 5-HT receptors named 5-HT_{1-7} and takes into account the information about intracellular signal transduction mechanisms and molecular structures besides the classical information on selective agonists and antagonists and ligand-binding affinities. Serotonin receptors can be grouped into three families: 5-HT_1 family, 5-HT_2 family, and that formed by 5-HT_4, 5-HT_6, and 5-HT_7 receptors. These three families are G-protein-coupled receptors. On the other hand, 5-HT_3 receptor is a ligand-gated ion channel and 5-HT_5 receptors are considered orphan receptors (Section 3.2.4). Each family of 5-HT receptors can have several subtypes, which will be discussed in detail in the next sections. Table 3.1 lists the currently accepted classification of serotonin receptors.

Table 3.1 Serotonin receptors[a,b].

Family	Subtype	Gene	Residues	Distribution in brain	Effector mechanism	Related neurodisorders
5-HT_1	5-HT_{1A}	*5-HT1A*	422	Amygdala, hippocampus, hypothalamus, raphe nuclei, entorhinal cortex, septum	Inhibition of adenylyl cyclase, opening of potassium channels, activation of MAP kinase pathway	Addiction, anxiety, memory, mood, sleep
	5-HT_{1B}	*5-HT1B*	390	Substantia nigra, basal ganglia	Inhibition of adenylyl cyclase	Addiction, anxiety, learning, memory, mood
	5-HT_{1D}	*5-HT1D*	377	Lateral geniculate, superior colliculus, frontal cortex, substantia nigra, hippocampus, cortex	Inhibition of adenylyl cyclase	Anxiety
	5-HT_{1E}	*5-HT1E*	365		Inhibition of adenylyl cyclase	
	5-HT_{1F}	*5-HT1F*	366	Striatum, cerebral cortex, nucleus accumbens, amygdala, hippocampus	Inhibition of adenylyl cyclase	
5-HT_2	5-HT_{2A}	*5-HT2A*	471	Cortex, olfactory tubercle, striatum, nucleus accumbens, claustrum	Activation of phospholipase C, closing of potassium channels	Addiction, anxiety, cognition, memory, mood, sleep
	5-HT_{2B}	*5-HT2B*	481	Cerebellum, septum, amygdala	Activation of phospholipase C	Anxiety, appetite
	5-HT_{2C}	*5-HT2C*	458	Globus pallidus, cortex, hypothalamus, substantia nigra, septum, median raphe, choroid plexus	Activation of phospholipase C	

5-HT_3	5-HT_3	*5-HT3A* *5-HT3B* *5-HT3C* *5-HT3D* *5-HT3E*	455 441 457 454 456	Hippocampus, amygdala, nucleus accumbens, striatum, substantia nigra, olfactory tubercle	Ligand-gated cation channel	Addiction, anxiety, learning, memory
5-HT_4	5-HT_4	*5-HT4*	388	Hippocampus, striatum, substantia nigra, olfactory tubercle	Activation of adenylyl cyclase	Anxiety, learning, memory, mood
5-HT_5	5-HT_{5A}	*5-HT5A*	357		Inhibition of adenylyl cyclase	Sleep
	5-HT_{5B}	*5-HT5B*	370			
5-HT_6	5-HT_6	*5-HT6*	440	Hippocampus, hypothalamus, striatum, nucleus accumbens, cortex, amygdala, olfactory tubercle	Activation of adenylyl cyclase	Anxiety, cognition, learning, memory, mood
5-HT_7	5-HT_7	*5-HT7*	479	Hippocampus, hypothalamus, amygdala	Activation of adenylyl cyclase	Anxiety, memory, mood, sleep

a) All the receptors are found in the brain except 5-HT_{5B}.
b) Corresponding to human receptors except 5-HT_{5B}, which corresponds to mouse (*Mus musculus*).

A wide-ranging treatise published in 2006 collects all the information about serotonin receptors thus far. The treatise includes sequences, molecular and structural biology as well as anatomical, cellular, and subcellular distribution of 5-HT receptors [23]. This information has been complemented with comprehensive reviews dedicated to chemical biology of serotonin receptors [24] and structural features [25].

3.2.2 Location

Serotonin receptors are widely distributed throughout the CNS and periphery. The distinct human 5-HT receptors have been cloned and sequenced facilitating identification of naturally occurring mutations and their relation to different neurodisorders. In the brain, 5-HT receptors are also widely distributed. The human 5-HT_{1A} receptor [26, 27] is located in cortical and limbic structures (hippocampus, entorhinal cortex, septum, amygdala, and frontal cortex). This distribution indicates its role in cognitive or integrative functions as well in emotional states. The 5-HT_{1B} receptor is found in the basal ganglia and substantia nigra [28]. The 5-HT_{1D} has been detected in brain neocortex and caudate nucleus [29]. The 5-HT_{1E} has also been detected in several areas of the brain [30]. The 5-HT_{2A} receptor has been detected in the brain cortex at protein level. The 5-HT_{2B} receptor has been found in cerebellum, occipital cortex, and frontal cortex [31]. The 5-HT_{2C} receptor is widely distributed in the brain, in particular, in the basal ganglia [32]. It is also found in the ventral tegmental area where contributes to a reduction of dopamine release into nucleus accumbens [33]. The 5-HT_{3A} subunit is expressed in the cerebral cortex, amygdala, and hippocampus [34]. The 5-HT_{3B} subunit is expressed in the brain cortex, caudate nucleus, hippocampus, thalamus, and amygdala [34, 35]. The 5-HT_4 receptor is found in the hippocampus, striatum, olfactory bulb, and substantia nigra. The expression of the 5-HT_{5A} receptor in raphe nuclei and in higher brain areas, such as the cerebral cortex and hippocampus, suggests an autoreceptor function. On the other hand, localization in the suprachiasmatic nucleus (SCN) suggests a role in circadian rhythm control. The 5-HT_6 receptor is expressed in several human brain regions including striatum olfactory tubercle, cortex, amygdala, hypothalamus, and hippocampus [36], most abundantly in the caudate nucleus [37]. The 5-HT_7 receptor is preferentially expressed in the caudate nucleus and hippocampus [38] although it is also found in the cerebral cortex, septum, thalamus, hypothalamus, and amygdala. Table 3.1 collects distribution of serotonin receptors in the brain and lists distribution of 5-HT receptors in the brain.

3.2.3 General Structural Considerations

Of the 14 known 5-HT receptors, 13 are G-protein-coupled receptors (GPCRs) exhibiting the common motif of a single-peptide chain looped back and forth through the cell membrane seven times (seven transmembrane domains identified from TM1 to TM7) with the extracellular N-terminus and the intracellular C-terminus. On the other hand, 5-HT_3 receptor is a ligand-gated cation channel comprising four transmembrane helical domains (Figure 3.2). The structure and

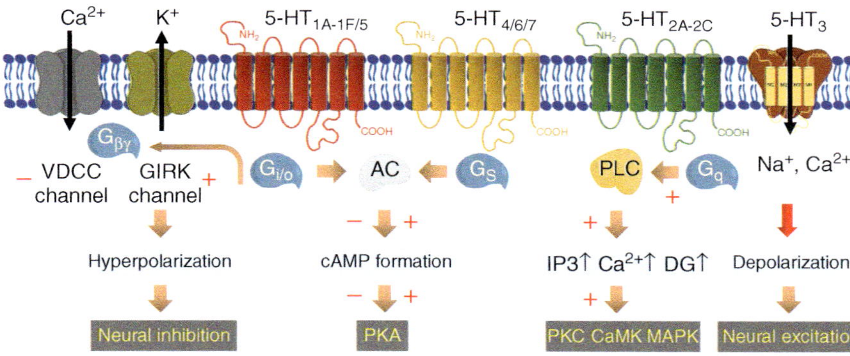

Figure 3.2 Signal transduction of 5-HT receptor subtypes. 5-$HT_{1,2}$ and 5-HT_{4-7} receptors are G-protein-coupled receptors with a seven transmembrane-spanning structure, mediating intracellular signal transduction via coupling to individual G proteins (e.g. $G_{i/o}$, G_s, and G_q). 5-HT_3 receptors form sodium- and calcium-permeable cation channels. AC: adenylate cyclase; CaMK: calcium/calmodulin-dependent protein kinase; DG: diacylglycerol; VDCC: voltage-dependent calcium channel; MAPK: mitogen-activated protein kinase; GIRK: G-protein-gated inwardly rectifying potassium; IP3: inositol triphosphate; PLC: phospholipase C; PKA: protein kinase A; PKC: protein kinase C.

function of serotonin receptors has been the objective of numerous studies [39] including invertebrate receptors [40], but only when some crystal structures have been resolved, full details at a molecular level have been determined [25]. To overcome the lack of structural information based on X-ray crystallography, computational methods have been used for determining pharmacophore models of some receptors including 5-HT_{1A}, 5-HT_{2A}, and 5-HT_7 [41]. Homology models based on rhodopsin structure or known X-ray structures of other receptors have also been used for predicting ligand interactions [42]. The structural features of each receptor, when available, are discussed in the Sections 3.2.6–3.2.12.

3.2.4 Effector Mechanism

Modulation of different biochemical signaling pathways in which 5-HT receptors are involved, leading to a variety of physiological effects, takes place by coupling with all three canonical G-protein signaling pathways through $G\alpha_i$, $G\alpha_{q/11}$, and $G\alpha_s$. The $G\alpha_i$-coupled serotonin receptors include 5-HT_1 and 5-HT_5 types. 5-HT_2 type couples to $G\alpha_{q/11}$, and $G\alpha_s$-coupled receptors comprise 5-HT_4, 5-HT_6, and 5-HT_7 receptors, also known as higher-end serotonin receptors [43]. Ligand binding to 5-HT receptors causes a conformational change that triggers signaling via guanine nucleotide-binding proteins (G proteins) and modulates the activity of downstream effectors, such as adenylate cyclase [44]. In fact, signaling from 5-HT_1 and 5-HT_2 receptors inhibits adenylate cyclase activity [45] and activates a phosphatidylinositol-calcium second messenger system that regulates the release of calcium ions from intracellular stores [46–48]. 5-HT_{1A} receptors are coupled to the opening of potassium channels. Examination of two related responses (upstream activation of $G\alpha_{q/11}$ proteins and the associated cascade of calcium mobilization) to diverse agonists at human 5-HT$_2$ receptors suggested a similar mechanism of functional selectivity for the three subtypes of receptors [47]. The mechanism is that corresponding to G$_q$-related pathways, consisting of activation phospholipase C producing inositol triphosphate (IP3) and diacylglycerol (DAG), inducing an increase in intracellular calcium. Stimulation of calcium release by 5-HT$_{2B}$ receptors takes place through a different pathway that is independent of phosphatidylinositol hydrolysis [49].

The 5-HT$_4$ receptor has been shown to couple positively to adenylyl cyclase, leading to increased levels of cAMP [50]. Isoforms of 5-HT$_4$ receptor interact in a different way with diverse kinases regulating G-protein signaling via phosphorylation of β-arrestin 1 [51]. Recent studies showed 5-HT$_{5B}$ receptor localizes to endosomes and interacts with Gα_i proteins [52]. 5-HT$_6$ and 5-HT$_7$ receptors are coupled to stimulation of adenylyl cyclase. Table 3.1 collects the effector mechanisms of 5-HT receptors, when known. Figure 3.2 illustrates the signal transduction of the different 5-HT receptor subtypes. Regarding effector mechanisms, it exists a clear parallelism between serotonergic, adrenergic, and dopaminergic receptors, in particular, between 5-HT$_{1A\text{-}1F/5}$, D$_2$-type and α_2; 5-HT$_{4/6/7}$, D$_1$-type and β, and 5-HT$_{2A\text{-}2C}$ and α_1 (see Sections 1.2.4 and 2.2.4).

3.2.5 Implication in Neurodisorders

Serotonin receptors are involved in multiple aspects of cognitive functions and are important pharmacological targets for neuropsychiatric disorders such as Alzheimer's disease and schizophrenia, among others [53]. Decreased serotonergic neurotransmission is associated with several mood disorders including bipolar illness, panic disorder, major depression, and suicidal thoughts. The role of 5-HT receptors in depression has been recently reviewed [54].

In particular, the 5-HT$_{1A}$ receptor has been implicated in affective disorders such as anxiety and depression [7]. Recent studies suggest that the 5-HT$_{1A}$ receptor G(-1019) allele is a risk allele that could be used as a marker for depression and related mood disorders [55]. Recent experimental evidences suggest that 5-HT$_{1A}$ and 5-HT$_{4A}$ receptors play critical roles in depression and anxiety [56]. Agonists of 5-HT$_{1A}$ receptors are being investigated for the treatment of schizophrenia and other behavioral disorders such as impulsivity and aggression. In fact, common variants in 5-HT$_{1A}$ and 5-HT$_{5A}$ receptors have been associated with such types of neurodisorders [57]. The 5-HT$_2$ receptors are molecular targets for drugs used in treating a variety of neurodisorders including schizophrenia, depression, anxiety, eating disorders, and obsessive-compulsive disorder (OCD) [58]. The 5-HT$_{2A}$ receptor is directly related to psychosis [59] and with neural activity, perception, cognition, and mood, playing a role in the regulation of behavior, including responses to anxiogenic situations and psychoactive substances [59]. The 5-HT$_{2B}$ receptor has been suggested to play a role in impulsivity [60]. The 5-HT$_{2C}$ receptor plays a role in the regulation of appetite and eating behavior, responses to anxiogenic stimuli and stress [61]. 5-HT$_{5A}$ receptor-selective ligands may be used as therapeutic agents in psychiatric disorders such as schizophrenia and unipolar depression [62]. The 5-HT$_6$ receptor is considered a viable drug target for treating cognitive deficits in Alzheimer's disease [63].

The 5-HT_7 receptor is considered as a potential therapeutic target for depression [64]. Table 3.1 lists main neurodisorders related to 5-HT receptors.

3.2.6 5-HT_1 Receptors

The 5-HT_1 subfamily comprises 5-HT_{1A}, 5-HT_{1B}, 5-HT_{1D}, 5-HT_{1E}, and 5-HT_{1F} subtypes [65] that share 40–63% overall sequence homology (Figure 3.3). There is no 5-HT_{1C} because it was renamed as 5-HT_{2C} based on its coupling to G_q. 5-HT_{1B}, 5-HT_{1D}, and 5-HT_{1F} are also receptors for various anxiolytic and antidepressant drugs and other psychoactive substances, such as lysergic acid diethylamide (LSD) [66].

The 5-HT_{1A} receptor has 422 residues that can be *N*-glycosylated with GlcNAc at Asn10, Asn11, and Asn24. Natural variants include replacement of Pro16, Pro184, and Arg220 by Leu and Gly273 by Asp [67], and Gly22 by Ser and Ileu28 by Val [68]. Sequence similarity with GPCRs suggests regions composed of residues 112–121 and 358–362 to be related to agonist binding. Motifs defined by residues 133–135 and 396 (DRY-motif) to 400 (NPxxY-motif) are important for ligand-induced conformational changes and signaling. A homology model based on the high-resolution crystal structure of the β_2-adrenergic receptor, further refined using molecular dynamics simulations in a 1-palmitoyl-2-oleoyl-*sn*-glycero-3-phosphocholine (POPC) lipid bilayer system, has been developed for the 5-HT_{1A} receptor [69]. Analysis of the binding pocket of the model revealed that the carboxylate group of Asp82 forms hydrogen bonds with the amino group of Asn396 (TM7), which is known to be conserved among bioamine GPCRs.

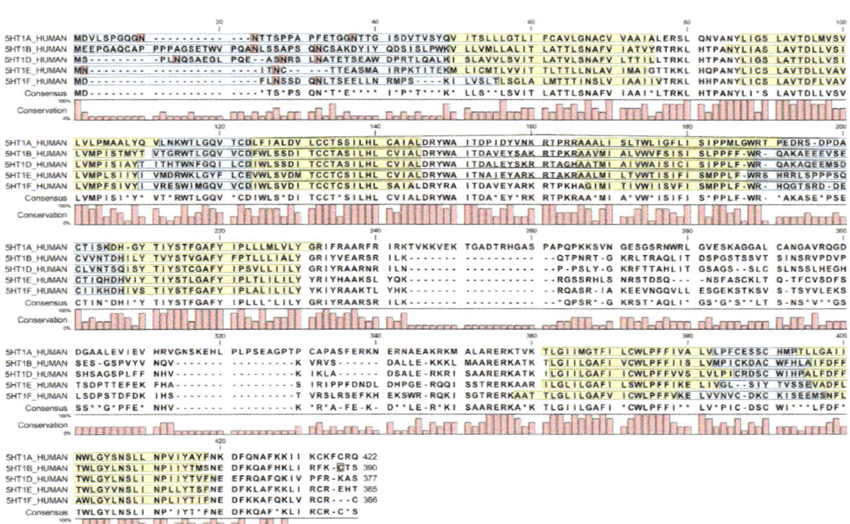

Figure 3.3 Sequences of the human 5-HT_1 receptor. Conservation graphic and consensus sequence are shown. Disulfide bonds (green), N-glycosylation sites (red), palmitoylation sites (brown), extracellular domains (light blue), intracellular domains (white), and transmembrane domains (yellow) are indicated on sequences.

Investigation of binding mode of serotonin shows H-bond interactions of the phenolic hydroxyl group with Ser199 and Thr200. The amino group of the neurotransmitter forms H-bonds with the carboxylate group of Asp116.

The human 5-HT$_{1B}$ receptor [70, 71] has 390 residues that can be *N*-glycosylated with GlcNAc at Asn24 and Asn32. Palmitoylation takes place at Cys388, and several sites have been suggested for phosphorylation [72]. Sequence similarity with GPCRs suggests regions comprising residues 125–134 and 327–331 to be related to agonist binding. Motifs defined by residues 146–148 (DRY-motif) and 365–369 (NPxxY-motif) are important for ligand-induced conformational changes and signaling.

The crystal structures of 5-HT$_{1B}$ receptor in complex with agonists dihydroergotamine (PDB ID: 4IAQ) and ergotamine (PDB ID: 4IAR) have been reported [73]. The structure in complex with ergotamine showed a canonical seven-transmembrane (7TM) α-helical bundle with the extracellular loop partially covering the ligand-binding pocket (Figure 3.4).

The N-terminus folds on top of the binding pocket, suggesting its role in ligand recognition. An orthosteric pocket embedded deep in the 7TM core and an extended binding pocket close to the extracellular entrance form a large binding cavity defined by residues from helices III, V, VI, VII, and el$_2$. Ergotamine adopts a binding mode with the ergoline ring system occupying the orthosteric binding pocket and the cyclic tripeptide moiety bound to the upper extended binding pocket. The ergoline ring is fixed through the salt-bridge interaction between its positively charged nitrogen and the carboxylate of Asp129 and a hydrogen bond between the indole NH group and Thr134. The ergoline ring moiety is also stabilized by a hydrophobic pocket formed by the side chains of Cys133, Ile130, Trp327, Phe330, and Phe331. Furthermore, a hydrogen bond is found between the side chain of Asp129 and the hydroxyl group of Tyr359. The 5-HT$_{1B}$/dihydroergotamine complex structure was very similar to complex with ergotamine showing essentially the same interactions. The resolution obtained for crystal structures in the case of 5-HT$_{1B}$ receptor resulted critical for predicting selective ligands, as it was demonstrated through structure-based docking screens of millions of compounds [74].

The human 5-HT$_{1D}$ receptor [75, 76] has 377 residues that can be *N*-glycosylated with GlcNAc at Asn5, Asn17, and Asn21. A natural variant consisting of replacement of Ser256 by Leu has been reported [77]. Sequence similarity with GPCRs suggests regions comprising residues 114–123 and 314–318 to be related to agonist binding. Motifs defined by residues 135–137 (DRY-motif) and 352–356 (NPxxY-motif) are important for ligand-induced conformational changes and signaling. Actually, 5-HT$_{1A}$ and 5-HT$_{5A}$ receptors can be considered as species variants of the same receptor due to their similarity in both pharmacological properties and distribution in the brain. However, the situation became more complicated after the discovery of two genes encoding the 5-HT$_{1D}$ receptor [78]. The resultant 5-HT$_{1D\alpha}$ and 5-HT$_{1D\beta}$ receptors showed different properties, but the rat 5-HT$_{1B}$ was the species homologue of the human 5-HT$_{1D}$ receptor [79]. Finally, 5-HT$_{1D\alpha}$ and 5-HT$_{1D\beta}$ receptors were associated with 5-HT$_{1D}$ and 5-HT$_{1B}$ receptors, respectively.

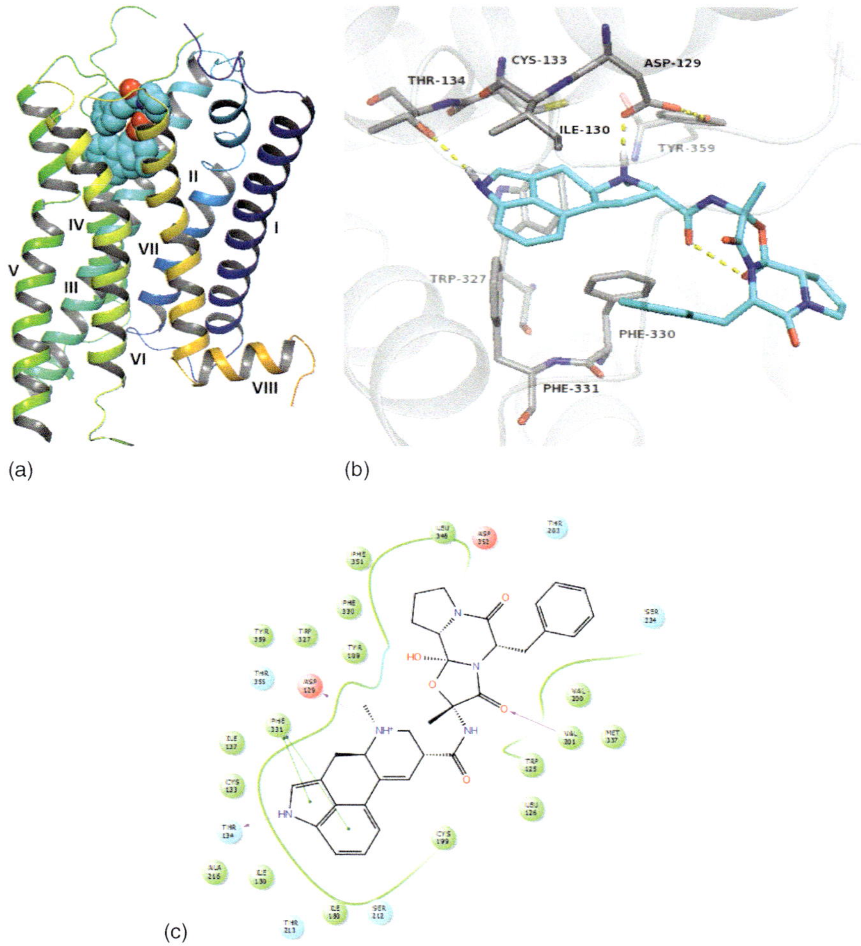

Figure 3.4 (a) X-ray structure of the 5-HT_{1B} receptor in complex with ergotamine (PDB ID: 4IAR). The receptor is shown as rainbow-colored ribbon cartoon; ligand is colored cyan and represented as CPK. (b) Details of ergotamine at the binding pocket. The ligand is colored cyan. H-bond interactions are colored in yellow. (c) Interactions between ergotamine and residues at binding site.

The human 5-HT_{1E} receptor [80, 81] has 365 residues that can be *N*-glycosylated with GlcNAc at Asn2 and Asn5. Sequence similarity with GPCRs suggests regions composed of residues 98–107 and 304–308 to be related to agonist binding. Motifs defined by residues 119–121 (DRY-motif) and 340–344 (NPxxY-motif) are important for ligand-induced conformational changes and signaling. Quantitative structure-activity and comparative molecular field analyses suggested that hydrogen bonding is crucial for binding interactions with the 5-HT_{1E} receptor [82]. 5-HT_{1D} and 5-HT_{1E} receptors have a 64% of homology and share c. 60% of residues in the transmembrane domains responsible for the different pharmacological properties observed for these two receptors. A detailed analysis showed that two residues (Ile333 and Ser334 in the 5-HT$_{1D}$ receptor, corresponding to Lys310 and Glu311 in the 5-HT$_{1E}$ receptor), located in the sixth transmembrane domain, are mostly responsible for the differential affinities of some, but not all, ligands for the 5-HT$_{1D}$ and 5-HT$_{1E}$ receptors [36].

The human 5-HT$_{1F}$ receptor [83] has 366 residues that can be *N*-glycosylated with GlcNAc at Asn5 and Asn10. Sequence similarity with GPCRs suggests regions comprising residues 99–108 and 306–310 are related to agonist binding. Motifs defined by residues 120–122 (DRY-motif) and 343–347 (NPxxY-motif) are important for ligand-induced conformational changes and signaling. The 5-HT$_{1F}$ receptor has a sequence homology of 61% with the 5-HT$_{1E}$ receptor.

3.2.7 5-HT$_2$ Receptors

The 5-HT$_2$ subfamily comprises 5-HT$_{2A}$, 5-HT$_{2B}$, and 5-HT$_{2C}$ subtypes presenting related features in their molecular structure, amino acid sequence, and signaling properties [84]. While 5-HT$_{2A}$ and 5-HT$_{2C}$ are widely distributed in the CNS, 5-HT$_{2B}$ receptors have a restricted expression. 5-HT$_{2A}$ and 5-HT$_{2C}$ receptors share very similar pharmacological profiles and cellular signaling pathways. Certain antipsychotic and antidepressant drugs are antagonists of 5-HT$_{2A}$ and 5-HT$_{2C}$ receptors. Also, they interact with specific sets of PDZ proteins [85]. Figure 3.5 collects the sequences of 5-HT$_2$ receptors.

The human 5-HT$_{2A}$ receptor [26, 27] has 471 residues that can be *N*-glycosylated with GlcNAc at Asn8, Asn38, Asn44, Asn51, and Asn54. Phosphorylation takes place at Ser280 [86]. Natural variants include replacement of Thr25 by Asn [87], Ile15 or Aa447 by Val [77], and His452 by Tyr [88]. An isoform with an alternative sequence to residues 1–138 has been reported [89]. Sequence similarity with GPCRs suggests regions comprising residues 155–160 and 336–340 are related to

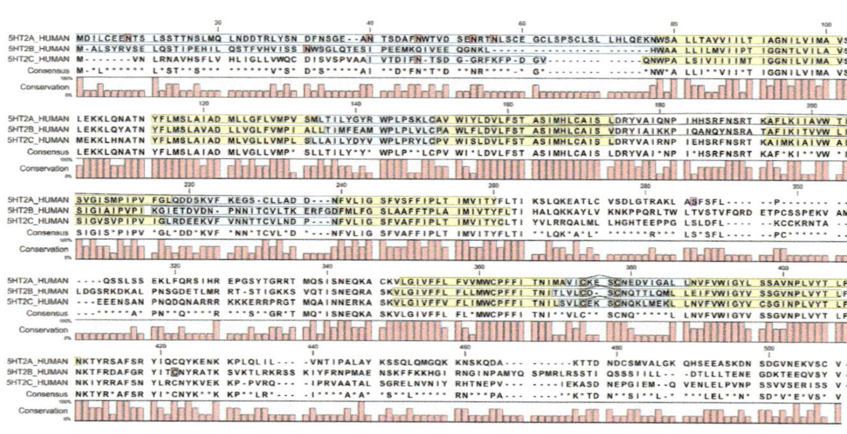

Figure 3.5 Sequences of the human 5-HT$_2$ receptor. Conservation graphic and consensus sequence are shown. Disulfide bonds (green), N-glycosylation sites (red), phosphorylation (magenta) sites, palmitoylation sites (brown), extracellular domains (light blue), intracellular domains (white), and transmembrane domains (yellow) are indicated on sequences.

agonist binding. Motifs defined by residues 172–174 (DRY-motif) and 376–380 (NPxxY-motif) are important for ligand-induced conformational changes and signaling. A PDZ-binding motif is composed of residues 469–471 [85, 90]. The 5-HT_{2A} receptor can also interact with the dopamine receptor D_2, forming a heteromer, which can affect signaling with implications for diseases involving both serotonin and dopamine systems such as schizophrenia [11, 91]. Several mutagenesis studies have been performed to investigate residues involved in different activities and responses. Replacement of Lys229 by Ala strongly increases dissociation of bound LSD, but binding affinity is not affected. Also, even though the signaling via arrestins is reduced, no effect on signaling via the phosphatidylinositol-calcium second messenger system is detected [92]. Interestingly, Cys470Ser and Asn465Ser mutations make the 5-HT_{2A} receptor acquire binding properties of the 5-HT_{2C} receptor [85]. The 5-HT_{2A} receptor is widely expressed in the cortex [93], and it is involved in schizophrenia, depression, and Tourette's syndrome [94].

A homology model of the same system previously reported for the 5-HT_{1A} receptor showed H-bond interactions between serotonin and Ser239 (phenolic hydroxyl group), Ser159 (indole nitrogen atom), and Asp155 (primary amino group) [69], in agreement with previous mapping of the binding site pocket [95]. The model is able to dock both agonists and antagonists, in good agreement with site-directed mutagenesis studies [96]. Homology models further refined by molecular dynamics simulations correlate very well with experimental observations [97]. The use of such homology models to determine key interaction sites with the receptor has allowed selecting antipsychotic analogues as therapeutic candidates targeting the 5-HT_{2A} receptor [98]. Very recently, the conformation of antagonist arylpiperazine **AP** in the binding site has been studied by molecular dynamics simulations (Figure 3.6a–c). The observed changes in the ligand–receptor complex 5-HT_{2A}·**AP** are in good agreement with the previous docking analysis, as demonstrated by comparison with the average structure derived from MD simulations (Figure 3.6d) [99]. The antagonist **AP** is anchored in a hydrophobic pocket formed by residues Val127, Trp151, Leu362, Val366, and Tyr370. An essential edge-to-face π–π stacking between Trp151 and ring C of **AP** is observed. H-bond interactions are predicted between the NH of the amide of compound **AP** and Asp155 (–NH···O–, 2.94 Å) and between the O-atom and Ser159 (–O···HO–, 3.37 Å). These results confirm the reliability of docking models for helping in the search of potent 5-HT_{2A} antagonists confirming the residues directly involved in binding and evidencing key interactions required for binding to the 5-HT_{2A} receptor (Figure 3.6e).

The human 5-HT_{2B} receptor [100] has 481 residues that can be *N*-glycosylated with GlcNAc at Asn30. Palmitoylation can take place at Cys397. Natural variants include replacement of Gln45 by Glu, Phe173 by Leu, Arg388 by Trp, and Met421 by Val [60]. Regions comprised of residues 135–140 and 337–341 have been identified with agonist binding. Motifs defined by residues 152–154 (DRY-motif) and 376–380 (NPxxY-motif) are important for ligand-induced conformational changes and signaling [92, 101]. A [DE]RFG motif defined by residues 212–215 may stabilize a conformation that preferentially activates signaling via β-arrestin family members. A PDZ binding motif is formed by residues 479–481 [90].

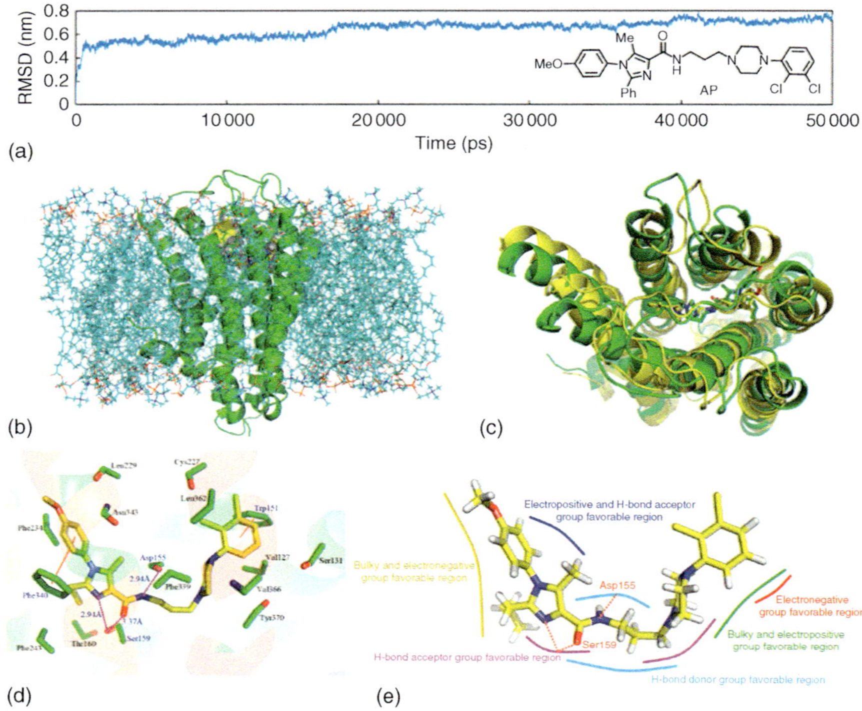

Figure 3.6 (a) The time evolutions of RMSD of the backbone atoms from the initial docked structure during the 50 000 ps MD simulations. Structure of compound **AP** is superimposed. (b) The ligand–receptor complex 5-HT_{2A}·**AP** embedded into the lipid bilayer after 50 000 ps of MD simulation. Ligand, receptor, and lipid molecules are shown as yellow spheres, green cartoon, and blue lines, respectively. (c) The projection of the superimposed backbone atoms of the average structure of the last 10 000 ps of the MD simulation (yellow) and the initial structure (green). Compound **AP** is respectively represented as carbon chain in green for the initial complex and yellow for the average structure. (d) Binding mode extracted from MD simulations in terms of π–π stacking, H-bonds, and hydrophobic interactions. Compound **AP** and the interaction groups of the key amino acids are depicted in sticks and highlighted with yellow and green carbons, respectively. H-bonds and π–π stacking interactions are represented as purple and orange dashed lines, respectively. (e) Key interaction features of template compound **AP** with 5-HT_{2A} receptor. Source: Reproduced with permission from Ref. [99]. https://creativecommons.org/licenses/by/4.0/

Biased signaling observed between 5-HT_{1B} and 5-HT_{2B} receptors with a series of ergolines including LSD and ergotamine has been investigated. Mutagenesis of Leu209 by Ala increases dissociation of LSD, but it does not affect binding affinity. Also, it reduces signaling via arrestins but without affecting signaling via the phosphatidylinositol-calcium second messenger system.

Several X-ray structures of 5-HT_{2B} receptor have been reported. The X-ray structure of the 5-HT2B receptor in complex with LSD (PDB ID: 5TVN) revealed conformational arrangements of the receptor for accommodating the ligand, thus explaining the conformational selectivity of LSD's diethylamine moiety (Figure 3.7) [92].

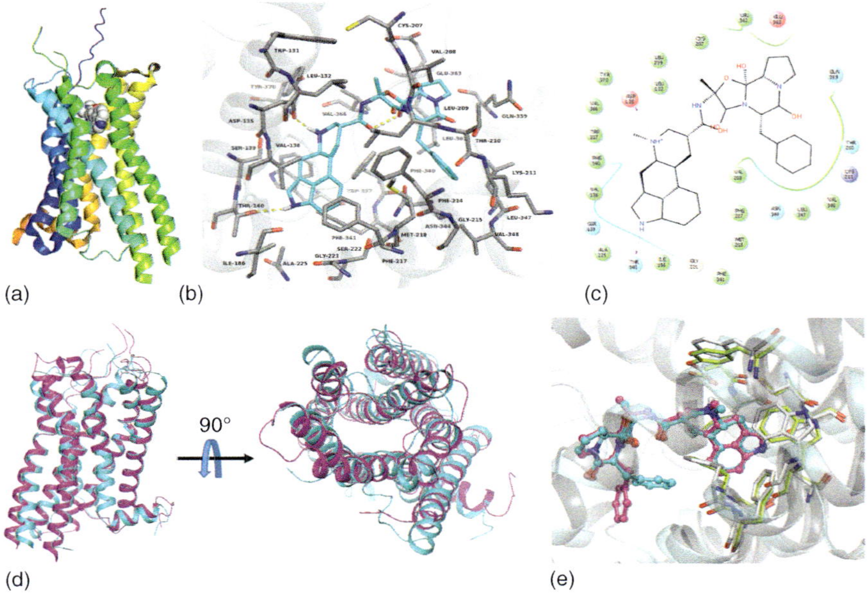

Figure 3.7 (a) X-ray structure of the 5-HT_{2B} receptor in complex with LSD (PDB ID: 5TVN). The receptor is shown as rainbow-colored ribbon cartoon; ligand is colored gray and represented as CPK. (b) Details of ergotamine at the binding pocket of the 5-HT_{2B} receptor (PDB ID: 4IB4). Residues at 5 Å are shown. The ligand is colored cyan. H-bond interactions are colored in yellow. (c) Interactions between ergotamine and residues at the binding site of the 5-HT_{2B} receptor (PDB ID: 4IB4). (d) Superposition of the structures of the 5-HT_{1B} receptor (PDB ID: 4IAR, cyan) and 5-HT_{2B} receptor (PDB ID: 4IB4, magenta). (e) Superposition of the ligand-binding pockets of the 5-HT_{1B} receptor (light blue) and 5-HT_{2B} receptor (gray). The carbons of ligand ergotamine in the structures of 5-HT_{1B} and 5-HT_{2B} receptors are shown as cyan and magenta, respectively. Residues forming the binding sites are shown as sticks and labeled in light green for the 5-HT_{1B} receptor and gray for the 5-HT_{2B} receptor.

The crystal structure of the human 5-HT_{2B} bound to ergotamine (PDB ID: 4IB4) with the 5-HT_{1B}/ergotamine structure was compared to study the structural basis of the different selectivities observed for both receptors [101]. Compared with the structure of the 5-HT_{2B} receptor, the 5-HT_{1B} receptor displays a 3 Å outward shift at the extracellular end of TM5, resulting in a more open extended pocket that explains subtype selectivity. Despite this difference, the cyclic tripeptide moiety of ergotamine maintains a rather similar orientation stabilized by an intramolecular hydrogen bond, except for the phenyl group. In the 5-HT_{2B} receptor it interacts with Leu347 and Val348, while in the 5-HT_{1B} receptor it occupies a cavity close to Thr209 at TM5. The crystal structure of the 5-HT_{2B} receptor in lipidic cubic phase (PDB ID: 4NC3) has been solved using serial femtosecond crystallography [102].

The human 5-HT_{2C} receptor [103] has 458 residues that can be *N*-glycosylated with GlcNAc at Asn39. Natural variants include replacement of Cys23 or Asn158 by Ser and Ile156 or Ile160 by Val [104]. Regions composed of residues 134–139 and 324–328 have been identified with agonist binding. Motifs defined by residues 151–153 (DRY-motif) and 364–368 (NPxxY-motif) are important for ligand-induced conformational changes and signaling. A PDZ binding motif is formed by residues 456–458 [85]. This domain-binding motif is involved in the interaction with MPDZ. In fact, Ser456Ala and Val458Ala mutations cause loss of interaction with MPDZ; on the other hand, Ser456Thr or Ser457Ala has no effect on the interaction [90]. An integrated approach combining experimental structure-activity relationship (SAR) studies of a series of 5-HT_{2C} agonists with hierarchical GPCR modeling protocol (consisting of homology models followed by optimization protocols and flexible docking) revealed that differences in interhelical interactions may be the major source of selectivity in G-protein activation by 5-HT_{2C} receptors [105]. Recent studies based on mutagenesis showed that some 5-HT_{2C} receptor residues important for pharmacology of one ligand are not necessarily important for another ligand [106].

3.2.8 5-HT_3 Receptors

5-HT_3 receptors are pentameric cation-selective transmitter-gated ion channels and, when activated, cause fast and depolarizing responses [107]. Five monomer subtypes, 5-$HT_{3A\text{-}E}$ subunits, have been identified, which show differences in the amino terminal and the transmembrane region. The sequences of these subunits are given in (Figure 3.8). The functional relevance of different receptor components is not yet clarified. Notably, the single-channel conductance of human 5-HT_3 receptors assembled as homomers of 5-HT_{3A} subunits (0.4 pS) [108], or heteromers of 5-HT_{3A} and 5-HT_{3B} subunits (16 pS) [109], are very different. A region within a large cytoplasmic loop of the receptor has been identified as

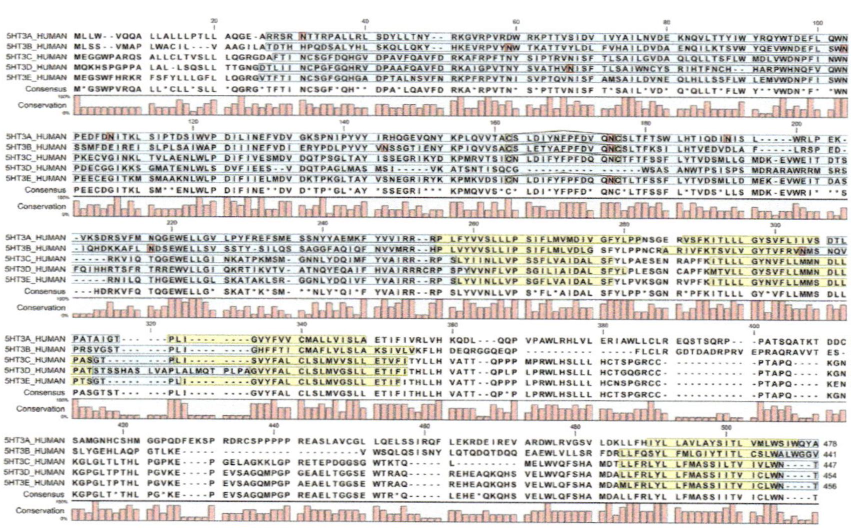

Figure 3.8 Sequences of the human 5-HT_3 receptors. Conservation graphic and consensus sequence are shown. Disulfide bonds (green), N-glycosylation sites (red), palmitoylation sites (brown), extracellular domains (light blue), intracellular domains (white), and transmembrane domains (yellow) are indicated on sequences.

responsible for such a different behavior [110]. 5-HT$_{3C}$, 5-HT$_{3D}$, and 5-HT$_{3E}$ receptors were initially considered as human specific. However, although absent in rodents, they are conserved in mammalians and expression profiles have been found in the brain [111]. These subunits form pentaheteromeric complexes with 5-HT$_{3A}$ and they are not functional as a homomeric complex [112]. On the other hand, some subtle differences exist; thus, novel heteromeric receptor interactions with 5-HT$_{3A}$ demonstrated that the expression or function of the receptor was modified by 5-HT$_{3C}$ and 5-HT$_{3E}$, but not 5-HT$_{3D}$ [113]. Studies correlating subunit structure with ionic selectivity and single channel conductance of 5-HT$_3$ receptors allowed identifying a novel domain, named intracellular MA-stretch, which contributes to ion permeation and selectivity [114].

The human 5-HT$_{3A}$ subunit has 455 residues that can be *N*-glycosylated with GlcNAc at Asn28, Asn104, Asn170, and Asn186. Natural variants include replacement of Ala33 by Thr [115], Ser253 by Asn, Arg344 by His, Pro391 by Arg and Arg409 by Gln [116]. Several isoforms have been described [117, 118]. Mutagenesis experiments including Arg432Gln, Arg436Asp, and Arg440Ala produced a substantial increase in conductance [110]. Formation of a pentahomomeric complex as well as pentaheteromeric complex with other 5-HT$_3$ receptors contributes to modulation of signaling [34, 112].

The human 5-HT$_{3B}$ subunit has 441 residues that can be *N*-glycosylated with GlcNAc at Asn52, Asn96, Asn138, Asn168, Asn203, and Asn287. *N*-Glycosylation is required for membrane localization [119]. Natural variants include replacement of Tyr129 by Ser [120], Ile143 by Thr [116], Ser156 by Arg, and Val183 by Ileu [115]. Mutagenesis experiments consisting of replacement of Asn52, Asn96, Asn138, Asn168, or Asn203 by Ser caused very little expression in the cell membrane [119].

The human 5-HT$_{3C}$ subunit [112] has 447 residues and no glycosylation residues have been identified. Natural variants include replacement of Val128 by Met [121], Asn163 by Lys, and Gly405 by Ala [111].

The human 5-HT$_{3D}$ subunit [112] has 454 residues that can be *N*-glycosylated with GlcNAc at Asn66. Natural variants include replacement of Ala171 by Gly and Arg225 or Arg435 by His. Four different isoforms with alternative sequences and/or missing fragments are found for this receptor, due to alternative splicing.

The human 5-HT$_{3E}$ subunit [112] has 456 residues that can be *N*-glycosylated with GlcNAc at Asn175. Natural variants include replacement of Ala71 or Ala430 by Thr. Five different isoforms with alternative sequences and/or missing fragments are found for this receptor due to alternative splicing [113].

The human 5-HT$_3$ receptor shares several structural and functional features with the nicotinic ACh receptor [122]. Once the structure of the latter was established [123], homology models were created for 5-HT$_3$ receptor [124, 125]. Mutagenesis experiments, consistent with these models, demonstrated that for 5-HT$_{3A}$ receptor subunits residues Arg436 and Arg440, located within the MA helix, act as rate-limiting determinants of single-channel conductance [126]. Moreover, a high-resolution structure of palonosetron, a potent 5-HT$_3$ receptor antagonist, bound to an engineered acetylcholine binding protein that mimics the 5-HT$_3$ receptor binding site, termed 5-HTBP (hydroxytryptamine-binding protein), led to identification of the molecular determinants of compound recognition in the receptor binding site [127]. The study demonstrated the particular importance of Trp183 (cation–π interaction and hydrogen bond) and Tyr153 (hydrogen bond).

The resolution of the crystal structure of the mouse 5-HT$_3$ receptor (PDB ID: 4PIRA) showing 89% of sequence identity for residues 30–474 with the human receptor [128] served for designing a more accurate homology model of the human 5-HT$_3$ receptor. Indeed, some differences are observed with homology model based on *n*AChR structure. In particular, shifts in register between the sequence alignment and the structural superimposition and distinct organizations of the MA helix bundle were found.

The crystallographic study showed a bullet-shaped structure with subunits arranged in fivefold symmetry around a central ion pathway perpendicular to the membrane plane (Figure 3.9). At the binding site, each of the five equivalent serotonin binding sites of the 5-HT$_3$ receptor is located between two subunits in an electronegative cleft. A hydrophobic pocket of c. 10 Å is formed by aromatic residues Trp156 (loop B), Phe199 and Tyr207 (loop C), Trp63 (loop D), and Tyr126 (loop E). The cation–π interaction observed for Trp156, initially postulated in homology models based on nACh receptors, is conserved. Other internal residues pointing toward the site include Asn101 (loop A), Thr152, and Thr154 (loop B), while peripheral ones include Asp42, Ile44 (loop G), Asp177, Ser179, Ile180 (loop F), and residues from loop C.

3.2.9 5-HT$_4$ Receptors

The human 5-HT$_4$ receptor has 388 residues distributed in seven transmembrane domains, four extracellular loops, and four cytoplasmic loops (Figure 3.10). It can be *N*-glycosylated with GlcNAc at Asn7 and palmitoylated at Cys329. Natural variants include replacement of Ser27 by Leu and Cys372 by Tyr [129]. Up to 10 different splice variants of unknown functional relevance can be identified [130].

Activation of 5-HT$_4$ receptors reduces the physiological drive to eat and can induce anorexia [9]. Potent and selective 5-HT$_4$ receptor agonists and antagonists have been discovered in the past and submitted to structure–activity relationships and pharmacological evaluation [131]. A recent review has discussed binding properties, pharmacological profiles, and clinical implications of an exhaustive compilation of antagonists and agonists for 5-HT$_4$ receptors [132]. A structural model for the human 5-HT$_4$ receptor has been developed by using structural mass spectrometry and computational and biochemical methods [133]. The predicted configuration of a docked antagonist (GR125487) revealed that the 5-HT$_4$ receptor shares several ligand binding characteristics typical of class A GPCRs.

3.2.10 5-HT$_5$ Receptors

The 5-HT$_5$ subfamily comprises 5-HT$_{5A}$ and 5-HT$_{5B}$ receptors [134]. These receptors, which occur in areas associated with learning and memory, have been suggested as therapeutic targets in psychiatric disorders [62], but no selective

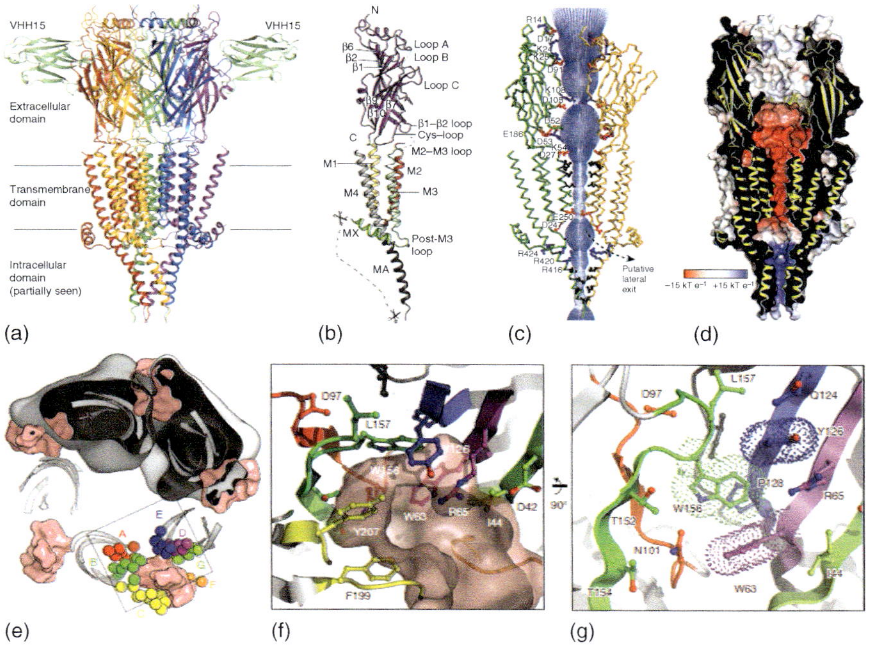

Figure 3.9 (a) Cartoon representation viewed parallel to the plane of the membrane. Only two of five VHH15 molecules are shown (pale green). (b) Intracellular view, perpendicular to the membrane, of the complex. (c) Global view of the ion permeation pathway. Two subunits are shown as ribbons with main residues lining the permeation pathway as sticks (negative in red, positive in blue, hydrophobic in black). The solvent-accessible surface is represented as blue dots. (d) Solvent-accessible electrostatic potential mapped on the receptor surface, color-coded from −15 to 15 kT e^{-1} (red to blue). (e) Extracellular view of the binding sites at the interface between subunits. Binding cleft surfaces are represented in pink, the extracellular domain core atoms are depicted in cartoon representation, with smooth surfaces for two of them. For one of the five binding sites, the binding loops A–G are indicated by colored spheres for Ca atoms. (f) Enlarged extracellular view with residues important for agonist binding represented as sticks, and the binding cleft surface depicted in pink. The image corresponds to the rectangle in panel. (e, g) Enlarged view parallel to the membrane plane. Van der Waals surface is depicted as dots for aromatic box residues and loop C is removed for clarity. Source: Reprinted with permission from Hassaine et al. [128]. Copyright 2014, Macmillan Publishers Ltd.

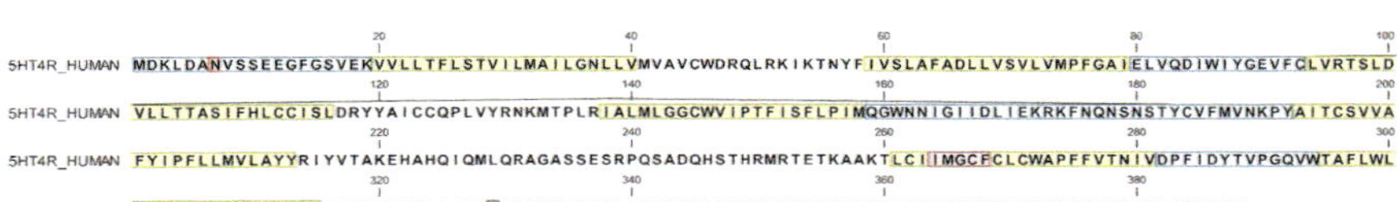

Figure 3.10 Sequence of the human 5-HT$_4$ receptor. Disulfide bonds (green), N-glycosylation sites (red), phosphorylation (magenta) sites, palmitoylation sites (brown), extracellular domains (light blue), intracellular domains (white), and transmembrane domains (yellow) are indicated on sequences.

agonists have been found to date. On the other hand, studies with a selective antagonist suggest that the blockade of 5-HT$_{5A}$ receptor appears to be able to impair memory, while its stimulation might facilitate memory [135]. Sequences for human 5-HT$_{5A}$ and mouse 5-HT$_{5B}$ receptors are given and compared in Figure 3.11.

The human 5-HT$_{5A}$ receptor has 357 residues and comprises seven transmembrane domains. *N*-glycosylation with GlcNAc can take place at Asn6 and Asn21. A natural variant in which Arg262 is replaced by Cys has been reported [121]. This subtype occurs in brain areas implicated in learning and memory [135].

The human 5-HT$_{5B}$ receptor has not been characterized. Instead, 5-HT$_{5B}$ receptors of rat (*Rattus norvegicus*) and mouse (*Mus musculus*) have been evidenced at transcript level, both of them having 370 residues. It is coexpressed in the raphe nuclei with serotonin transporter, and its presence in serotonergic neurons in the dorsal raphe and central superior nucleus suggests an autoreceptor role [136]. Both mouse and rat 5-HT$_{5B}$ receptors have seven transmembrane domains and can be *N*-glycosylated with GlcNAc at Asn5. No natural variants or isoforms have been reported.

3.2.11 5-HT$_6$ Receptors

The human 5-HT$_6$ receptor [137] has 440 residues and comprises the characteristic seven transmembrane domains of GPCRs (Figure 3.12). The interaction between the intracellular loops of the human 5-HT$_6$ and G$_{S/\alpha}$ protein has been studied at a molecular level [138].

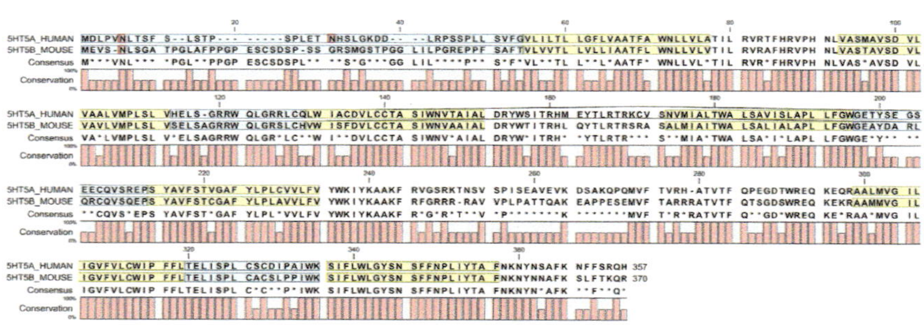

Figure 3.11 Aligned sequences of the human 5-HT$_5$ receptors. Conservation graphic and consensus sequence are shown. Disulfide bonds (green), N-glycosylation sites (red), palmitoylation sites (brown), extracellular domains (light blue), intracellular domains (white), and transmembrane domains (yellow) are indicated on sequences.

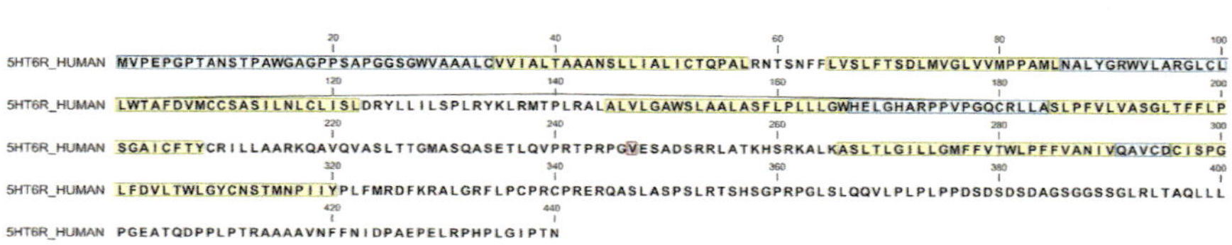

Figure 3.12 Sequence of the human 5-HT$_6$ receptor. Disulfide bonds (green), extracellular domains (light blue), intracellular domains (white), and transmembrane domains (yellow) are indicated on sequences.

N-glycosylation with GlcNAc has not been reported. Polymorphism by replacing Cys267 by Thr has been associated with schizophrenia although further studies concluded that it does not play a major role in susceptibility to the development of schizophrenia and is not related to cognitive impairment or age at onset in schizophrenic patients [139–141]. Association of polymorphisms of 5-HT_6 with other neurodisorders such as Alzheimer's [142, 143] and Parkinson [144] diseases has also been reported. The Cys267 polymorphism has also been associated with major depressive disorder [145]. It has been demonstrated that 5-HT_6 receptors physically interact with several proteins of the mammalian target of rapamycin (mTOR) pathway, including mTOR. Moreover, 5-HT_6 receptor activation increased mTOR signaling in rodent prefrontal cortex, suggesting that recruitment of mTOR by prefrontal 5-HT_6 receptors contributes to the perturbed cognition in schizophrenia [146]. Because of the therapeutic potential of acting on 5-HT6 receptors, several ligands have been discovered [147]. In this context, potential 5-HT_6 antagonists have been selected by combining a rapid 2D similarity search on multimillion compound databases with pharmacophore matching [148]. Hit validation of the study allowed definition of three pharmacophore models (Figure 3.13) that were in good agreement with those previously defined through a different approach [149].

3.2.12 5-HT_7 Receptors

The human 5-HT_7 receptor [150] has 479 residues and comprises the typical seven transmembrane domains of GPCRs (Figure 3.14). *N*-glycosylation with GlcNAc can take place at Asn5 and Asn66 [151]. Palmitoylation takes place at Cys401. Natural variants including replacement of Thr92 by Lys, Pro279 by Leu, and Pro448 by Gln have been associated with schizophrenia and bipolar affective disorder [152]. Studies on the Thr2Lys variant suggested involvement of the 5-HT_7 receptor in the regulation of circadian rhythm and, consequently, with sleep physiology and related disorders [153]. Four isoforms have been identified

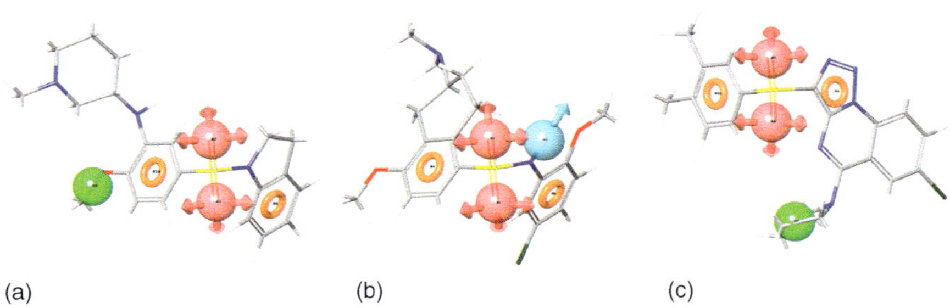

Figure 3.13 Pharmacophore hypotheses for 5-HT6 receptors; a superimposed seed compound is visualized. (a) AAHRR type for cluster#1. (b) AADRR type for cluster#2. (c) AAHRR type for cluster#3. Orange rings represent aromatic rings and green spheres hydrophobic pharmacophore sites. Pink spheres represent hydrogen-bond acceptor sites, while the cyan sphere is a hydrogen-bond donor site. The arrows define directionality. Types are defined according to the following pharmacophore features: H-bond acceptor (hin Source: Dobi et al. 2015 [148]. Reprinted with permission of John Wiley & Sons.

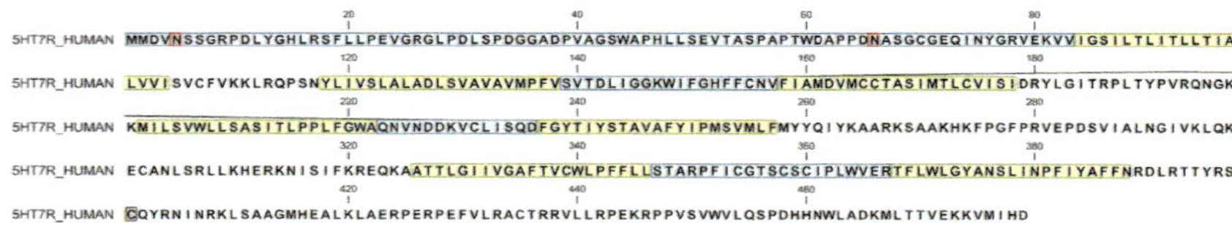

Figure 3.14 Sequences of the human 5-HT$_7$ receptor. Disulfide bonds (green), N-glycosylation sites (red), palmitoylation sites (brown), extracellular domains (light blue), intracellular domains (white), and transmembrane domains (yellow) are indicated on sequences.

showing tissue specificity [38]. No differences have been observed between splice variants in inverse agonist potencies, all of them showing similar constitutive activity [154]. A series of arylpiperazines have been evaluated as potential antagonists using homology models. The results were in line with experimental observations, and the studies revealed preferable binding to the 5-HT$_7$ receptor in respect to 5-HT$_{1A}$ and D$_2$ dopamine receptors [155]. The 5-HT$_7$ receptor has been recently used as a test case to ascertain suitability of a series of homology models of GPCRs [156]. The best 5HT$_7$ receptor–template sequence identities obtained for the models within transmembrane domain and binding site were 44.8 and 48.1% for different targeted proteins used as a template for modeling.

3.3 Serotonin Reuptake Transporter (SERT)

SERT is a transmembrane protein responsible for pumping the serotonin back into the presynaptic cell from the synaptic cleft (reuptake). It is closely related to other amine transporters such as dopamine (see Section 1.3) or norepinephrine (see Section 2.4) transporters [157]. There are several SERTs of molecular weights ranging from 60 to 80 kDA and all of them have 12 transmembrane domains with a large extracellular loop between TM3 and TM4. Both N- and C-termini are located within the cytoplasm. There are six potential sites of phosphorylation by protein kinase A (PKA) and protein kinase C (PKC); regulation of SERTs by protein kinases has been reported. The areas important for selective serotonin affinity are located within TMs 1–3 and TMs 8–12. The human SERT has 630 residues (Figure 3.15).

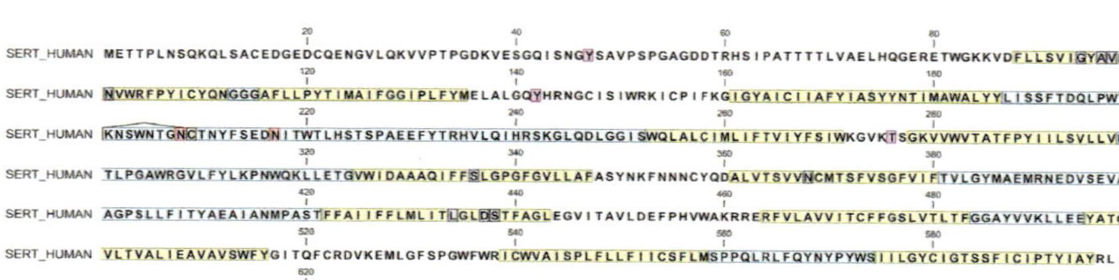

Figure 3.15 Sequence of the human serotonin transporter (SERT). Disulfide bonds (green), N-glycosylation sites (red), phosphorylation (magenta) sites, palmitoylation sites (brown), extracellular domains (light blue), intracellular domains (white), and transmembrane domains (yellow) are indicated on sequences.

N-glycosylation with GlcNAc can take place at Asn208 and Asn217. Phosphorylation can take place at Tyr47, Ty4142, Thr276, Ser611, Thr613, and Thr616. Phosphorylation at Thr276 [158] increases serotonin uptake and is required for cGMP (guanosine monophosphate)-mediated regulation of the transporter [159]. On the other hand, phosphorylation upon PKC stimulation modifies the SERT distribution and density in the membrane and diminishes the uptake capacity as observed for a naturally occurring mutation (Ileu425Val) associated with OCD and other neurodisorders [160]. Natural variants include replacement of Gly56 by Ala, Lys201 by Asn, Ile425 by Leu or Val, Phe 425 by Leu, Leu550 by Val, and Lys605 by Asn. Polymorphisms of SERT and their association with neurodisorders have been extensively studied [161, 162] because of their implication in emotion processing [163]. As an example, a polymorphism consisting of a 44-bp insertion or deletion involving repeat elements six to eight has been associated with suicidal thoughts [164, 165] and depression [166–168], which is the most studied neurodisorder related to serotonin [169]. The polymorphism Val425 seems to be linked to susceptibility to OCD [170, 171]. Other associations of SERT polymorphism include eating disorders [172, 173], psychoses (different from schizophrenia) [174], neuroticism and anxiety [175], bipolar disorder [176], sleep disorders [177], panic disorder [178], epilepsy [179], and Alzheimer's disease [180].

SERTs are dependent on extracellular sodium and chloride ions [181]. The reuptake of serotonin is energetically unfavorable. However, the driving force of the process is the concentration gradient of sodium [182]. The Na^+/K^+ pump (Na^+/K^+ ATPase) maintains the extracellular sodium concentration and the intracellular potassium concentration. The electrical potential produced with Na^+/K^+ exchange (Na^+/K^+ ATPase pumps three sodium ions out for each two potassium ions pumped into the cell) added to the sodium concentration used by SERT also leads to the loss of chloride ions from the cell. The mechanism of the transport initiates when SERT binds a sodium ion (Figure 3.16A). Serotonin then binds the transporter (Figure 3.16B) followed by chloride anion (Figure 3.16C), which is not necessary for binding but essential for net transport. When initial binding is completed (Figure 3.16D), a conformational change in SERT takes place (Figure 3.16E), moving from the neuron outside-facing orientation to an inward position that facilitates release of ions into the cytoplasm (Figure 3.16F,G). Intracellular potassium ion then binds SERT displacing serotonin (Figure 3.16G) inducing reorientation of the transporter (Figure 3.16H); the unoccupied binding site becomes exposed to the outside of the cell and another transport cycle can start again. In summary, SERT catalyzes the influx of serotonin together with Na^+ and Cl^- in a 1 : 1 : 1 stoichiometry. In the same catalytic cycle, there is coupled efflux of one K^+ ion. The catalytic cycle shown in Figure 3.16 corresponds to a full transport cycle, but SERT (like other monoamine transporters) can also operate in an exchange mode where the transporter seesaws through half-cycles. Recent studies showed that the N-terminus of SERTs plays a key role in the mechanism of action, acting as a switch between transport modes [183].

Residues Ser611, Thr613, and Thr616 have been identified to be crucial for serotonin uptake by mutagenesis studies. Replacement Ser611 by Asp produces a loss of 61% of serotonin uptake activity. If associated with mutations of Thr613

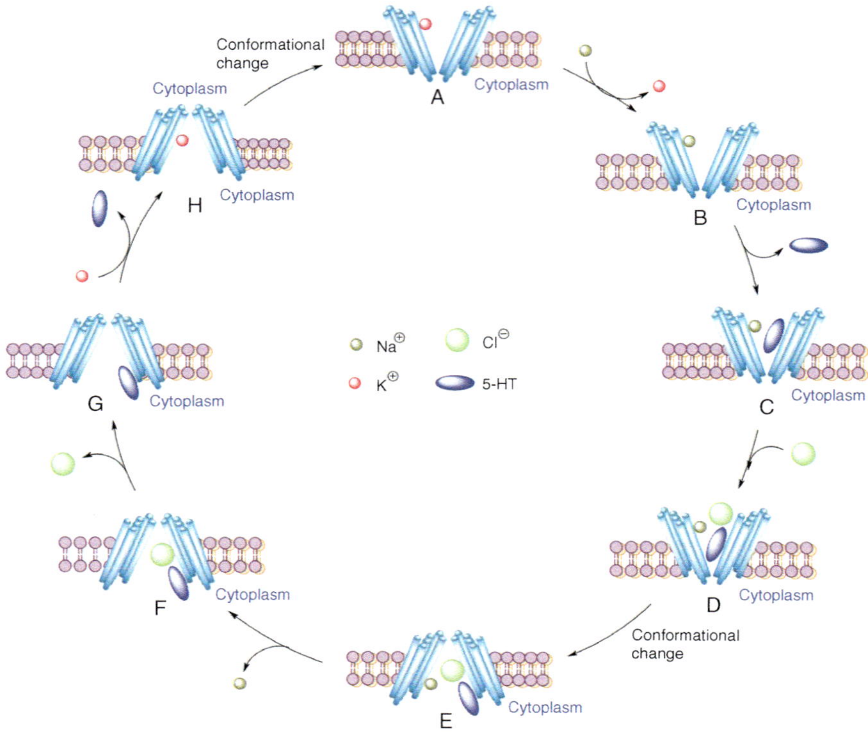

Figure 3.16 Model for reuptake of serotonin into serotonergic neurons by SERT. (A–D) outward-facing conformation; (E-H) inward-facing conformation.

and Th616, the loss of uptake activity reaches 95%. Similarly, when Thr613 and Thr616 are replaced by Asp, a loss of 26% of uptake activity (Thr613) or no change (Thr616) is detected, but associated with mutations of the other two residues, loss of 95% of uptake activity is observed [184]. The existence of an allosteric modulator site whose mechanism remains unclear is also known [185].

A variety of computational approaches including homology models [186], quantum-chemistry, pharmacophore approach, and quantitative structure-activity relationship (QSAR) studies complemented with docking [187], have been employed for the study of SERTs [188]. Homology models further improved by molecular dynamics simulations helped to identify key residues in substrate binding [189, 190]. Models based on the crystal structure of a bacterial transporter revealed a general structural basis of SERTs [191]. In 2016, the X-ray structure of the human serotonin transporter has been resolved alone (PDB ID: 5I6Z) and is bound to the antidepressants (*S*)-citalopram (PDB ID: 5I71) and its brominated derivative (PDB ID: 5I74) and paroxetine (PDB ID: 5I6X). Allosteric sites have been additionally identified for (*S*)-citalopram (PDB ID: 5I73) and its brominated derivative (PDB ID: 5I75) [192].

The structure of human SERT bound to either (*S*)-citalopram or paroxetine exhibits an outward-open conformation with the ligand bound to the central site,

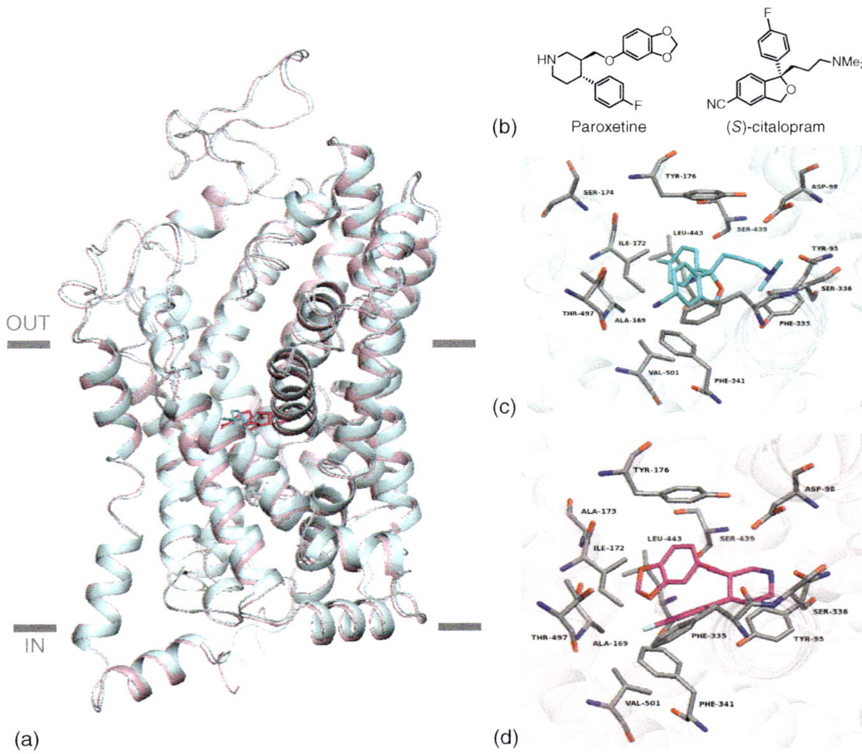

Figure 3.17 X-ray structure of human SERT. (a) Superimposed structures of the human SERT bound to (*S*)-citalopram (cyan) and paroxetine (pink). (b) Chemical structures of (*S*)-citalopram and paroxetine. (c) Interactions of (*S*)-citalopram in the central binding site. (d) Interactions of paroxetine in the central binding site

halfway across the membrane and located in a cavity made up of residues from TM1, TM3, TM6, TM8, and TM10 (Figure 3.17).

At the central binding site, three subsites A, B, and C are defined. In subsite A, the amine groups of the antidepressant drugs interact with Asp98 and with Tyr95, forming a cation–π interaction. Tyr95 also forms a hydrogen bond with the oxygen atom of (*S*)-citalopram. Ser336 participates in several interactions with ligands and ions. In subsite B, Tyr176 is involved in hydrophobic interactions and Asp98 in hydrogen bonding. Ile172 and Phe341 define a nonpolar ridge that cradles the hydrophobic groups of the ligands. Ser439, Leu443, Ala169, and Ala173 define a hydrophobic cavity into which the fluorine and dioxol ring groups of (*S*)-citalopram and paroxetine are inserted. In subsite C, Phe335 defines the extracellular gate, but the interactions of the two crystallized antidepressants with that residue are different. While for (*S*)-citalopram the cyanophtalane moiety forms an edge-to-face aromatic interaction, the fluorophenyl group of paroxetine stacks parallel to the aromatic ring. Val501 and Thr497 form a mixed nonpolar/polar surface into which the fluoro and cyano groups of paroxetine and (*S*)-citalopram are found. All these interactions are in concordance with the observed differences in activity for the different mutants studied and drugs evaluated.

Interactions with the different ions involved in transport could also be identified. One sodium ion is found in a site made up of residues from TM1, TM6, and TM7, and the ion is coordinated by Ala96, Asn101, Ser336, and Asn368. The second sodium cation is coordinated by Gly94, Val97, Leu434, Asp437, and Ser438 from TM1 and TM8. The chloride ion is coordinated by Tyr121, Gln332, Ser336, and Ser372 from TM2, TM6, and TM7.

In the structure of (*S*)-citalopram, a second molecule was found in the allosteric site, within the extracellular vestibule of the (*S*)-citalopram cocrystal structure, approximately 13 Å from the central site. The allosteric binding site of

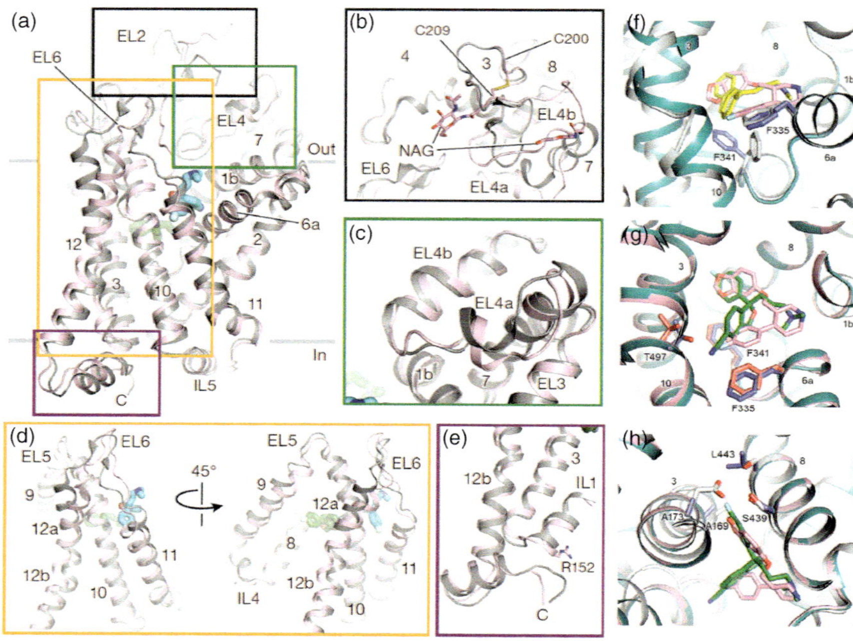

Figure 3.18 Comparison of serotonin and dopamine transporters. (a) Overall alignment of SERT (pink) versus dDAT (gray) using TM1–TM12; regions in SERT with structural differences are boxed. (*S*)-citalopram bound to the central (green) and allosteric (cyan) sites shown as sticks. (b) Close-up view of EL2, *N*-acetylglucosamine (NAG; SERT) and the disulfide bridge between Cys200 and Cys209 are shown as sticks. (c) View of EL4. (d) Structural differences at the SERT allosteric site showing TM9, TM10, TM11, TM12, EL6, and IL4. (e) Conformation of the C-terminal helix and IL1. (f) Comparison of SERT bound to paroxetine with dDAT (PDB ID: 4M48) bound to nortriptyline (yellow); superposition based on TM1–TM12. SERT is shown in blue and DAT in gray. (g) Alignment of paroxetine (blue) and (*S*)-citalopram (pink) structures using all atoms in superposition. Residues interacting with the antidepressant molecules are shown as sticks. Paroxetine (pink) and (*S*)-citalopram (green) are shown as sticks. (h) Insertion of benzodioxole and fluorophenyl groups of paroxetine and (*S*)-citalopram into a cavity in subsite B made up of Leu443, Ala169, Ala173, and Ser439. Note that Ser439 is equivalent to Thr439 in wild-type SERT. Equivalent residues in dDAT are shown in gray. EL: external loop, IL: internal loop. Source: Reprinted with permission from Coleman et al. [192]. Copyright 2016, Macmillan Publishers Ltd.

(*S*)-citalopram is formed by residues located in TM1b, TM6a, TM10, and TM11 and in el4 and el6. Among them, Glu494 and Arg104 are located at an interaction distance of the amino group, the latter presumably participating also in a cation–π interaction. Other residues involved in interactions with the ligand are Ala331, Gln332, Phe556, Pro560, and Pro561.

The crystallographic study was complemented with a comparison of the structure of the dopamine transporter (see Section 1.3) (Figure 3.18). The data suggest that the antidepressant molecules locked the transporter in an outward-open conformation, similar to the inhibitor-bound outward-open conformations of drosophila dopamine transporter (dDAT).

SERTs are sensitive to inhibition. By blocking the action of SERTs, the amount of serotonin in the synaptic cleft increases and, consequently, overactivation of the postsynaptic receptors takes place. Inhibitors directly bind to SERTs although it is unclear whether the inhibitors bind to the same SERT domain as serotonin or operate through more indirect mechanisms. Most known SERT inhibitors are citalopram, fluoxetine (Prozac), fluvoxamine, paroxetine (Paxil), and sertraline (Zoloft). Moreover, certain drugs of abuse such as cocaine, fenfluramine, and (3,4-methylenedioxy) methamphetamine (MDMA or ecstasy) are also inhibitors of SERTs although less selective than the above-mentioned inhibitors. The molecular mechanism of serotonin transporter inhibition has been elucidated by a new flexible docking protocol [193]. The docking results suggest that the inhibitors interact with Tyr176 and Phe335 through hydrophobic interactions, preventing the closure of the extracellular gate. These types of studies based on flexible docking were used for designing several 6-nitroquipazine analogues with high SERT affinity, three of them showing moderate antidepressant activity [194]. Moreover, by adding virtual screening, 46 compounds with K_i values of 1 μM, belonging to 16 structural classes, have been identified [195].

References

1 Törk, I. (1990). Anatomy of the serotonergic system. *Ann. N.Y. Acad. Sci.* 600 (1): 9–34.

2 Graeff, F.G. (1997). Serotonergic systems. *Psychiatr. Clin. North Am.* 20 (4): 723–739.

3 Charnay, Y. and Leger, L. (2010). Brain serotonergic circuitries. *Dialogues Clin. Neurosci.* 12 (4): 471–487.

4 Lei, S. (2012). Serotonergic modulation of neural activities in the entorhinal cortex. *Int. J. Physiol., Pathophysiol. Pharmacol.* 4 (4): 201–210.

5 Wassink, T.H., Hazlett, H.C., Epping, E.A. et al. (2007). Cerebral cortical gray matter overgrowth and functional variation of the serotonin transporter gene in autism. *Arch. Gen. Psychiatry* 64 (6): 709–717.

6 Huang, C.H. and Santangelo, S.L. (2008). Autism and serotonin transporter gene polymorphisms: a systematic review and meta-analysis. *Am. J. Med. Genet. B Neuropsychiatr. Genet.* 147b (6): 903–913.

7 Akimova, E., Lanzenberger, R., and Kasper, S. (2009). The serotonin-1A receptor in anxiety disorders. *Biol. Psychiatry* 66 (7): 627–635.

8 Lesemann, A., Reinel, C., Huehnchen, P. et al. (2013). MPTP-induced hippocampal effects on serotonin, dopamine, neurotrophins, adult neurogenesis and depression-like behavior are partially influenced by fluoxetine in adult mice [Erratum to document cited in CA157:510547]. *Brain Res.* 1510: 89.
9 Jean, A., Conductier, G., Manrique, C. et al. (2007). Anorexia induced by activation of serotonin 5-HT_4 receptors is mediated by increases in CART in the nucleus accumbens. *Proc. Natl. Acad. Sci. U.S.A.* 104 (41): 16335–16340.
10 Jhoo, J.H., Yoon, I.Y., Kim, Y.K. et al. (2010). Availability of brain serotonin transporters in patients with restless legs syndrome. *Neurology* 74 (6): 513–518.
11 Rasmussen, H., Erritzoe, D., Andersen, R. et al. (2010). Decreased frontal serotonin2A receptor binding in antipsychotic-naive patients with first-episode schizophrenia. *Arch. Gen. Psychiatry* 67 (1): 9–16.
12 Mossner, R., Schmitt, A., Syagailo, Y. et al. (2000). The serotonin transporter in Alzheimer's and Parkinson's disease. *Adv. Res. Neurodegener.* 8: 345–350.
13 Huot, P., Fox, S.H., and Brotchie, J.M. (2011). The serotonergic system in Parkinson's disease. *Prog. Neurobiol. (Oxford, UK)* 95 (2): 163–212.
14 Politis, M. and Niccolini, F. (2015). Serotonin in Parkinson's disease. *Behav. Brain Res.* 277: 136–145.
15 Huot, P. and Fox, S.H. (2013). The serotonergic system in motor and non-motor manifestations of Parkinson's disease. *Exp. Brain Res.* 230 (4): 463–476.
16 Tan, S.K.H., Hartung, H., Sharp, T., and Temel, Y. (2011). Serotonin-dependent depression in Parkinson's disease: a role for the subthalamic nucleus? *Neuropharmacology* 61 (3): 387–399.
17 Blum, R. and Lesch, K.-P. (2015). Parkinson's disease, anxious depression and serotonin - zooming in on hippocampal neurogenesis. *J. Neurochem.* 135 (3): 441–444.
18 Cheshire, P.A. and Williams, D.R. (2012). Serotonergic involvement in levodopa-induced dyskinesias in Parkinson's disease. *J. Clin. Neurosci.* 19 (3): 343–348.
19 Rylander, D. and Lindgren, H.S. (2013). The serotonergic component in Parkinson's disease symptoms and therapy side-effects. In: *Serotonergic Systems: Evolution, Functions and Roles in Disease* (ed. R.O. Villanueva), 107–130. New York, NY: Nova Science Publishers.
20 Politis, M. and Loane, C. (2011). Serotonergic dysfunction in Parkinson's disease and its relevance to disability. *ScientificWorldJournal* 11: 1726–1734.
21 Ohno, Y., Shimizu, S., Tokudome, K. et al. (2015). New insight into the therapeutic role of the serotonergic system in Parkinson's disease. *Prog. Neurobiol.* 134: 104–121.
22 Hilfiker, S., Pieribone, V.A., Czernik, A.J. et al. (1999). Synapsins as regulators of neurotransmitter release. *Philos. Trans. R. Soc. B: Biol. Sci.* 354 (1381): 269–279.
23 Roth, B.L. (2006). *The Receptors*, 637. Totowa, NJ: Humana Press.
24 Nichols, D.E. and Nichols, C.D. (2008). Serotonin receptors. *Chem. Rev.* 108 (5): 1614–1641.
25 McCorvy, J.D. and Roth, B.L. (2015). Structure and function of serotonin G protein-coupled receptors. *Pharmacol. Ther.* 150: 129–142.
26 Fargin, A., Raymond, J.R., Lohse, M.J. et al. (1988). The genomic clone G-21 which resembles a beta-adrenergic receptor sequence encodes the 5-HT_{1A} receptor. *Nature* 335 (6188): 358–360.

27 Aune, T.M., McGrath, K.M., Sarr, T. et al. (1993). Expression of $5HT_{1A}$ receptors on activated human T cells. Regulation of cyclic AMP levels and T cell proliferation by 5-hydroxytryptamine. *J. Immunol.* 151 (3): 1175–1183.
28 Middlemiss, D.N. and Hutson, P.H. (1990). The 5-HT_{1B} receptors. *Ann. N.Y. Acad. Sci.* 600: 132–147.
29 Gonzalez-Heydrich, J. and Peroutka, S.J. (1991). Postsynaptic localization of 5-HT_{1D} receptor binding sites in human caudate. *Exp. Neurol.* 113 (1): 28–30.
30 Bai, F., Yin, T., Johnstone, E.M. et al. (2004). Molecular cloning and pharmacological characterization of the guinea pig 5-HT_{1E} receptor. *Eur. J. Pharmacol.* 484 (2–3): 127–139.
31 Kursar, J.D., Nelson, D.L., Wainscott, D.B., and Baez, M. (1994). Molecular cloning, functional expression, and mRNA tissue distribution of the human 5-hydroxytryptamine2B receptor. *Mol. Pharmacol.* 46 (2): 227–234.
32 De Deurwaerdere, P., Lagiere, M., Bosc, M., and Navailles, S. (2013). Multiple controls exerted by 5-HT_{2C} receptors upon basal ganglia function: from physiology to pathophysiology. *Exp. Brain Res.* 230 (4): 477–511.
33 Di Giovanni, G., Di Matteo, V., Di Mascio, M., and Esposito, E. (2000). Preferential modulation of mesolimbic vs. nigrostriatal dopaminergic function by $serotonin_{2C/2B}$ receptor agonists: a combined in vivo electrophysiological and microdialysis study. *Synapse* 35 (1): 53–61.
34 Dubin, A.E., Huvar, R., D'Andrea, M.R. et al. (1999). The pharmacological and functional characteristics of the serotonin 5-HT_{3A} receptor are specifically modified by a 5-HT_{3B} receptor subunit. *J. Biol. Chem.* 274 (43): 30799–30810.
35 Davies, P.A., Pistis, M., Hanna, M.C. et al. (1999). The 5-HT_{3B} subunit is a major determinant of serotonin-receptor function. *Nature* 397 (6717): 359–363.
36 Parker, E.M., Izzarelli, D.G., Lewis-Higgins, L. et al. (1996). Two amino acid differences in the sixth transmembrane domain are partially responsible for the pharmacological differences between the 5-HT_{1D} beta and 5-HT_{1E} 5-hydroxytryptamine receptors. *J. Neurochem.* 67 (5): 2096–2103.
37 Kohen, R., Metcalf, M.A., Khan, N. et al. (1996). Cloning, characterization, and chromosomal localization of a human 5-HT_6 serotonin receptor. *J. Neurochem.* 66 (1): 47–56.
38 Heidmann, D.E., Metcalf, M.A., Kohen, R., and Hamblin, M.W. (1997). Four 5-hydroxytryptamine7 (5-HT_7) receptor isoforms in human and rat produced by alternative splicing: species differences due to altered intron-exon organization. *J. Neurochem.* 68 (4): 1372–1381.
39 Kroeze, W.K., Kristiansen, K., and Roth, B.L. (2002). Molecular biology of serotonin receptors - structure and function at the molecular level. *Curr. Top. Med. Chem. (Hilversum, Netherlands)* 2 (6): 507–528.
40 Tierney, A.J. (2001). Structure and function of invertebrate 5-HT receptors: a review. *Comp. Biochem. Physiol. A: Mol. Integr. Physiol.* 128A (4): 791–804.
41 Bielenica, A., Koziol, A.E., and Struga, M. (2013). Computational methods in determination of pharmacophore models of 5-HT_{1A}, 5-HT_{2A} and 5-HT_7 receptors. *Mini Rev. Med. Chem.* 13 (7): 933–951.
42 Rataj, K., Witek, J., Mordalski, S. et al. (2014). Impact of template choice on homology model efficiency in virtual screening. *J. Chem. Inf. Model.* 54 (6): 1661–1668.

43 Glennon, R.A. (2003). Higher-end serotonin receptors: 5-HT_5, 5-HT_6, and 5-HT_7. *J. Med. Chem.* 46 (14): 2795–2812.

44 Edvinsson, L., Uddman, E., Wackenfors, A. et al. (2005). Triptan-induced contractile (5-HT_{1B} receptor) responses in human cerebral and coronary arteries: relationship to clinical effect. *Clin. Sci. (London)* 109 (3): 335–342.

45 Zgombick, J.M., Schechter, L.E., Macchi, M. et al. (1992). Human gene S31 encodes the pharmacologically defined serotonin 5-hydroxytryptamine1E receptor. *Mol. Pharmacol.* 42 (2): 180–185.

46 Harrington, M.A., Shaw, K., Zhong, P., and Ciaranello, R.D. (1994). Agonist-induced desensitization and loss of high-affinity binding sites of stably expressed human 5-HT_{1A} receptors. *J. Pharmacol. Exp. Ther.* 268 (3): 1098–1106.

47 Cussac, D., Boutet-Robinet, E., Ailhaud, M.C. et al. (2008). Agonist-directed trafficking of signalling at serotonin 5-HT_{2A}, 5-HT_{2B} and 5-HT_{2C}-VSV receptors mediated Gq/11 activation and calcium mobilisation in CHO cells. *Eur. J. Pharmacol.* 594 (1–3): 32–38.

48 Singh, P., Paila, Y.D., and Chattopadhyay, A. (2012). Role of glycosphingolipids in the function of human serotonin(1)A receptors. *J. Neurochem.* 123 (5): 716–724.

49 Ullmer, C., Boddeke, H.G., Schmuck, K., and Lubbert, H. (1996). 5-HT_{2B} receptor-mediated calcium release from ryanodine-sensitive intracellular stores in human pulmonary artery endothelial cells. *Br. J. Pharmacol.* 117 (6): 1081–1088.

50 Bockaert, J., Sebben, M., and Dumuis, A. (1990). Pharmacological characterization of 5-hydroxytryptamine4(5-HT_4) receptors positively coupled to adenylate cyclase in adult guinea pig hippocampal membranes: effect of substituted benzamide derivatives. *Mol. Pharmacol.* 37 (3): 408–411.

51 Barthet, G., Carrat, G., Cassier, E. et al. (2009). Beta-arrestin1 phosphorylation by GRK5 regulates G protein-independent 5-HT_4 receptor signalling. *EMBO J.* 28 (18): 2706–2718.

52 Vogelgesang, S., Niebert, S., Renner, U. et al. (2017). Analysis of the serotonergic system in a mouse model of Rett syndrome reveals unusual upregulation of serotonin receptor 5b. *Front. Mol. Neurosci.* 10: 61.

53 Švob Štrac, D., Pivac, N., and Mück-Šeler, D. (2016). The serotonergic system and cognitive function. *Transl. Neurosci.* 7 (1): 35–49.

54 Yohn, C.N., Gergues, M.M., and Samuels, B.A. (2017). The role of 5-HT receptors in depression. *Mol. Brain* 10 (1): 28.

55 Le Francois, B., Czesak, M., Steubl, D., and Albert, P.R. (2008). Transcriptional regulation at a HTR_{1A} polymorphism associated with mental illness. *Neuropharmacology* 55 (6): 977–985.

56 Samuels, B.A., Mendez-David, I., Faye, C. et al. (2016). Serotonin 1A and serotonin 4 receptors: essential mediators of the neurogenic and behavioral actions of antidepressants. *Neuroscientist* 22 (1): 26–45.

57 Guan, F., Lin, H., Chen, G. et al. (2016). Evaluation of association of common variants in HTR_{1A} and HTR_{5A} with schizophrenia and executive function. *Sci. Rep.* 6: 38048.

58 Roth, B.L. and Shapiro, D.A. (2001). Insights into the structure and function of 5-HT_2 family serotonin receptors reveal novel strategies for therapeutic target development. *Expert Opin. Ther. Targets* 5 (6): 685–695.

59 Gonzalez-Maeso, J., Ang, R.L., Yuen, T. et al. (2008). Identification of a serotonin/glutamate receptor complex implicated in psychosis. *Nature* 452 (7183): 93–97.

60 Bevilacqua, L., Doly, S., Kaprio, J. et al. (2010). A population-specific HTR_{2B} stop codon predisposes to severe impulsivity. *Nature* 468 (7327): 1061–1066.

61 Xie, E., Zhu, L., Zhao, L., and Chang, L.S. (1996). The human serotonin 5-HT_{2C} receptor: complete cDNA, genomic structure, and alternatively spliced variant. *Genomics* 35 (3): 551–561.

62 Thomas, D.R. (2006). 5-HT_{5A} receptors as a therapeutic target. *Pharmacol. Ther.* 111 (3): 707–714.

63 Geldenhuys, W.J. and Van der Schyf, C.J. (2009). The serotonin 5-HT_6 receptor: a viable drug target for treating cognitive deficits in Alzheimer's disease. *Expert Rev. Neurother.* 9 (7): 1073–1085.

64 Abbas, A.I., Hedlund, P.B., Huang, X.P. et al. (2009). Amisulpride is a potent 5-HT_7 antagonist: relevance for antidepressant actions in vivo. *Psychopharmacology (Berlin)* 205 (1): 119–128.

65 Lanfumey, L. and Hamon, M. (2004). 5-HT_1 receptors. *Curr. Drug Targets CNS Neurol. Disord.* 3 (1): 1–10.

66 Veldman, S.A. and Bienkowski, M.J. (1992). Cloning and pharmacological characterization of a novel human 5-hydroxytryptamine1D receptor subtype. *Mol. Pharmacol.* 42 (3): 439–444.

67 Kawanishi, Y., Harada, S., Tachikawa, H. et al. (1998). Novel mutations in the promoter and coding region of the human 5-HT_{1A} receptor gene and association analysis in schizophrenia. *Am. J. Med. Genet.* 81 (5): 434–439.

68 Nakhai, B., Nielsen, D.A., Linnoila, M., and Goldman, D. (1995). Two naturally occurring amino acid substitutions in the human 5-HT_{1A} receptor: glycine 22 to serine 22 and isoleucine 28 to valine 28. *Biochem. Biophys. Res. Commun.* 210 (2): 530–536.

69 Yap, B.K., Buckle, M.J.C., and Doughty, S.W. (2012). Homology modeling of the human 5-HT_{1A}, 5-HT_{2A}, D_1, and D_2 receptors: model refinement with molecular dynamics simulations and docking evaluation. *J. Mol. Model.* 18 (8): 3639–3655.

70 Hamblin, M.W., Metcalf, M.A., McGuffin, R.W., and Karpells, S. (1992). Molecular cloning and functional characterization of a human 5-HT_{1B} serotonin receptor: a homologue of the rat 5-HT_{1B} receptor with 5-HT_{1D}-like pharmacological specificity. *Biochem. Biophys. Res. Commun.* 184 (2): 752–759.

71 Mochizuki, D., Yuyama, Y., Tsujita, R. et al. (1992). Cloning and expression of the human 5-HT_{1B}-type receptor gene. *Biochem. Biophys. Res. Commun.* 185 (2): 517–523.

72 Ng, G.Y., George, S.R., Zastawny, R.L. et al. (1993). Human $serotonin_{1B}$ receptor expression in Sf9 cells: phosphorylation, palmitoylation, and adenylyl cyclase inhibition. *Biochemistry* 32 (43): 11727–11733.

73 Wang, C., Jiang, Y., Ma, J. et al. (2013). Structural basis for molecular recognition at serotonin receptors. *Science* 340 (6132): 610–614.

74 Rodriguez, D., Brea, J., Loza, M.I., and Carlsson, J. (2014). Structure-based discovery of selective serotonin 5-HT_{1B} receptor ligands. *Structure* 22 (8): 1140–1151.

75 Hamblin, M.W. and Metcalf, M.A. (1991). Primary structure and functional characterization of a human 5-HT_{1D}-type serotonin receptor. *Mol. Pharmacol.* 40 (2): 143–148.
76 Weinshank, R.L., Zgombick, J.M., Macchi, M.J. et al. (1992). Human serotonin 1D receptor is encoded by a subfamily of two distinct genes: 5-HT_{1D} alpha and 5-HT_{1D} beta. *Proc. Natl. Acad. Sci. U.S.A.* 89 (8): 3630–3634.
77 Cargill, M., Altshuler, D., Ireland, J. et al. (1999). Characterization of single-nucleotide polymorphisms in coding regions of human genes. *Nat. Genet.* 22 (3): 231–238.
78 Harwood, G., Lockyer, M., Giles, H., and Fairweather, N. (1995). Cloning and characterisation of the rabbit 5-HT_{1D} alpha and 5-HT_{1D} beta receptors. *FEBS Lett.* 377 (1): 73–76.
79 Adham, N., Romanienko, P., Hartig, P. et al. (1992). The rat 5-hydroxytryptamine1B receptor is the species homologue of the human 5-hydroxytryptamine1D beta receptor. *Mol. Pharmacol.* 41 (1): 1–7.
80 Levy, F.O., Gudermann, T., Birnbaumer, M. et al. (1992). Molecular cloning of a human gene (S31) encoding a novel serotonin receptor mediating inhibition of adenylyl cyclase. *FEBS Lett.* 296 (2): 201–206.
81 McAllister, G., Charlesworth, A., Snodin, C. et al. (1992). Molecular cloning of a serotonin receptor from human brain ($5HT_{1E}$): a fifth $5HT_1$-like subtype. *Proc. Natl. Acad. Sci. U.S.A.* 89 (12): 5517–5521.
82 Dukat, M., Smith, C., Herrick-Davis, K. et al. (2004). Binding of tryptamine analogs at h5-HT_{1E} receptors: a structure-affinity investigation. *Bioorg. Med. Chem.* 12 (10): 2545–2552.
83 Adham, N., Kao, H.T., Schecter, L.E. et al. (1993). Cloning of another human serotonin receptor (5-HT_{1F}): a fifth 5-HT_1 receptor subtype coupled to the inhibition of adenylate cyclase. *Proc. Natl. Acad. Sci. U.S.A.* 90 (2): 408–412.
84 Leysen, J.E. (2004). 5-HT2 receptors. *Curr. Drug Targets CNS Neurol. Disord.* 3 (1): 11–26.
85 Becamel, C., Gavarini, S., Chanrion, B. et al. (2004). The serotonin 5-HT_{2A} and 5-HT_{2C} receptors interact with specific sets of PDZ proteins. *J. Biol. Chem.* 279 (19): 20257–20266.
86 Karaki, S., Becamel, C., Murat, S. et al. (2014). Quantitative phosphoproteomics unravels biased phosphorylation of serotonin 2A receptor at Ser280 by hallucinogenic versus nonhallucinogenic agonists. *Mol. Cell. Proteomics* 13 (5): 1273–1285.
87 Erdmann, J., Shimron-Abarbanell, D., Rietschel, M. et al. (1996). Systematic screening for mutations in the human serotonin-2A (5-HT_{2A}) receptor gene: identification of two naturally occurring receptor variants and association analysis in schizophrenia. *Hum. Genet.* 97 (5): 614–619.
88 Marshall, S.E., Bird, T.G., Hart, K., and Welsh, K.I. (1999). Unified approach to the analysis of genetic variation in serotonergic pathways. *Am. J. Med. Genet.* 88 (6): 621–627.
89 Ota, T., Suzuki, Y., Nishikawa, T. et al. (2004). Complete sequencing and characterization of 21,243 full-length human cDNAs. *Nat. Genet.* 36 (1): 40–45.

90 Becamel, C., Figge, A., Poliak, S. et al. (2001). Interaction of serotonin 5-hydroxytryptamine type 2C receptors with PDZ10 of the multi-PDZ domain protein MUPP1. *J. Biol. Chem.* 276 (16): 12974–12982.
91 Albizu, L., Holloway, T., Gonzalez-Maeso, J., and Sealfon, S.C. (2011). Functional crosstalk and heteromerization of serotonin 5-HT_{2A} and dopamine D_2 receptors. *Neuropharmacology* 61 (4): 770–777.
92 Wacker, D., Wang, S., McCorvy, J.D. et al. (2017). Crystal structure of an LSD-bound human serotonin receptor. *Cell* 168 (3): 377–389.e312.
93 Willins, D.L., Deutch, A.Y., and Roth, B.L. (1997). Serotonin 5-HT_{2A} receptors are expressed on pyramidal cells and interneurons in the rat cortex. *Synapse* 27 (1): 79–82.
94 Meltzer, H.Y. and Roth, B.L. (2013). Lorcaserin and pimavanserin: emerging selectivity of serotonin receptor subtype–targeted drugs. *J. Clin. Invest.* 123 (12): 4986–4991.
95 Almaula, N., Ebersole, B.J., Zhang, D. et al. (1996). Mapping the binding site pocket of the serotonin 5-Hydroxytryptamine2A receptor. Ser3.36(159) provides a second interaction site for the protonated amine of serotonin but not of lysergic acid diethylamide or bufotenin. *J. Biol. Chem.* 271 (25): 14672–14675.
96 Kristiansen, K., Kroeze, W.K., Willins, D.L. et al. (2000). A highly conserved aspartic acid (Asp-155) anchors the terminal amine moiety of tryptamines and is involved in membrane targeting of the 5-HT_{2A} serotonin receptor but does not participate in activation via a "salt-bridge disruption" mechanism. *J. Pharmacol. Exp. Ther.* 293 (3): 735–746.
97 Duan, X., Zhang, M., Zhang, X. et al. (2015). Molecular modeling and docking study on dopamine D2-like and serotonin 5-HT_{2A} receptors. *J. Mol. Graphics Modell.* 57: 143–155.
98 Mongeau, E., Yuan, G., Minden, Z. et al. (2017). Homology modeling inspired synthesis of 5-HT_{2A} receptor inhibitors: a diazepine analogue of the atypical antipsychotic JL13. *Cent. Nerv. Syst. Agents Med. Chem.* 17: 239–244.
99 Lin, F., Li, F., Wang, C. et al. (2017). Mechanism exploration of arylpiperazine derivatives targeting the 5-HT_{2A} receptor by in silico methods. *Molecules* 22 (7): 1064.
100 Schmuck, K., Ullmer, C., Engels, P., and Lubbert, H. (1994). Cloning and functional characterization of the human 5-HT_{2B} serotonin receptor. *FEBS Lett.* 342 (1): 85–90.
101 Wacker, D., Wang, C., Katritch, V. et al. (2013). Structural features for functional selectivity at serotonin receptors. *Science* 340 (6132): 615–619.
102 Liu, W., Wacker, D., Gati, C. et al. (2013). Serial femtosecond crystallography of G protein-coupled receptors. *Science* 342 (6165): 1521–1524.
103 Stam, N.J., Vanderheyden, P., van Alebeek, C. et al. (1994). Genomic organisation and functional expression of the gene encoding the human serotonin 5-HT_{2C} receptor. *Eur. J. Pharmacol.* 269 (3): 339–348.
104 Jahnsen, J.A. and Uhlen, S. (2012). The N-terminal region of the human 5-HT_{2C} receptor has as a cleavable signal peptide. *Eur. J. Pharmacol.* 684 (1–3): 44–50.

105 Heifetz, A., Storer, R.I., McMurray, G. et al. (2016). Application of an integrated GPCR SAR-modeling platform to explain the activation selectivity of human 5-HT_{2C} over 5-HT_{2B}. *ACS Chem. Biol.* 11 (5): 1372–1382.

106 Liu, Y., Canal, C.E., Cordova-Sintjago, T.C. et al. (2017). Mutagenesis analysis reveals distinct amino acids of the human serotonin 5-HT_{2C} receptor underlying the pharmacology of distinct ligands. *ACS Chem. Neurosci.* 8 (1): 28–39.

107 Faerber, L., Drechsler, S., Ladenburger, S. et al. (2007). The neuronal 5-HT_3 receptor network after 20 years of research - evolving concepts in management of pain and inflammation. *Eur. J. Pharmacol.* 560 (1): 1–8.

108 Brown, A.M., Hope, A.G., Lambert, J.J., and Peters, J.A. (1998). Ion permeation and conduction in a human recombinant 5-HT_3 receptor subunit (h5-HT_{3A}). *J. Physiol.* 507 (Pt. 3): 653–665.

109 Hussy, N., Lukas, W., and Jones, K.A. (1994). Functional properties of a cloned 5-hydroxytryptamine ionotropic receptor subunit: comparison with native mouse receptors. *J. Physiol.* 481 (Pt. 2): 311–323.

110 Kelley, S.P., Dunlop, J.I., Kirkness, E.F. et al. (2003). A cytoplasmic region determines single-channel conductance in 5-HT_3 receptors. *Nature* 424 (6946): 321–324.

111 Niesler, B., Frank, B., Kapeller, J., and Rappold, G.A. (2003). Cloning, physical mapping and expression analysis of the human 5-HT_3 serotonin receptor-like genes HTR_{3C}, HTR_{3D} and HTR_{3E}. *Gene* 310: 101–111.

112 Niesler, B., Walstab, J., Combrink, S. et al. (2007). Characterization of the novel human serotonin receptor subunits 5-HT_{3C}, 5-HT_{3D}, and 5-HT_{3E}. *Mol. Pharmacol.* 72 (1): 8–17.

113 Holbrook, J.D., Gill, C.H., Zebda, N. et al. (2009). Characterisation of 5-HT_{3C}, 5-HT_{3D} and 5-HT_{3E} receptor subunits: evolution, distribution and function. *J. Neurochem.* 108 (2): 384–396.

114 Barnes, N.M., Hales, T.G., Lummis, S.C.R., and Peters, J.A. (2008). The 5-HT_3 receptor - the relationship between structure and function. *Neuropharmacology* 56 (1): 273–284.

115 Frank, B., Niesler, B., Bondy, B. et al. (2004). Mutational analysis of serotonin receptor genes: HTR_{3A} and HTR_{3B} in fibromyalgia patients. *Clin. Rheumatol.* 23 (4): 338–344.

116 Yamada, K., Hattori, E., Iwayama, Y. et al. (2006). Distinguishable haplotype blocks in the HTR_{3A} and HTR_{3B} region in the Japanese reveal evidence of association of HTR_{3B} with female major depression. *Biol. Psychiatry* 60 (2): 192–201.

117 Bruss, M., Gothert, M., Hayer, M., and Bonisch, H. (1998). Molecular cloning of alternatively spliced human 5-HT_3 receptor cDNAs. *Ann. N.Y. Acad. Sci.* 861: 234–235.

118 Jin, P., Fu, G.K., Wilson, A.D. et al. (2004). PCR isolation and cloning of novel splice variant mRNAs from known drug target genes. *Genomics* 83 (4): 566–571.

119 Massoura, A.N., Dover, T.J., Newman, A.S., and Barnes, N.M. (2011). The identification of *N*-glycosylated residues of the human 5-HT_{3B} receptor subunit: importance for cell membrane expression. *J. Neurochem.* 116 (6): 975–983.

120 Tzvetkov, M.V., Meineke, C., Oetjen, E. et al. (2007). Tissue-specific alternative promoters of the serotonin receptor gene HTR_{3B} in human brain and intestine. *Gene* 386 (1–2): 52–62.

121 Sjoblom, T., Jones, S., Wood, L.D. et al. (2006). The consensus coding sequences of human breast and colorectal cancers. *Science* 314 (5797): 268–274.

122 Thompson, A.J. and Lummis, S.C. (2006). 5-HT_3 receptors. *Curr. Pharm. Des.* 12 (28): 3615–3630.

123 Brejc, K., van Dijk, W.J., Klaassen, R.V. et al. (2001). Crystal structure of an ACh-binding protein reveals the ligand-binding domain of nicotinic receptors. *Nature* 411 (6835): 269–276.

124 Maksay, G., Bikadi, Z., and Simonyi, M. (2003). Binding interactions of antagonists with 5-hydroxytryptamine3A receptor models. *J. Recept. Signal Transduct. Res.* 23 (2–3): 255–270.

125 Reeves, D.C., Sayed, M.F., Chau, P.L. et al. (2003). Prediction of 5-HT_3 receptor agonist-binding residues using homology modeling. *Biophys. J.* 84 (4): 2338–2344.

126 Carland, J.E., Cooper, M.A., Livesey, M.R. et al. (2013). Mutagenic analysis of the intracellular portals of the human 5-HT_{3A} receptor. *J. Biol. Chem.* 288 (44): 31592–31601.

127 Price, K.L., Lillestol, R.K., Ulens, C., and Lummis, S.C. (2016). Palonosetron-5-HT_3 receptor interactions as shown by a binding protein cocrystal structure. *ACS Chem. Neurosci.* 7 (12): 1641–1646.

128 Hassaine, G., Deluz, C., Grasso, L. et al. (2014). X-ray structure of the mouse serotonin 5-HT_3 receptor. *Nature* 512 (7514): 276–281.

129 Rademakers, R., Baker, M., Nicholson, A.M. et al. (2011). Mutations in the colony stimulating factor 1 receptor (CSF1R) gene cause hereditary diffuse leukoencephalopathy with spheroids. *Nat. Genet.* 44 (2): 200–205.

130 Ray, A.M., Kelsell, R.E., Houp, J.A. et al. (2009). Identification of a novel 5-HT_4 receptor splice variant (r5-HT_{4c1}) and preliminary characterisation of specific 5-HT_{4a} and 5-HT_{4b} receptor antibodies. *Eur. J. Pharmacol.* 604 (1–3): 1–11.

131 Clark, R.D. (1998). Medicinal chemistry of 5-HT4 receptor ligands. In: *5-HT4 Receptors in the Brain and Periphery* (ed. R.M. Eglen), 1–48. New York, NY: Springer.

132 Bureau, R., Boulouard, M., Dauphin, F. et al. (2010). Review of 5-HT_{4R} ligands: state of art and clinical applications. *Curr. Top. Med. Chem.* 10 (5): 527–553.

133 Padayatti, P.S., Wang, L., Gupta, S. et al. (2013). A hybrid structural approach to analyze ligand binding by the serotonin type 4 receptor (5-HT_4). *Mol. Cell. Proteomics* 12 (5): 1259–1271.

134 Nelson, D.L. (2004). 5-HT_5 receptors. *Curr. Drug Targets CNS Neurol. Disord.* 3 (1): 53–58.

135 Gonzalez, R., Chavez-Pascacio, K., and Meneses, A. (2013). Role of 5-HT_{5A} receptors in the consolidation of memory. *Behav. Brain Res.* 252: 246–251.

136 Serrats, J., Raurich, A., Vilaro, M.T. et al. (2004). 5-HT_{5B} receptor mRNA in the raphe nuclei: coexpression with serotonin transporter. *Synapse* 51 (2): 102–111.

137 Woolley, M.L., Marsden, C.A., and Fone, K.C. (2004). 5-HT_6 receptors. *Curr. Drug Targets CNS Neurol. Disord.* 3 (1): 59–79.

138 Kang, H., Lee, W.K., Choi, Y.H. et al. (2005). Molecular analysis of the interaction between the intracellular loops of the human serotonin receptor type 6 (5-HT_6) and the alpha subunit of G_S protein. *Biochem. Biophys. Res. Commun.* 329 (2): 684–692.

139 Chiu, H.J., Wang, Y.C., Liou, J.H. et al. (2001). Serotonin 6 receptor polymorphism in schizophrenia: frequency, age at onset and cognitive function. *Neuropsychobiology* 43 (3): 113–116.

140 Masellis, M., Basile, V.S., Meltzer, H.Y. et al. (2001). Lack of association between the T-->C 267 serotonin 5-HT_6 receptor gene (HTR_6) polymorphism and prediction of response to clozapine in schizophrenia. *Schizophr. Res.* 47 (1): 49–58.

141 Ohmori, O., Shinkai, T., Hori, H., and Nakamura, J. (2001). Novel polymorphism in the 5'-upstream region of the human 5-HT_6 receptor gene and schizophrenia. *Neurosci. Lett.* 310 (1): 17–20.

142 Thome, J., Retz, W., Baader, M. et al. (2001). Association analysis of HTR_6 and HTR_{2A} polymorphisms in sporadic Alzheimer's disease. *J. Neural. Transm.* 108 (10): 1175–1180.

143 Orlacchio, A., Kawarai, T., Paciotti, E. et al. (2002). Association study of the 5-hydroxytryptamine(6) receptor gene in Alzheimer's disease. *Neurosci. Lett.* 325 (1): 13–16.

144 Messina, D., Annesi, G., Serra, P. et al. (2002). Association of the 5-HT_6 receptor gene polymorphism $C26_{7T}$ with Parkinson's disease. *Neurology* 58 (5): 828–829.

145 Lee, S.H., Lee, K.J., Lee, H.J. et al. (2005). Association between the 5-HT_6 receptor $C26_{7T}$ polymorphism and response to antidepressant treatment in major depressive disorder. *Psychiatry Clin. Neurosci.* 59 (2): 140–145.

146 Meffre, J., Chaumont-Dubel, S., Mannoury la Cour, C. et al. (2012). 5-HT_6 receptor recruitment of mTOR as a mechanism for perturbed cognition in schizophrenia. *EMBO Mol. Med.* 4 (10): 1043–1056.

147 Witty, D., Ahmed, M., and Chuang, T.T. (2009). Advances in the design of 5-HT_6 receptor ligands with therapeutic potential. *Prog. Med. Chem.* 48: 163–224.

148 Dobi, K., Flachner, B., Pukancsik, M. et al. (2015). Combination of pharmacophore matching, 2D similarity search, and in vitro biological assays in the selection of potential 5-HT_6 antagonists from large commercial repositories. *Chem. Bio. Drug Des.* 86 (4): 864–880.

149 Bojarski, A.J. (2006). Pharmacophore models for metabotropic 5-HT receptor ligands. *Curr. Top. Med. Chem.* 6 (18): 2005–2026.

150 Bard, J.A., Zgombick, J., Adham, N. et al. (1993). Cloning of a novel human serotonin receptor (5-HT_7) positively linked to adenylate cyclase. *J. Biol. Chem.* 268 (31): 23422–23426.

151 Gellynck, E., Andressen, K.W., Lintermans, B. et al. (2012). Biochemical and pharmacological study of N-linked glycosylation of the human serotonin 5-HT_{7A} receptor. *FEBS J.* 279 (11): 1994–2003.

152 Erdmann, J., Nothen, M.M., Shimron-Abarbanell, D. et al. (1996). The human serotonin 7 (5-HT_7) receptor gene: genomic organization and systematic mutation screening in schizophrenia and bipolar affective disorder. *Mol. Psychiatry* 1 (5): 392–397.

153 Bruss, M., Kiel, S., Bonisch, H. et al. (2005). Decreased agonist, but not antagonist, binding to the naturally occurring $Thr^{92}Lys$ variant of the H5-$HT_{7(a)}$ receptor. *Neurochem. Int.* 47 (3): 196–203.

154 Krobert, K.A. and Levy, F.O. (2002). The human 5-HT_7 serotonin receptor splice variants: constitutive activity and inverse agonist effects. *Br. J. Pharmacol.* 135 (6): 1563–1571.

155 Kucwaj-Brysz, K., Warszycki, D., Podlewska, S. et al. (2016). Rational design in search for 5-phenylhydantoin selective 5-HT_{7R} antagonists. Molecular modeling, synthesis and biological evaluation. *Eur. J. Med. Chem.* 112: 258–269.

156 Shahaf, N., Pappalardo, M., Basile, L. et al. (2016). How to choose the suitable template for homology modelling of GPCRs: 5-HT_7 receptor as a test case. *Mol. Inf.* 35 (8–9): 414–423.

157 Focke, P.J., Wang, X., and Larsson, H.P. Neurotransmitter transporters: structure meets function. *Structure* 21 (5): 694–705.

158 Zhang, Y.W., Turk, B.E., and Rudnick, G. (2016). Control of serotonin transporter phosphorylation by conformational state. *Proc. Natl. Acad. Sci. U.S.A.* 113 (20): E2776–E2783.

159 Annamalai, B., Mannangatti, P., Arapulisamy, O. et al. (2012). Tyrosine phosphorylation of the human serotonin transporter: a role in the transporter stability and function. *Mol. Pharmacol.* 81 (1): 73–85.

160 Zhang, Y.W., Gesmonde, J., Ramamoorthy, S., and Rudnick, G. (2007). Serotonin transporter phosphorylation by cGMP-dependent protein kinase is altered by a mutation associated with obsessive compulsive disorder. *J. Neurosci.* 27 (40): 10878–10886.

161 Landrø, N.I., Jonassen, R., Clark, L. et al. (2015). Serotonin transporter polymorphisms predict response inhibition in healthy volunteers. *Neurosci. Lett.* 584: 109–112.

162 Iurescia, S., Seripa, D., and Rinaldi, M. (2016). Role of the 5-HTTLPR and SNP promoter polymorphisms on serotonin transporter gene expression: a closer look at genetic architecture and in vitro functional studies of common and uncommon allelic variants. *Mol. Neurobiol.* 53 (8): 5510–5526.

163 Jonassen, R. and Landrø, N.I. (2014). Serotonin transporter polymorphisms (5-HTTLPR) in emotion processing. Implications from current neurobiology. *Prog. Neurobiol.* 117: 41–53.

164 Caspi, A., Sugden, K., Moffitt, T.E. et al. (2003). Influence of life stress on depression: moderation by a polymorphism in the 5-HTT gene. *Science* 301 (5631): 386–389.

165 Hung, C.F., Lung, F.W., Chen, C.H. et al. (2011). Association between suicide attempt and a tri-allelic functional polymorphism in serotonin transporter gene promoter in Chinese patients with schizophrenia. *Neurosci. Lett.* 504 (3): 242–246.

166 Matsumoto, Y., Fabbri, C., Pellegrini, S. et al. (2014). Serotonin transporter gene: a new polymorphism may affect response to antidepressant treatments in major depressive disorder. *Mol. Diagn. Ther.* 18 (5): 567–577.

167 Arpawong, T.E., Lee, J., Phillips, D.F. et al. (2016). Effects of recent stress and variation in the serotonin transporter polymorphism (5-HTTLPR) on depressive symptoms: a repeated-measures study of adults age 50 and older. *Behav. Genet.* 46 (1): 72–88.

168 Wang, Y., Sun, N., Liu, Z. et al. (2016). Psychosocial mechanisms of serotonin transporter's genetic polymorphism in susceptibility to major depressive disorder: mediated by trait coping styles and interacted with life events. *Am. J. Translat. Res.* 8 (2): 1281–1292.

169 Peitl, V., Karlović, D., Štefanović, M., and Vrkić, N. (2015). Serotonin transporter polymorphism in relation to depression. *Alcoholism* 51 (2): 107–126.

170 Kilic, F., Murphy, D.L., and Rudnick, G. (2003). A human serotonin transporter mutation causes constitutive activation of transport activity. *Mol. Pharmacol.* 64 (2): 440–446.

171 Cengiz, M., Okutan, S.N., Bayoglu, B. et al. (2015). Genetic polymorphism of the serotonin transporter gene, *SLC6A4* rs16965628, is associated with obsessive compulsive disorder. *Genet. Test. Mol. Biomarkers* 19 (5): 228–234.

172 Munn-Chernoff, M.A., McQueen, M.B., Stetler, G.L. et al. (2012). Examining associations between disordered eating and serotonin transporter gene polymorphisms. *Int. J. Eat. Disorders* 45 (4): 556–561.

173 Rozenblat, V., Ong, D., Fuller-Tyszkiewicz, M. et al. (2017). A systematic review and secondary data analysis of the interactions between the serotonin transporter 5-HTTLPR polymorphism and environmental and psychological factors in eating disorders. *J. Psychiatr. Res.* 84: 62–72.

174 Golimbet, V., Korovaitseva, G., Lezheiko, T. et al. (2017). The serotonin transporter gene 5-HTTLPR polymorphism is associated with affective psychoses but not with schizophrenia: a large-scale study in the Russian population. *J. Affective Disorders* 208: 604–609.

175 Chang, C.C., Chang, H.A., Fang, W.H. et al. (2017). Gender-specific association between serotonin transporter polymorphisms (5-HTTLPR and rs25531) and neuroticism, anxiety and depression in well-defined healthy Han Chinese. *J. Affective Disorders* 207: 422–428.

176 Mohamed Saini, S., Nik Jaafar, N.R., Sidi, H. et al. (2014). Serotonin transporter gene polymorphism and its association with bipolar disorder across different ethnic groups in Malaysia. *Compr. Psychiatry* 55 (Suppl. 1): S76–S81.

177 Hartmann, J.A., Wichers, M., van Bemmel, A.L. et al. (2014). The serotonin transporter 5-HTTLPR polymorphism in the association between sleep quality and affect. *Eur. Neuropsychopharmacol.* 24 (7): 1086–1090.

178 Schumacher, J. and Deckert, J. (2010). Serotonin transporter polymorphisms and panic disorder. *Genome Med.* 2 (6): 40.

179 Schenkel, L.C., Bragatti, J.A., Torres, C.M. et al. (2011). Serotonin transporter gene (5HTT) polymorphisms and temporal lobe epilepsy. *Epilepsy Res.* 95 (1–2): 152–157.

180 Fehér, Á., Juhász, A., László, A. et al. (2013). Serotonin transporter and serotonin receptor 2A gene polymorphisms in Alzheimer's disease. *Neurosci. Lett.* 534 (1): 233–236.

181 Rudnick, G. (2006). Serotonin transporters--structure and function. *J. Membr. Biol.* 213 (2): 101–110.

182 Hasenhuetl, P.S., Freissmuth, M., and Sandtner, W. (2016). Electrogenic binding of intracellular cations defines a kinetic decision point in the transport cycle of the human serotonin transporter. *J. Biol. Chem.* 291 (50): 25864–25876.

183 Kern, C., Erdem, F.A., El-Kasaby, A. et al. (2017). The N terminus specifies the switch between transport modes of the human serotonin transporter. *J. Biol. Chem.* 292 (9): 3603–3613.

184 Ahmed, B.A., Bukhari, I.A., Jeffus, B.C. et al. (2009). The cellular distribution of serotonin transporter is impeded on serotonin-altered vimentin network. *PLoS One* 4 (3): e4730.

185 Loland, C.J., Sanchez, C., Plenge, P. et al. (2017). Allosteric binding in the serotonin transporter - pharmacology, structure, function and potential use as a novel drug target. In: *Allosterism in Drug Discovery* (ed. D. Doller), 360–380. London: RSC Publishing.

186 Jorgensen, A.M., Tagmose, L., Jorgensen, A.M. et al. (2007). Homology modeling of the serotonin transporter: insights into the primary escitalopram-binding site. *ChemMedChem* 2 (6): 815–826.

187 Zare, S., Fereidoonnezhad, M., Afshar, D., and Ramezani, Z. (2017). A comparative QSAR analysis and molecular docking studies of phenyl piperidine derivatives as potent dual NK_1R antagonists/serotonin transporter (SERT) inhibitors. *Comput. Biol. Chem.* 67: 22–37.

188 Pratuangdejkul, J., Schneider, B., Launay, J.M. et al. (2008). Computational approaches for the study of serotonin and its membrane transporter SERT: implications for drug design in neurological sciences. *Curr. Med. Chem.* 15 (30): 3214–3227.

189 Gabrielsen, M., Westrheim Ravna, A., Kristiansen, K., and Sylte, I. (2012). Substrate binding and translocation of the serotonin transporter studied by docking and molecular dynamics simulations. *J. Mol. Model.* 18 (3): 1073–1085.

190 Haddad, Y., Heger, Z., and Adam, V. (2016). Guidelines for homology modeling of dopamine, norepinephrine, and serotonin transporters. *ACS Chem. Neurosci.* 7 (11): 1607–1613.

191 Rudnick, G. (2006). Structure/function relationships in serotonin transporter: new insights from the structure of a bacterial transporter. In: *Neurotransmitter Transporters*, Handbook of Experimental Pharmacology, vol. 175 (ed. H.H. Sitte and M. Freissmuth), 59–73. Berlin, Heidelberg: Springer-Verlag.

192 Coleman, J.A., Green, E.M., and Gouaux, E. (2016). X-ray structures and mechanism of the human serotonin transporter. *Nature* 532 (7599): 334–339.

193 Gabrielsen, M., Kurczab, R., Ravna, A.W. et al. (2012). Molecular mechanism of serotonin transporter inhibition elucidated by a new flexible docking protocol. *Eur. J. Med. Chem.* 47 (1): 24–27.

194 Gabrielsen, M., Wolosewicz, K., Zawadzka, A. et al. (2013). Synthesis, antidepressant evaluation and docking studies of long-chain alkylnitroquipazines as serotonin transporter inhibitors. *Chem. Biol. Drug Des.* 81 (6): 695–706.

195 Gabrielsen, M., Kurczab, R., Siwek, A. et al. (2014). Identification of novel serotonin transporter compounds by virtual screening. *J. Chem. Inf. Model.* 54 (3): 933–943.

4

The Cholinergic System

4.1 Introduction

The cholinergic system has been associated with several cognitive functions, including memory, selective attention, and emotional processing [1]. The cholinergic system is formed by organized nerve cells that combines two groups of major cholinergic projections (Figure 4.1): (i) the **magnocellular basal forebrain** cholinergic system [2] and (ii) the **brainstem** cholinergic system[3]. The magnocellular basal forebrain cholinergic system includes the medial septal (MS) nucleus, the vertical and horizontal limbs of the diagonal band (DB) of Broca, and the nucleus basalis magnocellularis (nBM). The horizontal limb of the DB and nBM has extensive diffuse projections to neocortex as well as projections to the basolateral amygdala and olfactory bulb. The MS and vertical limb of the DB project to the hippocampus and entorhinal cortices. The brainstem cholinergic system includes the pedunculopontine tegmental nucleus (PPT) [4] and laterodorsal pontine tegmentum (LPT) and projects predominantly to the thalamus as well as to the basal forebrain region. The forebrain system is involved in enhancing signal-to-noise ratios so it is closely related with arousal, in particular vigilance processes. Also, it contributes to memory-encoding and working-memory processes. The brainstem system is involved in the sleep/waking cycle.

Cholinergic system dysfunction has been observed in several neurodegenerative diseases such as Alzheimer's disease [5, 6], Lewy body dementia [7, 8], Huntington's disease [9], and myasthenia gravis [10]. Recent studies have correlated nervous system dysfunction in multiple sclerosis, with inflammation and cholinergic system alterations [11]. A reduction in cholinergic activity has also been observed in Parkinson's disease [12], and although initially there was some controversy [13], schizophrenia has also been related to cholinergic dysfunction [14]. Identification of cause/effect relationship between those neurodisorders and the cholinergic system is of current interest [15].

The cholinergic system uses acetylcholine as neurotransmitter. Acetylcholine is synthesized from choline and activated acetate as acetyl coenzyme A, in a single-step reaction catalyzed by the enzyme choline acetyl transferase (ChAT). The reaction starts with a calcium-dependent transfer of mitochondrial acetyl coenzyme A (formed from D-glucose through intermediate pyruvate) into the cholinergic neuron cytoplasm. Once acetylcholine is formed, it is rapidly absorbed

Chemical Biology of Neurodegeneration: A Molecular Approach, First Edition. Pedro Merino.

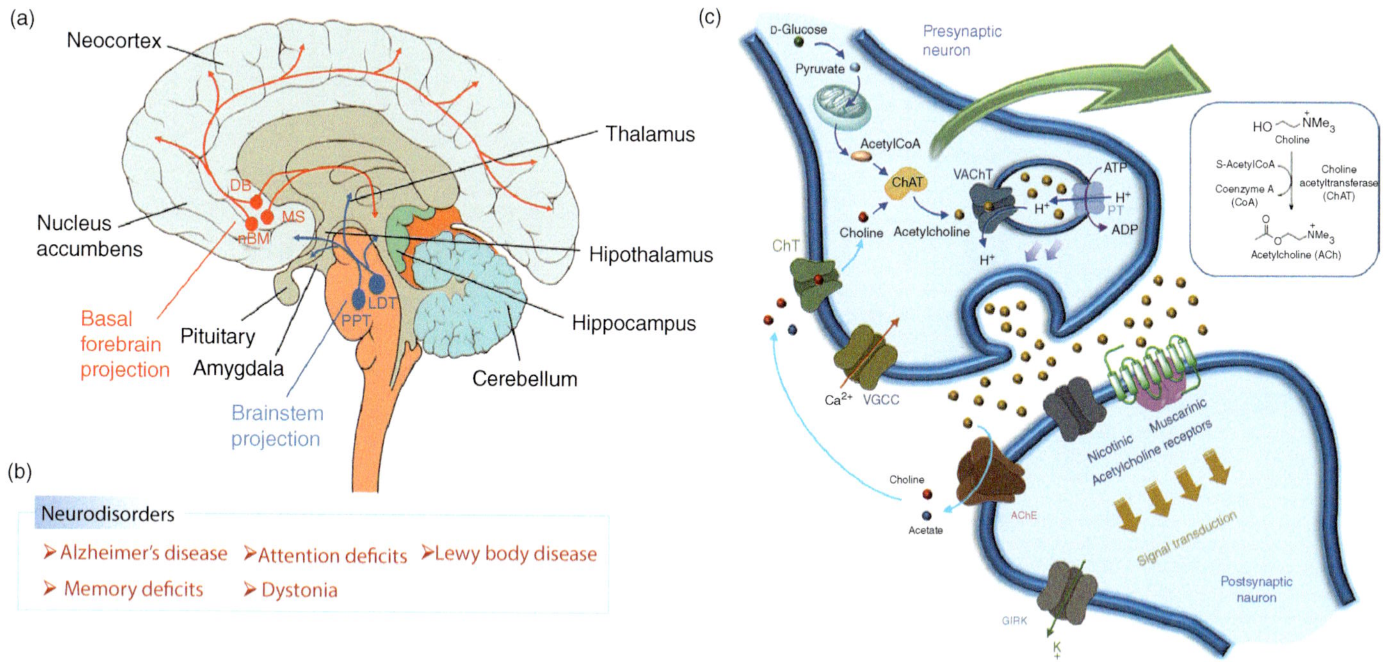

Figure 4.1 The cholinergic system. (a) Major projections of cholinergic system. (b) Neurological disorders associated to cholinergic system. (c) Detail of the cholinergic synapse: synthesis of acetylcholine by ChAT, absorption into synaptic vesicles by VAChT, release of acetylcholine, binding of acetylcholine to receptors, hydrolysis of acetylcholine by AChE, and reuptake of choline by ChT. VGCC: voltage-gated calcium channel; GIRK: G-protein-coupled inwardly rectifying potassium channel; VAChT: vesicular acetylcholine transporter; ChT: choline transporter; ChAT: choline acetyl transferase; AChE: acetylcholine esterase; PT: proton transporter.

into the synaptic vesicles. The uptake of acetylcholine (ACh) into storage vesicles occurs through an ATP-dependent pump that acidifies the vesicle. The acidified vesicle then uses a vesicular acetylcholine transporter (VAChT) to exchange protons for ACh molecules. Acetylcholine uptake into the vesicle by VAChT is driven by a proton-pumping ATPase. Coupled countertransport of a proton and acetylcholine allows the vesicle to remain isoosmotic and electroneutral. Then, upon the arrival of an action potential at the presynaptic neuron terminal, voltage-dependent calcium channels (VDCCs) open and calcium ions flow into the presynaptic neuron's cytosol. This influx of calcium causes neurotransmitter-containing vesicles to dock and fuse to the presynaptic neuron's cell membrane and release acetylcholine into the synaptic cleft. Acetylcholine diffuses into the synaptic cleft and binds to the cholinergic receptors. There are two types of receptors – nicotinic and muscarinic. Nicotinic receptors are ligand-gated ion channels, so when they bind acetylcholine, they allow sodium ions to flow in and potassium ions to flow out. Muscarinic receptors are G-protein-coupled receptors (GPCRs). Actually, acetylcholine binds briefly to receptors, and after dissociation acetylcholine is degraded by the enzyme acetylcholinesterase (AChE) producing choline and an acetate group. The choline produced by the action of AChE is transported, through reuptake by choline transporter (ChT), back into the presynaptic terminal, where it is used to synthesize new acetylcholine.

4.2 Cholinergic Receptors

4.2.1 Classification

Two cholinergic receptor subtypes, nicotinic and muscarinic, have been identified based on early studies of pharmacologic selectivity of agonists. Nicotinic receptors (nAChRs) are stimulated by nicotine; they are excitatory and show a rapid response to stimulation. Muscarinic receptors (mAChRs) are stimulated by muscarine, have either excitatory or inhibitory effects, and show a slower response to stimulation [16, 17]. Further studies showed that all receptors are not identical and different subtypes exist within the nicotinic and muscarinic classes.

Nicotinic receptors are subdivided according to whether they are neuromuscular (N_1) or ganglionic (N_2). Neuronal nicotinic receptors show ligand specificities distinct from the muscle-type nicotinic receptors in the neuromuscular junction. A book dedicated to nicotinic receptors in which most of the previous reported reviews [18–23] are referenced was published in 2014 [24]. Nicotinic receptors are formed by different subunits designated α, β, γ, and δ in order of increasing molecular weight. Neuronal nAChR subunits are classified as "α" when they have the vicinal cysteines required for agonist binding. The α_1 subunit is in muscle and α_2–α_{10} are neuronal subunits. The "β" subunits lack the vicinal cysteines; of them, β_1 is in the muscle, whereas β_2–β_4 are neuronal.

Ligand binding assays using radioactive agonists [^{3}H]-acetylcholine and [^{3}H]-nicotine and antagonist [^{125}I]-α-Bungarotoxin (α-Bgtx), a neurotoxic protein, were employed for further classification of nAChRs [25]. Two main classes of nAChR subtypes were identified: (i) α-Bgtx-sensitive receptors, made up of α_7,

α_8, α_9, and/or α_{10} subunits, which can form homomeric or heteromeric structures, and (ii) α-Bgtx-insensitive receptors, composed of α and β subunits that only form heteromeric structures. These receptors bind agonists with high affinity but not α-Bgtx [26].

Muscarinic receptors are subdivided into five subtypes (M1–M5) on the basis of pharmacological studies and molecular cloning [27, 28]. The M1 subtype was identified to be pirenzepine sensitive, whereas M2 subtype was pirenzepine insensitive. The affinity for pirenzepine, AF-DX 116 (an analogue of pirenzepine), HHSiD, and 4-DAMP allowed identification of M3 subtype. Differences in affinity by methoctramine provided evidences of M4 subtype (Figure 4.2).

Finally, M5 subtype was revealed by molecular cloning, since it has no distinct pharmacological profile different from the other four subtypes. For details on pharmacological and physiological properties of muscarinic receptors, the reader is directed to a published review [29]. A comprehensive treatise covering pharmacology and signaling of muscarinic receptors including those in the central nervous system (CNS) was published in 2012 [30].

4.2.2 Location

Nicotinic receptors (nAChRs) are located at the skeletal neuromuscular junction, adrenal medulla, and autonomic ganglia. They are divided into muscle-type (N_1) and neuronal (N_2) nAChR. Neuronal nicotinic acetylcholine receptors are widely distributed in the human brain [31] and operate at presynaptic locations to regulate the release of several CNS transmitters [32]. In this respect, several studies demonstrated that glutamatergic [33], dopaminergic [34], serotonergic, peptidergic, and cholinergic pathways are under the control of presynaptic nicotinic receptors [35]. Ionotropic neuronal nicotinic acetylcholine receptors are mainly located at the preterminal and presynaptic sites contributing to cognitive function [36]. At least 12 distinct genes encoding neuronal nicotinic receptor subunits α_2–α_{10} and β_2–β_4 have been identified in the central and peripheral nervous systems [37]. The α_3 subunit is prevalent in peripheral ganglia, usually

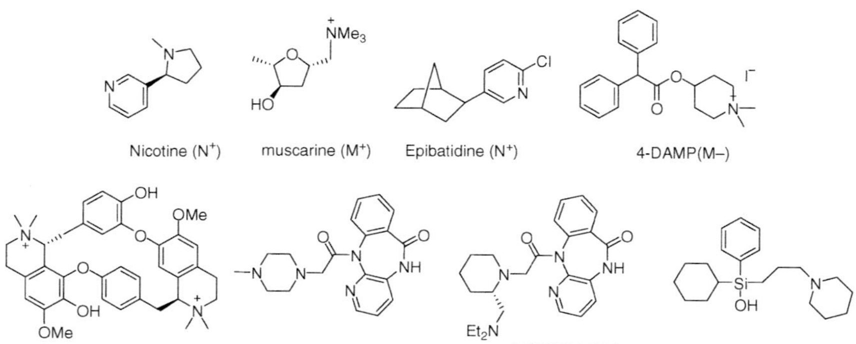

Figure 4.2 Cholinergic agonists and antagonists. Nicotinic (N) and muscarinic (M) specificity and agonist (+) and antagonist (–) activity are indicated between brackets.

with β_2, β_4, and α_5 subunits; the $\alpha_4\beta_2$ subunit combination predominates in the CNS. Four permutations of α_4, α_5, α_6, β_2, and β_3 subunits are found in the striatum, which mediate dopamine release [38]. The α_4 subunit is almost exclusively found in CNS neurons, whereas α_9 and α_{10} are not expressed in the CNS but are located in some sensory neurons.

Muscarinic receptors are located in the brain depending on the subtype. M1 and M4 receptors are located in the striatum, cortex, and hippocampus [39, 40], M2 receptors in the forebrain and thalamus [41], M3 receptors in the hypothalamus, and M5 receptors in the basal ganglia, brain vasculature, and dopaminergic neurons [42]. Notably, M5 receptors are the only muscarinic receptor observed (based on mRNA detection) in the substantia nigra pars compacta [43]. Recently, it has been demonstrated that all muscarinic receptors are expressed in mouse brain microvascular endothelial cells [44].

4.2.3 Nicotinic Receptors

Nicotinic receptors have been the subject of intense research, mainly because of the availability of nAChRs from the electric organs of electric eel (*Electrophorus electricus*) and the marine ray (*Torpedo californica*). In this section, we focus on nicotinic receptors in the brain (N_2). The structure of neuronal nicotinic receptors has been reviewed recently [45]. The structure and function of neuronal nicotinic acetylcholine receptors has also been reviewed [36, 46]. The subunit diversity has been reviewed and studied in detail [47]. These studies show that the number of subunit combinations generating functional nAChRs is considerably lower than all possible subunit combinations. For details on N_1 receptors, the reader is directed to specialized bibliography [48–50] and general treatises [24].

Nicotinic receptors are ligand-gated ion channel receptors that are structurally and functionally closely related to other ligand-gated ion channel receptors, such as those for γ-aminobutyric acid (GABA), and serotonin 5-HT$_3$ [18]. Several structural studies have been carried out with nicotinic muscle-type receptors. On the other hand, the source of neuronal nAChRs is scarce, limiting the initial studies on this subtype. The muscle-type receptors comprise the four distinct subunits α, β, γ (or ε), and δ [51], which are arranged as $\alpha_2\beta\gamma\delta$ for the embryonic form and $\alpha_2\beta\varepsilon\delta$ for the adult form. The nAChRs have a pentameric structure composed of five membrane spanning subunits (α_2 through α_{10} and β_2–β_4) arranged around a central pore. The α subunits are similar in sequence to the muscle α_1 subunit and contribute to the ligand-binding interface. The α_7–α_{10} subunits are homomeric, whereas α_2–α_6 and β_2–β_4 subunits are heteromeric.

4.2.3.1 Effector Mechanisms

Nicotinic receptors are ionotropic receptors that contribute to synaptic plasticity by virtue of their calcium permeability. In the brain, nicotinic receptors are closely involved in a variety of calcium-dependent processes [52]. Downstream signaling events mediated by nAChRs can be grouped into (i) instantaneous effects (e.g. neurotransmitter release), (ii) short-term effects mediated by kinases and phosphatases (such as the recovery of nAChR desensitization through cellular signaling cascades), and (iii) long-term effects mediated by transcriptional regulators (such as neuroprotection via gene expression) [53].

4.2.3.2 Implication in Neurodisorders

The loss of cholinergic neurons, involved in attention, learning, memory, perception, and consciousness, is particularly important in AD [6, 54]. Neuronal nicotinic acetylcholine receptors are involved in a number of physiological and behavioral processes. They are implicated in many neuropsychiatric disorders [55] and pathological conditions of adulthood and developmental age [56] such as Alzheimer's [57], and Parkinson's diseases [58]. A significant reduction in nAChRs in the brain of Alzheimer's patients is correlated with [11C] nicotine through positron emission tomography (PET) studies. Neuronal nAChR receptors are closely related to neurodegeneration [59]; they have been implicated in schizophrenia [60], Tourette's syndrome, anxiety, depression [61], epilepsy [62], and autoimmune diseases. Naturally occurring polymorphisms in neuronal nAChR subunits contribute to the risk factors for neurological and psychiatric diseases [63]. α_7-Containing nicotinic receptors are closely related to schizophrenia [64], and their dysfunction or dysregulated expression is found in various pathophysiological conditions of the nervous system including neurodegenerative diseases and neurodevelopmental disorders [65]. Given the role of nAChRs in regulating dopamine neuron activity [66], nicotinic receptors containing the α_7 subunit have been recently identified as therapeutic targets for Parkinson's disease [67].

4.2.3.3 Structure of Subunits

Neuronal nicotinic receptors are formed by subunits α_2–α_{10} and β_2–β_4. Each subunit is formed by the extracellular N-terminus portion followed by three hydrophobic transmembrane domains (M1–M3), a large intracellular loop, and a fourth hydrophobic transmembrane domain (M4) ending in the extracellular C-terminus. Major differences in sequences in α subunits are located in the first part of the extracellular N-terminus and in the large intracellular loop, the rest of the residues showing a high degree of conservation (Figure 4.3).

The X-ray crystal structure of the extracellular domain of the human α_2 subunit in complex with the agonist epibatidine has been resolved at 3.2 Å [68]. The subunit crystallized as a pentamer in which the full ligand binding pocket is conferred by two adjacent subunits (Figure 4.4).

In the case of β subunits, a high sequence identity is also observed (Figure 4.5).

4.2.3.4 Structure of Receptors

As other ligand-gated ion channel receptors, nAChRs are composed of five subunits arranged around a central pore. Figure 4.6 illustrates an example.

Assembly of subunits can be homomeric (five identical subunits) and heteromeric (combination of subunits). Among all the possible combinations of subunits, Table 4.1 lists the combinations, corresponding to neuronal nicotinic receptors, for which evidences derived from either native receptors or heterologous expression studies have been obtained. Subunit combinations have been grouped into those insensitive to α-bungarotoxin (α_2–α_6 and β_2–β_4) and those sensitive (α_7 and α_8).

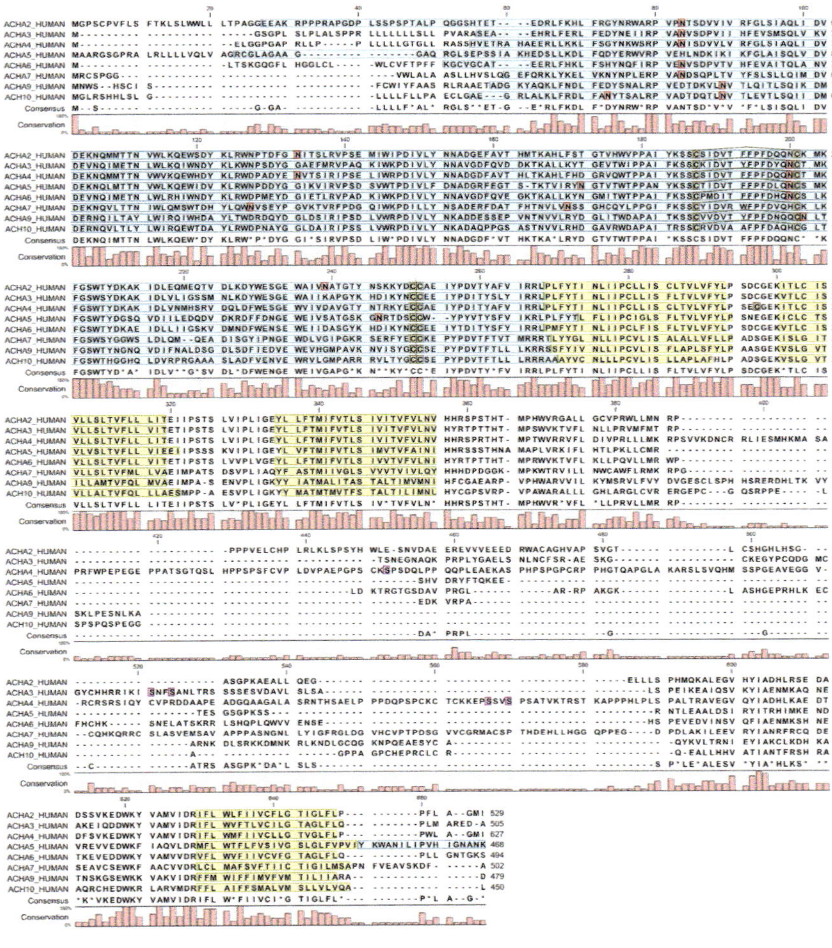

Figure 4.3 Sequences of the human α subunits α_2–α_7, α_9, and α_{10}. Conservation graphic and consensus sequence are shown. Disulfide bonds (green), N-glycosylation sites (red), phosphorylation (magenta) sites, palmitoylation sites (brown), extracellular domains (light blue), intracellular domains (white), and transmembrane domains (yellow) are indicated on sequences.

The homomeric receptors have five identical (orthosteric) ACh binding sites per receptor molecule (Figure 4.6). Each binding site is located at the interface between two adjacent subunits [70] as it has been recently demonstrated by X-ray crystallography (Figure 4.7) [68]. Heteromeric receptors have two (or three) α subunits and three (or two) β subunits, with two orthosteric binding sites per receptor molecule located at the interface between α and β subunits (Figure 4.7d) [22]. In these receptors, the fifth subunit does not contribute to the orthosteric site (the subunit is called accessory subunit). Each orthosteric ACh binding site has a principal and a complementary component. In homomeric receptors, each subunit contributes to both the principal and complementary components. In heteromeric receptors, α subunits contribute to the principal component,

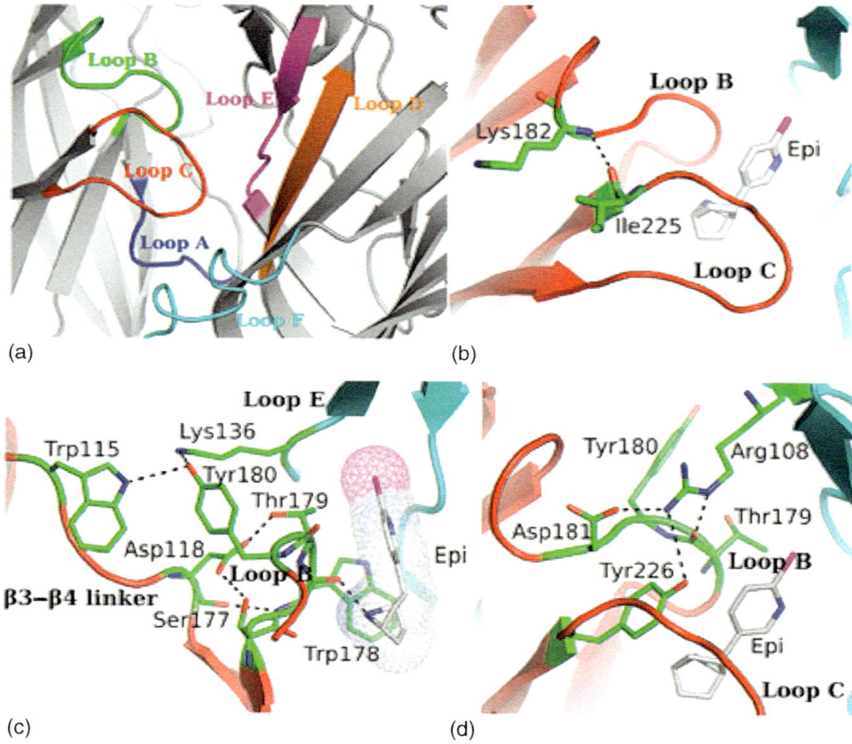

Figure 4.4 Loops and residues involved in the formation of the $\alpha_2(+)/\alpha_2(-)$ binding site. (a) Close-up view of the binding site in the α_2 Epi structure. Different colors are used to highlight loops A–F residing in two adjacent subunits. (b) Intrasubunit interaction (black dashed line) of Lys182 (loop B) and Ile225 (loop C) involved in ligand potency in nAChRs. Principal side backbone is colored in red, residues are shown in green sticks and epibatidine in white sticks. (c) Intrasubunit interaction of β_3–β_4 linker with loop B. Complementary side backbone is colored in cyan. (d) Intersubunit interactions involving the side chain of Arg108 (complementary side) and residues on loops B and C (primary side). Source: Reproduced with permission from Ref. [68]. Copyright 2016.

whereas β subunits contribute to the complementary one [18]. In particular, α_5 and β_3 subunits only form functional channels when they are coexpressed with primary and complementary subunits. On the other hand, α_3 or α_4 and β_2 or β_4 subunits can form orthosteric ligand binding sites or assemble in the accessory position to produce receptors with different stoichiometries. Receptors containing α_7 subunits have high calcium permeability and share structural and pharmacological properties of 5-HT$_3$ serotonergic receptors [71]. Studies have pointed out a higher complexity in binding sites and the involved subunits. The subtype combining α_4 and β_2 subunits is the most abundant in the brain, and it can assemble in two alternate stoichiometries – $(\alpha_4)_3(\beta_2)_2$ and $(\alpha_4)_2(\beta_2)_3$ [72]. The $(\alpha_4)_3(\beta_2)_2$ subtype showed an additional binding site at the α_4/α_4 interface in addition to the two orthosteric binding sites (Figure 4.7e). Both subtypes can be modulated by zinc(II) but in an opposite way. In fact, zinc(II) exerts an inhibitory modulatory

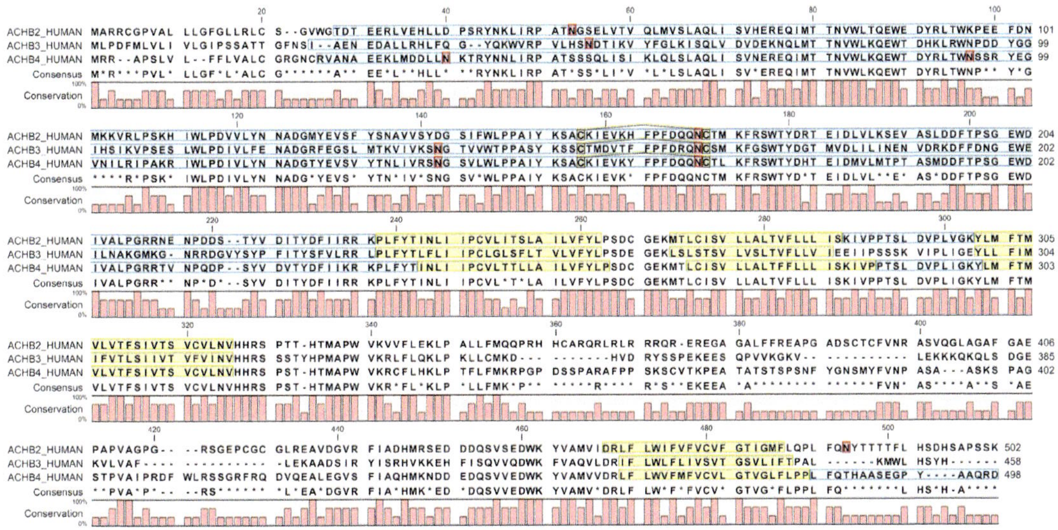

Figure 4.5 Sequences of the human β subunits β_2–β_4. Conservation graphic and consensus sequence are shown. Disulfide bonds (green), N-glycosylation sites (red), extracellular domains (light blue), intracellular domains (white), and transmembrane domains (yellow) are indicated on sequences.

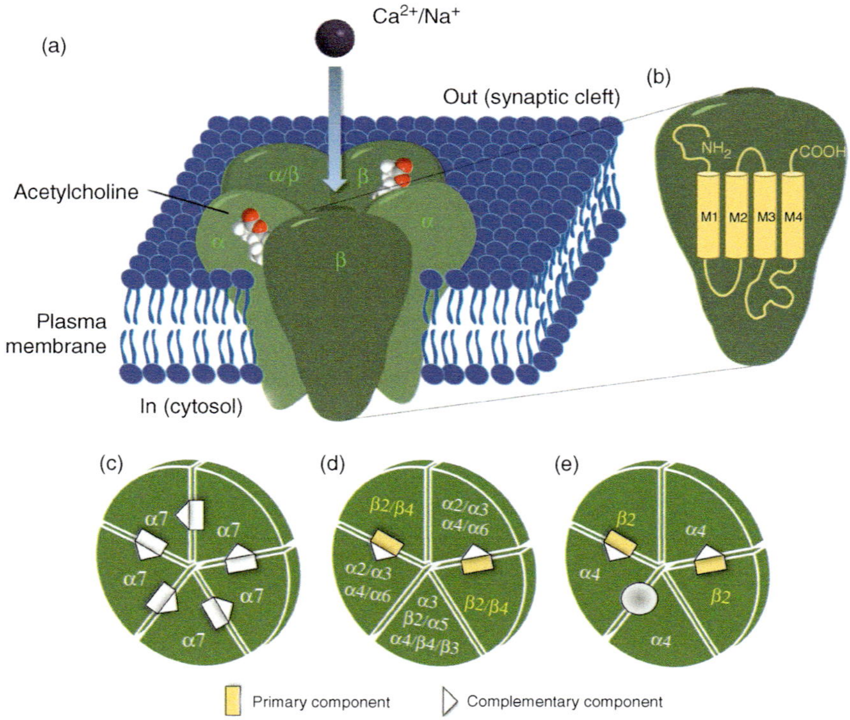

Figure 4.6 (a) General pentameric arrangement of nAChR subunits in an assembled receptor. (b) Detail of the putative transmembrane topology of nAChR subunits consisting of four hydrophobic transmembrane domains (M1–M4), a large intracellular loop between M3 and M4 and extracellular amino and carboxyl terminal portions. (c–e) Structure of hetero- and homopentameric neuronal subtypes. Pentameric arrangement of nAChR subunits in an α_7 homopentameric subtype (c), heteromeric receptor subtypes (d), and the $(\alpha_4)_3(\beta_2)_2$ subtype (e). The localization of the subunit interfaces of the orthosteric binding sites is indicated, with the primary component (rectangle) carried by the α subunits, and the complementary component (triangle) carried by α or non-α subunits. In addition to the two orthosteric sites, the $(\alpha_4)_3(\beta_2)_2$ subtype has a binding site at the α_4/α_4 interface (gray circle).

Table 4.1 Subunit combinations for neuronal cholinergic receptors.

Subunits	Combinations
α_2–α_6; β_2–β_4	$\alpha_2\beta_2$; $\alpha_2\beta_4$; $\alpha_3\beta_2$; $\alpha_3\beta_4$; $\alpha_4\beta_2$; $\alpha_4\beta_4$; $\alpha_6\beta_2$; $\alpha_6\beta_4$; $\alpha_2\alpha_5\beta_2$; $\alpha_3\alpha_5\beta_2$; $\alpha_3\alpha_5\beta_4$; $\alpha_3\alpha_6\beta_2$; $\alpha_3\alpha_6\beta_4$; $\alpha_3\alpha_3\beta_4$; $\alpha_4\alpha_5\beta_2$; $\alpha_5\alpha_6\beta_2$; $\alpha_6\beta_3\beta_4$; $\alpha_3\alpha_5\beta_2\beta_4$; $\alpha_3\alpha_6\beta_3\beta_4$; $\alpha_4\alpha_5\alpha_6\beta_2$; $\alpha_4\beta_2\beta_3\beta_4$
α_7 and α_8	α_7; α_8; $\alpha_7\beta_2$; $\alpha_7\beta_3$; $\alpha_7\alpha_8$; $\alpha_5\alpha_7\beta_2$; $\alpha_5\alpha_7\beta_4$

effect on $(\alpha_4)_2(\beta_2)_3$ receptors, but it potentiates or inhibits, depending on its concentration, the function of $(\alpha_4)_3(\beta_2)_2$ receptors [72]. For zinc(II) potentiation, the α_4/α_4 interface is required. Expression of $(\alpha_4)_2(\beta_2)_2(\alpha_5)$ pentamers showed that the α_5 subunit can occupy different positions, either as a nonbinding subunit or replacing a β_2 subunit participating in one of the orthosteric binding sites [73].

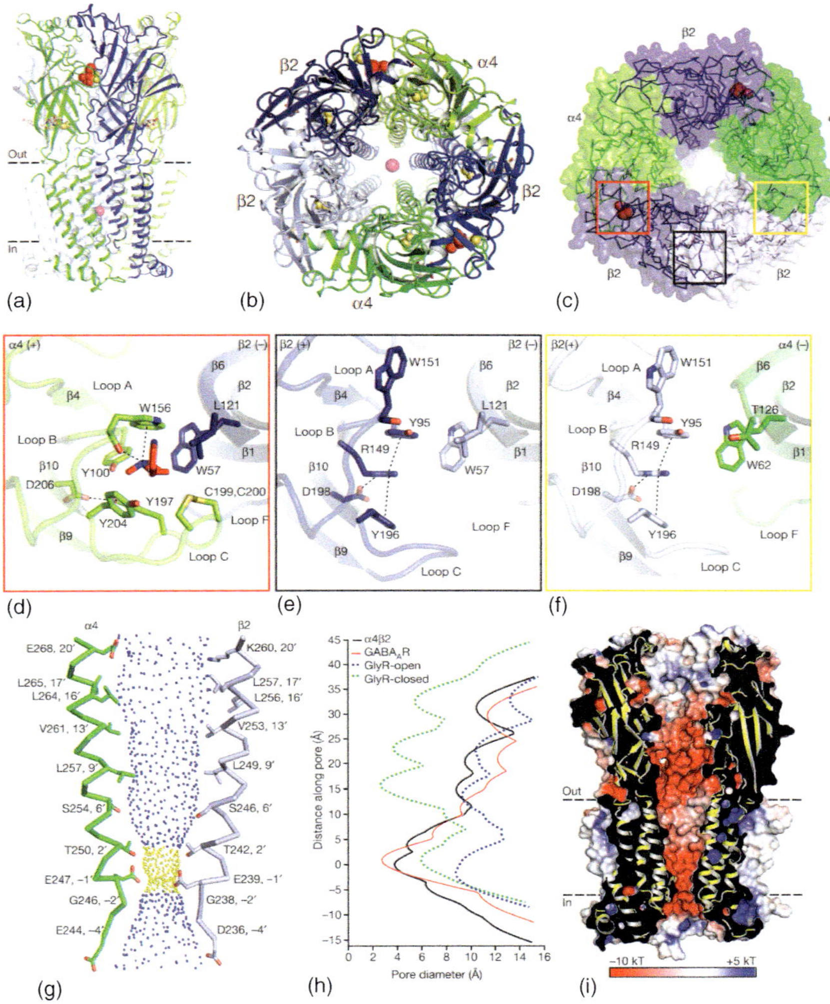

Figure 4.7 (a) View parallel to the plasma membrane. α_4 subunits are in green and β_2 in blue. Nicotine (red) and sodium (pink) are represented as spheres. The Cys-loop and loop C disulfide bonds are shown as yellow spheres. N-linked glycans (brown) are shown as sticks. Dashed lines indicate approximate membrane position. (b) View perpendicular to the plasma membrane looking from the extracellular side. (c) Extracellular view, with colored boxes indicating the three different interface classes. (d–f) Architectural details of interfaces boxed in (c) from the same orientation. Nicotine and interacting residues are shown as sticks. Potential hydrogen bonding and cation–π interactions are represented as dashed lines (2.7–5 Å). (g) M2 α-helices from opposing α_4 and β_2 subunits with side chains shown for pore-lining residues. Blue spheres indicate pore diameters >5.6 Å; yellow are >2.8 Å and <5.6 Å. (h) Pore diameter for the $\alpha_4\beta_2$ receptor and representative Cys-loop receptors in distinct functional states: desensitized/closed ($GABA_AR$ plus benzamidine; PDB ID: 4COF), activated/open (GlyR plus glycine; PDB ID: 3JAE) and resting/closed (GlyR plus strychnine; PDB ID: 3JAD). Structures were aligned using the M2 helix 9′ leucine, which occurs at $y \approx 15$ Å. The zero value along the y axis in the plot is aligned with the α-carbon of the M2 helix-1′ glutamate residue in $\alpha_4\beta_2$. (i) Cutaway of the receptor showing the permeation pathway colored by electrostatic potential. Source: Reproduced with permission from Morales-Perez et al. 2016 [69]. Copyright 2016, Springer Nature.

The X-ray structure of the $\alpha_4\beta_2$ receptor in complex with nicotine has been resolved, showing a structure resembling a cylinder formed by five subunits (Figure 4.7a,b) [69]. The subunits are arranged pseudosymmetrically around the pentameric ring in the order α-β-β-α-β confirming previous studies [74]. Nicotine is placed at the α-β interfaces almost fully buried from solvent (Figure 4.7c,d).

The X-ray structure also revealed a reorganization of the conserved residues in the β-β and β-α interfaces that precludes nicotine binding (Figure 4.7e,f). Crystallographic data also revealed the transmembrane channel in which a constriction point is formed at the interface with cytosol (Figure 4.7g). The narrowest point in the pore is defined by glutamate side chains at precise positions, which give rise to a constriction of 3.8 Å in diameter (Figure 4.7h,i). A comparison of the $\alpha_4\beta_2$ pore conformation with the three typical receptor states extracted from known data (resting/closed: glycine receptor plus strychnine [75]; activated/open: glycine receptor plus glycine [75]; desensitized/closed ($GABA_AR$ [76])) suggested that the $\alpha_4\beta_2$ receptor most closely resembles the desensitized $GABA_AR$, where the gate is at the cytosolic end of the pore (Figure 4.7i).

4.2.4 Muscarinic Receptors

Muscarinic receptors, designated M1–M5, are a part of the large family of GPCRs that occur principally in the CNS and present the typical seven transmembrane domains characteristic of GPCRs [77]. The gene family as a whole (Figure 4.8) shows 26.3% overall amino acid identity, with a larger variation between the receptor subtypes within the intracellular loops. The most variable third intracellular loop shows only 2.7% identity between receptors, while an average of 66% identity is found in the conserved transmembrane domains. Acetylcholine binds to the outer regions on the binding pocket where a crucial Asp105 stabilizes the positive charged nitrogen. Asp105 is conserved in most GPCRs, but there are six residues unique to the muscarinic receptors: Thr231, Thr234, Tyr148, Tyr506, Ty529, and Tyr533. Binding of acetylcholine to a muscarinic receptor causes a conformational change responsible for the activation of coupled G proteins [78].

Muscarinic ligands bind to a highly conserved pocket deep within the transmembrane regions TM3, TM5, TMS, and TM7. In addition to the orthosteric site, muscarinic receptors have several allosteric sites where ligands can modulate agonist function. The allosteric modulation of GPCRs offers several advantages over classic orthosteric ligands as therapeutic ligands, particularly selectivity and safety [79, 80]. The different allosteric binding sites of muscarinic receptors offer a great potential for selective targeting [79, 81, 82], and research is carried out on developing allosteric activators including positive allosteric modulators and allosteric agonists [82]. The study of both orthosteric and allosteric sites by X-ray crystallography illustrates the structural basis of GPCR function [83].

Structural details, including X-ray structures that might have implications in drug development [84], are discussed in the next sections for each receptor. Several general reviews [85–89] and treatises [30, 90, 91] on muscarinic receptors have been published elsewhere.

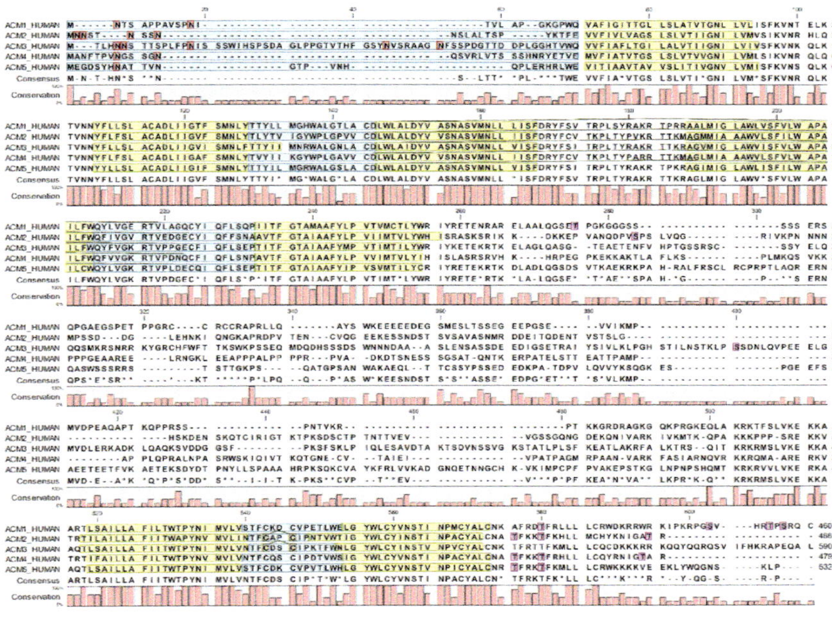

Figure 4.8 Sequences of the human muscarinic receptors M1–M5. Conservation graphic and consensus sequence are shown. Disulfide bonds (green), N-glycosylation sites (red), phosphorylation (magenta) sites, extracellular domains (light blue), intracellular domains (white), and transmembrane domains (yellow) are indicated on sequences.

4.2.4.1 Effector Mechanisms

Muscarinic receptors activate G proteins in a different manner (Figure 4.9). The M1, M3, and M5 mAChRs couple to G proteins of the $G_{q/11}$ family, increasing the activity of phosphoinositide-specific phospholipase C (PLC). Such activity results in the formation of inositol 1,4,5-triphosphate and diacylglycerol, second messengers responsible for the mobilization of intracellular calcium and activation of protein kinase C (PKC) and mitogen-activated protein kinase (MAPK). Coupling of M2 and M4 mAChRs with $G_{i/o}$ units results in inhibition of adenylyl cyclase. The subsequent reduction of cAMP leads to a decrease in the activity of protein kinase A. M2 and M4 receptors can also activate G-protein-coupled inwardly rectifying potassium channels (GIRKs), which causes membrane hyperpolarization. Those receptors are also capable of modulating negatively calcium currents VDCC, whereas they activate MAPK.

Muscarinic receptors M1, M3, and M5 parallel with 5-$HT_{2A\text{-}2C}$ serotonergic receptors and α_1 adrenergic receptors. M2 and M4 receptors couple with G proteins like 5-$HT_{1A\text{-}1F/5}$, serotonergic receptors, D_2-type dopaminergic receptors, and α_2 adrenergic receptors (see Sections 1.2.4, 2.2.4, and 3.2.4). This coupling efficiency to G-proteins has been the origin for the division of muscarinic receptors into two broad groups: M_1, M_3, and M_5-muscarinic receptors coupling to $G_{q/11}$-proteins and M2 and M4 muscarinic receptors coupling $G_{i/o}$-proteins [92]. Nevertheless, muscarinic receptors can couple to a wide range of diverse signaling pathways [93].

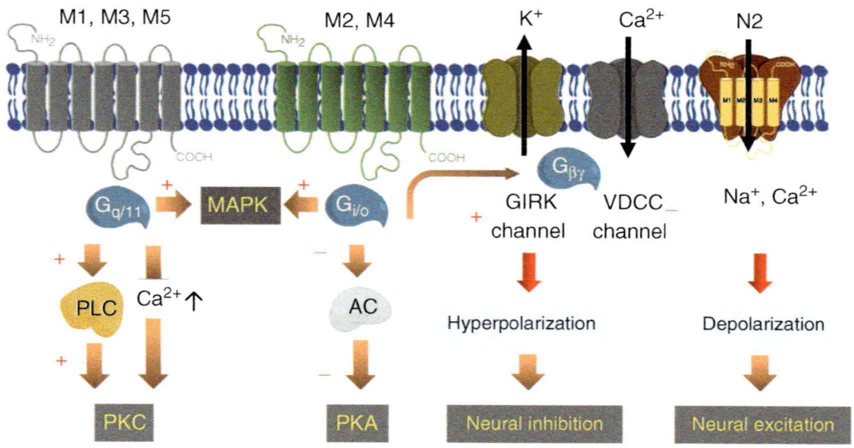

Figure 4.9 Signal transduction of cholinergic receptor subtypes. Muscarinic receptors are G-protein-coupled receptors with a seven transmembrane-spanning structure, mediating intracellular signal transduction via coupling to individual G proteins (e.g. $G_{i/o}$ and $G_{q/11}$). Nicotinic receptors form sodium and calcium-permeable cation channels. AC, adenylyl cyclase; MAPK, mitogen-activated protein kinase; GIRK, G-protein-coupled inwardly rectifying potassium; PLC, phospholipase C; PKA, protein kinase A; PKC, protein kinase C.

4.2.4.2 Implication in Neurodisorders

Muscarinic receptors have been implicated in the pathophysiology of psychiatric and neurological disorders [94]. In particular, they are involved in the pathophysiology of mood disorders [95] and psychiatric diseases such as schizophrenia [96]. M1 and M4 receptors are associated with learning, memory [97], and cognition [98]. Activation of those receptors has been investigated as potential treatments of Alzheimer's disease [99]. Schizophrenia [100] and drug addiction have also been related to M1 and M4 receptors. Genetic variations in the M2 receptor have been related to alcoholism and major depressive disorder [101]. The M5 receptor has been considered as a possible target for treatment of drug abuse due to their involvement in the regulation of striatal dopamine release [102] at postsynaptic plasma membranes distinct from dopamine transport [42].

4.2.4.3 M1 Receptor

The human M1 receptor has 460 residues that can be *N*-glycosylated with GlcNAc at Asn2 and Asn12. Phosphorylation takes place at Thr230, Thr428, Ser451, Thr455, and Ser457. An isoform in which residues 455–460 are missing has been described, but there is not experimental confirmation. The X-ray structures of M1 and M4 receptors have been resolved at the same time [103]. Both structures share important similarities with previously resolved structures of M2 and M3 receptors. For these reasons, the structural discussion for this receptor will be made in parallel to that of the M4 receptor (Section 4.2.4.6).

4.2.4.4 M2 Receptor

The human M2 receptor has 466 residues that can be *N*-glycosylated with GlcNAc at Asn2, Asn3, Asn6, and Asn9. Phosphorylation takes place at Ser232, Thr446, Thr450, and Thr465. The X-ray structure of the M2 receptor bound to the antagonist 3-quinuclidyl-benzilate (QNB) (PDB ID: 3UON) was resolved for the first time [104]. The agonist binding site is located in the middle of a long aqueous channel extending two-thirds through the membrane (Figure 4.10). The orthosteric binding pocket is conserved in all five muscarinic receptor subtypes. Dissociation of bound ligand is restricted by a layer of tyrosine residues that form a hydrophobic cap. Close to it and at the entrance to the binding pocket, an allosteric binding site has also been mapped.

Mutagenesis studies [105] showed that mutations Asp13Ala and Asn404Gln reduce affinity for acetylcholine. While the latter reduces signaling via downstream effectors, the former abolishes it completely. The same effect was found for mutations Asn58Ala and Tyr206Phe. The motifs involving residues 120–122 and 436–440 have also been found to be important for signaling. In the same paper, the mechanistic details of GPCR activation have been investigated by solving the structures of the M2 receptor bound to the agonist iperoxo alone (PDB ID: 4MQS) and combined with allosteric modulator LY2119620 (PDB ID: 4MQT) [105].

These studies suggest that, in contrast to other GPCRs such as rhodopsin and the β_2 adrenergic receptor, notable changes are observed in the orthosteric binding site and in the extracellular vestibule upon receptor activation where the allosteric modulator LY211962 is bound. This binding induces a slight contraction of the outer binding pocket of the iperoxo ligand. This structure with bound agonist and modulator definitively identifies the extracellular vestibule as an allosteric binding site. Notably, the extracellular loops and the extracellular end of TM segments that form the allosteric binding site show the greatest diversity between different receptors. The implication of the corresponding residues in the binding of several allosteric modulators has been also demonstrated through site-directed mutagenesis and chimeric receptor studies [106]. Thus, the observed diversity can be exploited for the design of more selective ligands that might modulate the binding of orthosteric ligands.

4.2.4.5 M3 Receptor

The human M2 receptor has 590 residues that can be *N*-glycosylated with GlcNAc at Asn5, Asn6, Asn15, Asn41, and Asn48. Phosphorylation takes place at Ser385. The M3 receptor is found as a homodimer that can form tetramers [107]. Natural variants corresponding to Val65Ile and Leu431Pro mutations have been found. The structure of the M3 receptor basolateral sorting signal corresponding to positions 271–289 has been resolved by NMR spectroscopy (PDB ID: 2CSA). Mutations Glu276Ala, Glu276Asp, Phe280Ala, and Val281Ala promote the loss of basolateral sorting, but no effect has been observed with mutations Phe280Leu and Val281Leu. Indeed, when these mutations are present, no effect is observed with Glu276Asp [108]. Regions containing residues 148–152 and 507–530 are involved in agonist binding.

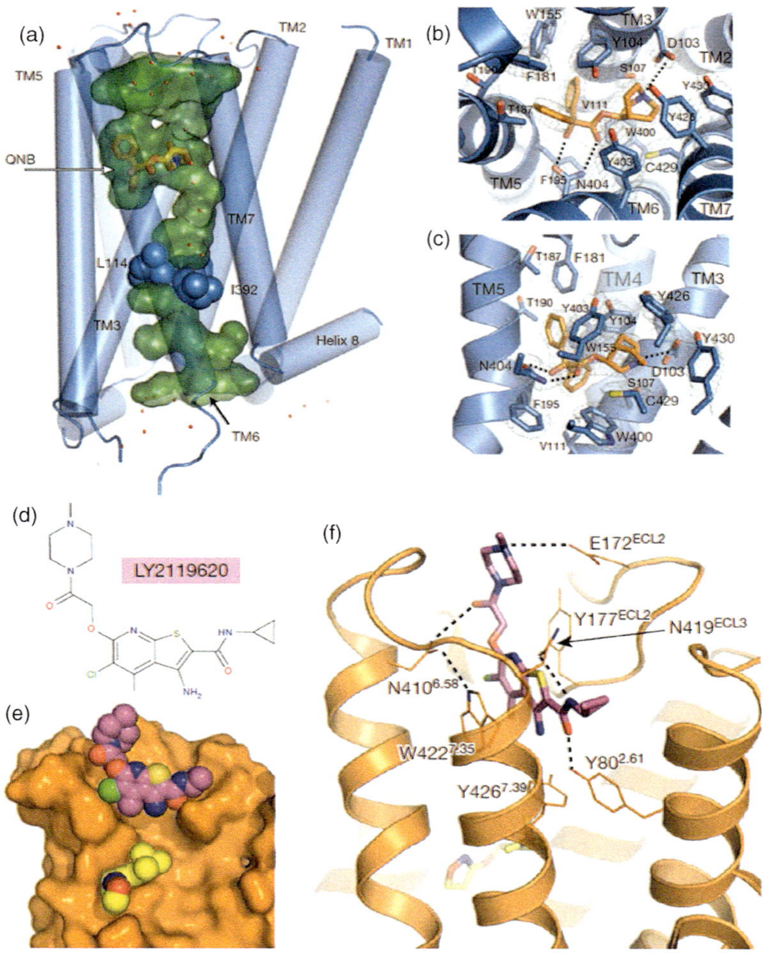

Figure 4.10 (a) Aqueous channel (green) extending from the extracellular surface into the transmembrane core is interrupted by a layer of three hydrophobic residues (blue spheres). Well-ordered water molecules are shown as red dots. (b,c) Two views of the QNB binding pocket. Amino acids within 4 Å of the ligand are shown as light blue sticks, with QNB in orange. Nitrogen and oxygen atoms are colored dark blue and red, respectively. Polar interactions are indicated by dashed lines. A $2F_o - 2F_c$ map is shown in wire at 1.5σ contour. (d) Positive allosteric modulator LY2119620. (e) The allosteric ligand binds to the extracellular vestibule just above the orthosteric agonist. A cross section through the membrane plane shows the relative positions of the two ligands. (f) Several polar contacts are involved in LY2119620 binding, in addition to extensive aromatic stacking interactions with Trp422 and Tyr177. Source: Haga et al. 2012 [104] and Kruse et al. 2013 [105]. Reproduced with permission of MacMillan Publishers.

Resolution of the X-ray structure of the M3 receptor from rat (human and rat M3 receptors show a 92% of residue identity, particularly within the transmembrane receptor core) bound to the bronchodilator drug tiotropium, a potent antagonist derived from atropine, showed that the ligand binds in a very similar

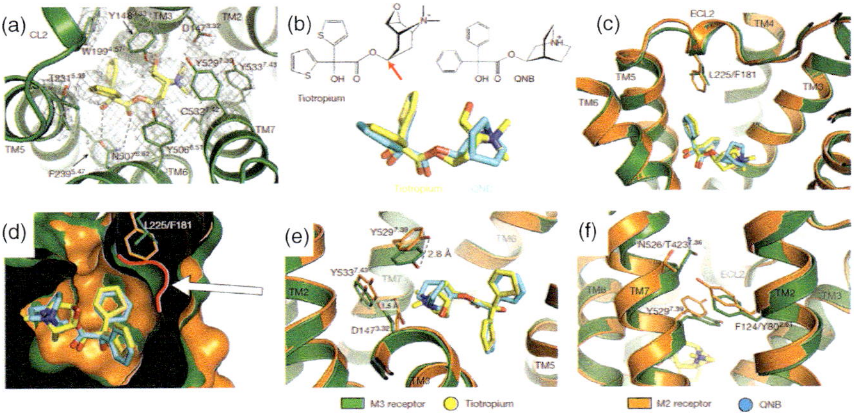

Figure 4.11 Orthosteric binding sites of the M2 and M3 receptors. In all panels, the M3 receptor is shown green with its ligand tiotropium in yellow, while the M2 receptor and its ligand QNB are shown in orange and cyan, respectively. (a) Tiotropium-binding site in the M3 receptor. A $2F_o$–F_c map contoured at 2 s is shown as mesh. (b) Chemical structures of ligands. A red arrow indicates the tropane C3 atom used as a tracking landmark. Superimposing the receptor structures reveals that the two ligands adopt highly similar poses (bottom). (c) There is a Phe (M2)/Leu (M3) sequence difference between the M2 and M3 receptors near the binding site. (d) This produces an enlarged binding pocket in the M3 receptor, outlined in red and indicated with an arrow. (e) Displacements of M3 Y529 and D147 are seen (black dashed lines). (f) The displacement of Y529 may arise from a sequence difference at position 2.61 (Tyr80 in M2 and Phe124 in M3). Source: Kruse et al. 2012 [109]. Reproduced with permission of MacMillan Publishers.

pose to that of QNB to M2 receptor (Figures 4.10 and 4.11 [92]) [109]. The residues interacting with the ligand are located on transmembrane helices 3–7 and are completely conserved among all muscarinic receptors, making extremely challenging to design subtype-selective muscarinic ligands targeting the orthosteric binding site [110]. Key interactions between the ligand and M3 receptor are the hydrogen bond between the hydroxyl group and Asn507 and a salt bridge between the ammonium group and Asp147, which is conserved among all biogenic amine GPCRs. This ionic interaction is surrounded by three conserved tyrosine residues Tyr148, Tyr506, and Tyr529. In a similar way to that observed for the M2 receptor, the X-ray structure of the M3 receptor also revealed a solvent-accessible vestibule separated from the orthosteric binding site by the tyrosine lid. The residues comprising that vestibule are less conserved among the five muscarinic receptor subtypes and, consequently, represent an attractive target for the design of selective-subtype ligands [83, 84]. In this context, several selective allosteric agents have been predicted to bind that extracellular vestibule [82].

To facilitate crystallogenesis, a fusion protein strategy, previously adopted with a series of GPCRs [111] was applied to the M3 receptor. The flexible third intracellular loop was replaced by several variants of T4 lysozyme (T4L) [112]. The original M3–T4L fusion protein produced twinned crystals, whereas TL4 variants gave rise to not twinned lattices that could be resolved up to 2.8 Å resolution.

Molecular dynamics simulations provided additional information suggesting that the ligand binds transiently to the allosteric binding site *en route* to the orthosteric binding site involving a metastable state in the extracellular vestibule (Figure 4.12a) [109]. When the ligand is placed in solvent, it binds to the same site (Figure 4.12b). The simulations are insufficiently long for the ligand to proceed into the orthosteric binding pocket; acetylcholine, a much smaller molecule, bound spontaneously to the orthosteric site in similar simulations. During these simulations, M2 and M3 receptors adopted a series of conformations included those observed in the crystal structure (Figure 4.12c), suggesting that the observed conformations are unlikely to be artifacts of the crystallization methodology; crystal structures only represent one conformation among many other that might be adopted (Figure 4.12d).

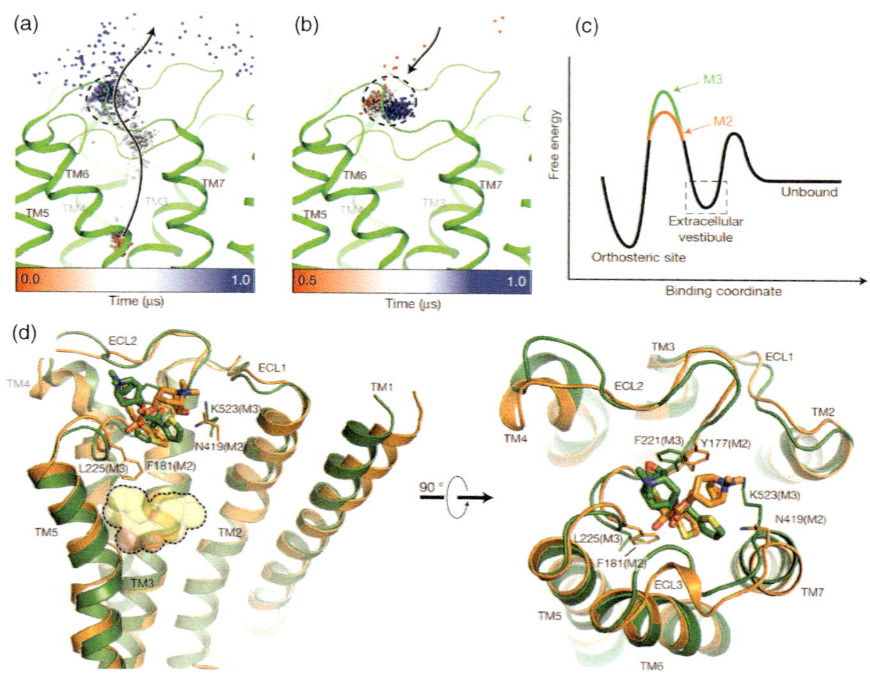

Figure 4.12 Molecular dynamics of ligand binding. (a) Tiotropium passes in the extracellular vestibule in the region outlined with a dashed circle. Spheres represent positions of the ligands C3 tropane atom at successive points in time. The direction of motion is indicated by an arrow. (b) Tiotropium binds to the same site in the extracellular vestibule when placed in solvent. (c) Schematic free-energy landscape for binding/dissociation. Differences between M2 and M3 are shown in orange and green, respectively, with the rest of the curve in black. (d) Common binding poses for tiotropium in the extracellular vestibule of M2 (orange) and of M3 (green). Nonconserved residues that contact the ligand are shown as thin sticks. The location of the orthosteric site is indicated by tiotropium (as spheres). Source: Kruse et al. 2012 [109]. Reproduced with permission of MacMillan Publishers.

4.2.4.6 M4 Receptor

The human M2 receptor has 479 residues that can be *N*-glycosylated with GlcNAc at Asn8 and Asn13. Phosphorylation takes place at Thr459, Thr463, and Thr477.

As mentioned earlier, the structure of M4 muscarinic receptor (PDB ID: 5DSG) was resolved by X-ray crystallography concurrently with that of M1 receptor (PDB ID: 5CXV) [103]. Both M1 and M4 receptors were crystallized in complex to tiotropium (as M3 receptor). The structures of M1 and M4 receptors are similar to those of M2 and M3 receptors. Slight differences are found on the extracellular and intracellular sides of largely conserved regions (Figure 4.13a–c). Given the almost complete conservation of residues corresponding to the orthosteric

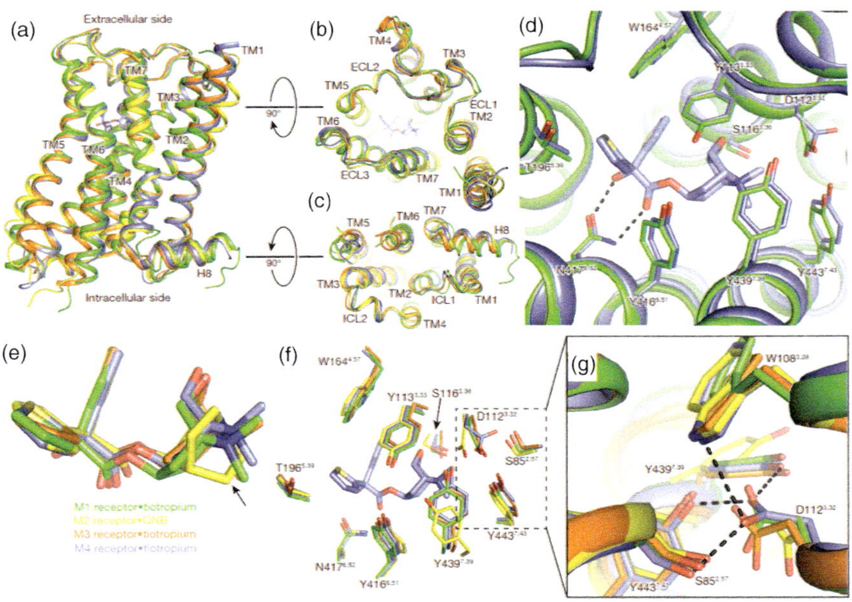

Figure 4.13 Structural comparison of M1–M4 receptors. (a) The overall view of the muscarinic structures is shown as cartoons aligned to the M3 receptor, with M1 colored in green, M2 in yellow (PDB ID: 3UON), M3 (PDB ID: 4U15, chain A) in orange, and M4 (chain A) in blue. Root mean squared deviations for the alignment (excluding T4L fusions) of M1, M2, and M4 versus the M3 receptor are 0.86, 0.81, and 0.62 Å, respectively. The ligand, tiotropium, for the M4 receptor is shown as sticks and colored according to element: carbon, light blue; oxygen, red; nitrogen, dark blue; sulfur, yellow. (b, c) Comparison of the views from (b) the extracellular side and (c) the intracellular side. (d) M1 and M4 residues involved in tiotropium binding are shown as sticks (several residues are omitted for clarity). The black dashed line indicates a bidentate hydrogen bond between Asn and tiotropium. (e) Superposition of tiotropium from the M1, M3, and M4 structures and QNB from the M2 structure. The arrow indicates the main structural difference between tiotropium and QNB. (f) Comparison of the orthosteric binding site of the M1–M4 receptors with orthosteric site residues shown as sticks. (g) The rotameric change of Asp112is stabilized by a network of hydrogen bonds. Source: Thal et al. 2016 [103]. Reproduced with permission of MacMillan Publishers.

binding site among muscarinic receptors, it is not surprising that the observed poses of the ligand are almost identical (Figure 4.13d–f). However, differences in the tertiary structure can exist and, in fact, a change in the rotamer of Asp112 of the M4 receptor is observed (Figure 4.13f,g). Interestingly, this change produces slight movements of Tyr439 and Tyr443 that induce the formation of a network of hydrogen bond interactions between Asp112 and Ser85, Trp108, Tyr439, and Tyr443, which is different in the M1, M2, and M3 receptor structures.

Important differences are found around residues Asp112, Tyr439, and Tyr443, which surround the amine group of the ligand. The importance of these residues had been pointed out by previous mutagenesis studies [113]. These studies showed that mutations Tyr439Ala and Ty443Ala produced 25- and 48-fold loss of affinity for ligand NMS (*N*-methylscopolamine, very similar to tiotropium), whereas little effect was observed for 3-quinuclidinyl-benzilate (QNB).

Figure 4.14a,b illustrates a high degree of conservation of the residues comprising the orthosteric binding site of the four resolved muscarinic receptors. On the other hand, the receptors have an extracellular vestibule constituting an

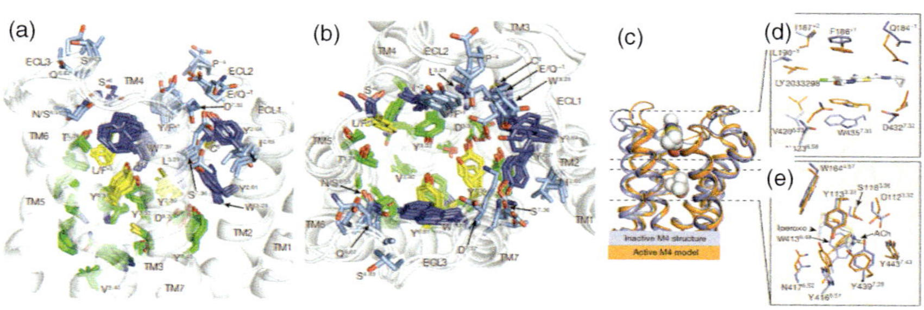

Figure 4.14 (a,b) Comparison of orthosteric and allosteric binding site residues across M1–M4 receptors. M1–M4 receptors are aligned to the M3 muscarinic receptor (PDB ID: 4U15, chain A) and are shown as gray colored cartoons with views from the (a) membrane and (b) extracellular side. Orthosteric-site residues are indicated in green, shared-site residues in yellow, allosteric-site conserved residues in blue and allosteric site nonconserved residues in cyan. Residues in ECL2 are numbered relative to the position of the disulfide-bonded cysteine. Carbon atoms are colored by site with the orthosteric residues in green, dark blue for allosteric conserved residues, light blue for allosteric nonconserved residues, and yellow for residues that contribute to both sites. Oxygen atoms are colored red, nitrogen blue, and sulfur yellow. Nonconserved residues are labeled according to either the most common residue or by the residue at the M4 receptor. (c–e) A model of an active M4 receptor bound to acetylcholine and LY2033298. (c) Comparison of the M4•tiotropium receptor structure (blue) versus an active-state model (orange) bound to acetylcholine and LY2033298 on the basis of the active M2•iperoxo•LY2119620 structure (PDB ID: 4MQT) as viewed from the membrane. The active M4 model was aligned to the M4•tiotropium structure (chain A, excluding T4L) with a root mean squared deviation of 0.9 Å. (d,e) Cross-sectional views of the (d) allosteric site and (e) orthosteric site as viewed from the extracellular side with a 90° rotation relative to the membrane from (a). Residues surrounding each site are shown as sticks (several removed for clarity). Acetylcholine and LY2033298 are shown as (c) spheres or (d,e) sticks and colored according to element: carbon, cyan; oxygen, red; nitrogen, blue; sulfur, yellow; chlorine, green. Acetylcholine is in the trans conformation and aligns in a similar pose to iperoxo (e, transparent yellow sticks). Source: Thal et al. 2016 [103]. Reproduced with permission of MacMillan Publishers.

allosteric binding site with differences in both amino acid sequence and tertiary structure, the latter as a result of the dynamic nature of the extracellular loop regions. As mentioned, these differences are potential targets for designing subtype-selective ligands that can act as allosteric modulators.

The resolved structures correspond to inactive forms of the receptor. A comparison with the active state model (Figure 4.14c–e) allows rationalizing the molecular bases of the mode of action of muscarinic receptors. A contraction of the extracellular vestibule consisting of an inward displacement of Asn423, Phe186, and Trp435 induces π-stacking interactions with the ligand in the active state. The large difference in size between acetylcholine and tiotropium is accommodated by a significant movement of residues Tyr416, Asn417, Trp413, and Tyr439, resulting in a contraction of the binding site. These studies identified for the first time a connection between the allosteric and orthosteric sites, involving the interface between transmembrane domains TM2, 3, 6, and 7, and extending along the top of the extracellular loop ECL2 (Figure 4.14, orange-colored residues). Definitively, the comparison between structures of muscarinic receptors evidences a divergence in residues forming the allosteric sites, demonstrating the importance of this region for designing novel selective ligands [44].

4.2.4.7 M5 Receptor

The human M2 receptor has 532 residues that can be *N*-glycosylated with GlcNAc at Asn8. Phosphorylation takes place at Thr501 and Thr505. Although there is X-ray crystallographic information of the M5 receptor, the sequence conservation and models allow foreseeing similar features to M1–M4 receptors. Despite this lack of structural information, positive allosteric modulators have been designed [114, 115].

4.3 Cholinergic Transporters

4.3.1 Vesicular Acetylcholine Transporter (VAChT)

The VAChT, responsible for the accumulation of acetylcholine into presynaptic vesicles of cholinergic neurons [116], has been cloned [117] and expressed from a "cholinergic" gene locus [118]. The distribution of VAChT mRNA coincides with that reported for ChAT in the peripheral and central cholinergic nervous system. The expression of VAChT is correlated with the severity of dementia and has been considered a significant diagnostic target for Alzheimer's disease [119].

The VAChT has a sequence of 532 residues, and it can be *N*-glycosylated with GlcNAc at Asn89 and Asn96 (Figure 4.15). Natural variants Gly186Ala and Asp398His have been reported to be a cause of congenital myasthenic syndrome in humans [120]. Other natural variants include Arg11Gln, Ala13Pro, Arg29Trp, and Ala520Glu.

X-ray structure is not yet available, but computer-based models predict 12 transmembrane domains with both intracellular N- and C-termini and a large extracellular loop connecting TM1 and TM2 [121], as other amine transporters found in adrenergic neurons [122].

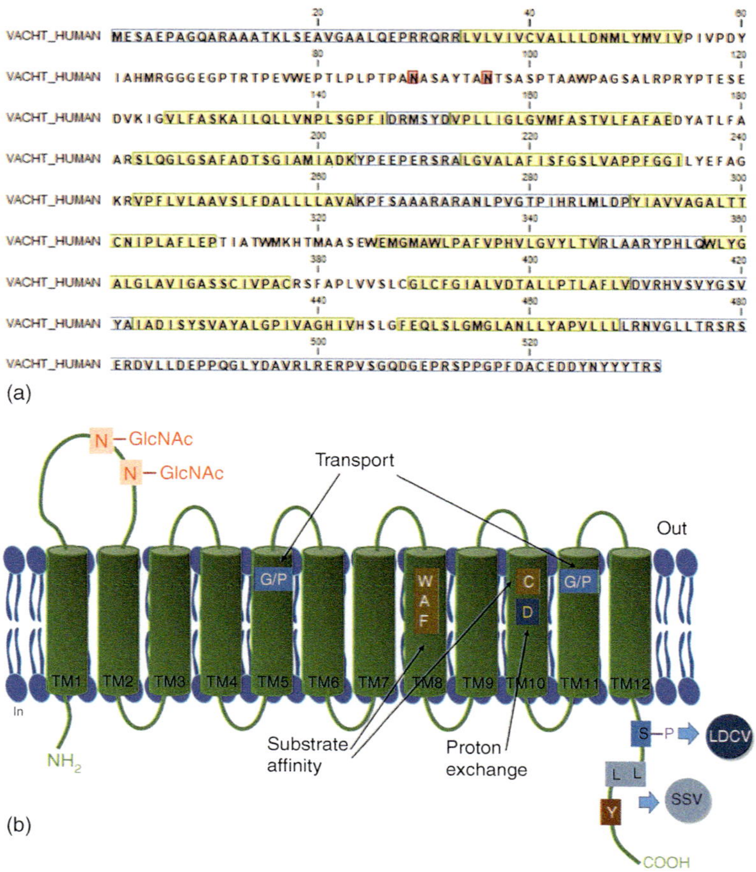

Figure 4.15 (a) Sequence of the human vesicular acetylcholine transporter. N-glycosylation sites (red) intracellular domains (light blue), extracellular domains (white), and transmembrane domains (yellow) are indicated on sequences. (b) Structure, function and trafficking of VAChT. Amino acids are indicated by one-letter code. SSV: small synaptic vesicles; LDCV: large-density core vesicles.

Important residues include Gly or Pro residues in TM5 and TM11 involved in transport; Trp331, Ala334, Phe335 (in TM8), and Cys391 (in TM10) involved in substrate affinity, and Asp398 residue (in TM10) involved in proton exchange [123]. Site-directed mutagenesis studies suggest the formation of cation-π interactions between the charged nitrogen of the acetylcholine and aromatic residues. The C-terminal region of VAChT, located in small synaptic vesicles [124], contains a di-Leu motif required for endocytosis [125] and sorting to small synaptic vesicles [126]. A tyrosine-based motif can also influence this trafficking. The di-Leu motif is regulated by phosphorylation of a Ser residue, located before in the peptide chain, that can vary the ratio of VAChT between small synaptic vesicles and large-dense core vesicles by driving a portion of VAChT onto the latter [127].

The VAChT forms a part of the vesicular monoamine transporter family SCL18 that also includes VMAT1 and VMAT2 [128]; all of them are structurally related [129]. The VAChT is predominantly present in synaptic vesicles, while VNATs are mostly in large-dense core vesicles [130]. These transporters accumulate biogenic amines into secretory vesicles used during transmission [124]. The transport of ACh, as the other positively charged amines (see previous chapters) is coupled to the countertransport of protons. The VAChT is able to concentrate acetylcholine in vesicles up to 100-fold higher than the cytosolic acetylcholine concentration. The current model accepted for VAChT proposes that the exchange of one luminal proton is used to transport substrate into the vesicle lumen whereas, a second proton is used to reorient the substrate binding site to allow the cycle to begin again. Thermodynamically, the two proton-dependent movements of the transporter are thought to be rate limiting, acetylcholine binding taking place at equilibrium. On the other hand, the reorientation of the empty acetylcholine-binding site is thought to be slower than the two translocation events [121].

Vesamicol, a selective inhibitor of ACh transport, inhibits vesicular ACh uptake in a noncompetitive way, suggesting that this inhibitor acts on a site other than the ACh-binding site on the transporter. Vesamicol also inhibits the release of newly synthesized ACh without affecting high-affinity choline uptake, ACh synthesis, or calcium influx. This loss of ACh release, secondary to the blockade of uptake by the vesicle, supports the fact that the vesicle is the site of ACh release. Structural analogues of vesamicol in which the phenyl ring was replaced by a benzoyl group showed increased affinity for VAChT and high selectivity [131].

4.3.2 High-Affinity Choline Transporter (ChT)

The high-affinity ChT is responsible for taking up choline from the extracellular space into neuron. Choline uptake is the rate-limiting step in the synthesis of acetylcholine. ChT cloned and identified in 2000 [132, 133] belongs to the sodium/glucose cotransporter family SLC5 [134], although it has a low analogy (20–25%) with other members of the family [133].

The ChT has a sequence of 580 residues, and it can be *N*-glycosylated with GlcNAc at extracellular Asn301. A putative glycosylation site has been predicted at Asn69 but experimental evidences point to only glycosylation at Asn301 [135]. ChT has 13 transmembrane domains with an extracellular N-terminus and an intracellular C-terminus containing several putative phosphorylation sites (Figure 4.16). Phosphorylation serves to regulate the transporter in response to altered synaptic activity. The carboxyl-terminal tail also contains a di-Leu internalization motif (Leu531, Val532) required for the clathrin-mediated endocytosis [136]. Deletion at residues 510 or 550 enhanced the activity to 120% of the wild-type activity supporting the hypothesis that C-terminal residues 560–577 are also involved in the internalization and recycling of the transporter internalized. A tyrosine-based motif (Tyr111-Val112-Thr113-Met114) in the second intracellular loop is also hypothesized to be involved in internalization. Tyr111Ala and Met114Ala mutants do not have uptake activity supporting that hypothesis [135]. Transmembrane domains 4, 11, 12, and 13 contain CRAC-like and

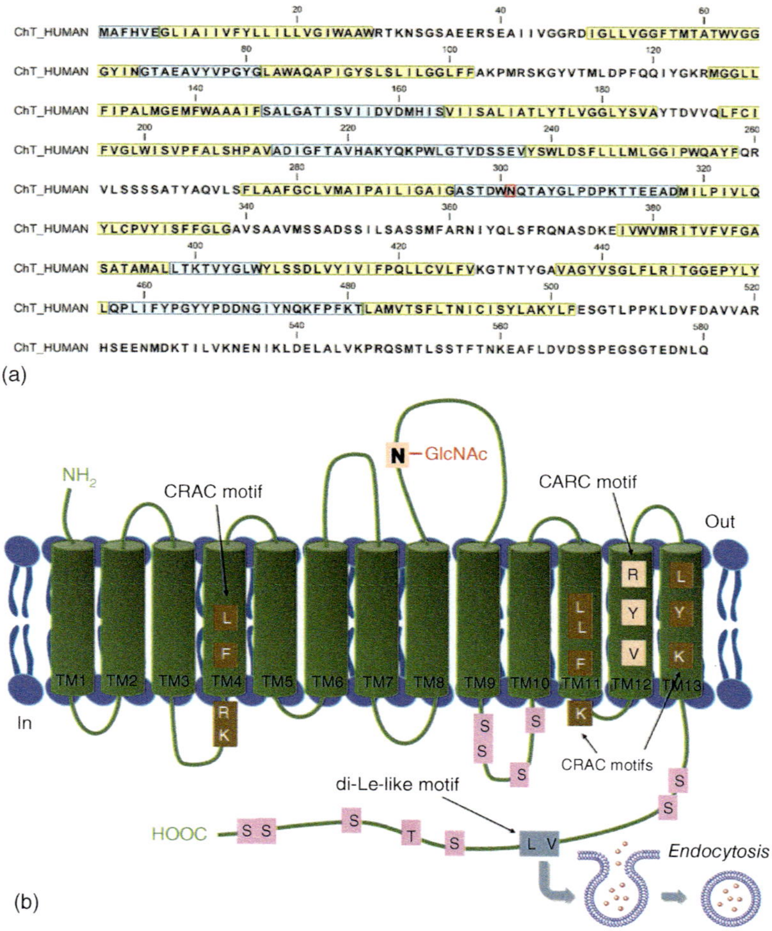

Figure 4.16 (a) Sequence of the human choline transporter. N-glycosylation sites (red), extracellular domains (light blue), intracellular domains (white), and transmembrane domains (yellow) are indicated on sequence. (b) Structure, function, and trafficking of ChT. Amino acids are indicated by one-letter code. Putative phosphorylation sites are indicated in pink. Putative CRAC and CARC motifs correspond to V/L-$X_{[1-5]}$-Y/F-$X_{[1-5]}$-K/R and K/R-$X_{[1-5]}$-Y/F-$X_{[1-5]}$-VL sequences, respectively.

CARC cholesterol binding sequence motifs [137]. Putative dimerization motifs Gly-x-x-x-Gly and Gly-x-x-x-Gly-x-x-x-Gly similar to other sodium-dependent neurotransmitter transporters are present in TM2 and TM12, respectively [138].

A number of natural variants found for ChT have important implications. Asp48Gly, Pro105Ser, Arg361Gln, and Arg446Gly lead to a decrease in the choline transmembrane transporter activity, whereas Gly65Glu produces a complete loss of activity [139]. Ile89Val results in 40% reduction in choline uptake rate [140]. Other natural variants are Tyr111His, Tyr175Cys, Ile291Thr, Val344Leu, and Phe418Val. Mutants Ile81Ala and Glu451Gln resulted in 20% and 5%, respectively, of wild-type choline uptake activity, suggesting a key role of those residues [141]. Polymorphisms in ChT are linked to neurological and psychological disorders, revealing the critical role of the transporter in maintaining cholinergic transmission [142].

Studies with antibodies revealed localization of ChT at presynaptic sites and exclusively in cholinergic neurons. However, a major part of the observed immunoreactivity was associated with intracellular vesicles involved in ACh synthesis, storage and release rather than at the presynaptic plasma membrane, suggesting that ChT is colocalized with the vesicular ACh transporter (VAChT) described above. Experimental evidences suggest that the majority of ChT is found intracellularly within endocytic and synaptic vesicles [143]. Consequently, since depolarization of neurons increases the exocytosis of synaptic vesicles, release of acetylcholine into the synaptic cleft enhances ChT trafficking to the cell surface. Therefore, the increased choline uptake during depolarization is responsible for maintaining high levels of acetylcholine release during neuronal stimulation [144]. Blocking endocytosis leads to ChT accumulation at the plasma membrane and an enhanced capacity for choline uptake [145]. According to these studies, ChT is normally removed from the plasma membrane and recycled. Mutations affecting the gene of ChT causing a defective presynaptic choline transport are the origin of hereditary motor neuropathy [146].

References

1 Hrabovska, A. and Krejci, E. (2014). Reassessment of the role of the central cholinergic system. *J. Mol. Neurosci.* 53 (3): 352–358.
2 Nagasaka, K., Watanabe, Y., and Takashima, I. (2017). Topographical projections from the nucleus basalis magnocellularis (Meynert) to the frontal cortex: a voltage-sensitive dye imaging study in rats. *Brain Stimul.* 10 (5): 977–980.
3 Xiao, C., Cho, J.R., Zhou, C. et al. (2016). Cholinergic mesopontine signals govern locomotion and reward through dissociable midbrain pathways. *Neuron* 90 (2): 333–347.
4 Vitale, F., Mattei, C., Capozzo, A. et al. (2016). Cholinergic excitation from the pedunculopontine tegmental nucleus to the dentate nucleus in the rat. *Neuroscience* 317: 12–22.
5 Mufson, E.J., Counts, S.E., Perez, S.E., and Ginsberg, S.D. (2008). Cholinergic system during the progression of Alzheimer's disease: therapeutic implications. *Expert Rev. Neurother.* 8 (11): 1703–1718.
6 Ferreira-Vieira, T.H., Guimaraes, I.M., Silva, F.R., and Ribeiro, F.M. (2016). Alzheimer's disease: targeting the cholinergic system. *Curr. Neuropharmacol.* 14 (1): 101–115.
7 Londos, E., Brun, A., Gustafson, L., and Passant, U. (2003). Lewy body dementia. Clinical challenges in diagnosis and management. In: *Alzheimer's Disease and Related Disorders: Research Advances* (ed. K. Iqbal and B. Winblad), 133–142. Bucharest, Romania: Ana Asian International Academy of Aging.
8 Förstl, H., Gratz, S., Hahn, U. et al. (2008). Dementia with Lewy bodies and reduced dopamine transporter binding indicates significant cholinergic deficit: therapeutic implications of abnormal dopamine transporter imaging with I123-labelled Ioflupane (DaTSCAN™). *Dtsch. Med. Wochenschr.* 133 (18 Suppl. 1): S11–S14.

9 D'Souza, G.X. and Waldvogel, H.J. (2016). Targeting the cholinergic system to develop a novel therapy for Huntington's disease. *J. Huntington's Dis.* 5 (4): 333–342.
10 Evoli, A., Iorio, R., and Bartoccioni, E. (2016). Overcoming challenges in the diagnosis and treatment of myasthenia gravis. *Expert Rev. Clin. Immunol.* 12 (2): 157–168.
11 Di Bari, M., Di Pinto, G., Reale, M. et al. (2017). Cholinergic system and neuroinflammation: implication in multiple sclerosis. *Cent. Nerv. Syst. Agents Med. Chem.* 17 (2): 109–115.
12 Bohnen, N.I. and Albin, R.L. (2011). The cholinergic system and Parkinson disease. *Behav. Brain Res.* 221 (2): 564–573.
13 Tandon, R., Taylor, S.F., DeQuardo, J.R. et al. (1999). The cholinergic system in schizophrenia reconsidered: anticholinergic modulation of sleep and symptom profiles. *Neuropsychopharmacology* 21 (S2): S189–S202.
14 Sterin-Borda, L. and Borda, E. (2010). Novel insight into cholinergic factors associated with schizophrenia and autoimmunity. *Curr. Immunol. Rev.* 6 (4): 273–286.
15 Tata, A.M., Velluto, L., D'Angelo, C., and Reale, M. (2014). Cholinergic system dysfunction and neurodegenerative diseases: cause or effect? *CNS Neurol. Disord. Drug Targets* 13 (7): 1294–1303.
16 Feldman, R.S., Meyer, J.S., and Quenzer, L.F. (1997). Acetylcholine. In: *Principles of Neuropsychopharmacology*, 246–249. Sunderland, MA: Sinauer Associates.
17 Iversen, L.L., Iversen, S.D., Bloom, F.E., and Roth, R.H. (2009). Acetylcholine. In: *Introduction to Neuropsychopharmacology*, 128–149. New York, NY: Oxford University Press.
18 Corringer, P.-J., Le Novere, N., and Changeux, J.-P. (2000). Nicotinic receptors at the amino acid level. *Annu. Rev. Pharmacol. Toxicol.* 40: 431–458, 432 plates.
19 O'Neill, M.J., Astles, P., Dell, C.P. et al. (2005). Neuronal nicotinic receptors and neuroprotection: newer ligands may help us understand their role in neurodegeneration. *Med. Chem. Rev.-Online* 2 (6): 455–470.
20 Dani, J.A. and Bertrand, D. (2007). Nicotinic acetylcholine receptors and nicotinic cholinergic mechanisms of the central nervous system. *Annu. Rev. Pharmacol. Toxicol.* 47: 699–729.
21 Lippiello, P.M., Bencherif, M., Hauser, T.A. et al. (2007). Nicotinic receptors as targets for therapeutic discovery. *Expert Opin. Drug Discovery* 2 (9): 1185–1203.
22 Taly, A., Corringer, P.-J., Guedin, D. et al. (2009). Nicotinic receptors: allosteric transitions and therapeutic targets in the nervous system. *Nat. Rev. Drug Discovery* 8 (9): 733–750.
23 Hurst, R., Rollema, H., and Bertrand, D. (2013). Nicotinic acetylcholine receptors: from basic science to therapeutics. *Pharmacol. Ther.* 137 (1): 22–54.
24 Lester, R.A.J. (ed.) (2014). *Nicotinic Receptors*, The Receptors, 468. New York, NY: Humana Press, Springer.
25 Clarke, P.B., Schwartz, R.D., Paul, S.M. et al. (1985). Nicotinic binding in rat brain: autoradiographic comparison of [3H]acetylcholine, [3H]nicotine, and [125I]-alpha-bungarotoxin. *J. Neurosci.* 5 (5): 1307–1315.
26 Gotti, C., Clementi, F., Fornari, A. et al. (2009). Structural and functional diversity of native brain neuronal nicotinic receptors. *Biochem. Pharmacol.* 78 (7): 703–711.

27 Caulfield, M.P. and Birdsall, N.J. (1998). International union of pharmacology. XVII. Classification of muscarinic acetylcholine receptors. *Pharmacol. Rev.* 50 (2): 279–290.

28 Eglen, R.M. (2012). Overview of muscarinic receptor subtypes. In: Handbook of Experimental Pharmacology, vol. 208 (ed. A.D. Fryer, N.M. Nathanson and A. Christopoulos), 3–28. Heidelberg: Springer-Verlag https://www.springer.com/us/book/9781493911660.

29 Eglen, R.M. (2005). Muscarinic receptor subtype pharmacology and physiology. In: *Progress in Medicinal Chemistry*, vol. 43 (ed. F.D. King and G. Lawton), 105–136. Amsterdam: Elsevier.

30 Fryer, A.D., Nathanson, N.M., and Christopoulos, A. (2012). *Muscarinic Receptors*, Handbook of Experimental Pharmacology, vol. 208, 513. Heidelberg: Springer-Verlag https://www.amazon.com/Muscarinic-Receptors-Handbook-Experimental-Pharmacology/dp/3642232736.

31 Paterson, D. and Nordberg, A. (2000). Neuronal nicotinic receptors in the human brain. *Prog. Neurobiol.* 61 (1): 75–111.

32 MacDermott, A.B., Role, L.W., and Siegelbaum, S.A. (1999). Presynaptic ionotropic receptors and the control of transmitter release. *Annu. Rev. Neurosci.* 22: 443–485.

33 Molas, S. and Dierssen, M. (2014). The role of nicotinic receptors in shaping and functioning of the glutamatergic system: a window into cognitive pathology. *Neurosci. Biobehav. Rev.* 46 (Pt. 2): 315–325.

34 Maskos, U. (2010). Role of endogenous acetylcholine in the control of the dopaminergic system via nicotinic receptors. *J. Neurochem.* 114 (3): 641–646.

35 Miwa, J., Freedman, R., and Lester, H. (2011). Neural systems governed by nicotinic acetylcholine receptors: emerging hypotheses. *Neuron* 70 (1): 20–33.

36 Albuquerque, E.X., Pereira, E.F., Alkondon, M., and Rogers, S.W. (2009). Mammalian nicotinic acetylcholine receptors: from structure to function. *Physiol. Rev.* 89 (1): 73–120.

37 Gotti, C. and Clementi, F. (2004). Neuronal nicotinic receptors: from structure to pathology. *Prog. Neurobiol.* 74 (6): 363–396.

38 Sakmann, B. (1992). Elementary steps in synaptic transmission revealed by currents through single ion channels. *Biosci. Rep.* 12 (4): 237–262.

39 Nastuk, M.A. and Graybiel, A.M. (1989). Ontogeny of M1 and M2 muscarinic binding sites in the striatum of the cat: relationships to one another and to striatal compartmentalization. *Neuroscience* 33 (1): 125–147.

40 Wang, Y., Gu, Q., Mao, F. et al. (1994). Activity-dependent expression and distribution of M1 muscarinic ACh receptors in visual cortex neuronal cultures. *J. Neurosci.* 14 (7): 4147–4158.

41 Carden, W.B. and Bickford, M.E. (1999). Location of muscarinic type 2 receptors within the synaptic circuitry of the cat visual thalamus. *J. Comp. Neurol.* 410 (3): 431–443.

42 Garzon, M. and Pickel, V.M. (2013). Somatodendritic targeting of M5 muscarinic receptor in the rat ventral tegmental area: implications for mesolimbic dopamine transmission. *J. Comp. Neurol.* 521 (13): 2927–2946.

43 Vilaro, M.T., Palacios, J.M., and Mengod, G. (1990). Localization of m5 muscarinic receptor mRNA in rat brain examined by in situ hybridization histochemistry. *Neurosci. Lett.* 114 (2): 154–159.

44 Radu, B.M., Osculati, A.M.M., Suku, E. et al. (2017). All muscarinic acetylcholine receptors (M1-M5) are expressed in murine brain microvascular endothelium. *Sci. Rep.* 7 (1): 5083.

45 Fasoli, F. and Gotti, C. (2015). Structure of neuronal nicotinic receptors. In: *The Neurobiology and Genetics of Nicotine and Tobacco* (ed. D.J.K. Balfour and M.R. Munafo), 1–17. Heidelberg: Springer-Verlag.

46 Dani, J.A. (2015). Neuronal nicotinic acetylcholine receptor structure and function and response to nicotine. *Int. Rev. Neurobiol.* 124: 3–19.

47 Millar, N.S. (2003). Assembly and subunit diversity of nicotinic acetylcholine receptors. *Biochem. Soc. Trans.* 31 (Pt. 4): 869–874.

48 Taylor, P., Osaka, H., Molles, B. et al. (2000). Contributions of studies of the nicotinic receptor from muscle to defining structural and functional properties of ligand-gated ion channels. In: *Neuronal Nicotinic Receptors*, Handbook of Experimental Pharmacology, vol. 144 (ed. F. Clementi, D. Fornasari and C. Gotti), 79–100. Berlin, Heidelberg: Springer-Verlag.

49 Lindstrom, J.M. (2003). Nicotinic acetylcholine receptors of muscles and nerves: comparison of their structures, functional roles, and vulnerability to pathology. *Ann. N.Y. Acad. Sci.* 998: 41–52 (Myasthenia Gravis and Related Disorders).

50 Sine, S.M., Wang, H.-L., and Gao, F. (2004). Toward atomic-scale understanding of ligand recognition in the muscle nicotinic receptor. *Curr. Med. Chem.* 11 (5): 559–567.

51 Arias, H.R. (2000). Localization of agonist and competitive antagonist binding sites on nicotinic acetylcholine receptors. *Neurochem. Int.* 36 (7): 595–645.

52 Shen, J.-x. and Yakel, J.L. (2009). Nicotinic acetylcholine receptor-mediated calcium signaling in the nervous system. *Acta Pharmacol. Sin.* 30 (6): 673–680.

53 Dajas-Bailador, F. and Wonnacott, S. (2004). Nicotinic acetylcholine receptors and the regulation of neuronal signalling. *Trends Pharmacol. Sci.* 25 (6): 317–324.

54 Bartus, R.T. (2000). On neurodegenerative diseases, models, and treatment strategies: lessons learned and lessons forgotten a generation following the cholinergic hypothesis. *Exp. Neurol.* 163 (2): 495–529.

55 Bacher, I., Rabin, R., Woznica, A. et al. (2010). Nicotinic receptor mechanisms in neuropsychiatric disorders: therapeutic implications. *Primary Psychiatry* 17 (1): 35–41.

56 Unti, E. and Baiamonte, P. (2013). Nicotinic receptors in neuropsychiatric pathologies of adulthood and developmental age. *EuroMediterranean Biomed. J.* 8: 54–59.

57 Ahmed, T., Zahid, S., Mahboob, A., and Farhat, S.M. (2017). Cholinergic system and post-translational modifications: an insight on the role in Alzheimer's disease. *Curr. Neuropharmacol.* 15 (4): 480–494.

58 Aquilonius, S.M. (1980). Cholinergic mechanisms in the CNS related to Parkinson's disease. *Parkinson's Disease Current Progress, Problems Management Processes North Europe Symposium Parkinson's Disease*, 17–27.

59 Posadas, I., Lopez-Hernandez, B., and Cena, V. (2013). Nicotinic receptors in neurodegeneration. *Curr. Neuropharmacol.* 11 (3): 298–314.

60 Berman, J.A., Talmage, D.A., and Role, L.W. (2007). Cholinergic circuits and signaling in the pathophysiology of schizophrenia. *Int. Rev. Neurobiol.* 78: 193–223, 191 plate (Integrating the Neurobiology of Schizophrenia).

61 Woznica, A.A. and George, T.P. (2009). Exploiting nicotinic receptor mechanisms for the treatment of schizophrenia and depression. *J. Dual Diagn.* 5 (2): 159–167.

62 Steinlein, O.K. and Bertrand, D. (2010). Nicotinic receptor channelopathies and epilepsy. *Pflugers Archiv. Eur. J. Physiol.* 460 (2): 495–503.

63 Steinlein, O.K. and Bertrand, D. (2008). Neuronal nicotinic acetylcholine receptors: from the genetic analysis to neurological diseases. *Biochem. Pharmacol.* 76 (10): 1175–1183.

64 Leonard, S., Mexal, S., and Freedman, R. (2007). Smoking, genetics and schizophrenia: evidence for self medication. *J. Dual Diagn.* 3 (3–4): 43–59.

65 de Jaco, A., Bernardini, L., Rosati, J., and Tata, A.M. (2017). Alpha-7 nicotinic receptors in nervous system disorders: from function to therapeutic perspectives. *Cent. Nerv. Syst. Agents Med. Chem.* 17 (2): 100–108.

66 Faure, P., Tolu, S., Valverde, S., and Naudé, J. (2014). Role of nicotinic acetylcholine receptors in regulating dopamine neuron activity. *Neuroscience* 282: 86–100.

67 Quik, M., Zhang, D., McGregor, M., and Bordia, T. (2015). Alpha7 nicotinic receptors as therapeutic targets for Parkinson's disease. *Biochem. Pharmacol.* 97 (4): 399–407.

68 Kouvatsos, N., Giastas, P., Chroni-Tzartou, D. et al. (2016). Crystal structure of a human neuronal nAChR extracellular domain in pentameric assembly: ligand-bound alpha2 homopentamer. *Proc. Natl. Acad. Sci. U.S.A.* 113 (34): 9635–9640.

69 Morales-Perez, C.L., Noviello, C.M., and Hibbs, R.E. (2016). X-ray structure of the human alpha4beta2 nicotinic receptor. *Nature* 538 (7625): 411–415.

70 Palma, E., Bertrand, S., Binzoni, T., and Bertrand, D. (1996). Neuronal nicotinic alpha 7 receptor expressed in Xenopus oocytes presents five putative binding sites for methyllycaconitine. *J. Physiol.* 491 (Pt. 1): 151–161.

71 Zwart, R., Bodkin, M., Broad, L.M. et al. (2004). Common structural and pharmacological properties of serotonin 5-HT3 receptors and α7 nicotinic acetylcholine receptors. In: *Cholinergic Mechanisms* (ed. L. Fuchs), 309–315. Boca Raton: CRC Press.

72 Moroni, M., Vijayan, R., Carbone, A. et al. (2008). Non-agonist-binding subunit interfaces confer distinct functional signatures to the alternate stoichiometries of the alpha4beta2 nicotinic receptor: an alpha4-alpha4 interface is required for $Zn2^{+}$ potentiation. *J. Neurosci.* 28 (27): 6884–6894.

73 Jin, X., Bermudez, I., and Steinbach, J.H. (2014). The nicotinic alpha5 subunit can replace either an acetylcholine-binding or nonbinding subunit in the alpha4beta2* neuronal nicotinic receptor. *Mol. Pharmacol.* 85 (1): 11–17.

74 Zhou, Y., Nelson, M.E., Kuryatov, A. et al. (2003). Human alpha4beta2 acetylcholine receptors formed from linked subunits. *J. Neurosci.* 23 (27): 9004–9015.

75 Du, J., Lu, W., Wu, S. et al. (2015). Glycine receptor mechanism elucidated by electron cryo-microscopy. *Nature* 526 (7572): 224–229.

76 Miller, P.S. and Aricescu, A.R. (2014). Crystal structure of a human GABAA receptor. *Nature* 512 (7514): 270–275.

77 Hulme, E.C., Lu, Z.L., Saldanha, J.W., and Bee, M.S. (2003). Structure and activation of muscarinic acetylcholine receptors. *Biochem. Soc. Trans.* 31 (1): 29–34.

78 Hulme, E.C., Lu, Z.L., Bee, M. et al. (2001). The conformational switch in muscarinic acetylcholine receptors. *Life Sci.* 68 (22/23): 2495–2500.

79 May, L.T., Avlani, V.A., Sexton, P.M., and Christopoulos, A. (2004). Allosteric modulation of G protein-coupled receptors. *Curr. Pharm. Des.* 10 (17): 2003–2013.

80 Keov, P., Sexton, P.M., and Christopoulos, A. (2011). Allosteric modulation of G protein-coupled receptors: a pharmacological perspective. *Neuropharmacology* 60 (1): 24–35.

81 Conn, P.J., Christopoulos, A., and Lindsley, C.W. (2009). Allosteric modulators of GPCRs: a novel approach for the treatment of CNS disorders. *Nat. Rev. Drug Discovery* 8 (1): 41–54.

82 Digby, G.J., Shirey, J.K., and Conn, P.J. (2010). Allosteric activators of muscarinic receptors as novel approaches for treatment of CNS disorders. *Mol. Biosyst.* 6 (8): 1345–1354.

83 Suno, R., Asada, H., and Kobayashi, T. (2015). Towards the crystal structure determination of muscarinic acetylcholine receptors. *Neuromethods* 107: 1–13 (Muscarinic Receptor: From Structure to Animal Models).

84 Kruse, A.C., Hu, J., Kobilka, B.K., and Wess, J. (2014). Muscarinic acetylcholine receptor X-ray structures: potential implications for drug development. *Curr. Opin. Pharmacol.* 16: 24–30.

85 Abrams, P., Andersson, K.E., Buccafusco, J.J. et al. (2006). Muscarinic receptors: their distribution and function in body systems, and the implications for treating overactive bladder. *Br. J. Pharmacol.* 148 (5): 565–578.

86 Ishii, M. and Kurachi, Y. (2006). Muscarinic acetylcholine receptors. *Curr. Pharm. Des.* 12 (28): 3573–3581.

87 Kow, R.L. and Nathanson, N.M. (2012). Structural biology. Muscarinic receptors become crystal clear. *Nature (London, UK)* 482 (7386): 480–481.

88 Yeomans, J.S. (2012). Muscarinic receptors in brain stem and mesopontine cholinergic arousal functions. *Handb. Exp. Pharmacol.* 208: 243–259 (Muscarinic Receptors).

89 El-Fakahany, E.E. and Jakubik, J. (2015). Radioligand binding at muscarinic receptors. *Neuromethods* 107: 37–68 (Muscarinic Receptor: From Structure to Animal Models).

90 Brown, J.H. (1990). *The Receptors*, 478. Clifton, NJ: Humana Press.

91 Myslivecek, J. and Jakubik, J. (2016). *The Receptors*, 287. Clifton, NJ: Humana Press.

92 Wess, J. (1996). Molecular biology of muscarinic acetylcholine receptors. *Crit. Rev. Neurobiol.* 10 (1): 69–99.

93 Nathanson, N.M. (2000). A multiplicity of muscarinic mechanisms: enough signaling pathways to take your breath away. *Proc. Natl. Acad. Sci. U.S.A.* 97 (12): 6245–6247.

94 Scarr, E. (2012). Muscarinic receptors: their roles in disorders of the central nervous system and potential as therapeutic targets. *CNS Neurosci. Ther.* 18 (5): 369–379.

95 Jeon, W.J., Dean, B., Scarr, E., and Gibbons, A. (2015). The role of muscarinic receptors in the pathophysiology of mood disorders: a potential novel treatment? *Curr. Neuropharmacol.* 13 (6): 739–749.

96 Dean, B. and Scarr, E. (2015). Possible involvement of muscarinic receptors in psychiatric disorders: a focus on schizophrenia and mood disorders. *Curr. Mol. Med.* 15 (3): 253–264.

97 Hasselmo, M.E. (2006). The role of acetylcholine in learning and memory. *Curr. Opin. Neurobiol.* 16 (6): 710–715.

98 Hasselmo, M.E. and Giocomo, L.M. (2006). Cholinergic modulation of cortical function. *J. Mol. Neurosci.* 30 (1–2): 133–135.

99 Foster, D.J., Choi, D.L., Jeffrey Conn, P., and Rook, J.M. (2014). Activation of M1 and M4 muscarinic receptors as potential treatments for Alzheimer's disease and schizophrenia. *Neuropsychiatr. Dis. Treat.* 10: 183–191.

100 Foster, D.J., Jones, C.K., and Conn, P.J. (2012). Emerging approaches for treatment of schizophrenia: modulation of cholinergic signaling. *Discov. Med.* 14 (79): 413–420.

101 Wang, J.C., Hinrichs, A.L., Stock, H. et al. (2004). Evidence of common and specific genetic effects: association of the muscarinic acetylcholine receptor M2 (CHRM2) gene with alcohol dependence and major depressive syndrome. *Hum. Mol. Genet.* 13 (17): 1903–1911.

102 Raffa, R.B. (2009). The M5 muscarinic receptor as possible target for treatment of drug abuse. *J. Clin. Pharm. Ther.* 34 (6): 623–629.

103 Thal, D.M., Sun, B., Feng, D. et al. (2016). Crystal structures of the M1 and M4 muscarinic acetylcholine receptors. *Nature* 531 (7594): 335–340.

104 Haga, K., Kruse, A.C., Asada, H. et al. (2012). Structure of the human M2 muscarinic acetylcholine receptor bound to an antagonist. *Nature* 482 (7386): 547–551.

105 Kruse, A.C., Ring, A.M., Manglik, A. et al. (2013). Activation and allosteric modulation of a muscarinic acetylcholine receptor. *Nature* 504 (7478): 101–106.

106 Gregory, K.J., Sexton, P.M., and Christopoulos, A. (2007). Allosteric modulation of muscarinic acetylcholine receptors. *Curr. Neuropharmacol.* 5 (3): 157–167.

107 Patowary, S., Alvarez-Curto, E., Xu, T.R. et al. (2013). The muscarinic M3 acetylcholine receptor exists as two differently sized complexes at the plasma membrane. *Biochem. J.* 452 (2): 303–312.

108 Iverson, H.A., Fox, D. III, Nadler, L.S. et al. (2005). Identification and structural determination of the M(3) muscarinic acetylcholine receptor basolateral sorting signal. *J. Biol. Chem.* 280 (26): 24568–24575.

109 Kruse, A.C., Hu, J., Pan, A.C. et al. (2012). Structure and dynamics of the M3 muscarinic acetylcholine receptor. *Nature* 482 (7386): 552–556.

110 Kruse, A.C., Li, J., Hu, J. et al. (2014). Novel insights into M3 muscarinic acetylcholine receptor physiology and structure. *J. Mol. Neurosci.* 53 (3): 316–323.

111 Rosenbaum, D.M., Cherezov, V., Hanson, M.A. et al. (2007). GPCR engineering yields high-resolution structural insights into β2-adrenergic receptor function. *Science (Washington, DC)* 318 (5854): 1266–1273.

112 Thorsen, T.S., Matt, R., Weis, W.I., and Kobilka, B.K. (2014). Modified T4 lysozyme fusion proteins facilitate G protein-coupled receptor crystallogenesis. *Structure* 22 (11): 1657–1664.

113 Lu, Z.L., Saldanha, J.W., and Hulme, E.C. (2001). Transmembrane domains 4 and 7 of the M(1) muscarinic acetylcholine receptor are critical for ligand binding and the receptor activation switch. *J. Biol. Chem.* 276 (36): 34098–34104.

114 Gentry, P.R., Bridges, T.M., Lamasal, A. et al. (2013). Discovery of ML326: the first sub-micromolar, selective M(5) PAM. *Bioorg. Med. Chem. Lett.* 23 (10): 2996–3000.

115 Gentry, P.R., Kokubo, M., Bridges, T.M. et al. (2014). Development of a highly potent, novel M(5) positive allosteric modulator (PAM) demonstrating CNS exposure: 1-([1*H*-indazol-5-yl]sulfoneyl)-*N*-ethyl-*N*-(2-[trifluoromethyl] benzyl)piperidine-4-carboxamide (ML380). *J. Med. Chem.* 57 (18): 7804–7810.

116 Prado, V.F., Roy, A., Kolisnyk, B. et al. (2013). Regulation of cholinergic activity by the vesicular acetylcholine transporter. *Biochem. J.* 450 (2): 265–274.

117 Arvidsson, U., Riedl, M., Elde, R., and Meister, B. (1997). Vesicular acetylcholine transporter (VAChT) protein: a novel and unique marker for cholinergic neurons in the central and peripheral nervous systems. *J. Comp. Neurol.* 378 (4): 454–467.

118 Erickson, J.D., Varoqui, H., Schafer, M.K. et al. (1994). Functional identification of a vesicular acetylcholine transporter and its expression from a "cholinergic" gene locus. *J. Biol. Chem.* 269 (35): 21929–21932.

119 Giboureau, N., Som, I.M., Boucher-Arnold, A. et al. (2010). PET radioligands for the vesicular acetylcholine transporter (VAChT). *Curr. Top. Med. Chem.* 10 (15): 1569–1583.

120 O'Grady, G.L., Verschuuren, C., Yuen, M. et al. (2016). Variants in SLC18A3, vesicular acetylcholine transporter, cause congenital myasthenic syndrome. *Neurology* 87 (14): 1442–1448.

121 Parsons, S.M. (2000). Transport mechanisms in acetylcholine and monoamine storage. *FASEB J.* 14 (15): 2423–2434.

122 Wimalasena, K. (2011). Vesicular monoamine transporters: structure-function, pharmacology, and medicinal chemistry. *Med. Res. Rev.* 31 (4): 483–519.

123 Khare, P., Mulakaluri, A., and Parsons, S.M. (2010). Search for the acetylcholine and vesamicol binding sites in vesicular acetylcholine transporter: the region around the lumenal end of the transport channel. *J. Neurochem.* 115 (4): 984–993.

124 Varoqui, H. and Erickson, J.D. (1998). The cytoplasmic tail of the vesicular acetylcholine transporter contains a synaptic vesicle targeting signal. *J. Biol. Chem.* 273 (15): 9094–9098.

125 Ferreira, L.T., Santos, M.S., Kolmakova, N.G. et al. (2005). Structural requirements for steady-state localization of the vesicular acetylcholine transporter. *J. Neurochem.* 94 (4): 957–969.

126 Colgan, L., Liu, H., Huang, S.Y., and Liu, Y.-J. (2007). Dileucine motif is sufficient for internalization and synaptic vesicle targeting of vesicular acetylcholine transporter. *Traffic (Oxford, UK)* 8 (5): 512–522.

127 Krantz, D.E., Waites, C., Oorschot, V. et al. (2000). A phosphorylation site regulates sorting of the vesicular acetylcholine transporter to dense core vesicles. *J. Cell Biol.* 149 (2): 379–395.

128 Eiden, L.E., Schaefer, M.K.H., Weihe, E., and Schuetz, B. (2004). The vesicular amine transporter family (SLC18): amine/proton antiporters required for vesicular accumulation and regulated exocytotic secretion of monoamines and acetylcholine. *Pfluegers Arch.* 447 (5): 636–640.

129 Lawal, H.O. and Krantz, D.E. (2013). SLC18: Vesicular neurotransmitter transporters for monoamines and acetylcholine. *Mol. Aspects Med.* 34 (2–3): 360–372.

130 Liu, Y. and Edwards, R.H. (1997). The role of vesicular transport proteins in synaptic transmission and neural degeneration. *Annu. Rev. Neurosci.* 20: 125–156.

131 Tu, Z., Efange, S.M.N., Xu, J. et al. (2009). Synthesis and in vitro and in vivo evaluation of 18F-labeled positron emission tomography (PET) ligands for imaging the vesicular acetylcholine transporter. *J. Med. Chem.* 52 (5): 1358–1369.

132 Apparsundaram, S., Ferguson, S.M., George, A.L. Jr., and Blakely, R.D. (2000). Molecular cloning of a human, hemicholinium-3-sensitive choline transporter. *Biochem. Biophys. Res. Commun.* 276 (3): 862–867.

133 Okuda, T., Haga, T., Kanai, Y. et al. (2000). Identification and characterization of the high-affinity choline transporter. *Nat. Neurosci.* 3 (2): 120–125.

134 Wright, E.M. (2013). Glucose transport families SLC5 and SLC50. *Mol. Aspects Med.* 34 (2–3): 183–196.

135 Haga, T. (2014). Molecular properties of the high-affinity choline transporter CHT1. *J. Biochem.* 156 (4): 181–194.

136 Ribeiro, F.M., Black, S.A.G., Cregan, S.P. et al. (2005). Constitutive high-affinity choline transporter endocytosis is determined by a carboxyl-terminal tail dileucine motif. *J. Neurochem.* 94 (1): 86–96.

137 Cuddy, L.K., Winick-Ng, W., and Rylett, R.J. (2014). Regulation of the high-affinity choline transporter activity and trafficking by its association with cholesterol-rich lipid rafts. *J. Neurochem.* 128 (5): 725–740.

138 Sitte, H.H., Farhan, H., and Javitch, J.A. (2004). Sodium-dependent neurotransmitter transporters: oligomerization as a determinant of transporter function and trafficking. *Mol. Interventions* 4 (1): 38–47.

139 Bauche, S., O'Regan, S., Azuma, Y. et al. (2016). Impaired presynaptic high-affinity choline transporter causes a congenital myasthenic syndrome with episodic apnea. *Am. J. Hum. Genet.* 99 (3): 753–761.

140 Okuda, T., Okamura, M., Kaitsuka, C. et al. (2002). Single nucleotide polymorphism of the human high affinity choline transporter alters transport rate. *J. Biol. Chem.* 277 (47): 45315–45322.

141 Okuda, T., Osawa, C., Yamada, H. et al. (2012). Transmembrane topology and oligomeric structure of the high-affinity choline transporter. *J. Biol. Chem.* 287 (51): 42826–42834.

142 Black, S.A.G. and Rylett, R.J. (2012). Choline transporter CHT regulation and function in cholinergic neurons. *Cent. Nerv. Syst. Agents Med. Chem.* 12 (2): 114–121.

143 Nakata, K., Okuda, T., and Misawa, H. (2004). Ultrastructural localization of high-affinity choline transporter in the rat neuromuscular junction: enrichment on synaptic vesicles. *Synapse* 53 (1): 53–56.

144 Ribeiro, F.M., Black, S.A.G., Prado, V.F. et al. (2006). The "ins" and "outs" of the high-affinity choline transporter CHT1. *J. Neurochem.* 97 (1): 1–12.

145 Ferguson, S.M. and Blakely, R.D. (2004). The choline transporter resurfaces: new roles for synaptic vesicles? *Mol. Interventions* 4 (1): 22–37.

146 Barwick, K.E., Wright, J., Al-Turki, S. et al. (2012). Defective presynaptic choline transport underlies hereditary motor neuropathy. *Am. J. Hum. Genet.* 91 (6): 1103–1107.

5

The Glutamatergic System

5.1 Introduction

The glutamatergic system can be considered ubiquitous in the central nervous system (CNS) having a special role in the modulation of neuronal activity related to processing limbic and cognitive information. In particular, it plays an important role in the regulation of cell activity in the ventral tegmental area (VTA) including dopaminergic outflow [1]. The glutamatergic systems combines three main pathways that are linked to many other neurotransmitter pathways (Figure 5.1): (i) the **corticostriatal** projection, which is organized into two networks, i.e. the *orbital network,* involved in processing sensory information relating to food and feeding, and the *medial network*, related to emotional motor functions [2, 3]; (ii) the **amygdalostriatal** projection arising from various components of the basal and accessory basal nuclear complexes [4] and mostly targeting ventromedial striatum [5], and (iii) the **hippocampostriatal** projection for which the main source is the subiculum although additional minor inputs can come from parasubiculum, prosubiculum, and CA1 and CA3 regions [6].

There are other glutamate projections between the cortex, substantia nigra, subthalamic nucleus, and pallidum. The glutamate projections from the VTA to the lateral habenula are involved in aversive conditioning [7]. The VTA contains largely segregated populations of dopaminergic and nondopaminergic neurons that project to various cortical and subcortical brain structures, including the nucleus accumbens and the prefrontal cortex. The ventral hippocampus, entorhinal cortex, and amygdala, most likely via polysynaptic pathways that involve projections to the ventral striatum, are an additional source of glutamatergic projections that influence VTA neurons. These routes allow glutamate exerting direct control on midbrain dopaminergic neurons.

Glutamate is synthesized from glutamine in the mitochondria by the action of glutaminase. The synthesized glutamate is internalized in the vesicles by the ATP-dependent vesicular glutamate transporter (VGLUT). The VGLUT uses a vacuolar ATPase to cleave ATP to provide the energy required to produce and maintain a high proton gradient that allows to reach an extremely high concentration of glutamate in the vesicles. Like other neurotransmitters, vesicular release of glutamate takes place in a calcium-dependent manner following depolarization of the presynaptic neuron. Once released into the synapse following an

Chemical Biology of Neurodegeneration: A Molecular Approach, First Edition. Pedro Merino.

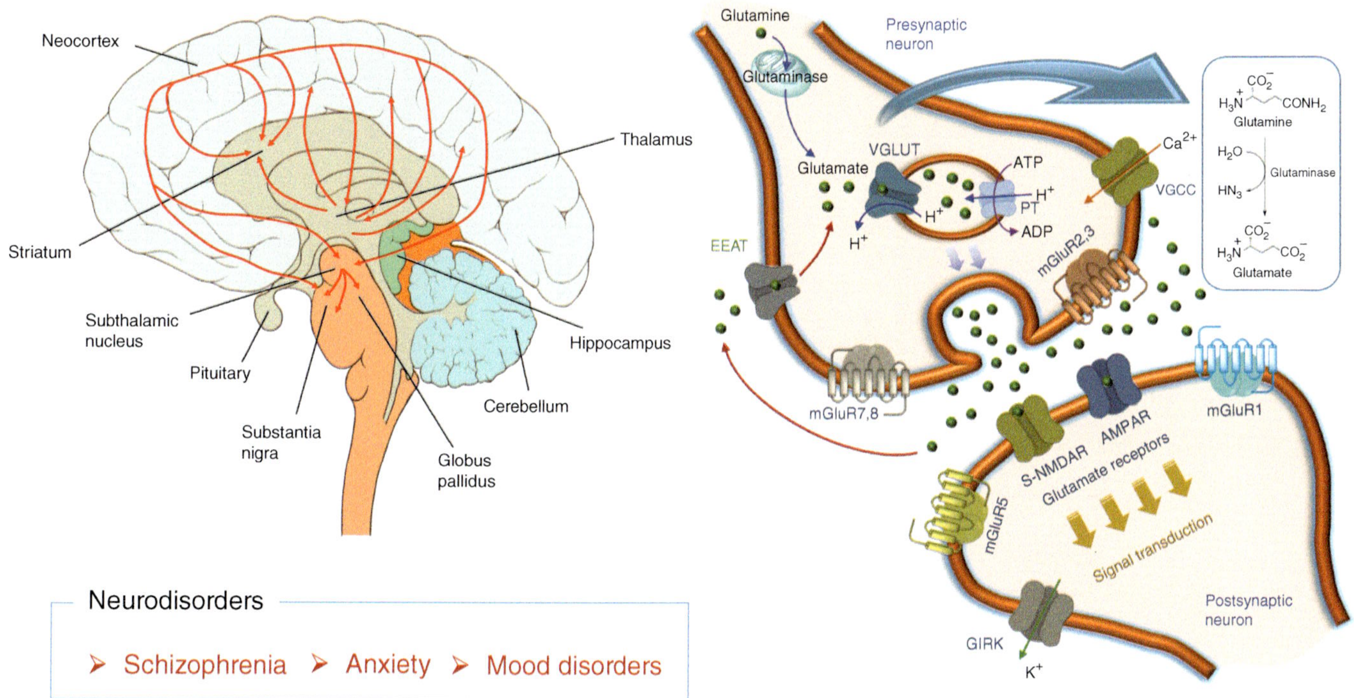

Figure 5.1 The glutamatergic system. Top right: Major projections of glutamatergic system. Top left: Neurological disorders associated with glutamatergic system. Bottom: Details of the glutamatergic synapse: synthesis of glutamate from glutamine in the mitochondria, absorption into synaptic vesicles by VGLUT, release of glutamate, binding of glutamate to receptors, and reuptake of choline by EEAT. VGCC: Voltage-gated calcium channel; GIRK: G-protein-coupled inwardly rectifying potassium channel; VGLUT: Vesicular glutamate transporter; EAAT: Excitatory amino acid transporter; NMDAR: *N*-methyl-D-aspartate receptor; AMPAR: α-amino-3-hydroxy-5-methyl-4-isoxazolepropionic acid receptor; mGluR: metabotropic glutamate receptor; PT: proton transporter.

action potential, glutamate binds to a number of postsynaptic and presynaptic targets within the synapse that comprises various glutamate receptors. Among them, ionotropic receptors NMDA and α-amino-3-hydroxy-5-methyl-4-isoxazole propionic acid (AMPA) facilitates the influx of sodium and calcium ions into the cell. Metabotropic receptors (mGluRs) modulate glutamate signaling by reuptaking glutamate from the synaptic cleft (Figure 5.1). To maintain homeostasis within the glutamate synapse, neighboring astrocytes, modulated by glycine (not shown in the figure), remove glutamate via excitatory amino acid transporters (EAATs) that internalize glutamate and coupled to the cotransport of a proton and three sodium ions within the astrocyte and extrusion of one potassium ion. After the conversion of glutamate into glutamine by the action of glutamine synthetase within the astrocytes, glutamine is released into the extracellular space, then it enters the neuron and starts a new synaptic cycle.

Several neurological and psychiatric diseases such as schizophrenia are related to an abnormal regulation of the glutamatergic synapse [8]. Glutamatergic cortico-cortical connections may be affected in schizophrenia. Abnormal activity of prefrontal glutamatergic plays a key role in various psychiatric disorders [9]. Perturbations or dysfunction of the aforementioned mechanisms of the glutamatergic synapse to control the glutamate concentrations in the extracellular space can lead to the release of glutamate in excess, inducing hyperexcitability in postsynaptic neurons and causing cytotoxicity [10]. The overactivity of the glutamatergic system, termed excitotoxicity, is responsible for death of neurons observed after ischemia (stroke) or epileptic episodes. The neuronal death is a consequence of the massive calcium entry resulting from depolarization. Excitotoxicity is suggested to be involved in the pathogenesis of Huntington's disease [11, 12] and amyotrophic lateral sclerosis (ALS).

5.2 Glutamate Receptors

5.2.1 Classification

Glutamate receptors belong to a large family of EAARs. More than 20 glutamate receptors have been identified in the mammalian CNS, which are classified into two main categories: ionotropic (voltage sensitive) and metabotropic (ligand sensitive). For a comprehensive description of glutamate receptors, the reader is directed to a published treatise [13]. Several reviews dedicated to glutamate receptors have been published [14–18], most of which are devoted to one of the two types, so they are indicated in the appropriate sections.

Each ionotropic or metabotropic receptor has three types, depending on binding specificity, ion permeability, conductance properties, and other factors [19]. The different types and subunits are listed in Table 5.1 using the current nomenclature and old names.

The classification of ionotropic receptors is based on the affinity for the ligands N-methyl-D-aspartate (NMDA), AMPA, and kainic acid (Scheme 5.1). Ionotropic receptors are thus divided into NMDA receptors and non-NMDA receptors, which include AMPA and kainate receptors (KARs). AMPA receptors include

Table 5.1 Glutamate receptors[a].

Type	Receptor	Subunits	Common names
Ionotropic	NMAD	GluN1	NMDAR1, NR1, GluRε1
		GluN2A	NMDAR2A, NR2A, GluRε1
		GluN2B	NMDAR2B, NR2B, GluRε2
		GluN2C	NMDAR2C, NR2C, GluRε3
		GluN2D	NMDAR2D, NR2D, GluRε4
		GluN3A	NR3A
		GluN3B	NR3B
	AMPA	GluA1	GluR1, GluRA
		GluA2	GluR2. GluRB
		GluA3	GluR3, GluRC
		GluA4	GluR4, GluRD
	Kainate	GluK1	GluR5
		GluK2	GluR6
		GluK3	GluR7
		GluK4	KA1
		GluK5	KA2
	Delta	GluD1	δ1, GluR delta-1
		GluD2	δ2, GluR delta-2
Metabotropic	Group 1	mGluR1	
		mGluR5	
	Group 2	mGluR2	
		mGluR3	
	Group 3	mGluR4	
		mGluR6	
		mGluR7	
		mGluR8	

a) The currently accepted IUPHAR nomenclature is used along this book. For detailed discussions on nomenclature of ionotropic glutamate receptors, see Refs [20–22].

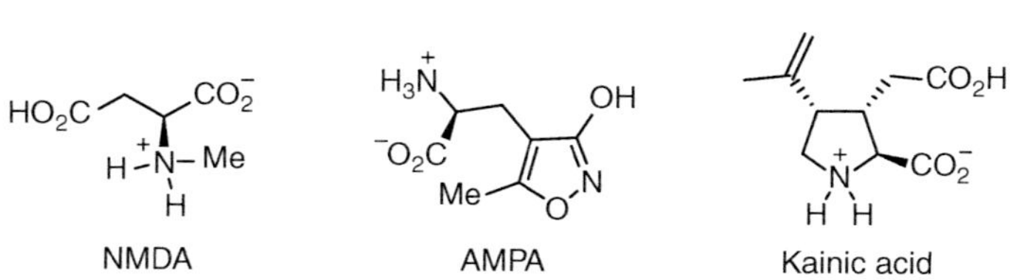

Scheme 5.1 Selective ligands for glutamate receptors.

subunits GluR1, GluR2, GluR3, and GluR4. KARs include subunits GluR5, GluR6, GluR7, GluR8, KA-1, and GluR KA-2, but they do not contribute significantly to synaptic transmission and thus will not be discussed in detail in this chapter.

AMPA-selective receptors are tetrameric assemblies of the corresponding subunits. Two groups have been traditionally defined based on the inclusion or exclusion of the Glu2 receptor subunit. A third subclass based on the variation of stoichiometry of existing subunits has been suggested [23].

Eight different metabotropic receptors categorized in three groups have been identified to date. Classification of metabotropic receptors is based on pharmacology, sequence similarity, and intracellular effector signaling systems. Receptors of Group 1 include mGlu1 and mGlu5; receptors of Group 2 are mGlu2 and mGlu3, and Group 3 comprises receptors mGlu4, mGlu6, mGlu7, and mGlu8.

5.2.2 Ionotropic Receptors

Ionotropic receptors have been receiving great attention. A book on the main aspects of ionotropic receptors was published in 1997 [24]. Two years later, a book devoted to ionotropic receptors in the CNS was published [25]. The last collective volume on both types of glutamate receptors was published in 2008 [13]. Several reviews covering general [22] and specific aspects such as origin and molecular evolution [26], pharmacology [27], genomics [28], molecular structure [29], regulation [30], and activation [31], among others, have been reported.

The ligand-gated ion channel receptors (ionotropic) [32] are composed of homo- or heterotetramers of individual subunits that confer cation selectivity. Functional receptors are exclusively formed subunits within the same functional receptor class. Each subunit has an extracellular N-terminus with three transmembrane domains (TMDs), an intramembrane reentrant "p-loop" between the first and third TMDs, and an intracellular C-terminus (Figure 5.2). These receptors comprise the glutamate receptors family, which is different from the receptor Cys-loop superfamily that includes nicotinic, 5-HT_3, and $GABA_A$ and glycine receptors. The main difference between the two families is that ionotropic glutamate receptors form tetramers and the receptors belonging to the Cys-loop family form pentamers. Other important differences include the presence of a disulfide bond in the N-terminus of the Cys-loop superfamily and the shorter length of the second transmembrane (TM2) domain (involved in ion conduction and selectivity) in glutamate ionotropic receptors [20]. Ionotropic glutamate receptor subunits have four defined domains: (i) the extracellular N-terminal domain, (ii) the intracellular C-terminal domain, (iii) the extracellular ligand-binding domain (LBD), and (iv) the TMD. Detailed crystal structures are discussed in the following sections.

5.2.2.1 Location

Ionotropic receptors are widely distributed in the CNS not only in neurons but also in astrocytes [33]. Seven NMDA receptor subunits (N1, N2A, N2B, N2C, N2D, N3A, and N3B) have been found in astrocytes showing expression at

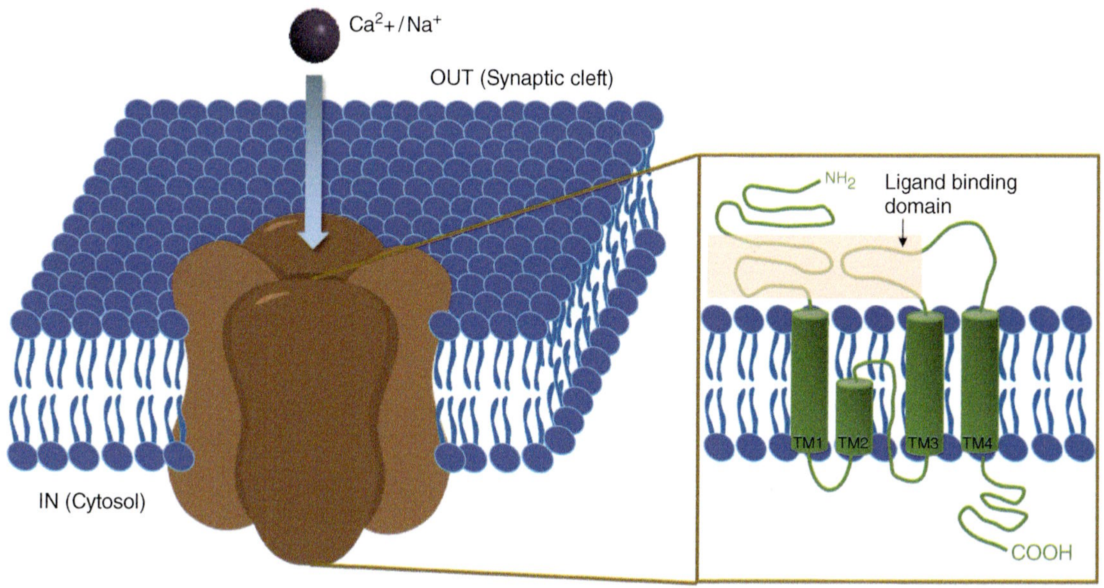

Figure 5.2 General structure of a glutamate ionotropic receptor formed by four subunits, each of which has three transmembrane domains and one intramembrane domain.

different levels [34]. NMDA receptors are widely distributed in the cerebral cortex [35]; in the brain, they are also localized in primary visual cortices [35] and the lower auditory pathway [36].

AMPA receptors are predominantly postsynaptic receptors, widely distributed in the cortex and ventral striatum and in temporal lobe structures such as the hippocampus and amygdala, with lower levels in the thalamus. Of the three ionotropic receptors in the CNS, AMPA receptors occur at the greatest density mainly in the cerebral cortex and hippocampus. They are also located in the anterior olfactory nucleus, caudate putamen, nucleus accumbens, and the lateral septum, and, in a lesser extent, in the olfactory bulb, inferior colliculus, and thalamus [37]. mRNAs for GluR1, GluR2, and GluR3 are widely expressed in the CNS. mRNA for GluR4 is expressed in the cerebellar cortex, dentate gyrus, CA1 region of the hippocampus, and the olfactory bulb [38].

KARs are predominantly presynaptic and regulate glutamate release. They are distributed mainly in the hippocampus and the infragranular layers of the cortex. KAR agonists include kainic and domoic acid.

5.2.2.2 Effector Mechanisms

Ionotropic glutamatergic receptors open cation-permeable channels to mediate sodium (Na^+), potassium (K^+), or calcium (Ca^{2+}) ions flow by increasing the sodium influx and potassium efflux. In addition, NMDA receptors also increase calcium influx due to their high calcium permeability [39]. In general, NMDA receptors are composed of N1 and N2 subunits. Those containing N3 subunits play a negative modulatory role in channel function [40]. NMDA receptors containing N2A subunits are characterized by fast response kinetics (lasting tens of milliseconds), while those containing N2B subunits display slow response kinetics (lasting hundreds of milliseconds).

Activation of the NMDA receptors requires binding of two agonists: glutamate to the N2 subunit and glycine to a binding site on the N1 subunit [41]. Although posttranslational modifications (operating via intracellular C-terminal domains) are the major regulators of glutamate trafficking of NMDA receptors, mutagenesis experiments have demonstrated that gating features of ionotropic receptors also control glutamate trafficking relevant to development and plasticity [42]. The activation of NMDA receptors is associated with divergent signaling to neuronal survival or death depending on whether the receptor is located synaptically or extrasynaptically [43]. Indeed, both areas correlate, and there are various mechanisms shared between synaptic plasticity and neuronal survival and death [44]. The different activation of the mitogen-activated protein (MAP) kinases extracellular signal-regulated kinase (ERK) and p38 by synaptic and extrasynaptic NMDA receptors is one of the mechanisms responsible for such dichotomy. The subunit diversity of NMDA receptors has a high impact on receptor activities and effector mechanisms [45]. The intracellular C-terminal domain of NMDA subunits is involved in receptor trafficking and couples receptors to signaling cascades.

In AMPA receptors, the presence of the subunit GluR2 prevents the transport of calcium. On the other hand, when the subunit GluR2 is not present, calcium is efficiently conducted through the ion channel. In neurons, GluR2 exists only in the edited form. KARs have low calcium permeability. The NMDA and AMPA ionotropic glutamate receptors are closely related [46]. At negative membrane potentials, NMDA receptors are blocked by extracellular magnesium, requiring AMPA-R depolarization to relieve this block.

5.2.2.3 Implication in Neurodisorders

The association between ionotropic receptors and CNS disorders has been reviewed [47]. Specifically, NMDA receptors are involved in neurodegeneration following cerebral ischemia, Alzheimer's disease (AD), and Huntingdon's disease (HD), as well as a range of CNS dysfunctional conditions such as epilepsy, Parkinson's disease (PD), schizophrenia, and anxiety [48]. The high calcium permeability of NMDA receptors is suggested to be responsible for the involvement of those receptors in several neurodisorders associated with learning and memory, including AD [49]. Glutamate excitotoxicity is also an important factor in vascular dementia [50] and ALS [51]. NMDA receptors play an important role in schizophrenia [52, 53], and glycine modulation has emerged as a therapeutic target [54]. NMDA receptors are also involved in the pathophysiology of bipolar disorder [55]. NMDA receptors located in the lower auditory pathway play different roles in auditory synaptic development, plasticity, and temporal processing, thus being involved in neurological auditory diseases.

Defects in AMPA receptor trafficking are important to fragile X mental retardation, and ectopic expression of KAR synapses contributes to the pathology of temporal lobe epilepsy [47].

5.2.2.4 NMDA Receptors

NMDA receptors are heterotetramers composed of a number of subunit forms (GluN1, GluN2A, GluN2B, GluN2C, GluN2D, GluN3A, and GluN3B) that confer unique pharmacological properties to each receptor [56]. *N*-methyl-D-aspartate receptors (NMDARs) comprise two N1 subunits and a variety of GluN2A, B, C, and D subunits. Figure 5.3 collects the sequences of the subunits corresponding to the NMDA receptors. The total number of residues ranges from 938 (GluN1) to 1484 (GluN2B), and the size difference is almost entirely due to the length of the intracellular C-terminal domain.

The GluN1 subunit presents several natural variants with important consequences. Among them, Arg217Trp increases inhibition by zinc [57]. Pro557Arg, Tyr647Ser, and Gly815Arg induce activity loss in the glutamate-gated calcium ion channel [57, 58]. Gly618Arg and Gly620Arg cause loss of function in calcium ion transmembrane import into cytosol [59]. A significant calcium influx is produced by Glu662Lys [60]. Other natural variants that are not affecting glutamate-gated ion calcium ion channel activity include Ala645Ser and Serr688Tyr.

Many natural variants have also been observed in the subunit GluN2A. Reduced zinc affinity is observed for Ala243Val [61]. Cy2436Arr, Pro552Arg, Lys669Asn, and Pro699Ser induce increased glutamate potency, whereas Gly483Arg, Val685Gly, Ala716Thr, Ala727Thr, Val734Leu, and Lys772Glu have the opposite effect, decreasing glutamate potency [62]. Decreased protein abundance and localization to the cell membrane have been observed for Cys436Arg, Arg504Trp, Arg518His, Thr531Met, Ile694Thr, Met705Val, Ala716Thr, Ala727Thr,

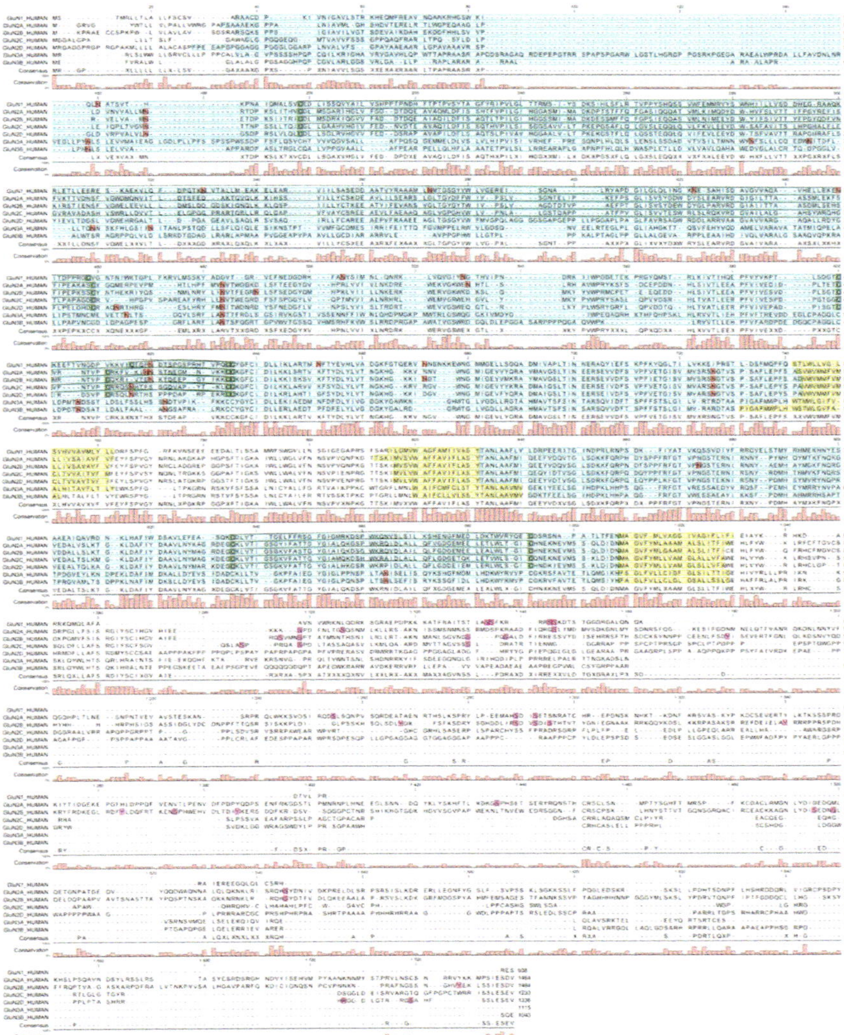

Figure 5.3 Sequences of the human NMDA receptor subunits. Conservation graphic and consensus sequence are shown. Disulfide bonds (green), N-glycosylation sites (red), phosphorylation (magenta) sites, extracellular domains (light blue), intracellular domains (white), and transmembrane domains (yellow) are indicated on sequences.

and Lys772Glu [61]. Some natural variants with unknown pathological significance such as Ile184Ser, Cys231Tyr, Ala290Val, Gly295Ser, Arg370Trp, Ile814Thr, Asp933Asn, Asn976Ser, and Asp1251Asn have also been observed.

For the subunit GluN2B, only the natural variant Arg696His has been associated with increased glutamate and glycine potency. Glu413Gly has been associated with decreased glutamate potency [62]. Decreased protein abundance and localization to the cell membrane in the case of GluN2B have been observed for Glu413Gly, Cys456Tyr, Cys456Phe, Arg540His, and Arg696His. Activity loss of glutamate-gated calcium ion channel has been associated with Pro553Leu [58] and increased calcium permeability with Asn615Ile and Val618Gly [63]. Other natural variants for GluN2A include Val15Met, Val18Ile, Ile50Asn, Ala271Val, Leu362Met, Ser407Asn, Gly820Asp, Gly1026Ser, Met1342Arg, Ser1415Leu, Pro1439Ala, and Ser1452Phe [64].

The subunit GluN2C presents less natural variants, most of which (Val90Ile, Ala166Val, Met573Thr, Ala641Thr, Arg679Cys, Ile863Thr, Gln871Arg, Ser911Ile, Pro982Ala, Ser989Pro, Ser995Leu, His1079Tyr, Pro1141Ala, Gly1180Arg, Thr1196Ile, and Arg1209Ser) are of unknown pathological significance. Similarly, GluN2D shows natural variants (Leu466Val, Met592Leu, Met733Val, Arg872His, Met883Ile, Ala922Val, and Ala926Thr) without a clear association to an effect with the exception of Val667Ile, which potentiates ionotropic glutamate receptor signaling [64].

The subunits GluN3A and GluN3B also present natural variants of unknown pathological significance. Those found for GluN3A are Ser72Ile, Arg111Gly, Arg313Trp, Val362Met, Gly373Arg, Ile375Thr, Ser434Pro, Arg480His, Gly487Arg, Tyr493Phe, Asn565Ser, Lys585Asn, Val641Leu, Arg793Ser, Asp835Asn, Thr921Met, Tyr956His, Ile1028Asn, and Arg1041Gln. For GluN3B, Thr157Met, Arg180Cys, Arg194Trp, Arg247Gln, Arg247Trp, Ala285Gly, Glu332Ly, Arg336Cys, Arg350His, Pro381Leu, Arg404Trp, Trp414Arg, Arg470is. Leu472Pro, Gly515Ser, Thr577Met, Ala583Val, Arg598Cys, Arg608His, Thr612Ala, Thr641Met, Ser678Leu, Ala845Thr, Glu919Asp, Val928Met, and Ala948Val.

As illustrated in Figure 5.3, the different NMDA receptors subunits have multiple glycosylation and phosphorylation sites (Table 5.2).

The X-ray structure of Ca^{+2}/calmodulin bound to GluN1 subunits (PDB ID: 2HDW) showed functional differences between homologous domains [65]. The X-ray structure of the human GluN1/GluN2A LBD has been resolved in the *apo*

Table 5.2 Glycosylation and phosphorylation sites for NMDA subunits.

Subunit	Glycosylation sites[a)]	Phosphorylation sites[b)]
GluN1	61/203/239/276/300/350/368/440/471/491	889/890/896/897
GluN2A	75/340/380/443/444/541	882/890/929/1025/1059/1062/1198/1291
GluN2B	74/341/348/444/491/542/688	882/886/917/920/962[c)]/1039[c)]/1058/1061/1064/1109[c)]/1133[c)]/1143/1155[c)]/1255/1259/1303/1474[c)]
GluN2C	70/337/438/539	875/881/912
GluN2D	92/352/366/467/569	1326
GluN3A	145/264/275/286/296/426/439/549/565/886	
GluN3B	69/344/451/465/786	

a) In all cases, residues are Asn.
b) In all cases, residues are Ser, unless otherwise indicated.
c) The residue is Thr.

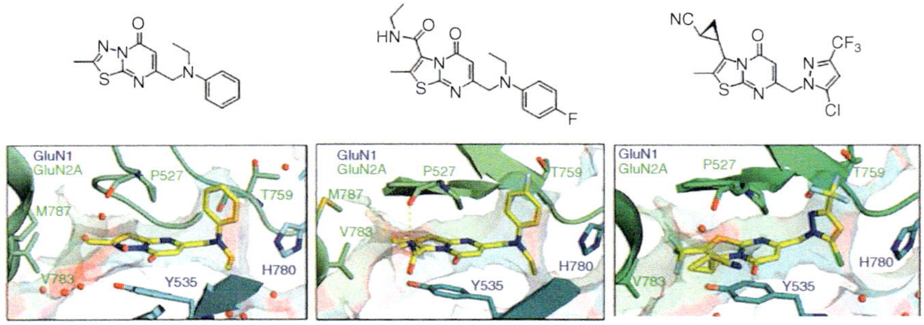

(a) (b) (c)

Figure 5.4 Crystal structure of ligands a, b, and c bound to the GluN1/GluN2A ligand-binding domain (PDB IDs: 5H8H and 5I2K, respectively). Solvent-accessible surface rendering depicts the binding site with relevant amino acids denoted by their single amino acid code. Blue coloring represents residues from GluN1 and green coloring represents residues from GluN2A. Source: Reproduced with permission from Yuen et al. [67]. Copyright 2016, American Chemical Society.

form (PDB ID: 5H8F) and served to identify a new modulatory site [66], which facilitated the resolution of other structures in complex with a variety of ligands (PDB IDs: 5H8H, 5H8N, 5TP9, 5KCJ, and 5H8Q) [66–68] (Figure 5.4).

X-ray crystal structures of the *Xenopus laevis* GluN1–GluN2B NMDA receptor in complex with the allosteric inhibitor, Ro25-6981, partial agonists, and the ion channel blocker, MK-801, have been resolved [69]. The structure of the receptor presents an overall twofold axis of symmetry, and the receptor domains are organized into three layers: (i) the amino terminal domain (ATD) layer at the top; (ii) the LBD layer in the middle, and the TMD layer at the bottom (Figure 5.5a,b). While GluN1 ATDs reside at the periphery of the receptor assembly, GluN2B ATDs are present in the R2 lobes of subunits proximal to each other and near the overall twofold axis of symmetry (Figure 5.5c,d).

The agonist-binding LBDs are organized as closely equivalent pair GluN1-GluN2B heterodimers (Figure 5.5e) resembling isolated GluN1-GluN2B LBDs, as well as homodimeric assemblies of AMPA and kainate receptors (see below). The X-ray structure provides some information on how allosteric inhibition promotes the closure of the ion channel gate despite the binding of agonists. The ATDs are placed in a position that allows mediating conformational changes at both LBD intradimer and interdimer interfaces. However, since the intradimer LBD interface is intact, the action of allosteric inhibitors should not necessarily involve rupture of that interface, and other conformational changes involving LBD layer may be involved in promoting the inactive conformation observed in the X-ray structures.

5.2.2.5 AMPA Receptors

The AMPA receptors are tetrameric complexes composed of either homo- or heteromeric combinations of four subunits: GluA1 GluA2, GluA3, and GluA4. Figure 5.6 collects the sequences of subunits corresponding to AMPA receptors.

As illustrated in Figure 5.6, the different AMPA receptors subunits have multiple glycosylation, phosphorylation, and palmitoylation sites (Table 5.3).

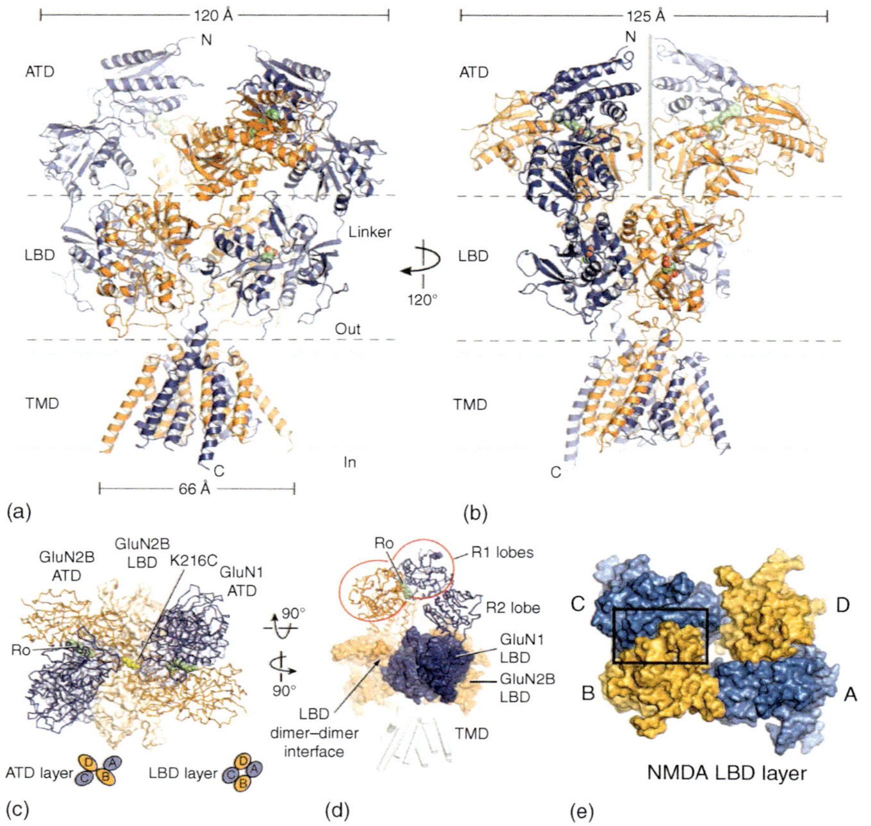

Figure 5.5 Architecture, symmetry and domain organization of the GluN1–GluN2B NMDA receptor. (a) View of the receptor complex, parallel to the membrane, with the GluN1 subunits in blue and the GluN2B subunits in orange. The ligands Ro25-6981, ACPC, and *t*-ACBD are in space-filling representation. (b) View of the complex rotated 120° around the overall twofold axis of the receptor. The approximate position of the overall twofold axis is shown by a vertical gray bar in the center of the ATD layer. (c) View of the ATD layer along the overall twofold axis, from the extracellular side of the membrane, centered on the overall twofold axis, and showing the relative location of the underlying LBD layer. Ro25-6981 (Ro) is green and the K216C disulfide is yellow. The arrangements of the subunits for ATD and LBD layers are shown as insets. (d) The inverted ATD heterodimeric "V" straddles GluN1 and GluN2B LBD subunits on different local LBD heterodimers. The ATDR2 lobes interact with the LBDs and the R1 lobes cradle bound Ro25-6981 at an ATD subunit interface. (e) View from the extracellular side of the membrane, along the overall twofold axis of the receptor, showing the LBDs of the GluN1 and GluN2B subunits, with the LBD heterodimer interface of the B–C subunits emphasized by a box. Source: Reproduced with permission from Lee et al. [69]. Copyright 2014, Springer Nature.

Natural variants of the GluA1 subunit include Asp487Asn, Pro521Thr, Ala536Ser, Ile548Met, Phe588Leu, and Ala636Thr – all of them are of unknown pathological significance. Variants of the GluA2 subunit include Gln607Arg and Glu608Arg [70]. Up to four different isoforms have been described for subunits GluA1 and GluA2. The subunit GluA3 presents natural variants (Arg631Ser and

Figure 5.6 Sequences of the human AMPA receptor subunits. Conservation graphic and consensus sequence are shown. Disulfide bonds (green), N-glycosylation sites (red), phosphorylation (magenta) sites, extracellular domains (light blue), intracellular domains (white), and transmembrane domains (yellow) are indicated on sequences.

Table 5.3 Glycosylation and phosphorylation sites for AMPA subunits.

Subunit	Glycosylation sites[a)]	Phosphorylation sites[b)]	Palmitoylation sites[c)]
GluA1	63/249/257/363/401/406	645/710/849/863	603
GluA2	256/370/406/413	683/717/860/863/876[d)]/880	610/836
GluA3	63/266/380/415/422	663/728/877[d)]/887	621
GluA4	56/258/371/407/414	862	611/837

a) In all cases, residues are Asn.
b) In all cases, residues are Ser, unless otherwise indicated.
c) In all cases, residues are Cys.
d) The residue is Tyr.

Met706Thr) for which homomers have minimal or no current but heteromers have altered desensitization. The Gly833Arg variant reduces receptor expression. Natural variants Arg450Gln and Phe525Leu are of unknown pathological significance [71]. This subunit presents two isoforms differing in residues 776-811 [72]. The subunit GluA4 can present two isoforms in which 434–902 residues are missing and 424–433 are replaced by an alternative sequence.

The different subunits have different residues involved in glutamate binding. In the subunit GluA1, the residues are 492–494 and 668–669. For the subunit GluA2, the residues involved in glutamate binding are 499–501 and 675 and 676. In the subunit GluA3, the residues are 508–510 and 686–687. In the subunit GluA4, the residues are 500–502 and 676–677. Residues 903–906 of the subunit GluA1 are involved in PDZ binding. The TM4 domain of the subunits mediates tetramerization and is required for cell surface expression.

The X-ray structure of the common 400-residue ATD has been resolved for the human GluA2 receptor in two crystal forms (PDB IDs: 2WJW and 2WJX) [73]. GluA2 is dimeric in solution but crystallizes as a tetramer (dimer–dimer). Positive modulators bind GluA2 to the twofold axis of the dimer as illustrated by the crystal structure of *N*-[(2*S*)-5-(6-fluoro-3-pyridinyl)-2,3-dihydro-1H-inden-2-yl]-2-propanesulfonamide (FPIPS) in complex with human GluA2 S1S2 LBD (PDB ID: 2XHD) (Figure 5.7a) [74]. Similarly, phenyl-1,4-bisalkylsulfonamide (CMPDA) and CMPDB, two positive modulators of GluA2, bind within the same modulator-binding pocket (PDB IDs: 3RNN and 3RN8, respectively) (Figure 5.7b) [75]. Quinazolinedione sulfonamides are also representative examples of AMPA receptor antagonists that also bind the LBD (PDB IDs: 3R7X and 3UA8) [76, 77]. A comparison of binding modes of positive allosteric modulators (PAMs), like GNE-6901 (PDB ID: 5H8S), and negative allosteric modulators (NAMs) revealed that NAMs bind to a distinct but overlapping region of the interdomain interface as the PAMs (Figure 5.7c–e) [66]. These studies showed some flexibility for the residues of the binding site, identified similar but unique binding sites for GluA2 allosteric modulators and resulted in the discovery of highly potent and brain-penetrant GluA2-selective allosteric agonists [67, 68].

5.2.2.6 Kainate Receptors

KARs are composed of subunits GluK1, GluK2, GluK3, GluK4, and GluK5. Figure 5.8 collects the sequences of the subunits corresponding to KARs.

As illustrated in Figure 5.8, the different kainate receptor subunits have multiple glycosylation and phosphorylation sites (Table 5.4).

Natural variants of the GluK1 subunit include Ala332Val, Gln636Arg, Ile751Val, Arg862Gln, Ala870Val, and Leu902Ser [78]. For this subunit, a second isoform has been reported with missing residues 402–416 and an alternative

Figure 5.7 (a) Crystal structure of FPIPS in complex with hGluA2 S1S2 ligand-binding domain. The compound (yellow carbons) binds at the dimer interface; a glutamate (red, solid) is bound at the core of each ligand-binding domain. Source: Reprinted (adapted) with permission from Ward et al. [74]. Copyright 2010, American Chemical Society. (b) FPIPS: *N*-[(2*S*)-5-(6-fluoro-3-pyridinyl)-2,3-dihydro-1H-inden-2-yl]-2-propanesulfonamide. CMPDA: phenyl-1,4-bisalkylsulfonamide. CMPDB: phenyl-1,4-bis-carboxythiophene. (c) View of the PAM-binding site from an unbound apo structure of the human GluN1/GluN2A LBD. Key interface residues are shown as stick models, and residues that move upon PAM binding (GluN1 Y535 and GluN2A E530) are highlighted in magenta. Solvent-accessible surface area in the interface is shown as a transparent surface. (d) View of GNE-6901 bound in the PAM-binding site of the GluN1/GluN2A LBD. (e) The GluN2A-selective NAM bound to GluN1/GluN2A LBD. Source: (c–e) Hackos et al. 2016 [66]. Reprinted with permission of Elsevier.

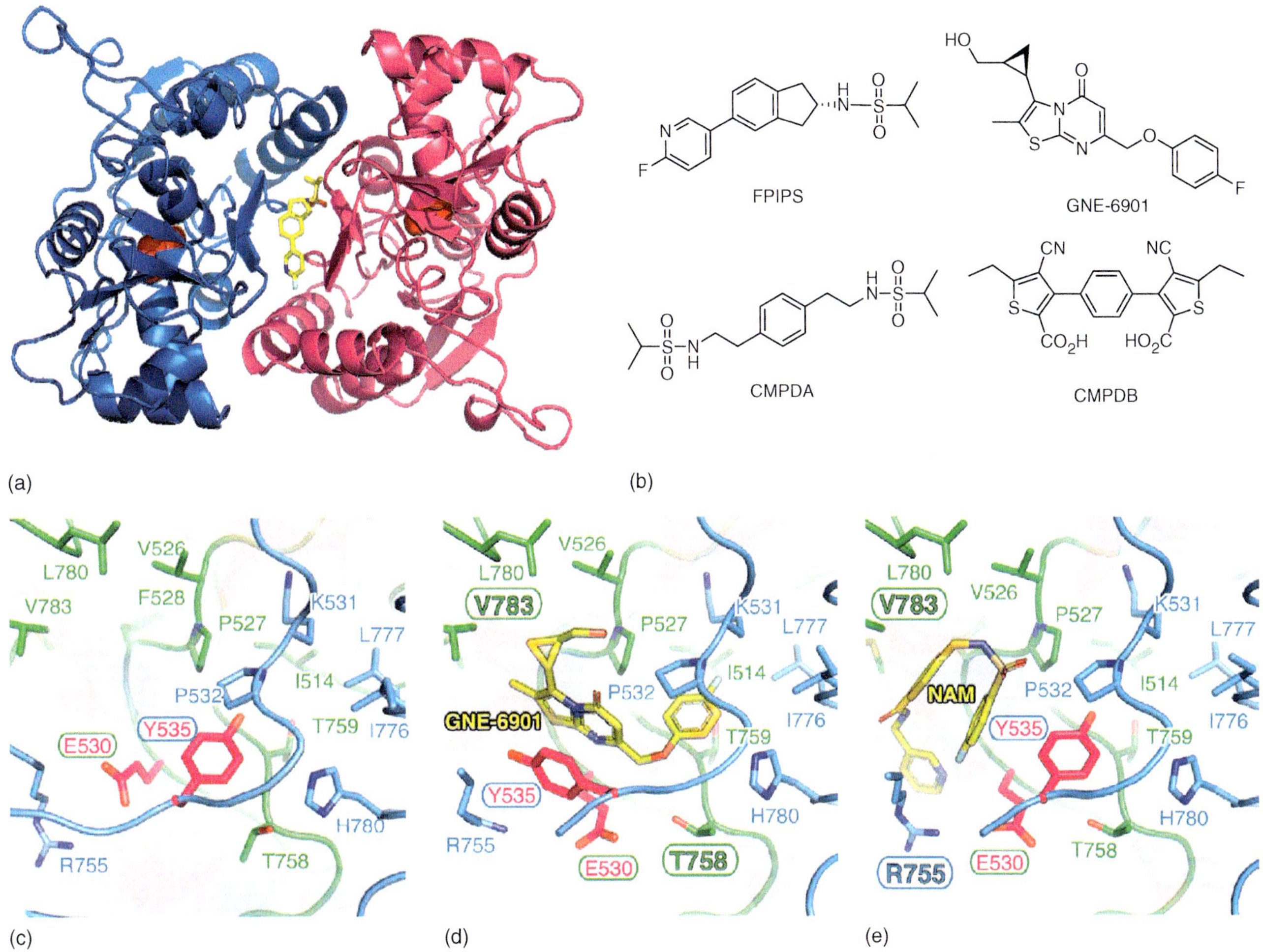

(a)
(b)
FPIPS
GNE-6901
CMPDA
CMPDB
(c)
L780
V783
V526
F528
P527
K531
L777
I514
P532
E530
Y535
T759
I776
H780
R755
T758
(d)
L780
V783
V526
P527
K531
L777
I514
P532
GNE-6901
I776
T759
Y535
H780
R755
E530
T758
(e)
L780
V783
V526
P527
K531
L777
I514
P532
NAM
Y535
I776
T759
H780
R755
E530
T758

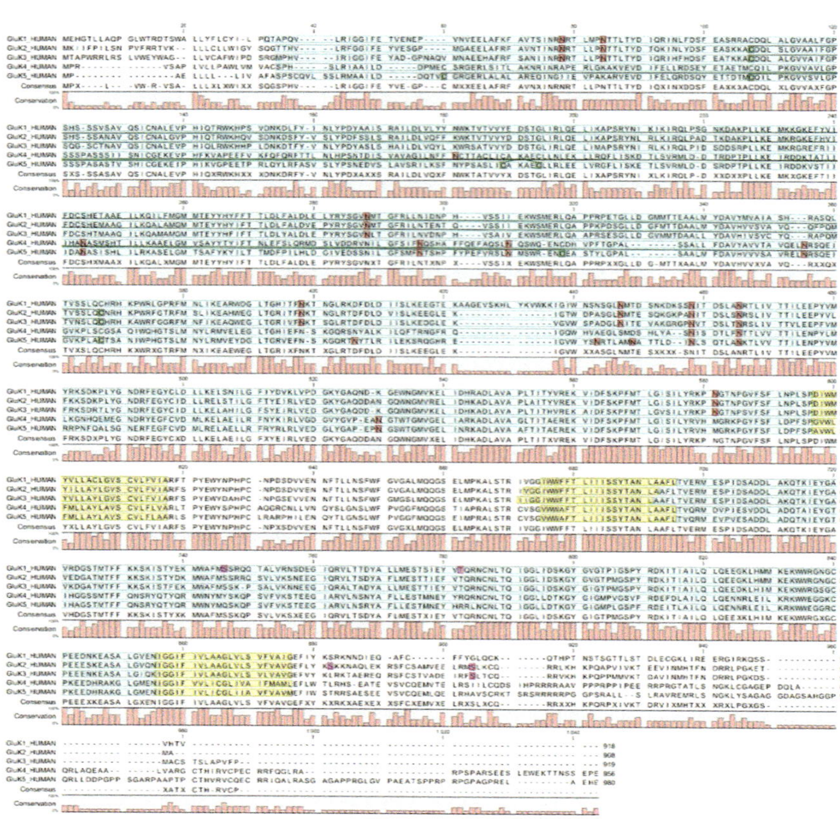

Figure 5.8 Sequences of the human kainate receptor subunits. Conservation graphic and consensus sequence are shown. Disulfide bonds (green), N-glycosylation sites (red), phosphorylation (magenta) sites, extracellular domains (light blue), intracellular domains (white), and transmembrane domains (yellow) are indicated on sequences.

Table 5.4 Glycosylation and phosphorylation sites for kainate subunits.

Subunit	Glycosylation sites[a)]	Phosphorylation sites[b)]
GluK1	68/74/276/379/428/439/446/561	725/761[c)]
GluK2	67/73 275/378/412/423 /430/546	868/886[c)]
GluK3	70/76/278/381/415/426/433/548	869
GluK4	158/220/272/286/323/408/415/479	
GluK5	219/271/285/322/372/394/400/407/414/478	

a) In all cases, residues are Asn.
b) In all cases, residues are Ser, unless otherwise indicated.
c) The residue is Thr.

sequence for residues 870–918 [79]. Glutamate-binding regions have been identified in residues 531–533 and 704–705. Natural variants of the GluK2 subunit include Glu187Gln, Ile567Val, Tyr571Cys, Gln621Arg, Ala657Thr, Val766Ile, and Met867Ile [80]. A total of seven isoforms have been identified for the subunit GluK2 with missing residues (585–908 for isoform 3; 547–622 for isoform 4; 509–695 for isoform 6; 510–714 for isoform 7) and alternative sequences (855–908 for isoforms 1 and 6; 856–908 for isoforms 5 and 7) [81]. Glutamate-binding regions have been identified in residues 516–518 and 689–690. Natural variants of the GluK3 subunit include Arg215His, Ser310Ala, Arg352Gln, and Asp391His [82]. A second isoform has been identified with an alternative sequence in residues 856–919. Glutamate-binding regions have been identified in residues 690–692. Natural variants of the GluK4 subunit include Val528Ile and Met824Thr [83]. For GluK5, a second isoform has been suggested with an alternative sequence in residues 839–980 [84].

The subunits GluK1, GluK2, and GluK3 form functional homo- and heteromeric receptors; GluK4 and GuK5 subunits require the presence of GluK1, GluK2, and GluK3 subunits to properly assemble into functional receptors. A crystallographic analysis of the mutant GluK2 Tyr490Cys Leu752Cys revealed some relaxation from the active conformation [85], although it is not sufficient to trigger desensitization, a universal feature of ligand-gated ion channels.

The binding sites of GluK1 and GluK2 have been characterized by crystallographic studies of bound forms with a series of dysiherbaine and neodysiherbaine analogues (PDB IDs: 2ZNS, 2ZNT, 2ZNU, 3FUZ, 3FV1, 3FV2, 3FVN, and 3FVO) [86]. The studies revealed that differences in three amino acids (Thr503, Ser706, and Ser726 in GluK1 and Ala487, Asn690 and Thr710 in GluK2) in the ligand-binding pocket generate differences in the binding modes of neodysiherbaine to GluK1 and GluK2. Binding sites of human GluK1 and GluK2 are very similar, sharing 87% sequence identity. Important differences include a larger binding cavity for GluK1 and diverse noncovalent interactions observed between ligand-bound GluK1 and GluK2. It was found that H-bond interactions between Glu409 and Asn690 for GluK1 (Glu411 and Asn721 for GluK2), associated to a positive relationship between channel opening and domain closure, are not so determinant of the degree of the channel opening since small ligands such as K-glutamate and kainate do not induce any conformational changes in the Asn residue. Quinoline derivatives showed similar binding modes for GluK1 [87].

The structural basis of kainate subunit desensitization has been established by determining the structure of the KAR GluK2 subtype in its desensitized state by cryo-electron microscopy (cryo-EM) at 3.8 Å resolution (Figure 5.9a,b) [88]. Desensitization is featured by a ring-like structure adopted in the LBD layer of the receptor. Adoption of such a disposition is promoted by staggered helix contacts between adjacent subunits. The arrangement produces subtle changes in symmetry, acting as a molecular switch, that are important for the gating mechanism. Disruption of the desensitization ring, upon agonist unbinding and opening of the individual LBDs, enables restoration of the receptor to its resting state, thereby completing the gating cycle (Figure 5.9c).

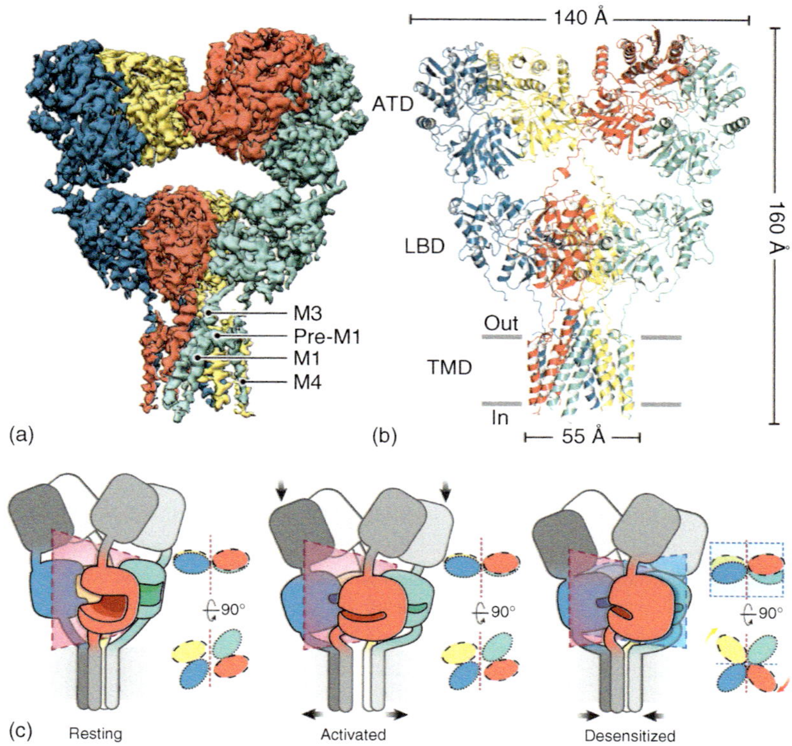

Figure 5.9 Cryo-EM density map (a) and atomic model (b) of 2*S*,4*R*-4-methylglutamate-bound GluK2, with each chain colored uniquely and (a) has features shown from two map contours. (c) Schematic view of structural rearrangements that are involved in transition from resting to activated to desensitized states of glutamate receptors. Source: Reprinted with permission from Meyerson et al. [88]. Copyright 2016, Springer Nature.

5.2.2.7 Delta Receptors

Delta receptors are comprised of subunits GluD1 and GluD2. Figure 5.10 collects the sequences of the subunits corresponding to delta receptors. The human GluD1 subunit has 1009 residues and comprises three helical domains. N-glycosylation with GlcNAc takes place at Asn200, Asn422, and Asn498. A second isoform has been reported with residues 1–429 missing. A natural variant in which Val529 is replaced by Ile has been found. The human GluD2 subunit has 1007 residues and comprises three helical domains. N-glycosylation with GlcNAc takes place at Asn293 and Asn426. Phosphorylation can take place at Ser883, Ser886, Ser890, and Ser1007. A second isoform has been reported with residues 82–176 missing. Natural variants include Thr68Met, Thr209Asn, Phe398Ser, Val490Ile, Ala654Asp, Ala654Thr, and Leu656Val [89]. A PDZ-binding motif is identified in residues 1005–1007.

The structural basis for the integration of delta receptors within synaptic organizer complexes has been elucidated through the crystal structure of Cbln1, a soluble synaptic organizer molecule belonging to the C1q–tumor necrosis factor α (C1q-TNFa) superfamily, in complex with GluD2 [90]. In contrast to previ-

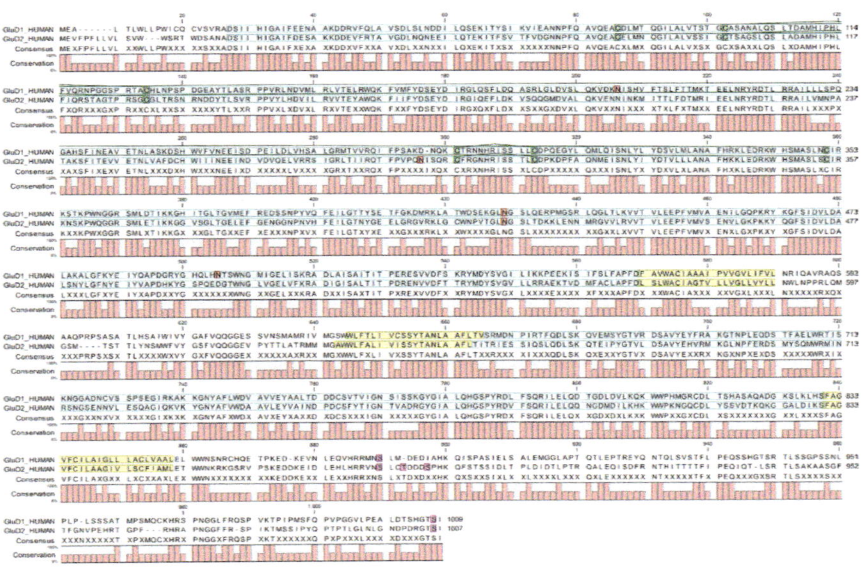

Figure 5.10 Sequences of the human delta receptor subunits. Conservation graphic and consensus sequence are shown. Disulfide bonds (green), N-glycosylation sites (red), phosphorylation (magenta) sites, extracellular domains (light blue), intracellular domains (white), and transmembrane domains (yellow) are indicated on sequences.

ous studies with GluA and GluK receptors suggesting a marked conformational variability in the ATD layer, this study proposes that, in a transsynaptic context, anchoring of GluD to the Cbln1 complex will limit or prevent large-scale motions of the ATD layer.

5.2.3 Metabotropic Receptors

The metabotropic glutamate receptors are G-protein-coupled receptors (GPCRs) that play crucial roles in the modulation of synaptic transmission and neuronal excitability. Readers interest in comprehensive description of these receptors and their functions are referred to reviews and treatises published elsewhere [91–95]. The physiological roles of these receptors have also been reviewed extensively [96–100].

5.2.3.1 Location

Metabotropic glutamate receptors are widely expressed in the CNS in both neurons and glia [101]. Group I metabotropic receptors are mostly expressed on the postsynaptic membrane and regulate learning and memory [102]. Group II metabotropic receptors are found on both postsynaptic and presynaptic cells. Group III metabotropic receptors are mostly presynaptic and found within the hippocampus and hypothalamus. In fact, all mGlu receptors exist in the hippocampus and entorhinal cortex except mGluR6 [103]. In particular, mGluR4 is located only in the brain, in locations such as the cerebellum, thalamus, hypothalamus, and caudate nucleus. mGluR7 is expressed in many areas of the brain, especially in the cerebral cortex, hippocampus, and cerebellum [104, 105].

5.2.3.2 Effector Mechanisms

Metabotropic glutamate receptors modulate synaptic transmission through coupling with G proteins, which regulates voltage-sensitive calcium channels, G-protein-regulated inwardly rectifying potassium channels, $GABA_A$ receptors, AMPA receptors, and NMDA receptors [106, 107]. In fact, the classification of three different groups is based on their different effector mechanisms. Group I mGlu receptors are activated by 3,5-dihydroxyphenylglycine and modulated by extracellular calcium [108]. They are coupled to $G_{q/11}$, which activates phosphoinositide-specific phospholipase C_β (PI-PLC) to produce inositol-1,4,5-trisphosphate (IP_3) and diacylglycerol. The increase of cytoplasmic IP_3 leads to the release of calcium from cytoplasmic stores, activating protein kinase C (PKC). The presence of diacylglycerol can also activate adenylate cyclase to increase the concentration of c-AMP. Different activities have been detected for the different subtypes of group I mGlu receptors and their splice variants. Indeed, these receptors can modulate additional signaling pathways including other cascades downstream of G_q as well as pathways stemming from $G_{i/o}$, G_s, and other molecules independent of G proteins. Depending on the cell type or neuronal population, group I mGluRs can activate a range of downstream effectors, including phospholipase D and protein kinase pathways [109]. Group I mGlu receptors are usually excitatory, causing more glutamate to be released from the presynaptic cell. They can also increase the inhibitory postsynaptic potential. Group II and group III mGlu receptors are coupled to $G_{i/o,}$ and their activation inhibits adenylate cyclase, thereby decreasing the concentration of c-AMP. Group II mGlu receptors are involved in presynaptic inhibition and do not appear to affect the postsynaptic membrane potential by themselves. Group II and group III mGlu receptors also couple to other signaling pathways, including activation of mitogen-activated protein kinase (MAPK) and phosphatidyl inositol 3-kinase PI3 kinase pathways [110].

Activation of presynaptic mGlu receptors blocks both excitatory glutamatergic and inhibitory GABAergic synaptic transmission in a variety of central structures including the hippocampus, olfactory tract, spinal cord, and thalamus. Presynaptic mGlu receptors regulate synaptic transmission through an unclear mechanism that could involve presynaptic calcium channels.

Postsynaptic mGlu receptors modulate several ligand- and voltage-gated ion channels expressed in central neurons. Activation of all three classes of mGlu receptor inhibits L-type voltage-dependent calcium channels, whereas activation of group I and II mGlu receptors inhibits N-type calcium channels. Depending on the component of the signal transduction mechanism is affected and the tissue in which takes place, the activation of mGlu receptors can inhibit or potentiate a given ionotropic receptor [111]. Thus, activation of mGlu receptors closes voltage-dependent, calcium-dependent potassium channels in the hippocampal and other cortical neurons through an unclear mechanism, whereas in cerebellar granule cells, mGlu receptor activation inhibits NMDA-receptor-induced increase of intracellular calcium through a mechanism involving PKC.

Several proteins interact directly with the C-terminal tails of different mGluR subtypes playing important roles in regulating mGluR signaling. The most well characterized are Homer proteins, containing PDZ1 domains that interact with the last amino acids, Pro-Pro-XX-Phe of mGluR1a, mGluR5a, and mGluR5b [112, 113]. A variety of interactions have also been identified between group II and III mGlu receptors and a variety of proteins. Among them, interactions with Ran binding protein in the microtubule-organizing center (RanBPM) have been identified [114], calmodulin and protein phosphatase 2C (PP2C) have been shown to interact with the C-terminus of mGluR3 [115], and isoforms of mGluR8 interact with a series of proteins reducing their ability to inhibit c-AMP production [116].

5.2.3.3 Implication in Neurodisorders

Metabotropic glutamate receptors are implicated in important neurodegenerative disorders such as AD, PD, and HD [117]. Metabotropic glutamate receptors are key modulators of glutamatergic dysfunction in PD [118], and they are considered targets for PD therapy [119]. For instance, allosteric modulations of mGLu5 offer neuroprotection that reduces L-DOPA induced dyskinesia in PD patients [120]. Group I metabotropic receptors are involved in problems with learning and memory, addiction, motor regulation, and, particularly, AD and fragile X syndrome [91, 102, 121]. Figure 5.11 illustrates dysfunctions on signaling

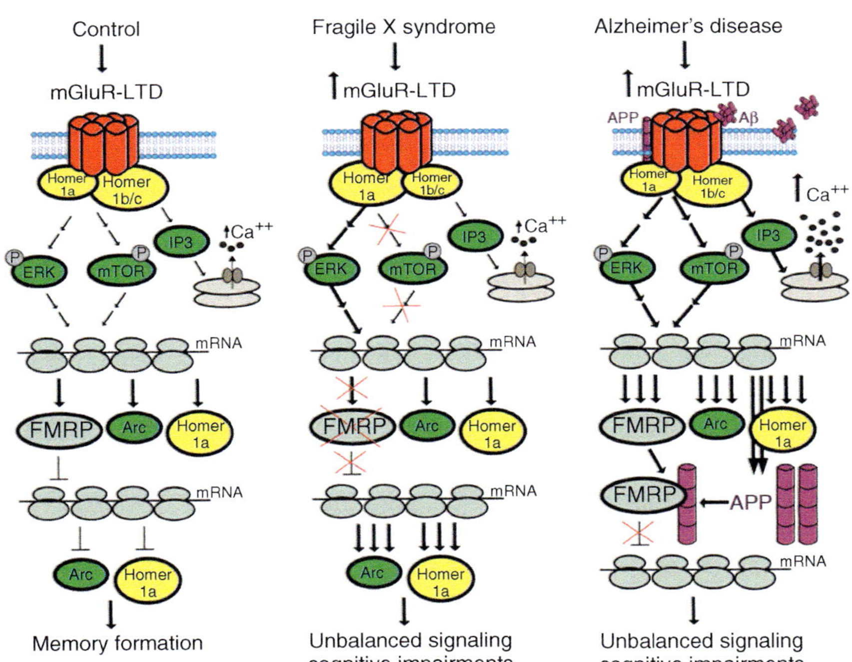

Figure 5.11 Comparison of mGluR-related signaling pathway activation in successful cognitive aging, fragile X syndrome, and Alzheimer's disease. Source: Reproduced with permission from Menard and Quirion [121]. Copyright 2012, https://creativecommons.org/licenses/by/4.0/.

pathways responsible for the last two aforementioned disorders. Activation of group 1 mGluR mediates calcium release from intracellular stores and protein translation through the activation of kinases ERK and mammalian target of rapamycin (mTOR). Once translated, Fragile X Mental Retardation Protein (FMRP) proteins act as a negative regulator of protein synthesis and reduce long term depression (LTD)-related proteins levels and immediate early gene (IEG) expression. A good balance of these mechanisms leads to memory formation in the aged brain. Excessive mGluR-LTD has been linked to the fragile X syndrome. While ERK activation is enhanced, protein synthesis related to the mTOR pathway is lost. Without FMRP to decrease the protein translation, LTD-related proteins are overexpressed and memory is impaired. Excessive mGluR-LTD has also been reported in Alzheimer's disease. Aβ oligomers cluster mGluRs on the plasma membrane promoting calcium signaling and release from intracellular stores. Stimulation of group 1 mGluR also increases amyloid precursor protein (APP) translation. FMRP can bind to APP possibly altering the protein synthesis negative control and creating memory deficits.

Dysfunction of group II metabotropic receptors has been implicated in anxiety, schizophrenia, and Alzheimer's disease [122]. Group III metabotropic receptors play a role in PD [123] and anxiety disorders [124]. Single nucleotide polymorphisms associated with altered activities of metabotropic receptors are closely related to bipolar disorder [125]. The signal transduction pathways of metabotropic glutamate receptors have been related to temporal lobe epilepsy [126]. The allosteric modulation of metabotropic receptors has demonstrated efficacy in animal models used for the evaluation of cognitive deficits that occur in schizophrenia [127].

5.2.3.4 Structural Features

As members of the superfamily of G-protein-coupled receptors, mGlu receptors have the typical structural features including an extracellular N-domain, seven transmembrane-spanning domains and an intracellular C-terminus. In particular, mGlu receptors belong to class C of GPCRs characterized by a large extracellular N-terminal domain, named Venus flytrap domain (VFD), that contains the endogenous glutamate-binding site [128]. Each VFD consists of two lobes that sit one atop the other and bind glutamate in a cleft between them [129]. Two VFDs dimerize together, back to back, and large conformational changes are induced when agonists bind to one or both VFDs; the relative orientation of these domains is of importance for receptor activation [130]. Three main states of the VFD dimer exist: open–open, open–closed, and closed–closed (Figure 5.12). The open–open (inactive) conformation is stabilized by antagonists; the open–closed and closed–closed conformations are induced by the binding of ligand to one or two protomers. The signal induced by ligand binding is transmitted from VFDs through cysteine-rich domains formed by nine Cys residues, eight of which are linked by disulfide bridges [128]. It has been suggested that intracellular loop i2 plays a role in regulating G-protein specificity [131, 132]. The i2 loop is also a site for regulation by kinases [133]. The C-termini of mGlu receptors are crucial regions for modu-

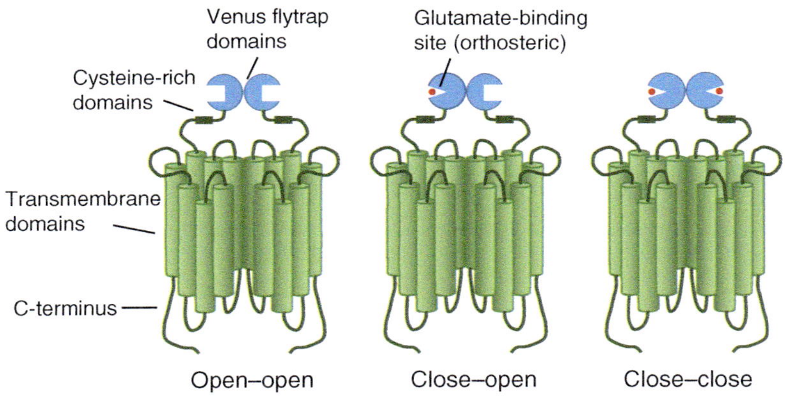

Figure 5.12 Metabotropic glutamate receptors (dimers) in the three possible activity states. The open–open state is the inactive one.

lating G-protein coupling. Indeed, several mGlu receptors' subtypes show alternative splicing generating diverse C-terminal tails that have different effects [134].

Metabotropic receptors are divided into three subfamilies consisting of eight subtypes as listed in Table 5.1. Figure 5.13 collects the sequences of the different metabotropic receptors.

As illustrated in Figure 5.13, the different metabotropic receptors have multiple glycosylation and phosphorylation sites (Table 5.5).

Group 1 Receptors Group 1 of metabotropic receptors is formed by mGluR1 and mGluR5 receptors.

Natural variants of mGluR1 receptors include Ser34Tyr, Tyr262Cys, Arg285Lys, Ser593Phe, Arg696Trp, Glu741Asp, Tyr792Cys, Gly884Glu, Val929Ile, and Ser993Phe. For this receptor, two isoforms have been reported with missing residues 907–1194 (isoform 2) and 909–1194 (isoform 3) and alternative sequences for residues 887–906 (isoform 2) and 887–908 (isoform 3). Glutamate-binding region has been identified in residues 186–188. For mGluR5, two isoforms (canonical is isoform 1) have been reported with missing residues 877–908 (isoform 1) and 896–1165 (isoform 3).

The structure of the human mGluR1 bound to the NAM 4-fluoro-*N*-(4-(6-(isopropylamino)pyrimidin-4-yl)thiazol-2-yl)-*N*-methylbenzamide (FITM) has been resolved (PDB ID: 4OR2) [135]. The crystallographic studies show some differences between mGluR1, a class C GPCR, and other families of GPCRs, although the architecture of the seven transmembrane bundle is generally preserved. The modulator binding site partially overlaps the orthosteric binding sites of class A GPCRs, but it is more restricted. Dimerization of mGluR1 is mediated by molecules of cholesterol; interactions between TMDs within the full-length receptor dimer might mediate the signaling initiated by glutamate recognition at the extracellular domain. The structure of mGluR1 complexed with antagonist LY341495 has also been deposited (PDB ID: 3KS9). Important

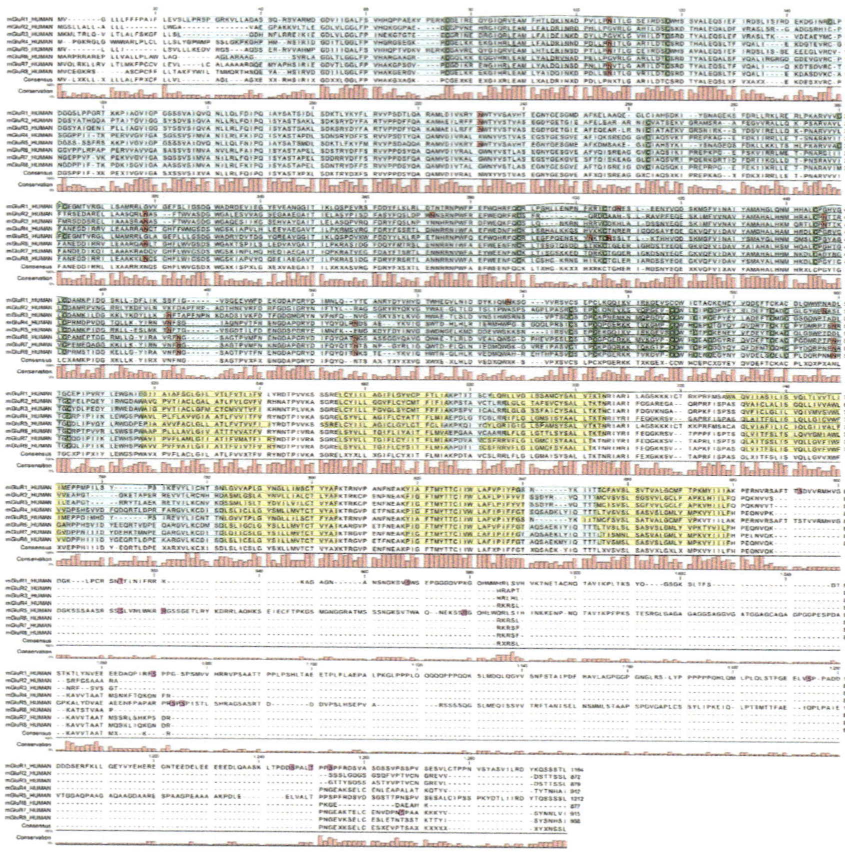

Figure 5.13 Sequences of the human metabotropic receptors. Conservation graphic and consensus sequence are shown. Disulfide bonds (green), N-glycosylation sites (red), phosphorylation (magenta) sites, extracellular domains (light blue), intracellular domains (white), and transmembrane domains (yellow) are indicated on sequences.

Table 5.5 Glycosylation and phosphorylation sites for metabotropic receptors.

Subunit	Glycosylation sites[a)]	Phosphorylation sites[b)]
mGluR1	98/397/515	853/871[c)]/894 969/1091/1142/1146[c)]/1149
mGluR2	203/286/338/402 /547	
mGluR3	2090/291/414/439	
mGluR4	98/301/454/484/569	
mGluR5	88/210/378/382/445/734	861/1018/1020
mGluR6	296/451/479/567	
mGluR7	98/458/486/572	900
mGluR8	95/298/452/489/565	

a) In all cases, residues are Asn.
b) In all cases, residues are Ser, unless otherwise indicated.
c) The residue is Thr.

information on the architecture of the TMD of class C GPCRs has been obtained by resolving the structure of such a domain for mGluR5 in complex with the antagonist mavoglurant (PDB ID: 4OO9) [136]. These studies are in close agreement with the structure of mGluR1 and show the precise location of the allosteric binding sites within the TMD and key microswitches that regulate receptor signaling (Figure 5.14a–d).

The structure of the LBD of mGluR5 complexed with glutamate has been resolved at 2.44 Å and deposited (PDB ID: 3LMK). The structures of mGluR5 complexed with the NAM HTL14242 (PDB ID: 5CGC) and an analogue (PDB ID: 5CGD) showed an allosteric pocket similar to that found for mavoglurant illustrated in Figure 5.14 [137]. The pocket is defined by residues from TM2, TM3, TM5, TM6, and TM7. The corresponding heterocyclic ring (pyrazole for HTL14242 and pyridine for the analogue) sits in a pocket between Ala810$^{7.40}$ and Pro655$^{3.40}$, bordered on one side by Ile625$^{2.46}$, Gly628$^{2.49}$, Ser654$^{3.39}$, Ser658$^{3.43}$ and on the other by Tyr6593^{44} and Ser8097^{39}, which makes a hydrogen bond to the nitrogen atom of the respective heterocyclic ring (Figure 5.14e–h).

Group 2 Receptors Group 2 metabotropic receptors are formed by mGluR2 and mGluR3 receptors. Glutamate-binding region for mGluR2 has been identified in residues 166–168. The only natural variant of mGluR3 receptor is Gly475Asp. For this receptor, an isoform has been reported with missing residues 538–879 and alternative sequences for residues 442–537. Glutamate-binding region has been identified in residues 172–174. The crystal structures of the entire extracellular LBD of human mGluR2 complexed with five different agonists, including Glu (PDB ID: 2E4U), DCG-IV (PDB ID: 2E4V), 1S,3S-ACPD (PDB ID: 2E4W), 1S,3R-ACPD (PDB ID: 2E4X), and 2R,4R-APDC (PDB ID: 2E4Y), showed identical packing [128]. These structures (Figure 5.15) include the cysteine-rich domain, which is linked to the LBD by a disulfide bridge. Such a disposition suggests a potential role in transmitting a ligand-induced conformational change into the downstream transmembrane region. The structure of mGluR3 in complex with antagonist LY341495 has also been described (PDB ID: 3SM9) [138].

The ATD of recombinant human mGluR2 has been cocrystallized with agonists LY354740 (PDB ID: 4XAQ), LY2934747 (PDB ID: 4XAS), and LY2812223 (PDB ID: 5CNJ). Recombinant human mGluR3 has also been cocrystallized with LY354740 (PDB ID: 4XAR) and LY2812223 (PDB ID: 5CNM). These structures showed key interactions between the ligand and the residues of the binding pocket (Figure 5.16), leading to the identification of LY22934747 [139] and LY2812223 [140] as a high-potent agonist of mGluR2 and mGluR3.

Further characterization of LY3020371 antagonist binding to human mGluR2 has also been achieved (PDB ID: 5KZQ). The observed interactions support the antagonist pharmacology [141]. Recently, the structure of mGluR3 in complex with agonist LY2794193 (PDB ID: 6B7H) revealed critical binding interactions with residues adjacent to the glutamate binding site (Figure 5.16) [142].

Group 3 Receptors Group 3 metabotropic receptors are formed by mGluR4, mGluR6, mGluR7, and mGluR8 receptors. Natural variants of mGluR4 receptor include Leu169Phe and Val797Ile. For this receptor, four additional isoforms

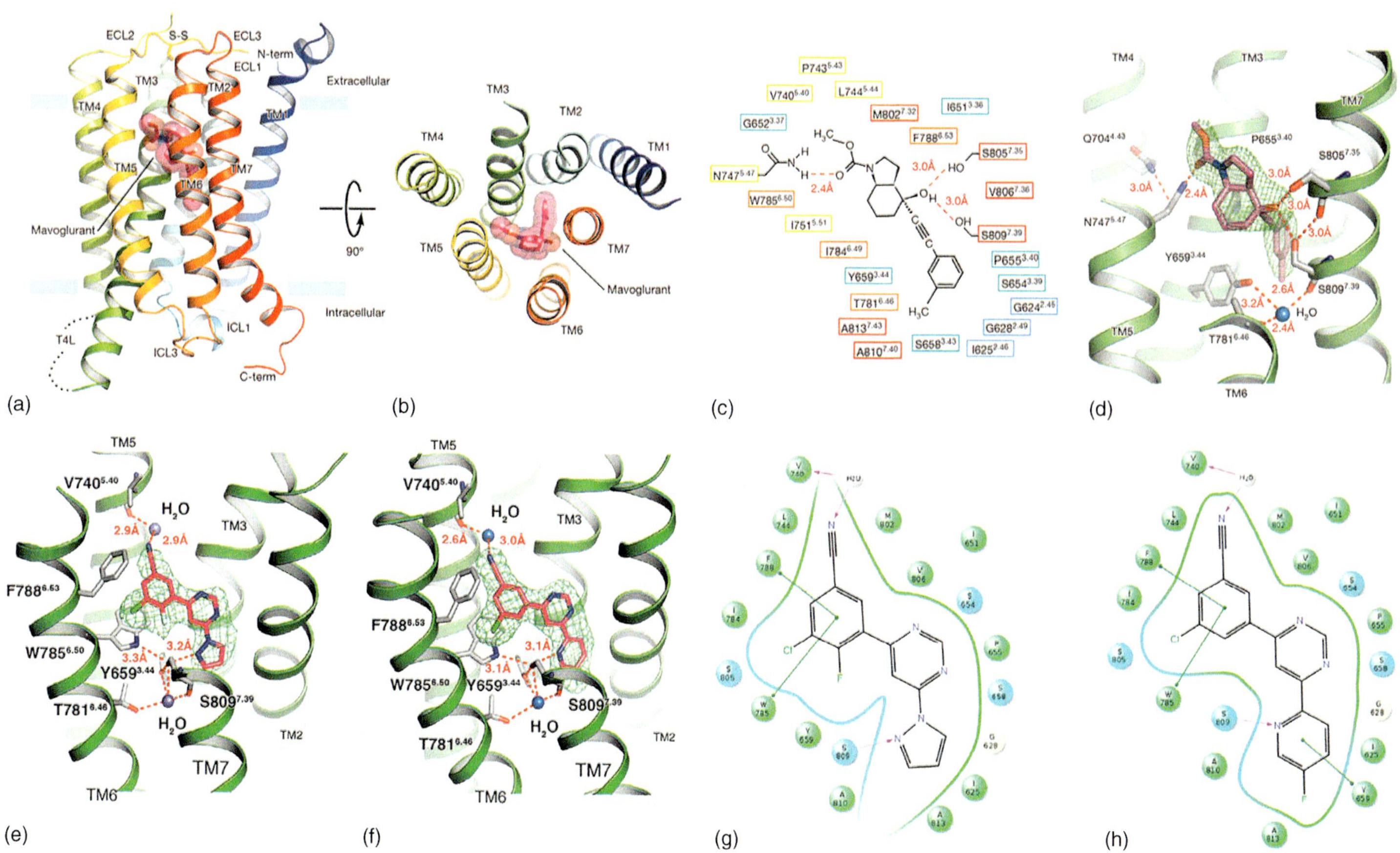
(a)
ECL2
S-S
ECL3
N-term
ECL1
Extracellular
TM3
TM2
TM4
TM1
TM5
TM7
TM6
Mavoglurant
90°
Intracellular
T4L
ICL1
ICL3
C-term
(b)
TM3
TM2
TM4
TM1
TM5
TM7
Mavoglurant
TM6
(c)
P743^5.43
V740^5.40
L744^5.44
M802^7.32
I651^3.36
G652^3.37
F788^6.53
N747^5.47
S805^7.35
3.0Å
2.4Å
V806^7.36
3.0Å
W785^6.50
I751^5.51
S809^7.39
I784^6.49
P655^3.40
Y659^3.44
S654^3.39
T781^6.46
G624^2.45
A813^7.43
G628^2.49
A810^7.40
S658^3.43
I625^2.46
(d)
TM4
TM3
TM7
Q704^4.43
P655^3.40
S805^7.35
3.0Å
2.4Å
3.0Å
3.0Å
3.0Å
N747^5.47
Y659^3.44
2.6Å
S809^7.39
3.2Å
H2O
2.4Å
TM5
T781^6.46
TM6
(e)
TM5
V740^5.40
H2O
2.9Å
2.9Å
TM3
F788^6.53
3.2Å
W785^6.50
3.3Å
Y659^3.44
T781^6.46
S809^7.39
H2O
TM2
TM7
TM6
(f)
TM5
V740^5.40
H2O
2.6Å
3.0Å
TM3
F788^6.53
3.1Å
W785^6.50
3.1Å
Y659^3.44
S809^7.39
T781^6.46
H2O
TM2
TM7
TM6
(g)
(h)

have been reported with missing residues 34–173 (isoform 2) and alternative sequences for residues 1–33 (isoform 2), 1–172 (isoforms 3 and 4), and 1–173 and 343–390 (isoform 3). Natural variants of the mGluR6 receptor include Phe46Leu, Gly58Arg, Gln59Phe, Gly150Ser, Glu227Val, Ile236Phe, Ile405Thr, Cys522Tyr, Met712Val, Glu781Lys, Ala807Val, and Thr817Ser. Natural variants of the mGluR7 receptor include Tyr433Phe, Ile495Val, and Gly745Glu. For this receptor, four additional isoforms have been reported with alternative sequences for residues 900–915 in all of them [105]. Natural variants of the mGluR8 receptor include Ser10Cys, Phe21Cys, Ile265Thr, Arg343Gln, Phe362Tyr, Gly368Asp, Arg392Gln, Leu430Phe, Val548Gly, Ile768Asn, and Ser902Ile. For this receptor, two additional isoforms have been reported with missing residues 502–908 (isoform 2) and alternative sequences for residues 893–908 (isoform B), and 454–501 (isoform C). Glutamate-binding region has been identified in residues 177–179. The crystal structure of the ligand-binding region of mGluR7 has been resolved (PDB ID: 2E4Z) [128], and it has been found to be essentially the same as those of the corresponding regions in mGluR1 and mGluR3. The structures of the extracellular LBD of human mGluR7 (PDB ID: 5C5C) [143] and that of the same receptor complexed with LY341495 antagonist (PDB ID: 3MQ4) [144] have also been determined. The human mGluR8 ATD has been resolved in complex with Glu (PDB ID: 6BSZ) and *L*-2-amino-4-phosphonobutyric acid (L-AP4), a potent and selective agonist for group 3 receptors (PDB ID: 6BT5) [145]. The molecular basis for L-AP4's group 3 mGluR functional potency and selectivity has been determined to be due to strong ionic interactions between the agonist and mGluR8 (Figure 5.17a). When *L*-AP4 interacts with mGluR8, it forms direct

Figure 5.14 (a, b) Ribbon representation of mGlu5 in rainbow coloration (N-terminus, blue; C-terminus, red) viewed parallel to the membrane and from the extracellular space, respectively. Mavoglurant is represented in translucent stick representation. Carbon, nitrogen, and oxygen atoms are colored magenta, blue, and red, respectively. The position of the T4L-insertion to ICL2 is indicated. (c) Diagram of ligand interactions in the binding pocket. The color scheme of the boxes follows the rainbow coloration as in a, b. Hydrogen bonds depicted as dashed red lines with distances between heavy atoms in angstroms. (d) View from the membrane of specific interactions in the binding pocket. Extracellular portions of TM5/TM6 removed for clarity. F_o–F_c OMIT density contoured at 2.0σ calculated before mavoglurant inclusion in the model, mavoglurant shown as sticks as in a. Source: (a–d) Reproduced with permission from Doré et al. [136]. Copyright 2014, Springer Nature.(e) Diagram of ligand interactions within the allosteric site of mGlu5 with HTL14242 bound. (f) Diagram of ligand interactions within the allosteric site of mGlu5 with pyridine analogue. mGlu5 in ribbon representation colored green. Ligands in stick representation with carbon, nitrogen, oxygen, chlorine, and fluorine atoms oxygen atoms colored magenta, blue, red, green, and white, respectively. Selected residues from mGlu5 in stick representation with carbon, nitrogen, and oxygen atoms colored gray, blue, and red, respectively. TM helices are labeled and selected hydrogen bonds depicted as dashed red lines with distances labeled in Å. In both figures, the extracellular portion of TM7 has been removed for clarity. F_o–F_c OMIT density for both ligands contoured at 2.0σ calculated prior to inclusion of the small molecule in the model. (g, h) 2D diagrams of ligand interactions in the mGlu5-StaR allosteric pocket, respectively. Green lines denote π-stack interactions, dashed purple arrows denote hydrogen bond (side chain), solid purple arrows denote hydrogen bond (main chain), residues in cyan spheres denote polar interactions, and residues in green spheres denote hydrophobic interactions. Source: (e–h) Reproduced with permission from Christopher et al. [137]. Copyright 2015, ACS.

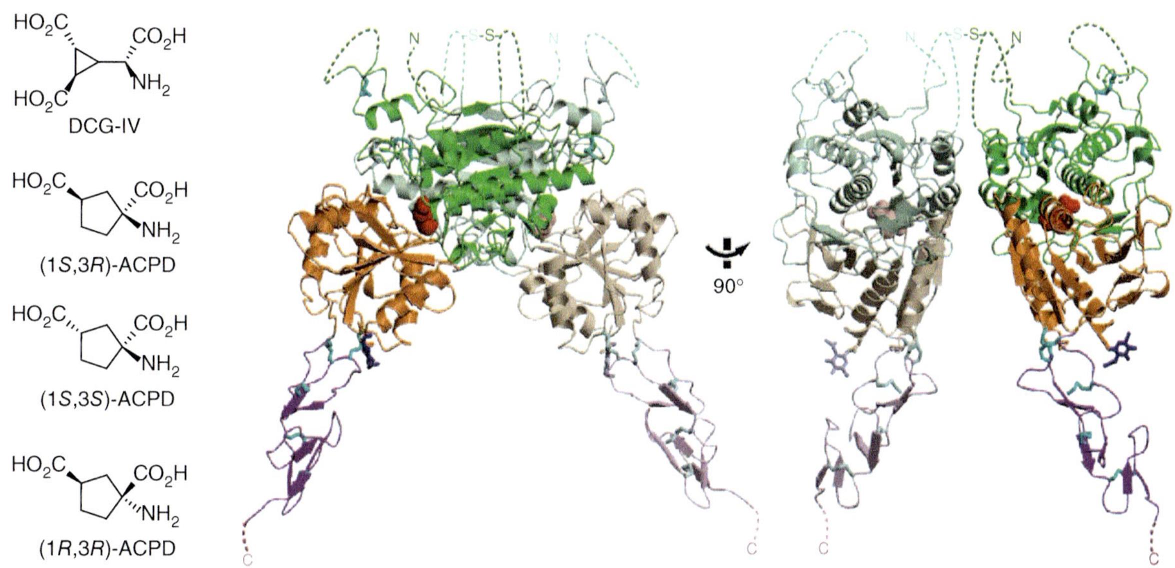

(a) (b)

Figure 5.15 (a) Structures of mGluR3 agonists. (b) Two orthogonal views of the mGluR3 EC region complexed with Glu. Source: Reproduced with permission from Muto et al. [128]. Copyright 2007, PNAS.

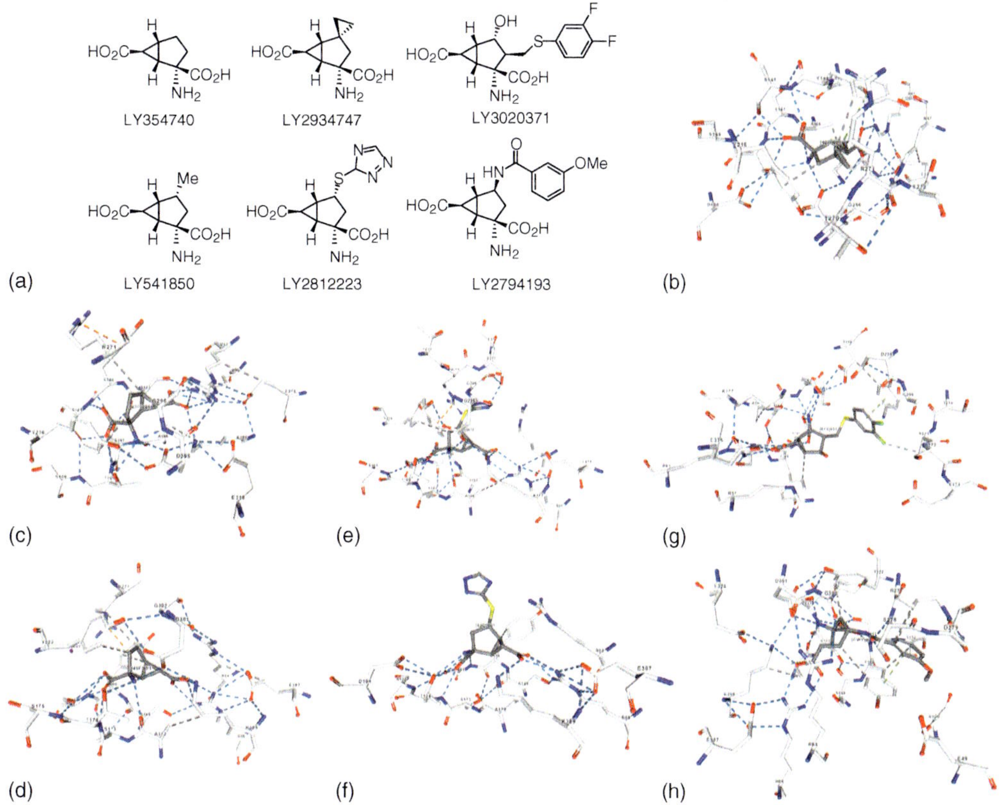

Figure 5.16 (a) Structures of mGluR2 and mGluR3 ligands. (b) Interactions of mGluR2 with LY2934747. (c, d) Details on interactions of LY354740 with mGluR2 (c) and mGluR3 (d). (e, f) Details on interactions of LY2812223 with mGluR2 (e) and mGluR3 (f). (g, h) Interactions of mGluR2 with LY3020371 (g) and LY2794192 (h).

ionic interactions with residues from both LB1 (Lys71, Arg75, and Lys401) and LB2 (Lys314). Accordingly, it is postulated that *L*-AP4 will form less productive interactions with LB1 residues in mGluR1 or no direct or indirect ionic interactions with LB2 in mGluR2 (Figure 5.17b). There are also important steric reasons that justify the selectivity of *L*-AP4 by group 3 receptors (Figure 5.17c). It seems plausible that the sterically imposing side chains of Trp110 of mGluR1 and Tyr144 of mGluR2 cause steric clashes with the bulkier phosphate group of *L*-AP4, leading to a less favored binding to these receptors. This hypothesis is supported by docking studies [145].

A comparison with other L-glu-bound ATD structures revealed that L-glu binds in a very different way in the mGluR1 ATD structure compared to its complex with mGluR2 or mGluR8 (Figure 5.17d–f). On the other hand, L-Glu adopts similar conformations in the mGluR2 and mGluR8 ATD structures.

5.3 Glutamate Transporters

There are three known vesicular transporters (VGLUT1, VGLUT2, and VGLUT3) that mediate glutamate uptake into synaptic vesicles [146]. All of them are driven by a proton electrochemical gradient generated by the vacuolar H^+-ATPase. The three transporters have a close resemblance at the sequence level (Figure 5.18), and they are highly similar with respect to substrate specificity, kinetics, and pharmacology [147].

On the other hand, they differ in their expression profiles. VGLUT1 and VGLUT2 are the predominant isoforms in the adult brain although they display complementary expression patterns. VGLUT1 predominates in the cerebral and cerebellar cortices and hippocampus, whereas VGLUT2 is mostly expressed in diencephalon, brainstem, and spinal cord [148]. However, there is no exclusivity for any of these brain regions, and glutamatergic neurons express either VGLUT1 or VGLUT2, as demonstrated in cerebellar mossy fiber terminals [149] and cerebellar unipolar brush cells [150]. The third isoform, VGLUT3, is less widely expressed than the other two and, notably, this transporter is not expressed in glutamatergic neurons but is rather found in cholinergic, serotoninergic, and even GABAergic neurons. VGLUT3 may therefore play a unique role in nonconventional glutamatergic neurotransmission [151]. To date, no differences in the

Figure 5.17 (a) Selected residues in the vicinity of the distal phosphate of *L*-AP4 in human mGlu8R ATD. The cyan dotted lines indicate favorable interactions between ligands and residues. (b, c) Electronic (b) and steric (c) environments of the distal acid in mGluR1 (yellow), mGluR2 (cyan), and mGluR8 (red). (d) Overlay of *L*-glu mGluR1 ATD (yellow), *L*-Glu mGluR2 ATD (cyan), *L*-glu mGluR8 ATD (gray), and *L*-AP4 mGluR8 ATD (red). (i) Conformation of the distal carboxylate of *L*-glu in mGluR1 ATD (yellow), mGluR2 ATD (cyan), and mGluR8 ATD (gray). (ii) Conformation of the distal phosphate of *L*-AP4 relative to that of the distal carboxylate of *L*-Glu in mGluR8 ATD. (e) Selected residues in the vicinity of the distal carboxylate of *L*-Glu in mGlur1 ATD. (f) Selected residues in the vicinity of the distal carboxylate of *L*-Glu in mGluR2 ATD. Source: Reproduced with permission from Schkeryantz et al. [145]. Copyright 2007, Elsevier.

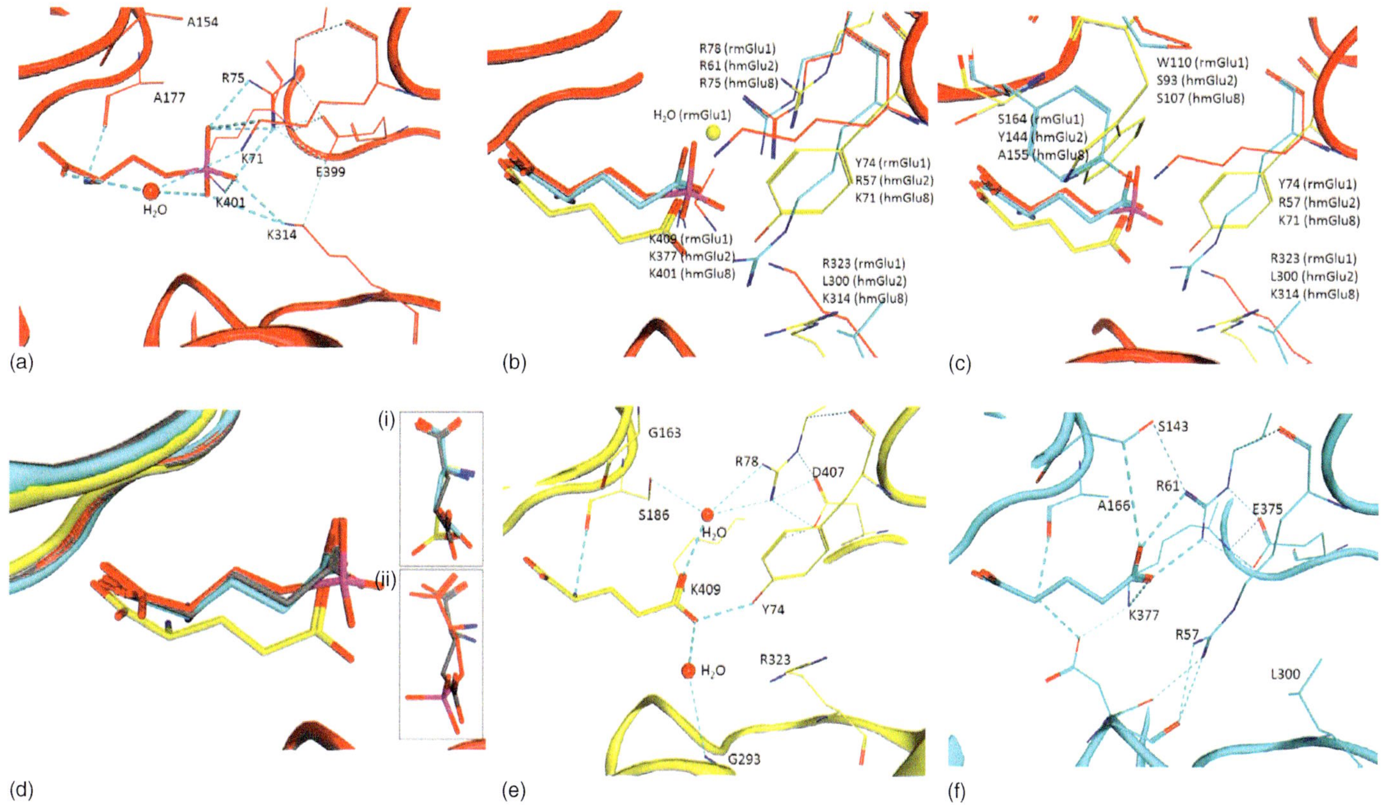

A154
A177
R75
K71
E399
H2O
K401
K314
(a)
R78 (rmGlu1)
R61 (hmGlu2)
R75 (hmGlu8)
H2O (rmGlu1)
Y74 (rmGlu1)
R57 (hmGlu2)
K71 (hmGlu8)
K409 (rmGlu1)
K377 (hmGlu2)
K401 (hmGlu8)
R323 (rmGlu1)
L300 (hmGlu2)
K314 (hmGlu8)
(b)
W110 (rmGlu1)
S93 (hmGlu2)
S107 (hmGlu8)
S164 (rmGlu1)
Y144 (hmGlu2)
A155 (hmGlu8)
Y74 (rmGlu1)
R57 (hmGlu2)
K71 (hmGlu8)
R323 (rmGlu1)
L300 (hmGlu2)
K314 (hmGlu8)
(c)
(i)
(ii)
(d)
G163
R78
D407
S186
H2O
K409
Y74
H2O
R323
G293
(e)
S143
R61
A166
E375
K377
R57
L300
(f)

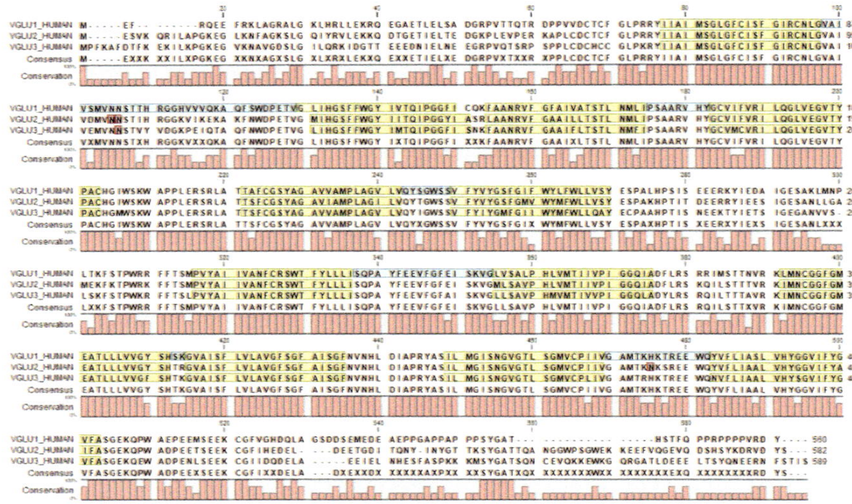

Figure 5.18 Sequences of the human vesicular glutamate transporters. N-Glycosylation sites (red), extracellular domains (light blue), intracellular domains (white), and transmembrane domains (yellow) are indicated on sequences.

transport characteristics of VGLUT1 and VGLUT2 have been identified. However, it should be expected that both transporters might be associated with different modes of synaptic glutamate release [152].

5.3.1 Implication of Glutamate Transporters in Neurodegeneration

Excitotoxicity, i.e. the excessive activation of glutamate receptors caused by an excess of extracellular concentrations of glutamate resulting in cell death, has been associated with a wide range of acute and chronic neurodegenerative disorders. Those neurodegenerative disorders can be associated with a change in localization and/or expression of glutamate transporters, so enhancing transporter expression might be therapeutically beneficial for the treatment of such disorders. The role of glutamate transporters in neurodegenerative diseases including Alzheimer's, Parkinson's, and Huntington's diseases as well as ALS, cerebral ischemia, and epilepsy among others, has been reviewed comprehensively [153].

The altered expression of glutamate transporters in ALS has been widely studied [154], and several evidences have been obtained linking alterations in glutamate handling to the pathophysiology of ALS. In this context, glutamate transporters are targets for therapeutic drug development [155]. As for ALS, reduced glutamate transporter expression and glutamate uptake are associated with AD. In fact, glutamate transporter expression declines with age but declines even more rapidly in AD as demonstrated in animal models [156]. Reduction in glutamate transporters may aggravate HD symptoms and the mutant huntingtin may decrease glutamate transporter levels, resulting in decreased glutamate uptake and exacerbating excitotoxicity [157]. On the other hand, recent studies indicate that glutamate transporter function in HD could be sufficient to maintain normal levels of glutamate [158]. As is observed in AD, glutamate uptake is reduced in platelets from patients with PD. Promotion of glutamate transporters by suppressing glutamate excitotoxicity has been recently considered a neuroprotective therapy for PD. [159]

5.3.2 Vesicular Glutamate Transporter 1 (VGLUT1)

The VGLUT1, expressed in several regions of the brain including amygdala, cerebellum, cerebral cortex, hippocampus, frontal lobe, medulla, occipital lobe, putamen, and temporal lobe [160], has a sequence of 560 residues, and it can be phosphorylated at Ser504. It can present two additional isoforms with residues 1–67 missing (isoform 2) and a different sequence in residues 1–21 (isoform 3). A natural variant Pro431Arg has been described. An X-ray structure of human VGLUT1 is not available yet, although the structures of homologues have been reported [161–163].

VGLUT1 exhibits a C-terminal dileucine-like motif sequence, known to be important for the trafficking of multispanning transmembrane proteins and consisting of acidic resides at the positions -4 and -5 upstream of two hydrophobic residues ((Glu/Asp)XXX(Leu/Ileu)). In VGLUT1, the sequence is Glu505-Glu-Lys-Cys-GlyPhe-Val511. VGLUT2 and VGLUT3 have similar sequences. The VGLUT1 C-terminus is crucial for the protein trafficking through its dileucine-like Phe-Val structure [164]. The VGLUT1 N-terminus also contains dileucine-like motif sequences [165], which are not well conserved in the other glutamate transporters. There is a Ser504 residue, which can be phosphorylated, flanking the dileucine-like motif upstream of the sequence that is also present in VGLUT2 and VGLUT3. To date, it is not known whether VGLUT1 is functional without its C-terminus.

5.3.3 Vesicular Glutamate Transporter 2 (VGLUT2)

The VGLUT2, expressed in amygdala, caudate nucleus, cerebral cortex, frontal lobe, hippocampus, medulla, occipital lobe, putamen, spinal cord, substantia nigra, subthalamic nucleus, temporal lobe, and thalamus [160], has a sequence of 582 residues, and it can be glycosylated with NAcGlc at Asn100, Asn101, and Asn470. Natural variants Thr40Ser and Asn551Ser have been described. An X-ray structure is not available yet. As for VGLUT1, the trafficking of VGLUT2 also depends on the conserved C-terminal dileucine-like motif sequence [165].

5.3.4 Vesicular Glutamate Transporter 3 (VGLUT3)

The VGLUT3, expressed in amygdala, cerebellum, hippocampus, medulla, spinal cord, and thalamus [166], has a sequence of 589 residues, and it can be glycosylated with NAcGlc at Asn106. It can present an additional isoform in which residues 302–351 are missing. Natural variants Thr8Iso, Ala211Val, Ala220Thr, and Gly246Asp have been reported [167]. An X-ray structure is not available yet.

Data about VGLUT3 and a role of its dileucine-like motif sequence are currently not available.

References

1 Geisler, S. and Wise, R.A. (2008). Functional implications of glutamatergic projections to the ventral tegmental area. *Rev. Neurosci.* 19 (4-5): 227–244.

2 Price, J.L., Carmichael, S.T., and Drevets, W.C. (1996). Networks related to the orbital and medial prefrontal cortex; a substrate for emotional behavior? *Prog. Brain Res.* 107: 523–536.

3 Ferry, A.T., Ongur, D., An, X., and Price, J.L. (2000). Prefrontal cortical projections to the striatum in macaque monkeys: evidence for an organization related to prefrontal networks. *J. Comp. Neurol.* 425 (3): 447–470.

4 Kita, H. and Kitai, S.T. (1990). Amygdaloid projections to the frontal cortex and the striatum in the rat. *J. Comp. Neurol.* 298 (1): 40–49.

5 Fudge, J.L., Kunishio, K., Walsh, P. et al. (2002). Amygdaloid projections to ventromedial striatal subterritories in the primate. *Neuroscience* 110 (2): 257–275.

6 Friedman, D.P., Aggleton, J.P., and Saunders, R.C. (2002). Comparison of hippocampal, amygdala, and perirhinal projections to the nucleus accumbens: combined anterograde and retrograde tracing study in the Macaque brain. *J. Comp. Neurol.* 450 (4): 345–365.

7 Root, D.H., Mejias-Aponte, C.A., Qi, J., and Morales, M. (2014). Role of glutamatergic projections from ventral tegmental area to lateral habenula in aversive conditioning. *J. Neurosci.* 34 (42): 13906–13910.

8 Volk, L., Chiu, S.-L., Sharma, K., and Huganir, R.L. (2015). Glutamate synapses in human cognitive disorders. *Annu. Rev. Neurosci.* 38 (1): 127–149.

9 Moghaddam, B. (2002). Stress activation of glutamate neurotransmission in the prefrontal cortex: implications for dopamine-associated psychiatric disorders. *Biol. Psychiatry* 51 (10): 775–787.

10 Lewis, D.A. and Gonzalez-Burgos, G. (2000). Intrinsic excitatory connections in the prefrontal cortex and the pathophysiology of schizophrenia. *Brain Res. Bull.* 52 (5): 309–317.

11 Estrada Sanchez, A.M., Mejia-Toiber, J., and Massieu, L. (2008). Excitotoxic neuronal death and the pathogenesis of Huntington's disease. *Arch. Med. Res.* 39 (3): 265–276.

12 Sepers, M.D. and Raymond, L.A. (2014). Mechanisms of synaptic dysfunction and excitotoxicity in Huntington's disease. *Drug Discovery Today* 19 (7): 990–996.

13 Gereau, R.W. and Swanson, G.T. (2008). *The Receptors* (ed. K.A. Neve), 609. Totowa: Humana Press.

14 Nakanishi, S. (1992). Molecular diversity of glutamate receptors and implications for brain function. *Science (Washington, D. C., 1883-)* 258 (5082): 597–603.

15 MacDonald, J.F., Wojtowicz, J.M., and Baskys, A. (1996). Glutamate receptors. In: *Brain Mechanisms and Psychotropic Drugs* (ed. A. Baskys and G. Remington), 117–130. Boca Raton, FL: CRC.

16 Oswald, R.E., Ahmed, A., Fenwick, M.K., and Loh, A.P. (2007). Structure of glutamate receptors. *Curr. Drug Targets* 8 (5): 573–582.

17 Rousseaux, C.G. (2008). A review of glutamate receptors I: current understanding of their biology. *J. Toxicol. Pathol.* 21 (1): 25–51.

18 Rousseaux, C.G. (2008). A review of glutamate receptors II: pathophysiology and pathology. *J. Toxicol. Pathol.* 21 (3): 133–173.

19 Kew, J.N. and Kemp, J.A. (2005). Ionotropic and metabotropic glutamate receptor structure and pharmacology. *Psychopharmacology* 179 (1): 4–29.

20 Collingridge, G.L., Olsen, R.W., Peters, J., and Spedding, M. (2009). A nomenclature for ligand-gated ion channels. *Neuropharmacology* 56 (1): 2–5.

21 Lodge, D. (2009). The history of the pharmacology and cloning of ionotropic glutamate receptors and the development of idiosyncratic nomenclature. *Neuropharmacology* 56 (1): 6–21.

22 Traynelis, S.F., Wollmuth, L.P., McBain, C.J. et al. (2010). Glutamate receptor ion channels: structure, regulation, and function. *Pharmacol. Rev.* 62 (3): 405–496.

23 Bowie, D. (2012). Redefining the classification of AMPA-selective ionotropic glutamate receptors. *J. Physiol.* 590 (1): 49–61.

24 Monaghan, D.T. and Wenthold, R. (1997). *The Receptors* (ed. K.A. Neve), 380. Totowa: Humana Press.

25 Jonas, P. and Monyer, H. (1999). *Handbook of Experimental Pharmacology*, vol. 141 (ed. G.Y.R. Born, P. Cuatrecasas, D. Ganten, et al.), 544. Berlin: Springer.

26 Tikhonov, D.B. and Magazanik, L.G. (2009). Origin and molecular evolution of ionotropic glutamate receptors. *Neurosci. Behav. Physiol.* 39 (8): 763–773.

27 Planells-Cases, R., Lerma, J., and Ferrer-Montiel, A. (2006). Pharmacological intervention at ionotropic glutamate receptor complexes. *Curr. Pharm. Des.* 12 (28): 3583–3596.

28 Lipsky, R.H. and Goldman, D. (2003). Genomics and variation of ionotropic glutamate receptors. *Ann. N. Y. Acad. Sci.* 1003 (Glutamate and Disorders of Cognition and Motivation): 22–35.

29 Kaczor, A.A. and Matosiuk, D. (2010). Molecular structure of ionotropic glutamate receptors. *Curr. Med. Chem.* 17 (24): 2608–2635.

30 Tomita, S. (2010). Regulation of ionotropic glutamate receptors by their auxiliary subunits. *Physiology* 25 (February): 41–49.

31 Maksay, G. (2005). Activation of ionotropic receptors and thermodynamics of binding. *Neurochem. Int.* 46 (4): 281–291.

32 Dingledine, R., Borges, K., Bowie, D., and Traynelis, S.F. (1999). The glutamate receptor ion channels. *Pharmacol. Rev.* 51 (1): 7–61.

33 Seifert, G. and Steinhauser, C. (2001). Ionotropic glutamate receptors in astrocytes. *Prog. Brain Res.* 132 (Glial Cell Function): 287–299.

34 Lee, M.-C., Ting, K.K., Adams, S. et al. (2010). Characterisation of the expression of NMDA receptors in human astrocytes. *PLoS One* 5 (11): e14123.

35 Huntley, G.W., Vickers, J.C., Janssen, W. et al. (1994). Distribution and synaptic localization of immunocytochemically identified NMDA receptor subunit proteins in sensory-motor and visual cortices of monkey and human. *J. Neurosci.* 14 (6): 3603–3619.

36 Sanchez, J.T., Ghelani, S., and Otto-Meyer, S. (2015). From development to disease: diverse functions of NMDA-type glutamate receptors in the lower auditory pathway. *Neuroscience* 285: 248–259.
37 Hawkins, L.M., Beaver, K.M., Jane, D.E. et al. (1995). Binding of the new radioligand (*S*)-[3H]AMPA to rat brain synaptic membranes: effects of a series of structural analogues of the non-NMDA receptor agonist willardiine. *Neuropharmacology* 34 (4): 405–410.
38 Michaelis, E.K. (1998). Molecular biology of glutamate receptors in the central nervous system and their role in excitotoxicity, oxidative stress and aging. *Prog. Neurobiol.* 54 (4): 369–415.
39 Jahr, C.E. and Stevens, C.F. (1993). Calcium permeability of the *N*-methyl-d-aspartate receptor channel in hippocampal neurons in culture. *Proc. Natl. Acad. Sci. U.S.A.* 90 (24): 11573–11577.
40 Das, S., Sasaki, Y.F., Rothe, T. et al. (1998). Increased NMDA current and spine density in mice lacking the NMDA receptor subunit NR3A. *Nature* 393: 377.
41 Erreger, K., Chen, P.E., Wyllie, D.J., and Traynelis, S.F. (2004). Glutamate receptor gating. *Crit. Rev. Neurobiol.* 16 (3): 187–224.
42 Petzoldt, A.G., Lee, Y.H., Khorramshahi, O. et al. (2014). Gating characteristics control glutamate receptor distribution and trafficking in vivo. *Curr. Biol.* 24 (17): 2059–2065.
43 Jiang, X., Knox, R., Pathipati, P., and Ferriero, D. (2011). Developmental localization of NMDA receptors, Src and MAP kinases in mouse brain. *Neurosci. Lett.* 503 (3): 215–219.
44 Bartlett, T.E. and Wang, Y.T. (2013). The intersections of NMDAR-dependent synaptic plasticity and cell survival. *Neuropharmacology* 74 (Supplement C): 59–68.
45 Paoletti, P., Bellone, C., and Zhou, Q. (2013). NMDA receptor subunit diversity: impact on receptor properties, synaptic plasticity and disease. *Nat. Rev. Neurosci.* 14: 383.
46 Lüscher, C. and Malenka, R.C. (2012). NMDA receptor-dependent long-term potentiation and long-term depression (LTP/LTD). *Cold Spring Harbor Perspect. Biol.* 4 (6): a005710.
47 Bowie, D. (2008). Ionotropic glutamate receptors and CNS disorders. *CNS Neurol. Disord. Drug Targets* 7 (2): 129–143.
48 Doble, A. (1999). The role of excitotoxicity in neurodegenerative disease: implications for therapy. *Pharmacol. Ther.* 81 (3): 163–221.
49 Wang, R. and Reddy, P.H. (2017). Role of glutamate and NMDA Receptors in Alzheimer's disease. *J. Alzheimer's Dis.* 57 (4): 1041–1048.
50 Tanović, A. and Alfaro, V. (2006). Glutamate-related excitotoxicity neuroprotection with memantine, an uncompetitive antagonist of NMDA-glutamate receptor, in Alzheimer's disease and vascular dementia. *Rev. Neurol.* 42 (10): 607–616.
51 Spalloni, A., Nutini, M., and Longone, P. (2013). Role of the *N*-methyl-d-aspartate receptors complex in amyotrophic lateral sclerosis. *Biochim. Biophys. Acta, Mol. Basis Dis.* 1832 (2): 312–322.
52 Stahl, S.M. (2007). The genetics of schizophrenia converge upon the NMDA glutamate receptor. *CNS Spectr.* 12 (8): 583–588.

53 Nishikawa, T. and Ishiwata, S. (2013). NMDA-type glutamate receptor and schizophrenia. *Jpn. J. Neuropsychopharmacology* 33 (5–6): 217–224.
54 Stahl, S.M. (2007). Novel therapeutics for schizophrenia: targeting glycine modulation of NMDA glutamate receptors. *CNS Spectr.* 12 (6): 423–427.
55 Fountoulakis, K.N. (2012). The possible involvement of NMDA glutamate receptor in the etiopathogenesis of bipolar disorder. *Curr. Pharm. Des.* 18 (12): 1605–1608.
56 Antonio, S.-C., Roger, A.N., and Katherine, W.R. (2012). Diversity in NMDA receptor composition: many regulators, many consequences. *Neuroscientist* 19 (1): 62–75.
57 Lemke, J.R., Geider, K., Helbig, K.L. et al. (2016). Delineating the GRIN1 phenotypic spectrum: a distinct genetic NMDA receptor encephalopathy. *Neurology* 86 (23): 2171–2178.
58 Ogden, K.K., Chen, W., Swanger, S.A. et al. (2017). Molecular mechanism of disease-associated mutations in the pre-M1 helix of NMDA receptors and potential rescue pharmacology. *PLoS Genet.* 13 (1): e1006536.
59 Zehavi, Y., Mandel, H., Zehavi, A. et al. (2017). De novo GRIN1 mutations: an emerging cause of severe early infantile encephalopathy. *Eur. J. Med. Genet.* 60 (6): 317–320.
60 Hamdan, F.F., Gauthier, J., Araki, Y. et al. (2011). Excess of de novo deleterious mutations in genes associated with glutamatergic systems in nonsyndromic intellectual disability. *Am. J. Hum. Genet.* 88 (3): 306–316.
61 Lemke, J.R., Lal, D., Reinthaler, E.M. et al. (2013). Mutations in GRIN2A cause idiopathic focal epilepsy with rolandic spikes. *Nat. Genet.* 45 (9): 1067–1072.
62 Swanger, S.A., Chen, W., Wells, G. et al. (2016). Mechanistic insight into NMDA receptor dysregulation by rare variants in the GluN2A and GluN2B agonist binding domains. *Am. J. Hum. Genet.* 99 (6): 1261–1280.
63 Lemke, J.R., Hendrickx, R., Geider, K. et al. (2014). GRIN2B mutations in West syndrome and intellectual disability with focal epilepsy. *Ann. Neurol.* 75 (1): 147–154.
64 Tarabeux, J., Kebir, O., Gauthier, J. et al. (2011). Rare mutations in *N*-methyl-d-aspartate glutamate receptors in autism spectrum disorders and schizophrenia. *Transl. Psychiatry* 1: e55.
65 Ataman, Z.A., Gakhar, L., Sorensen, B.R. et al. (2007). The NMDA receptor NR1 C1 region bound to calmodulin: structural insights into functional differences between homologous domains. *Structure (London, England: 1993)* 15 (12): 1603–1617.
66 Hackos, D.H., Lupardus, P.J., Grand, T. et al. (2016). Positive allosteric modulators of GluN2A-containing NMDARs with distinct modes of action and impacts on circuit function. *Neuron* 89 (5): 983–999.
67 Volgraf, M., Sellers, B.D., Jiang, Y. et al. (2016). Discovery of GluN2A-selective NMDA receptor positive allosteric modulators (PAMs): tuning deactivation kinetics via structure-based design. *J. Med. Chem.* 59 (6): 2760–2779.
68 Villemure, E., Volgraf, M., Jiang, Y. et al. (2017). GluN2A-selective pyridopyrimidinone series of NMDAR positive allosteric modulators with an improved in vivo profile. *ACS Med. Chem. Lett.* 8: 84–89.
69 Lee, C.-H., Lü, W., Michel, J.C. et al. (2014). NMDA receptor structures reveal subunit arrangement and pore architecture. *Nature* 511: 191.

70 Sun, W., Ferrer-Montiel, A.V., and Montal, M. (1994). Primary structure and functional expression of the AMPA/kainate receptor subunit 2 from human brain. *Neuroreport* 5 (4): 441–444.

71 Wu, Y., Arai, A.C., Rumbaugh, G. et al. (2007). Mutations in ionotropic AMPA receptor 3 alter channel properties and are associated with moderate cognitive impairment in humans. *Proc. Natl. Acad. Sci. U.S.A.* 104 (46): 18163–18168.

72 Rampersad, V., Elliott, C.E., Nutt, S.L. et al. (1994). Human glutamate receptor hGluR3 flip and flop isoforms: cloning and sequencing of the cDNAs and primary structure of the proteins. *Biochim. Biophys. Acta* 1219 (2): 563–566.

73 Clayton, A., Siebold, C., Gilbert, R.J. et al. (2009). Crystal structure of the GluR2 amino-terminal domain provides insights into the architecture and assembly of ionotropic glutamate receptors. *J. Mol. Biol.* 392 (5): 1125–1132.

74 Ward, S.E., Harries, M., Aldegheri, L. et al. (2010). Discovery of *N*-[(2*S*)-5-(6-fluoro-3-pyridinyl)-2,3-dihydro-1H-inden-2-yl]-2-propanesulfonamide, a novel clinical AMPA receptor positive modulator. *J. Med. Chem.* 53 (15): 5801–5812.

75 Timm, D.E., Benveniste, M., Weeks, A.M. et al. (2011). Structural and functional analysis of two new positive allosteric modulators of GluA2 desensitization and deactivation. *Mol. Pharmacol.* 80 (2): 267–280.

76 Koller, M., Lingenhoehl, K., Schmutz, M. et al. (2011). Quinazolinedione sulfonamides: a novel class of competitive AMPA receptor antagonists with oral activity. *Bioorg. Med. Chem. Lett.* 21 (11): 3358–3361.

77 Orain, D., Ofner, S., Koller, M. et al. (2012). 6-Amino quinazolinedione sulfonamides as orally active competitive AMPA receptor antagonists. *Bioorg. Med. Chem. Lett.* 22 (2): 996–999.

78 Shibata, H., Joo, A., Fujii, Y. et al. (2001). Association study of polymorphisms in the GluR5 kainate receptor gene (GRIK1) with schizophrenia. *Psychiatr. Genet.* 11 (3): 139–144.

79 Korczak, B., Nutt, S.L., Fletcher, E.J. et al. (1995). cDNA cloning and functional properties of human glutamate receptor EAA3 (GluR5) in homomeric and heteromeric configuration. *Receptors Channels* 3 (1): 41–49.

80 Guzman, Y.F., Ramsey, K., Stolz, J.R. et al. (2017). A gain-of-function mutation in the GRIK2 gene causes neurodevelopmental deficits. *Neurol. Genet.* 3 (1): e129.

81 Zhawar, V.K., Kaur, G., deRiel, J.K. et al. (2010). Novel spliced variants of ionotropic glutamate receptor GluR6 in normal human fibroblast and brain cells are transcribed by tissue specific promoters. *Gene* 459 (1-2): 1–10.

82 Nutt, S.L., Hoo, K.H., Rampersad, V. et al. (1994). Molecular characterization of the human EAA5 (GluR7) receptor: a high-affinity kainate receptor with novel potential RNA editing sites. *Receptors Channels* 2 (4): 315–326.

83 Kamboj, R.K., Schoepp, D.D., Nutt, S. et al. (1994). Molecular cloning, expression, and pharmacological characterization of humEAA1, a human kainate receptor subunit. *J. Neurochem.* 62 (1): 1–9.

84 Kamboj, R.K., Schoepp, D.D., Nutt, S. et al. (1992). Molecular structure and pharmacological characterization of humEAA2, a novel human kainate receptor subunit. *Mol. Pharmacol.* 42 (1): 10–15.

85 Weston, M.C., Schuck, P., Ghosal, A. et al. (2006). Conformational restriction blocks glutamate receptor desensitization. *Nat. Struct. Mol. Biol.* 13 (12): 1120–1127.
86 Unno, M., Shinohara, M., Takayama, K. et al. (2011). Binding and selectivity of the marine toxin neodysiherbaine A and its synthetic analogues to GluK1 and GluK2 kainate receptors. *J. Mol. Biol.* 413 (3): 667–683.
87 Martinez-Perez, J.A., Iyengar, S., Shannon, H.E. et al. (2013). GluK1 antagonists from 6-(tetrazolyl)phenyl decahydroisoquinoline derivatives: in vitro profile and in vivo analgesic efficacy. *Bioorg. Med. Chem. Lett.* 23 (23): 6463–6466.
88 Meyerson, J.R., Chittori, S., Merk, A. et al. (2016). Structural basis of kainate subtype glutamate receptor desensitization. *Nature* 537 (7621): 567–571.
89 Coutelier, M., Burglen, L., Mundwiller, E. et al. (2015). GRID2 mutations span from congenital to mild adult-onset cerebellar ataxia. *Neurology* 84 (17): 1751–1759.
90 Elegheert, J., Kakegawa, W., Clay, J.E. et al. (2016). Structural basis for integration of GluD receptors within synaptic organizer complexes. *Science* 353 (6296): 295–299.
91 Niswender, C.M. and Conn, P.J. (2010). Metabotropic glutamate receptors: physiology, pharmacology, and disease. *Annu. Rev. Pharmacol. Toxicol.* 50: 295–322.
92 Vaidya, A., Jain, S., Jain, A.K. et al. (2013). Metabotropic glutamate receptors: a review on prospectives and therapeutic aspects. *Mini-Rev. Med. Chem.* 13 (13): 1967–1981.
93 Sherman, S.M. (2014). The function of metabotropic glutamate receptors in thalamus and cortex. *Neuroscientist* 20 (2): 136–149, 114 pp.
94 Golubeva, A.V., Moloney, R.D., O'Connor, R.M. et al. (2016). Metabotropic glutamate receptors in central nervous system diseases. *Curr. Drug Targets* 17 (5): 538–616.
95 Collingridge, G.L., Manahan-Vaughan, D., Nicoletti, F., and Schoepp, D.D. (2017). Metabotropic glutamate receptors, 5 years on. *Neuropharmacology* 115: 1–3.
96 Anwyl, R. (1999). Metabotropic glutamate receptors: electrophysiological properties and role in plasticity. *Brain Res. Brain Res. Rev.* 29 (1): 83–120.
97 Coutinho, V. and Knopfel, T. (2002). Metabotropic glutamate receptors: electrical and chemical signaling properties. *Neuroscientist* 8 (6): 551–561.
98 Valenti, O., Conn, P.J., and Marino, M.J. (2002). Distinct physiological roles of the Gq-coupled metabotropic glutamate receptors Co-expressed in the same neuronal populations. *J. Cell Physiol.* 191 (2): 125–137.
99 Bellone, C., Luscher, C., and Mameli, M. (2008). Mechanisms of synaptic depression triggered by metabotropic glutamate receptors. *Cell Mol. Life Sci.* 65 (18): 2913–2923.
100 Pinheiro, P.S. and Mulle, C. (2008). Presynaptic glutamate receptors: physiological functions and mechanisms of action. *Nat. Rev. Neurosci.* 9 (6): 423–436.
101 D'Antoni, S., Berretta, A., Bonaccorso, C.M. et al. (2008). Metabotropic glutamate receptors in glial cells. *Neurochem. Res.* 33 (12): 2436–2443.

102 Ménard, C. and Quirion, R. (2012). Group 1 metabotropic glutamate receptor function and its regulation of learning and memory in the aging brain. *Front. Pharmacol.* 3, Article 182 (12 pages). DOI: 10.3389/fphar.2012.00182

103 Shigemoto, R., Kinoshita, A., Wada, E. et al. (1997). Differential presynaptic localization of metabotropic glutamate receptor subtypes in the rat hippocampus. *J. Neurosci.* 17 (19): 7503.

104 Makoff, A., Pilling, C., Harrington, K., and Emson, P. (1996). Human metabotropic glutamate receptor type 7: molecular cloning and mRNA distribution in the CNS. *Brain Res. Mol. Brain Res.* 40 (1): 165–170.

105 Schulz, H.L., Stohr, H., and Weber, B.H. (2002). Characterization of three novel isoforms of the metabotrobic glutamate receptor 7 (GRM7). *Neurosci. Lett.* 326 (1): 37–40.

106 Conn, P.J. and Pin, J.P. (1997). Pharmacology and functions of metabotropic glutamate receptors. *Annu. Rev. Pharmacol. Toxicol.* 37: 205–237.

107 Suh, Y.H., Chang, K., and Roche, K.W. (2018). Metabotropic glutamate receptor trafficking. *Mol. Cell Neurosci.* 10–24.

108 Zou, J., Jiang, J.Y., and Yang, J.J. (2017). Molecular basis for modulation of metabotropic glutamate receptors and their drug actions by extracellular Ca^{2+}. *Int. J. Mol. Sci.* 18 (3): 672.

109 Chung, G. and Kim, S.J. (2017). Sustained activity of metabotropic glutamate receptor: homer, arrestin, and beyond. *Neural. Plast.* 2017: 5125624.

110 Iacovelli, L., Bruno, V., Salvatore, L. et al. (2002). Native group-III metabotropic glutamate receptors are coupled to the mitogen-activated protein kinase/ phosphatidylinositol-3-kinase pathways. *J. Neurochem.* 82 (2): 216–223.

111 Rojas, A. and Dingledine, R. (2013). Ionotropic glutamate receptors: regulation by G-Protein-coupled receptors. *Mol. Pharmacol.* 83 (4): 746.

112 Tu, J.C., Xiao, B., Yuan, J.P. et al. (1998). Homer binds a novel proline-rich motif and links group 1 metabotropic glutamate receptors with IP3 receptors. *Neuron* 21 (4): 717–726.

113 Ehrengruber, M.U., Kato, A., Inokuchi, K., and Hennou, S. (2004). Homer/Vesl proteins and their roles in CNS neurons. *Mol. Neurobiol.* 29 (3): 213–227.

114 Seebahn, A., Rose, M., and Enz, R. (2008). RanBPM is expressed in synaptic layers of the mammalian retina and binds to metabotropic glutamate receptors. *FEBS Lett.* 582 (16): 2453–2457.

115 Flajolet, M., Rakhilin, S., Wang, H. et al. (2003). Protein phosphatase 2C binds selectively to and dephosphorylates metabotropic glutamate receptor 3. *Proc. Natl. Acad. Sci. U.S.A.* 100 (26): 16006–16011.

116 Rose, M., Dutting, E., and Enz, R. (2008). Band 4.1 proteins are expressed in the retina and interact with both isoforms of the metabotropic glutamate receptor type 8. *J. Neurochem.* 105 (6): 2375–2387.

117 Ribeiro, F.M., Vieira, L.B., Pires, R.G.W. et al. (2017). Metabotropic glutamate receptors and neurodegenerative diseases. *Pharmacol. Res.* 115: 179–191.

118 Amalric, M. (2015). Targeting metabotropic glutamate receptors (mGluRs) in Parkinson's disease. *Curr. Opin. Pharmacol.* 20: 29–34.

119 Gasparini, F., Di Paolo, T., and Gomez-Mancilla, B. (2013). Metabotropic glutamate receptors for Parkinson's disease therapy. *Parkinsons Dis.* 11. https://doi.org/10.1155/2013/196028.

120 Litim, N., Morissette, M., and Di Paolo, T. (2017). Metabotropic glutamate receptors as therapeutic targets in Parkinson's disease: an update from the last 5 years of research. *Neuropharmacology* 115: 166–179.

121 Malter, J. S., Westmark, P. R., Westmark, C. J. (2010). Fragile X Syndrome and Alzeimer's Disease: Just Another Story About APP and β-amyloid. *Curr. Alzheimer Res.* 7(3): 200–206.

122 Muguruza, C., Meana, J.J., and Callado, L.F. (2016). Group II metabotropic glutamate receptors as targets for novel antipsychotic drugs. *Front. Pharmacol.* 7 (May): 12. https://doi.org/10.3389/fphar.2016.00130.

123 Duty, S. (2010). Therapeutic potential of targeting group III metabotropic glutamate receptors in the treatment of Parkinson's disease. *Br. J. Pharmacol.* 161 (2): 271–287.

124 Dickerson, J.W. and Conn, P.J. (2012). Therapeutic potential of targeting metabotropic glutamate receptors for Parkinson's disease. *Neurodegener. Dis. Manag.* 2 (2): 221–232.

125 Blacker, C.J., Lewis, C.P., Frye, M.A., and Veldic, M. (2017). Metabotropic glutamate receptors as emerging research targets in bipolar disorder. *Psychiatry Res.* 257: 327–337.

126 Qian, F. and Tang, F.-R. (2016). Metabotropic glutamate receptors and interacting proteins in epileptogenesis. *Curr. Neuropharmacol.* 14 (5): 551–562.

127 Vinson, P.N. and Conn, P.J. (2012). Metabotropic glutamate receptors as therapeutic targets for schizophrenia. *Neuropharmacology* 62 (3): 1461–1472.

128 Muto, T., Tsuchiya, D., Morikawa, K., and Jingami, H. (2007). Structures of the extracellular regions of the group II/III metabotropic glutamate receptors. *Proc. Natl. Acad. Sci. U.S.A.* 104 (10): 3759–3764.

129 Jingami, H., Nakanishi, S., and Morikawa, K. (2003). Structure of the metabotropic glutamate receptor. *Curr. Opin. Neurobiol.* 13 (3): 271–278.

130 Bessis, A.S., Rondard, P., Gaven, F. et al. (2002). Closure of the Venus flytrap module of mGlu8 receptor and the activation process: insights from mutations converting antagonists into agonists. *Proc. Natl. Acad. Sci. U.S.A.* 99 (17): 11097–11102.

131 Pin, J.P., Joly, C., Heinemann, S.F., and Bockaert, J. (1994). Domains involved in the specificity of G protein activation in phospholipase C-coupled metabotropic glutamate receptors. *EMBO J.* 13 (2): 342–348.

132 Gomeza, J., Joly, C., Kuhn, R. et al. (1996). The second intracellular loop of metabotropic glutamate receptor 1 cooperates with the other intracellular domains to control coupling to G-proteins. *J. Biol. Chem.* 271 (4): 2199–2205.

133 Dhami, G.K., Babwah, A.V., Sterne-Marr, R., and Ferguson, S.S. (2005). Phosphorylation-independent regulation of metabotropic glutamate receptor 1 signaling requires g protein-coupled receptor kinase 2 binding to the second intracellular loop. *J. Biol. Chem.* 280 (26): 24420–24427.

134 Hermans, E. and Challiss, R.A. (2001). Structural, signalling and regulatory properties of the group I metabotropic glutamate receptors: prototypic family C G-protein-coupled receptors. *Biochem. J.* 359 (Pt 3): 465–484.

135 Wu, H., Wang, C., Gregory, K.J. et al. (2014). Structure of a class C GPCR metabotropic glutamate receptor 1 bound to an allosteric modulator. *Science* 344 (6179): 58–64.

136 Doré, A.S., Okrasa, K., Patel, J.C. et al. (2014). Structure of class C GPCR metabotropic glutamate receptor 5 transmembrane domain. *Nature* 511 (7511): 557–562.

137 Christopher, J.A., Aves, S.J., Bennett, K.A. et al. (2015). Fragment and structure-based drug discovery for a class C GPCR: discovery of the mGlu5 negative allosteric modulator HTL14242 (3-chloro-5-[6-(5-fluoropyridin-2-yl) pyrimidin-4-yl]benzonitrile). *J. Med. Chem.* 58 (16): 6653–6664.

138 Wernimont, A.K., Dong, A., Seitova, A. et al. (2011). Crystal structure of metabotropic glutamate receptor 3 precursor in presence of LY341495 antagonist. PDB ID: 3SM9; DOI: https://doi.org/10.2210/pdb3sm9/pdb.

139 Monn, J.A., Prieto, L., Taboada, L. et al. (2015). Synthesis and pharmacological characterization of C4-disubstituted analogs of 1*S*,2*S*,5*R*,6*S*-2-aminobicyclo[3.1.0]hexane-2,6-dicarboxylate: identification of a potent, selective metabotropic glutamate receptor agonist and determination of agonist-bound human mGlu2 and mGlu3 amino terminal domain structures. *J. Med. Chem.* 58 (4): 1776–1794.

140 Monn, J.A., Prieto, L., Taboada, L. et al. (2015). Synthesis and pharmacological characterization of C4-(thiotriazolyl)-substituted-2-aminobicyclo[3.1.0] hexane-2,6-dicarboxylates. Identification of (1*R*,2*S*,4*R*,5*R*,6*R*)-2-amino-4-(1*H*-1,2,4-triazol-3-ylsulfanyl)bicyclo[3.1.0]hexane-2,6-dicarboxylic acid (LY2812223), a highly potent, functionally selective mGlu2 receptor agonist. *J. Med. Chem.* 58 (18): 7526–7548.

141 Chappell, M.D., Li, R., Smith, S.C. et al. (2016). Discovery of (1*S*,2*R*,3*S*,4*S*,5*R*,6*R*)-2-amino-3-[(3,4-difluorophenyl)sulfanylmethyl]-4-hydroxy-bicyclo[3.1.0] hexane-2,6-dicarboxylic acid hydrochloride (LY3020371·HCl): a potent, metabotropic glutamate 2/3 receptor antagonist with antidepressant-like activity. *J. Med. Chem.* 59 (24): 10974–10993.

142 Monn, J.A., Henry, S.S., Massey, S.M. et al. (2018). Synthesis and pharmacological characterization of C4beta-amide-substituted 2-aminobicyclo[3.1.0]hexane-2,6-dicarboxylates. Identification of (1*S*,2*S*,4*S*,5*R*,6*S*)-2-amino-4-[(3-methoxybenzoyl)amino]bicyclo[3.1.0] hexane-2,6-dicarboxylic acid (LY2794193), a highly potent and selective mGlu3 receptor agonist. *J. Med. Chem.* 61 (6): 2303–2328.

143 Dobrovetsky, E., Dong, A., Hutchinson, A. et al. (2015). Human metabotropic glutamate receptor 7, extracellular ligand binding domain. *To be published,* RCSB Protein Data Bank, deposited June 19, 2015; DOI: https://doi.org/10.2210/pdb5c5c/pdb.

144 Dobrovetsky, E., Khutoreskaya, G., Seitova, A. et al. (2010). mGluR7 complexed with LY341495. *To be published,* RCSB Protein Data Bank, deposited April 27, 2010; DOI: https://doi.org/10.2210/pdb3mq4/pdb.

145 Schkeryantz, J.M., Chen, Q., Ho, J.D. et al. (2018). Determination of L-AP4-bound human mGlu8 receptor amino terminal domain structure and the molecular basis for L-AP4's group III mGlu receptor functional potency and selectivity. *Bioorg. Med. Chem. Lett.* 28 (4): 612–617.

146 Fremeau, R.T. Jr., Troyer, M.D., Pahner, I. et al. (2001). The expression of vesicular glutamate transporters defines two classes of excitatory synapse. *Neuron* 31 (2): 247–260.

147 Takamori, S., Rhee, J.S., Rosenmund, C., and Jahn, R. (2000). Identification of a vesicular glutamate transporter that defines a glutamatergic phenotype in neurons. *Nature* 407 (6801): 189–194.
148 Herzog, E., Bellenchi, G.C., Gras, C. et al. (2001). The existence of a second vesicular glutamate transporter specifies subpopulations of glutamatergic neurons. *J. Neurosci.* 21 (22): RC181–RC181.
149 Hisano, S., Sawada, K., Kawano, M. et al. (2002). Expression of inorganic phosphate/vesicular glutamate transporters (BNPI/VGLUT1 and DNPI/VGLUT2) in the cerebellum and precerebellar nuclei of the rat. *Brain Res. Mol. Brain Res.* 107 (1): 23–31.
150 Nunzi, M.G., Russo, M., and Mugnaini, E. (2003). Vesicular glutamate transporters VGLUT1 and VGLUT2 define two subsets of unipolar brush cells in organotypic cultures of mouse vestibulocerebellum. *Neuroscience* 122 (2): 359–371.
151 Fremeau, R.T., Burman, J., Qureshi, T. et al. (2002). The identification of vesicular glutamate transporter 3 suggests novel modes of signaling by glutamate. *Proc. Natl. Acad. Sci. U.S.A.* 99 (22): 14488–14493.
152 Varoqui, H., Schäfer, M.K.-H., Zhu, H. et al. (2002). Identification of the differentiation-associated Na^+/P_I transporter as a novel vesicular glutamate transporter expressed in a distinct set of glutamatergic synapses. *J. Neurosci.* 22 (1): 142–155.
153 Sheldon, A.L. and Robinson, M.B. (2007). The role of glutamate transporters in neurodegenerative diseases and potential opportunities for intervention. *Neurochem. Int.* 51 (6–7): 333–355.
154 Rattray, M. and Bendotti, C. (2006). Does excitotoxic cell death of motor neurons in ALS arise from glutamate transporter and glutamate receptor abnormalities? *Exp. Neurol.* 201 (1): 15–23.
155 Blasco, H., Mavel, S., Corcia, P., and Gordon, P.H. (2014). The glutamate hypothesis in ALS: pathophysiology and drug development. *Curr. Med. Chem.* 21 (31): 3551–3575.
156 Revett, T.J., Baker, G.B., Jhamandas, J., and Kar, S. (2013). Glutamate system, amyloid ss peptides and tau protein: functional interrelationships and relevance to Alzheimer disease pathology. *J. Psychiatry. Neurosci.* 38 (1): 6–23.
157 Dvorzhak, A., Vagner, T., Kirmse, K., and Grantyn, R. (2016). Functional indicators of glutamate transport in single striatal astrocytes and the influence of Kir4.1 in normal and huntington mice. *J. Neurosci.* 36 (18): 4959–4975.
158 Malkki, H. (2016). Huntington disease: new study challenges the hypothesis of glutamate transporter dysfunction in Huntington disease. *Nat. Rev. Neurol.* 12 (5): 251.
159 Zhang, Y.L., Liu, Y., Kang, X.P. et al. (2018). Ginsenoside Rb1 confers neuroprotection via promotion of glutamate transporters in a mouse model of Parkinson's disease. *Neuropharmacology* 131: 223–237.
160 Aihara, Y., Mashima, H., Onda, H. et al. (2000). Molecular cloning of a novel brain-type Na(+)-dependent inorganic phosphate cotransporter. *J. Neurochem.* 74 (6): 2622–2625.
161 Yernool, D., Boudker, O., Jin, Y., and Gouaux, E. (2004). Structure of a glutamate transporter homologue from Pyrococcus horikoshii. *Nature* 431: 811.

162 Akyuz, N., Georgieva, E.R., Zhou, Z. et al. (2015). Transport domain unlocking sets the uptake rate of an aspartate transporter. *Nature* 518 (7537): 68–73.
163 Scopelliti, A.J., Font, J., Vandenberg, R.J. et al. (2018). Structural characterisation reveals insights into substrate recognition by the glutamine transporter ASCT2/SLC1A5. *Nat. Commun.* 9 (1): 38.
164 Voglmaier, S.M., Kam, K., Yang, H. et al. (2006). Distinct endocytic pathways control the rate and extent of synaptic vesicle protein recycling. *Neuron* 51 (1): 71–84.
165 Foss, S.M., Li, H., Santos, M.S. et al. (2013). Multiple dileucine-like motifs direct VGLUT1 trafficking. *J. Neurosci.* 33 (26): 10647–10660.
166 Takamori, S., Malherbe, P., Broger, C., and Jahn, R. (2002). Molecular cloning and functional characterization of human vesicular glutamate transporter 3. *EMBO Rep.* 3 (8): 798–803.
167 Ruel, J., Emery, S., Nouvian, R. et al. (2008). Impairment of SLC17A8 encoding vesicular glutamate transporter-3, VGLUT3, underlies nonsyndromic deafness DFNA25 and inner hair cell dysfunction in null mice. *Am. J. Hum. Genet.* 83 (2): 278–292.

6

The GABAergic System

6.1 Introduction

The GABAergic system is constituted by a series of structures, including receptors and transporters, involved in release and binding of γ-Aminobutyric acid (GABA) as a neurotransmitter. GABAergic inhibition is located at all levels of the central nervous system (CNS), with more than 50% of the inhibitory synapses in the brain being mediated by GABA. It can be found at the hypothalamus, hippocampus, cerebral cortex, and cerebellar cortex. GABA pathways are well established and distributed by the whole brain (Figure 6.1). Cortical GABAergic neurons contribute to long-range projections in the septo-hippocampal system [1–4], between the hippocampus and entorhinal cortex [5], from the ventral hippocampal formation to the amygdala [6, 7], and within the hippocampal formation [8, 9] and the neocortex [10, 11]. GABA-releasing neurons located in the ventral tegmental area (VTA) provide local inhibition but also project to the nucleus accumbens inhibiting cholinergic interneurons and thus, enhancing stimulus-outcome learning [12]. Indeed, VTA GABA neurons modulate specific learning behaviors through the control of dopaminergic and cholinergic systems [13]. GABAergic neurons are critical for the network activity of the amygdala and hippocampus. Specific subpopulations of cortical GABAergic neurons send long-range axonal projections to subcortical and other cortical areas. A variety of GABAergic neurons in the hippocampus project to neighboring and remote areas including the subiculum and other retrohippocampal areas, the medial septum, and the contralateral dentate gyrus [9].

The GABAergic system plays a crucial role in neurodevelopmental disorders [14]. GABAergic dysfunction in the basolateral amygdala causes hyperexcitability, strongly associated with anxiety, hypervigilance, and inability to regulate emotions. It can also be related to Alzheimer's disease (AD) [15] and epilepsy [16]. The impairment between structure and function of the GABAergic circuit structure is strongly related to fragile X syndrome [17], the most common heritable cause of autism and intellectual disability [18]. Genetic and environmental risk factors that predispose to schizophrenia disrupt the development and normal functioning of the GABAergic system [19].

GABA is synthesized from glutamate by the action of glutamic acid decarboxylase (GAD) in presynaptic neurons. Even though glutamate is an

Chemical Biology of Neurodegeneration: A Molecular Approach, First Edition. Pedro Merino.

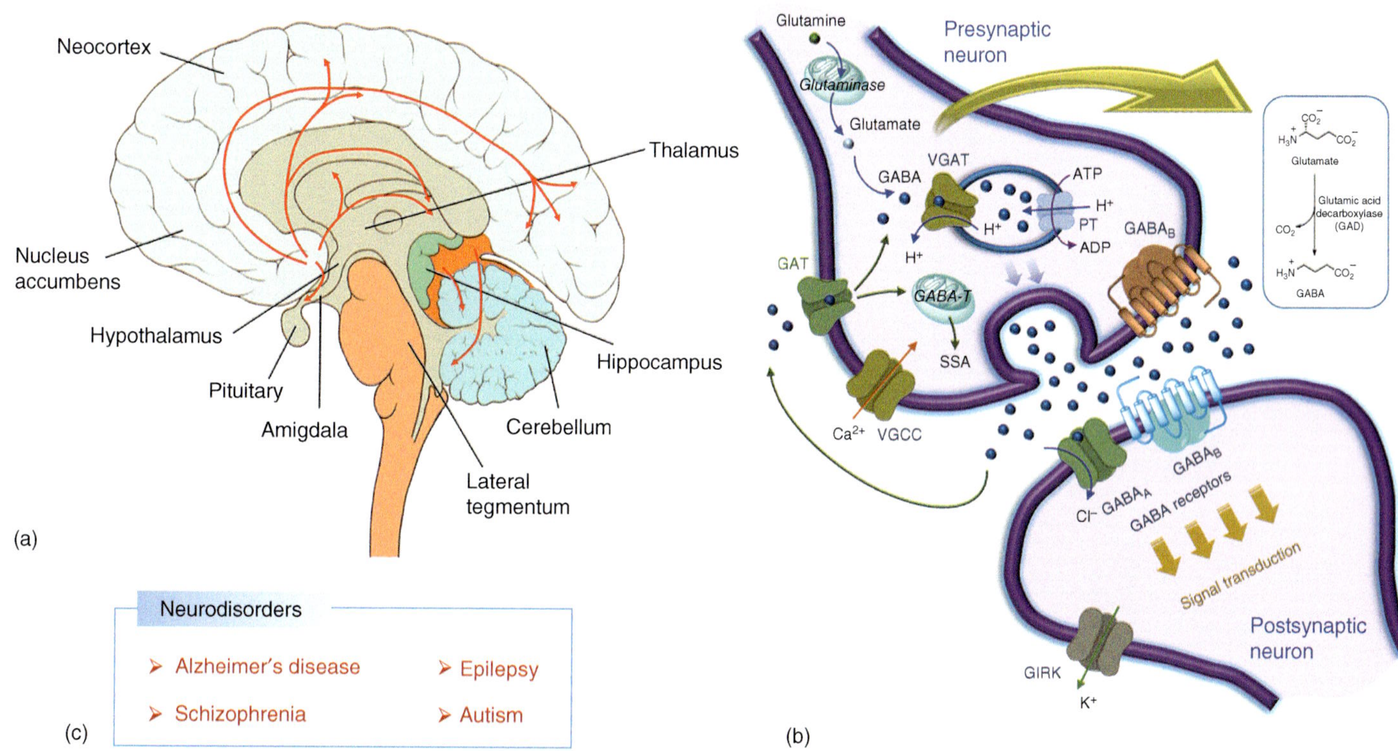

Figure 6.1 The GABAergic system. (a) Major projections of the GABAergic system. (b) Neurological disorders associated with the GABAergic system. VGCC: voltage-gated calcium channel; GIRK: G-protein-coupled inwardly rectifying potassium channel; VGAT: vesicular GABA transporter; GAT: GABA transporter; GABA-T: GABA transaminase; SSA: succinic semialdehyde; PT: proton transporter. (c) Details of the GABAergic synapse: synthesis of GABA from glutamate, absorption into synaptic vesicles by VGAT, release of GABA, binding of GABA to receptors, and reuptake of GABA by GAT.

important neurotransmitter by itself, the control of GABA synthesis depends on the activity of GAD and its cofactor, pyridoxal phosphate. Like other neurotransmitters, after being synthesized, GABA is stored in vesicles by the action of a specific vesicular GABA transporter (VGAT) [20]. Further calcium-mediated depolarization leads to exocytosis of vesicles releasing GABA in the synaptic cleft. Once released, GABA interacts with different receptors to produce the signal. Regulation of GABA release occurs via presynaptic $GABA_B$ autoreceptors. The termination of the GABA-mediated signal takes place through the reuptake transporter system that transports GABA back into the presynaptic neurons or into surrounding glial cells and astrocytes. This dual neuronal–glial reuptake process is distinct from that observed with the classical monoamines, for which neuronal reuptake largely predominates. Four transporters with 12 transmembrane-spanning segments have been identified and cloned: γ-Aminobutyric acid transporter (GAT-1), GAT-2, GAT-3, and betaine-γ-aminobutyric acid cotransporter (BGT-1) [21]. All four transporters are capable of transporting GABA, as well as other inhibitory amino acids (IAAs), particularly, glycine, β-alanine, and taurine. One of the transporters, BGT-1, is capable of transporting the dimethylated glycine derivative betaine and thus is also known as the betaine transporter.

6.2 GABA Receptors

6.2.1 Classification

GABA receptors have been classified into three types: $GABA_A$, $GABA_B$, and $GABA_C$ receptors [22]. $GABA_A$ and $GABA_C$ receptors are members of a superfamily of transmitter-gated ion channels such as 5-HT3 receptors. Both $GABA_A$ and $GABA_C$ are selective for chloride anions. While the former are pentameric complexes formed by different subunits, the latter are made up of a single type of protein subunits. Due to their similarity in pharmacology, some authors consider $GABA_C$ receptors as a subclass of $GABA_A$ receptors just with a different stoichiometry between subunits. Up to 19 different subunits (α1–6, β1–3, γ1–3, δ, ε, θ, π, and ρ1–3) have been identified for $GABA_A$ receptors [23, 24]. $GABA_B$ receptors are heteromeric receptors, comprising two subunits $GABA_{B1}$ and $GABA_{B2}$, belonging to the family of G-protein-coupled receptors that activate second messenger systems and calcium and potassium ion channels. $GABA_{B1}$ activates neurons, and $GABA_{B2}$ is involved in signaling and membrane targeting [25]. $GABA_{B1}$ receptors can be further subdivided into $GABA_{B1a}$ and $GABA_{B1b}$ subtypes (Table 6.1). The sequences of subunits comprising GABA receptors are collected in Figure 6.2.

6.2.2 Location

In the CNS, each $GABA_A$ subunit has a particular distribution pattern in the brain according to immunohistochemical studies and *in situ* hybridization studies (Table 6.2) [27]. For instance, α5 subunits are highly localized in the olfactory

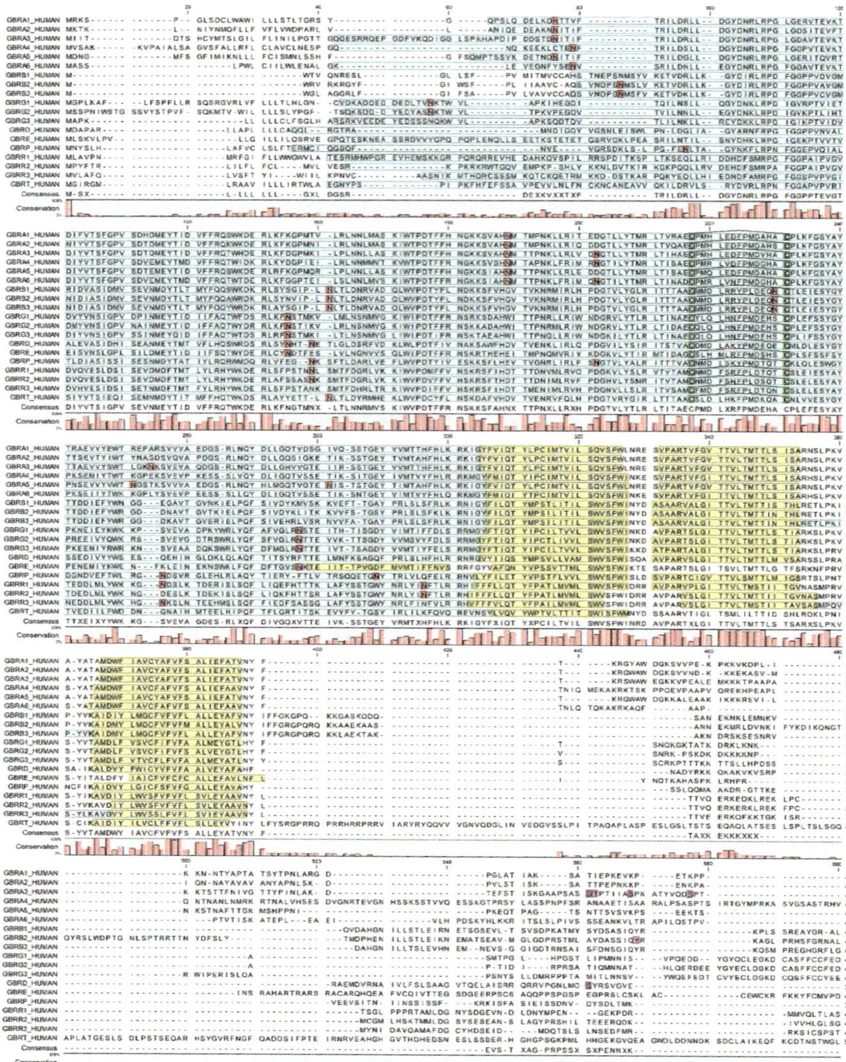

Figure 6.2 Sequences of the human subunits of $GABA_A$ receptors. Conservation graphic and consensus sequence are shown. Disulfide bonds (green), N-glycosylation sites (red), phosphorylation (magenta) sites, extracellular domains (light blue), intracellular domains (white), and transmembrane domains (yellow) are indicated on sequences.

Table 6.1 Classification of GABA receptors.

Receptor	Subunits[a)]	Subtypes
$GABA_A$	$\alpha_1\beta_2\gamma_2$, $\alpha_1\beta_3\gamma_2$, $\alpha_1\beta_2\delta$, $\alpha_1\beta_2$	
	$\alpha_2\beta_3\gamma_2$	
	$\alpha_3\beta_2\gamma_2$, $\alpha_3\beta_3\gamma_2$, $\alpha_3\beta_3\theta$, $\alpha_3\beta_3\varepsilon$, $\alpha_3\beta_3$	
	$\alpha_4\beta_3\gamma_2$, $\alpha_4\beta_2\delta$, $\alpha_4\beta_3\delta$, $\alpha_4\beta_3$	
	$\alpha_5\beta_3\gamma_2$	
	$\alpha_6\beta_2\gamma_2$, $\alpha_6\beta_3\gamma_2$, $\alpha_6\beta_2\delta$, $\alpha_6\beta_3\delta$, $\alpha_6\beta_3$	
$GABA_B$	$GABA_{B1}$	$GABA_{B1a}$, $GABA_{B1b}$
	$GABA_{B2}$	

a) Common stoichiometries are shown for $GABA_A$ receptors.

Table 6.2 Localization of major $GABA_A$ receptor subtypes [26].

Subunit	Abundance	Major sites of expression
$\alpha_1\beta_2\gamma_2$ $\alpha_1\beta_3\gamma_2$ $\alpha_1\beta_2\delta$ $\alpha_1\beta_2$	60%	Cerebral cortex (layers 1–6), hippocampus, amygdala, olfactory bulb, thalamus, basal forebrain, globus pallidus, substantia nigra pars reticulata, inferior colliculus, cerebellum, brainstem
$\alpha_2\beta_3\gamma_2$	15–20%	Cerebral cortex (layers 1–4), hippocampal formation, amygdala, striatum, olfactory bulb, hypothalamus, superior colliculus, inferior olive, motor nuclei, spinal cord dorsal horn
$\alpha_3\beta_2\gamma_2$ $\alpha_3\beta_3\gamma_2$ $\alpha_3\beta_3\theta$ $\alpha_3\beta_3\varepsilon$ $\alpha_3\beta_3$	<5%	Cerebral cortex (layers 5 and 6), amygdala, olfactory bulb, thalamic reticular and intralaminar nuclei, superior colliculus, brainstem, spinal cord, locus coeruleus, raphe, medial septum
$\alpha_4\beta_3\gamma_2$ $\alpha_4\beta_2\delta$ $\alpha_4\beta_3\delta$ $\alpha_4\beta_3$	<5%	Dentate gyrus, thalamus
$\alpha_5\beta_3\gamma_2$	<5%	Highest in hippocampus, considerably lower in deep cortical layers, amygdala, olfactory bulb, hypothalamus, superior colliculus, superior olivary nucleus, spinal trigeminal nucleus, spinal cord
$\alpha_6\beta_2\gamma_2$ $\alpha_6\beta_3\gamma_2$ $\alpha_6\beta_2\delta$ $\alpha_6\beta_3\delta$ $\alpha_6\beta_3$	<5%	Cerebellum, dorsal cochlear nucleus

bulb, the hypothalamus, and the hippocampus. The α4 subunits are concentrated in the thalamus, hippocampus, olfactory tubercle, and basal ganglia. The distribution of δ subunits parallels that of α4 subunits in the thalamus, striatum, outer layers of the cortex, and the hippocampus [28]. Autoradiography studies have revealed that α1β receptors (with or without δ units) are located in the thalamus and other forebrain regions, and receptors containing α6 subunits are expressed in the granule cell layer of the cerebellum [24].

$GABA_B$ receptors are located throughout the brain at both presynaptic and postsynaptic sites. The brain regions possessing the highest density of $GABA_B$ receptors are the thalamic nuclei, the molecular layer of the cerebellum, the cerebral cortex, and the interpeduncular nucleus. In the brainstem and certain thalamic nuclei, $GABA_B$ receptors can account for up to 90% of the total GABA binding sites. The $GABA_{B1}$ and $GABA_{B2}$ subunits show an overlapping pattern of distribution throughout the CNS although there are some exceptions such as caudate–putamen and cerebellum where $GABA_{B1}$ is in much greater abundance [21]. In particular, $GABA_{B1a}$ receptors are located at the presynaptic terminals of excitatory synapses and $GABA_{B1b}$ receptors are found at the postsynaptic terminals [25]. $GABA_{B2}$ receptors are highly expressed in the brain, in particular in cerebral cortex, thalamus, hippocampus, frontal, occipital and temporal lobe, occipital pole and cerebellum followed by corpus callosum, caudate nucleus, spinal cord, amygdala, and medulla [29–31].

6.2.3 Implication in Neurodegenerative Disorders

A disruption of the GABAergic system may contribute to important neurodegenerative disorders such as Alzheimer's [32] and Parkinson's [33] diseases. For instance, failure of the calcium/GABA control might promote a series of cascading processes leading to both weakened protective barriers, such as the blood–brain barrier, and an accumulation of intracellular deposits of calcium and Lewy bodies. There are documented aberrations in the GABAergic system in AD that demonstrate alterations of particular $GABA_A$ receptors [34]. $GABA_A$ receptors have also shown great clinical significance in epilepsy and anxiety [35]. In fact, they are well-known targets for anxiolytics [36]. $GABA_B$ receptors play a crucial role in the pathogenesis of Parkinson's [37] and Alzheimer's [15] diseases. Consequently, they are targets for therapies [38]. The role of these receptors and calcium channels has been studied in Parkinson's disease (PD) models [39], and there are evidences of decreased $GABA_B$ receptors in the putamen and external globus pallidus (GP) [40]. Currently, the GABAergic pharmacotherapy is being considered for motor disorders of the CNS [41] as well as for schizophrenia, neuropathic pain, and Alzheimer's and Parkinson's diseases [42]. Reduced GABA concentration is associated with physical disability caused by progressive multiple sclerosis [43].

6.2.4 $GABA_A$ Receptors

The $GABA_A$ receptor is a ligand-gated ion channel (ionotropic) that belongs to the Cys-loop (cysteine–cysteine loop located on the N-terminal region) family of receptors (such as nicotinic acetylcholine (nACh) and serotonin 5-HT3 receptors). It has a pentameric structure (Figure 6.3) composed of various combinations of subunits (α, β, γ, δ, ε, π, θ, and ρ) that are arranged in a circle, forming the

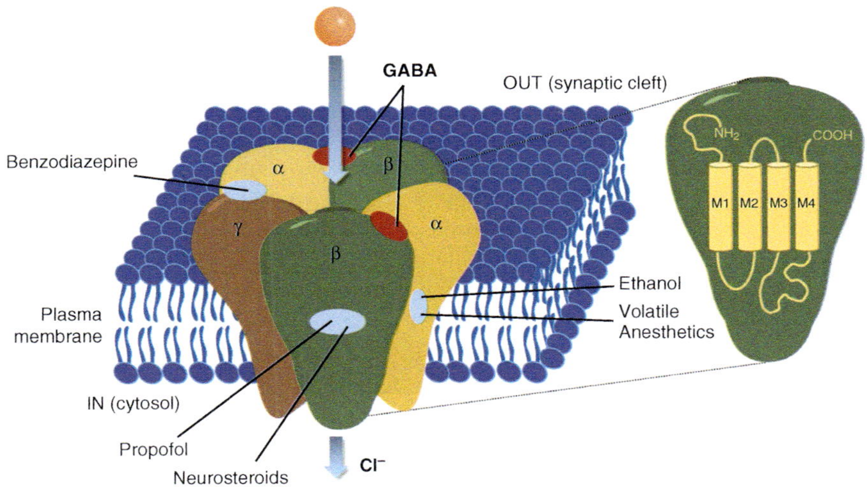

Figure 6.3 Schematic model of a pentameric $GABA_A$ receptor ($\alpha_2\beta_2\gamma$) comprising a chloride-selective pore. Specific sites for allosteric modulators are indicated in cyan. Active sites of GABA are indicated in red.

ion channel with a diameter of c. 8 nm in the same way as that of cholinergic ionotropic receptors (see Chapter 4). Various subunit isoforms have been identified with 6α, 3β, 3γ, 1δ, 1ε, 1θ, 1π, and 3ρ subunits [23]. Despite the potential for receptor heterogeneity, the majority of $GABA_A$ receptors will contain two α subunits, two β subunits, and one γ subunit [44].

Each subunit has four transmembrane domains (TMDs) and extracellular N- and C-termini. The most common arrangement in mammalian CNS consists of an $\alpha_2\beta_2\gamma$ stoichiometry although up to seventeen different combinations of subunits have been identified. Such a variability in subunit composition allows high levels of functional diversity showing subtle variations in binding affinity, channel activity, and degree of influence of modulators in GABA binding. Several general [45–47] and more dedicated reviews regarding structure [48], function [49, 50], pharmacology [51, 52], and drug interactions [23] have been reported elsewhere.

6.2.4.1 Effector Mechanisms

$GABA_A$ receptors, activated by binding of two molecules of GABA, allow the entry of chloride ions into the cell (Figure 6.3), thereby hyperpolarizing the neurons and making cell firing less likely to occur, and, in consequence, the cell becomes less reactive to excitatory neurotransmitters. $GABA_A$ receptors are responsible for fast inhibitory postsynaptic potentials on the cell soma of principal neurons in the cortex and hippocampus, and depolarization of primary afferent terminals (presynaptic inhibition). In neonatal neurons, the chloride gradient is opposite to that in mature neurons, and, therefore, activation of $GABA_A$ receptors leads to membrane depolarization [53].

6.2.4.2 Structure of Subunits

The regulation of $GABA_A$ receptors is determined by their heterogeneity but also by posttranslational phosphorylation that can affect the allosteric modulation of the receptors [54]. In addition, $GABA_A$ receptors can also be glycosylated at Asn residues located at different positions. Table 6.3 collects glycosylation and phosphorylation sites of the different $GABA_A$ receptor subunits.

The α1 subunit presents several natural variants. Arg112Gln, Gly251Ser, and Lys306Thr are found in Dravet syndrome [55]; Leu146Met [56], Asp219Asn [57], and Ala322Asp [58] are involved in epilepsy. The α2 subunit has an isoform that differs from the canonical sequence in the residues 1–85 and 353. For α4 and α6 subunits, three natural variants have been reported [59]: Ala19Thr, Leu26Met, and Ser516Arg for α4 and Pro180His, Thr187Met, and Pro404Ser for α6. No natural variants have been described for α3 and α5 subunits.

The natural variants Phe246Ser and Thr287Ile [60] of β1 subunits are related to epilepsy [61, 62]. The natural variant His421Gln is related to schizophrenia [63], and no effect has been reported for Ile429Asn. There is also a second iso-

Table 6.3 Glycosylation and phosphorylation sites for $GABA_A$ subunits.

Subunits	Glycosylation sites[a)]	Phosphorylation sites[b)]
α1, α2	38/138	
α3	63/163/176/228	426/427[c)]/433/442
α4	47/144/157	
α5	45/145/207/236	
α6	31/128/141	403
β1	105/174	
β2	32/104/173	441[d)]
β3	33/105/174	
γ1	50/127/245	
γ2	52/129/247	
γ2	110/228	
δ	103/106	390
ε	134/252	
π	43/102/145/196/228	
ρ1	140/234/274	
ρ2	120/254	
ρ3	220	
θ	127	

a) In all cases, residues are Asn.
b) In all cases, residues are Ser, unless otherwise indicated.
c) The residue is Thr.
d) The residue is Tyr.

form differing in residues 59–87, in which residues 88–474 are missing. In this subunit, residues 120–122 and 180–182 have been identified as agonist-binding sites; residues 290–311 have been identified for allosteric effector binding. For β2 subunits, residue 224 has been identified as the binding site for agonists as well as the regions composed of residues 119–121 and 179–181; residues 289–310 have been identified for allosteric effector binding. Three alternative isoforms have been described: isoform 2 is missing residues 360–397; isoform 3 is missing residues 1–63 and 377–512, and differs in residues 360–376 [64]; isoform 4 has residues 360–397 missing and Glu410 has been replaced by Asp [65]. The β3 subunits present three alternative isoforms. Isoform 2 differs in residues 1–26 and isoform 3 in residues 3–80. Isoform 4 is missing residues 1–85. The canonical isoform has a natural variant (Glu32Arg) that is related to epilepsy [66]. Other natural variants include [67]: Asp120Asn [61, 62], Leu124Phe [56], Thr157Met, Tyr182Phe, Gln249Lys, Ser254Phe, Leu256Gln, Thr287Ile, Leu293His, and Ala305Thr. Isoform 2 has two natural variants related to epilepsy [68]: Pro11Ser and Ser15Phe.

The crystal structure of human β3 homopentamer in complex with agonist benzamidine has been resolved, showing a closed gate at the base of the pore, characteristic of a desensitized state [69]. The structure revealed the forces responsible for assembly between subunits (Figure 6.4a–c). These interactions take place between extracellular domains (ECD) through hydrogen bonds, salt bridges, and van der Waals contacts. Interactions between the extracellular and TMD involve hydrophobic contacts (Figure 6.4d,e). These contacts ensure a considerable solvent-inaccessible surface. Mutations linked to epilepsy highlight the importance of this interface [61, 71]. Crystallization with the agonist benzamidine allowed the identification of residues involved in binding (Figure 6.4f,g). Stacking between side chains of Phe200 and Tyr62 is observed as well as H-bond interactions between the amidinium group and Glu155. Cation-π interactions with Tyr157 and Tyr205 complete the set of interactions fixing the agonist in the binding site.

The crystal structure of a chimeric β3–α5 $GABA_A$ receptor in apo form (PDB ID: 5OJM) and complexed with pregnanolone (PDB ID: 5O8F) has been resolved [70]. The chimeric human receptor was constructed by fusing the α5 subunit TMD to the β3 subunit ECD. β3-homopentamers were gated by histamine (natural agonist) rather than GABA. The architecture of the receptor in both apo and complexed forms is illustrated in Figure 6.5a,b. Each subunit comprises an N-terminal β3 ECD folded into a twisted β-sandwich followed by an α5 four-helix-bundle TMD. Binding of pregnanolone is facilitated by hydrophobic cavities containing the critical Gln245 residue, whose ε-oxygen contributes a putative hydrogen bond from the top of the site to anchor the pregnanolone A-ring 3α-hydroxyl group (Figure 6.5c). Binding of neurosteroid pregnanolone has a dramatic impact on TMD conformation (Figure 6.5d,e). A shift is produced that leads to an overall straightening of the subunits, with the ECD and TMD undergoing a rocking motion about their shared interface as α-5TMD switches between the apo and pregnanolone-bound states (Figure 6.5d). This observed motion is consistent in type and magnitude with that previously proposed for nACh receptors (Figure 6.5e) [72].

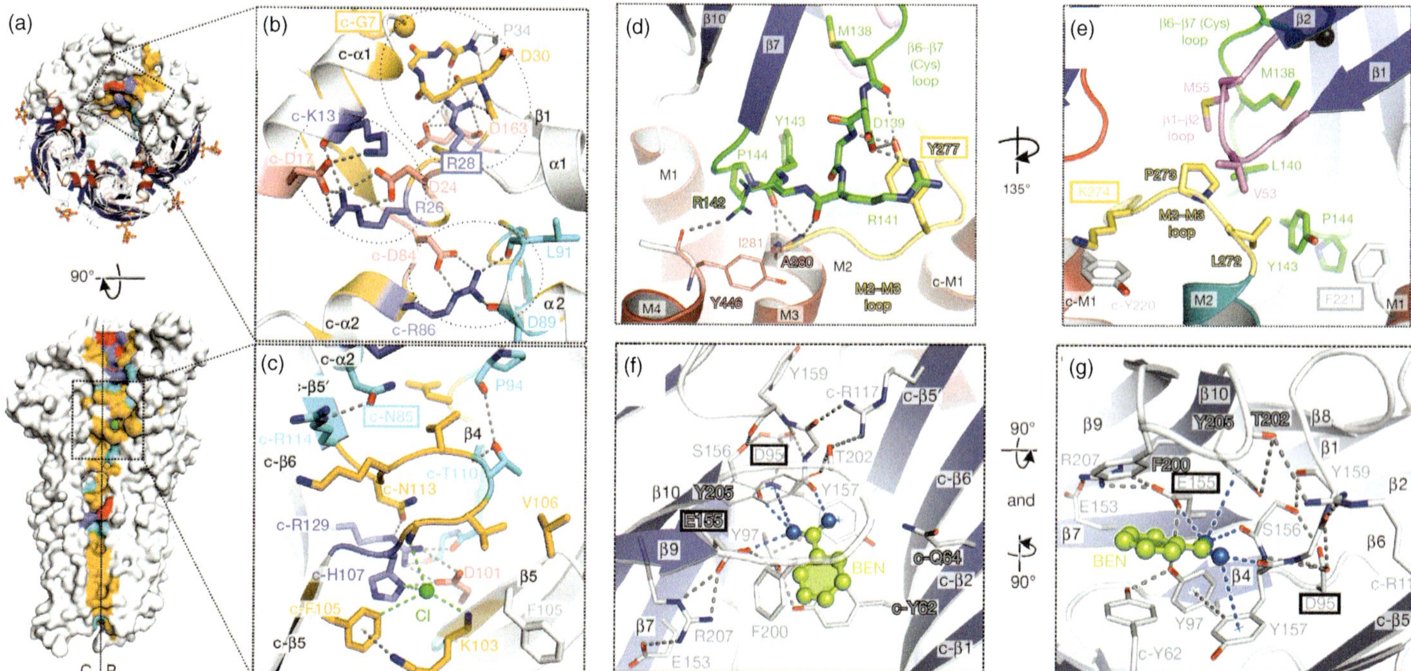

Figure 6.4 Structure of human GABA$_A$ β3 homopentamer. (a) Top-down view and side-on view of two neighboring subunits from the vestibule highlighting interactions between subunits. Salt-bridging residues are colored purple and red, those forming hydrogen bonds in cyan, and residues forming van der Waals contacts in orange. (b) Upper ECD close-up showing the intersubunit α1–β1 loop (upper dotted oval), the intersubunit salt bridges connecting a1 helices (middle dotted oval), and intersubunit a2–a2 interactions (lower dotted oval). (c) The ECD anion-binding site and surrounding intersubunit interface (chloride shown as a green sphere). Gray dashed lines indicate putative salt bridges and hydrogen bonds; green dashes indicate chloride coordination. (d, e) Structural coupling at the ECD and TMD interface. (f, g) Neurotransmitter pocket occupied by the agonist benzamidine. Gray dashed lines indicate putative hydrogen bonds, salt bridges, and cation-π interactions; blue dashed lines indicate the coordination sphere of benzamidine nitrogen atoms. Boxed residue labels indicate disease mutations. Source: Reprinted with permission from Miller and Aricescu 2014 [69]. Copyright 2014, Springer Nature.

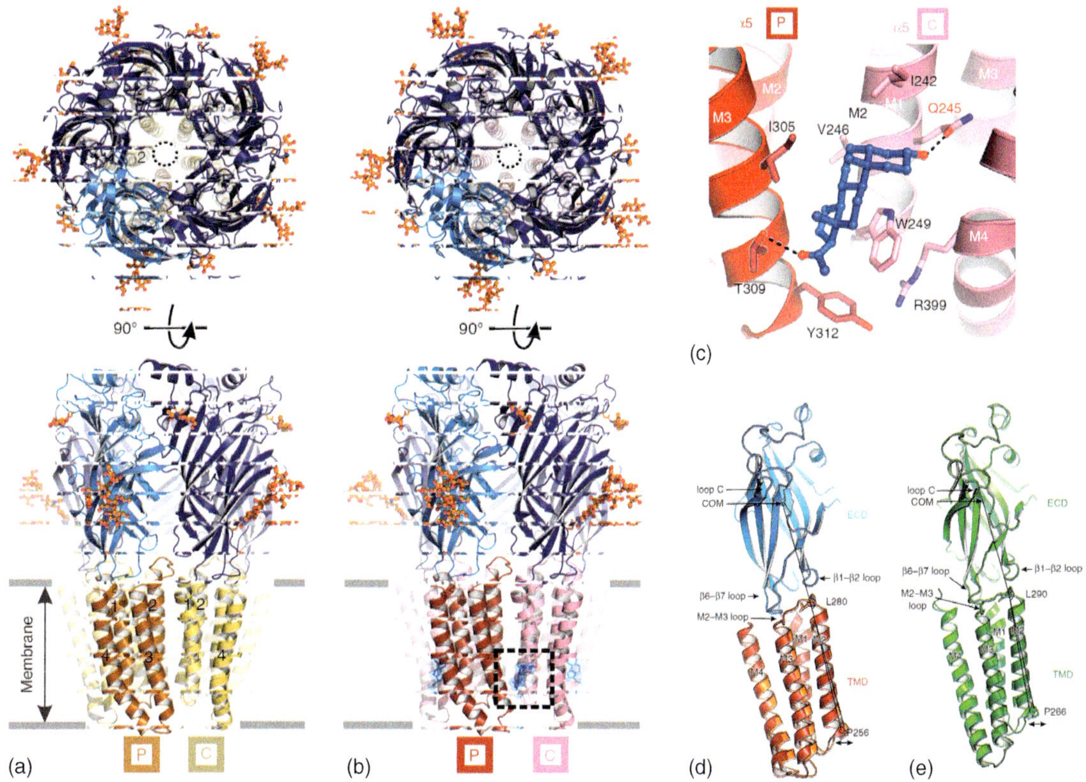

Figure 6.5 Structure of a chimeric β3–α5 $GABA_A$ receptor. (a,b) Architecture of $GABA_A$ α5-TMD. Top-down (top) and side-on (bottom) views of the α5TMD pentamer in apo (a) and pregnanolone-bound (b) forms. (c) The neurosteroid potentiation site. Pregnanolone (ball-and-stick representation; carbon atoms in blue, oxygen atoms in red) bound to an intersubunit site between M3 residues on the principal face of one subunit and M1 residues on the complementary face of the next. Putative hydrogen bonds between the pregnanolone 3α-hydroxyl and the Gln245 amide (2.4 Å) and the C20 ketone and Thr309 hydroxyl (3.0 Å) are indicated by dashed lines. (d, e) TMD motions relative to the ECD. (d) Side-on view of a single subunit (B chains), from apo α5TMD (orange TMD) and pregnanolone-bound α5TMD (red TMD), superposed by their ECDs (blue). (e) Equivalent view and superposition between open GlyR (light green, PDB ID: 3JAE) and desensitized (or partially open) GlyR (dark green, PDB ID: 3JAF). Source: Reprinted with permission from Miller et al. 2017 [70]. Copyright 2017, Springer Nature.

The subunit γ1 has neither natural variants nor additional isoforms. Several natural variants of the subunit γ2 (Asn79Ser [73], Arg82Gln [74], Pro83Ser [57], Tyr274Cys [56], Arg323Gln [75], and Lys328Met [71]) have been associated with epilepsy. Other natural variants of this subunit include Arg177Gly [76] and His35Arg. Two isoforms have been identified in which residue 376 has been replaced by a sequence of nine amino acids. In isoform 3, residue 211 has also been replaced by a sequence of 41 amino acids. For the subunit γ3, one natural variant (Thr352Ala) [77] and one additional isoform differing in residues 239–253 and having missing residues 254–467 have been identified.

The subunit δ has five natural variants including Pro62Ser [78] and those related to epilepsy, Glu177Ala, Arg220Cys, Arg220His, and Val370Ile [79].

The subunit ε has an alternative isoform in which residues 366–506 are missing and residue Ile365 has been replaced by Met. Two natural variants (Leu9Phe and Ser102Ala) have been reported [80–82]. The subunit π has two natural variants Phe391Leu and His416Arg [59]. The subunit ρ1 has two natural variants (Met26Val and His27Arg) and two alternative isoforms in which residues 41–57 (isoform 2) and 1–87 (isoform 3) are missing. The subunit ρ2 has an alternative isoform in which Met1 has been replaced by a chain of 26 residues. No natural variants have been reported for subunits ρ2 and ρ3. The subunit θ has two natural variants Leu15le and Phe478Ile [82].

6.2.5 $GABA_B$ Receptors

The $GABA_B$ receptor is a G-protein coupled receptors (GPCR) heterodimer composed of two subunits: $GABA_{B1}$ and $GABA_{B2}$. Each subunit has seven TMD with extracellular N-terminus and intracellular C-terminus, the latter connecting in a 1 : 1 stoichiometric fashion. The $GABA_{B1}$ subunit was also found to exist in two isoforms, $GABA_{B1a}$ and $GABA_{B1b}$. It has been suggested that the $GABA_{B1a}$ subunit is preferentially located presynaptically, whereas the $GABA_{B1b}$ subunit is located postsynaptically. The $GABA_{B1}$ subunits contain the GABA binding domain extracellularly, whereas the $GABA_{B2}$ subunit couples to the G-protein intracellularly. Both subunits are required for normal receptor function. The structure and function of metabotropic $GABA_B$ receptors have been comprehensively reviewed [83]. Sequences of human $GABA_B$ receptors are illustrated in Figure 6.6.

6.2.5.1 Effector Mechanisms

$GABA_B$ receptors can mediate both postsynaptic and presynaptic inhibition by coupling to a variety of effector systems. Presynaptic inhibition takes place through a decrease in the influx of calcium, which results in a reduced release of neurotransmitters. $GABA_B$ receptors are G-proteins coupled to a number of cellular effector mechanisms, including adenylyl cyclase, voltage-dependent calcium channels, and inwardly rectifying potassium channels. These different effectors enable $GABA_B$ receptors to produce not only inhibition but also a diverse range of other effects on neuronal function.

$GABA_B$ receptors form heterodimers of $GABA_{B1}$ and $GABA_{B2}$ subunits that interact with each other through the C-terminal coiled-coil domains (Figure 6.7) [84, 85].

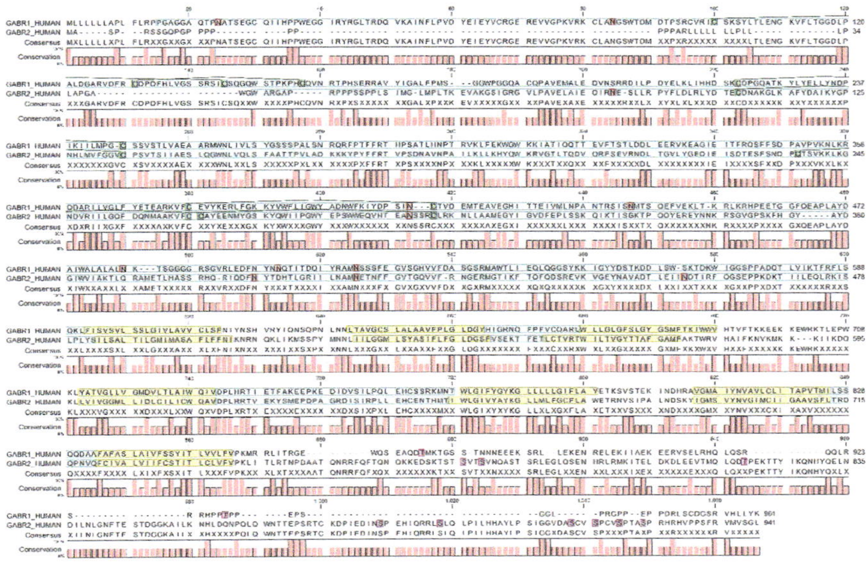

Figure 6.6 Sequences of human $GABA_B$ receptors. Conservation graphic and consensus sequence are shown. Disulfide bonds (green), N-glycosylation sites (red), phosphorylation (magenta) sites, palmitoylation sites (brown), extracellular domains (light blue), intracellular domains (white), and transmembrane domains (yellow) are indicated on sequences.

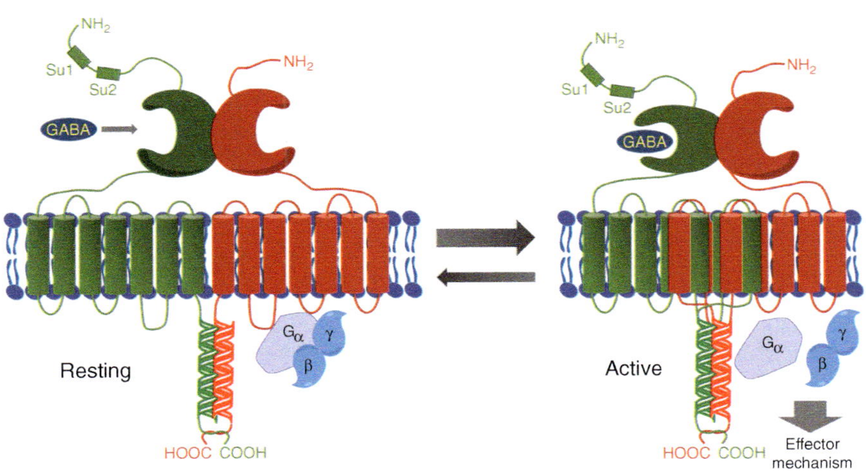

Figure 6.7 Effector mechanism of $GABA_B$ receptors. Subunits $GABA_{B1}$ and $GABA_{B2}$ are indicated in blue and red, respectively.

Heterodimerization is required for the formation of a functional receptor [86, 87]. The $GABA_{B1}$ subunit is responsible for GABA binding, and the $GABA_{B2}$ subunit plays a crucial role in signal transduction [88]. Homodimers may also be formed, but they are inactive [85].

In the resting conformation, the GABA-binding pocket is open, and TMD of each subunit are apart. Upon GABA binding, the active site closes and a conformational change is induced, causing the TMD of both subunits to move closer together. The conformational change initiates the effector signaling. The $GABA_{B1}$ subunit has two domains at the amino terminus denominated sushi (Su1 and Su2),which are the key for receptor targeting in the cell.

The activation of $GABA_B$ receptors inhibits adenylyl cyclase activity through the activation of $G_{i\alpha}$ and/or $G_{o\alpha}$ proteins, but, in contrast to this effector mechanism, $GABA_B$ receptors can have an opposite effect on the cyclic adenosine monophospate (cAMP) concentration. When adenylyl cyclase is stimulated to produce cAMP by another G_s-protein-coupled receptor, $GABA_B$ receptor activation enhances that effect contributing to increase the cAMP concentration [89]. The activation of $GABA_B$ receptors also inhibits voltage-dependent calcium channels but activates potassium channels [87]. These effects are dependent upon $G_{i/o}$ proteins through a direct interaction of βγ-subunits of the G-proteins with the corresponding channel.

Presynaptic GABA receptors inhibit neurotransmitter release by downregulating high-voltage-activated calcium channels, whereas postsynaptic GABA receptors decrease neuronal excitability by activating a prominent inwardly rectifying potassium conductance that underlies the late inhibitory postsynaptic potentials.

6.2.5.2 Structure of Receptors

The $GABA_{B1}$ subunits can be *N*-glycosylated at Asn residues located in positions 24, 84, 49, 440, 482, 502, and 514 [89]. Phosphorylation takes place at Thr873 and Thr930. Four isoforms have been identified: isoform 1B differing in residues 1–164, isoform 1C in which residues 98-159 are missing, isoform 1D differing in residues 905–961, and isoform 1E differing in residues 570–961 [90]. Three natural variants have been reported for $GABA_{B1}$ receptors, related to epilepsy [90–92]: Ala20Val, GLy498Ser, and Phe645Leu. Residues Phe247, Ser270, His287, Tyr367, Trp395, and Phe466 have been identified to be a part of agonist-binding sites. Mutagenesis studies showed that Tyr230Ala and Tyr234Ala decrease signaling via G-proteins, whereas His287Ala, Tyr367Ala, and Trp395Ala strongly reduce signaling via proteins. On the other hand, agonist binding is affected in a different way, whereas Tyr367Ala has no effect, Trp395Ala causes a strong decrease, and His287A abolishes the binding completely [93]. Furthermore, Trp182Ala abolishes both antagonist binding and signaling via G-proteins [89].

The $GABA_{B2}$ subunits can be glycosylated at Asn residues located in positions 90, 298, 389, 404, and 453. Phosphorylation takes place at Ser776, Ser779, Thr819, Ser884, Ser893, Ser913, Ser916, Ser920, and Ser924. Four natural variants have been reported for this receptor subtype: Leu162Phe, Ala567Thr [94], Tyr628Phe, and Thr869Ala [29]. Mutagenesis studies showed that Tyr118Ala decreases signaling via G-proteins [93].

The extracellular domain of human $GABA_B$ receptors has been crystallized (PDB IDs: 4F11 and 4F12) and the structure resolved [93]. The structure consists of two lobe-shaped domains (LB1 and LB2) connected by three short loops and has revealed a polar heterodimeric interface (Figure 6.8a). The observed architecture is similar to that of mGlu receptors and adopts an open conformation.

More recently, the crystal structures of the extracellular domain of human $GABA_B$ receptors in the apo form (PDB ID: 4MQE) and complexed with antagonists 2-hydroxysaclofen (PDB ID: 4MQF), CGP54626 (PDB ID: 4MR7), CGP35348 (PDB ID: 4MR8), SCH50911 (PDB ID: 4MR9), phaclofen (PDB ID:4MRM), CGP46381 (PDB ID: 4MS1), and agonist baclofen (PDB ID: 4MS4) as well as endogenous agonist GABA (PDB ID: 4MS3) have been resolved [89]. Agonist binding causes large conformational changes within the heterodimeric complex. In addition to domain closure of GBR_{1bVFT}, agonist binding brings the LB2 domain of GBR_{1bVFT} into close contact with the LB2 domain of GBR_{2VFT} to form a large heterodimer interface unique to the active site. The LB2 domain of GBR_{2VFT} moves toward the LB2 domain of GBR_{1bVFT} to form new heterodimeric contacts. These conformational changes from the apo to the agonist-bound state shorten the distance between the C termini of the two subunits from 45 to 32 Å (Figure 6.8b,c). Agonist binding induces the formation of an additional heterodimer interface between the LB2 domains of the GBR_{1bVFT} and GBR_{2VFT} subunits. The interaction between LB1 domains is mediated by the B and C helices of both subunits (Figure 6.8d). LB2 domains are mediated by two strand–loop–helix motifs (Figure 6.8f). In both cases, the interface can be divided into three adjacent areas from each domain on the basis of the observed interactions (Figure 6.8e,g).

All of the antagonists are derivatives of GABA and have the general structure of a γ-amino acid. The receptor–antagonist interactions are mainly based on H-bond interactions (Figure 6.8h,i), and $GABA_B$ receptors recognize both agonists GABA and (*R*)-baclofen AGO in essentially the same way (Figure 6.8j,k). Comparison of the binding sites of agonists and antagonists and evaluation of (un)favorable interactions with the residues of the binding site revealed the molecular basis of the agonist/antagonist activities (Figure 6.8l,m).

6.3 GABA Transporters

There is some heterogeneity between GABA transporters [95]. In principle, there are four distinct GABA transporters [96]: the betaine transporter (BGT1) and sodium- and chloride-dependent GABA transporters (GAT1–3), responsible for removing GABA from the synaptic cleft [97]. In addition, there is a fifth GABA transporter, termed VGAT – also known as vesicular inhibitory amino acid transporter (VIAAT) – responsible for the introduction of GABA into vesicles.

The sequences of VGAT, BGT1 and GAT1–3 are given in Figure 6.9.

GABA is mainly transported by GAT1 and GAT3 that are widely expressed in the brain [98]. GAT1 is located primarily on GABAergic nerve terminals where it recycles GABA; GAT3 is located predominantly on astrocytes around the GABAergic synapse [99]. GAT1 and GAT3 are the only transporters found in basal ganglia [100]. In particular, light and electron microscopic studies revealed very dense GAT3 immunoreactivity in the medial septum, diagonal band, basal

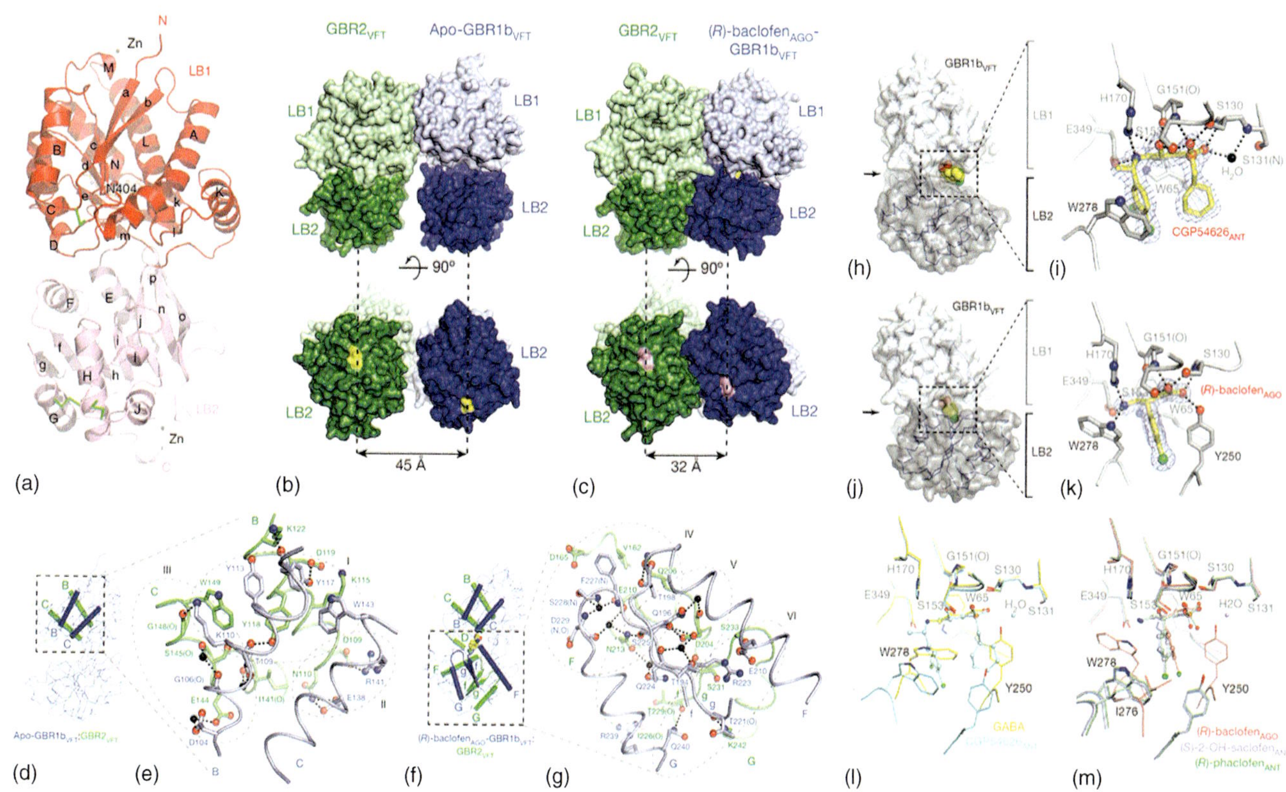
(a)
N
Zn
LB1
N404
LB2
Zn
(b)
GBR2$_{VFT}$
Apo-GBR1b$_{VFT}$
LB1
LB2
90°
45 Å
(c)
GBR2$_{VFT}$
(R)-baclofen$_{AGO}$-GBR1b$_{VFT}$
LB1
LB2
90°
32 Å
(d)
Apo-GBR1b$_{VFT}$:GBR2$_{VFT}$
(e)
(f)
(R)-baclofen$_{AGO}$-GBR1b$_{VFT}$:GBR2$_{VFT}$
(g)
(h)
GBR1b$_{VFT}$
LB1
LB2
(i)
H170
G151(O)
S130
E349
S131(N)
H_2O
W65
W278
CGP54626$_{ANT}$
(j)
GBR1b$_{VFT}$
LB1
LB2
(k)
H170
G151(O)
S130
E349
(R)-baclofen$_{AGO}$
W65
W278
Y250
(l)
H170
G151(O)
S130
E349
S153
W65
H_2O
S131
W278
Y250
GABA
(m)
H170
G151(O)
S130
E349
S153
W65
H2O
S131
W278
I276
Y250
(R)-baclofen$_{AGO}$
(S)-2-OH-saclofen$_{ANT}$
(R)-phaclofen$_{ANT}$

Figure 6.8 (a) Structure of the extracellular domain of human $GABA_B$ receptors. LB1 domain, red; LB2, pink; α-helices, uppercase; β-strands, lowercase. Observed N-linked carbohydrates at Asn404 are in yellow. Disulfide bonds are in green. Two zinc ions are depicted as black spheres. (b, c) Agonist-induced conformational changes. Superposition of apo (cyan) and (*R*)-baclofen AGO-bound (red) complexes based on the LB1 domain of GBR_{1bVFT} (b) or GBR_{2VFT} (c). Side view is shown on the right for the superimposed GBR_{1bVFT} (b) and GBR2VFT (c) subunits. (d–g) Heterodimer interface. (d) Structure of apo GBR_{1bVFT}:GBR_{2VFT} with the elements involved in heterodimer formation highlighted by ribbons (LB1–LB1: B and C helices). (e) Specific contacts at the LB1–LB1 heterodimer interface of apo-GBR_{1bVFT}:GBR_{2VFT}. The interface area is divided into three regions, I, II, and III (thin dashed lines). (f) Structure of (*R*)-baclofen AGO-GBR_{1bVFT}:GBR_{2VFT} showing the elements involved in heterodimer formation (LB1–LB1: B and C helices; LB2–LB2: F and G helices, f and g strands, and connecting loops). (g) Specific contacts at the LB2–LB2 heterodimer interface of (*R*)-baclofen AGOGBR$_{1bVFT}$:GBR_{2VFT}. The interface area is divided into three regions: IV, V, and VI (thin dashed lines). Thick dashed lines indicate hydrogen bonds. Ligand recognition by GBR_{1bVFT}. (h, j) Molecular surface of GBR1bVFT bound to antagonist CGP54626ANT (h) or agonist (*R*)-baclofen AGO (j). Ligand is displayed as a space-filling model. (i, k) Specific contacts between GBR_{1bVFT} (gray) and CGP54626ANT (yellow) (i) or (*R*)-baclofen AGO (j), viewed in the direction of the arrow in (i) panel (h) or (k) panel (j). Hydrogen bonds are represented by thick dashed lines. (l) Comparison of the binding sites of agonist GABA and antagonist CGP54626ANT. (m) Comparison of the binding sites of agonist (*R*)-baclofen AGO and two related antagonists, (*S*)-2-OH-saclofen ANT and (*R*)-phaclofen ANT.

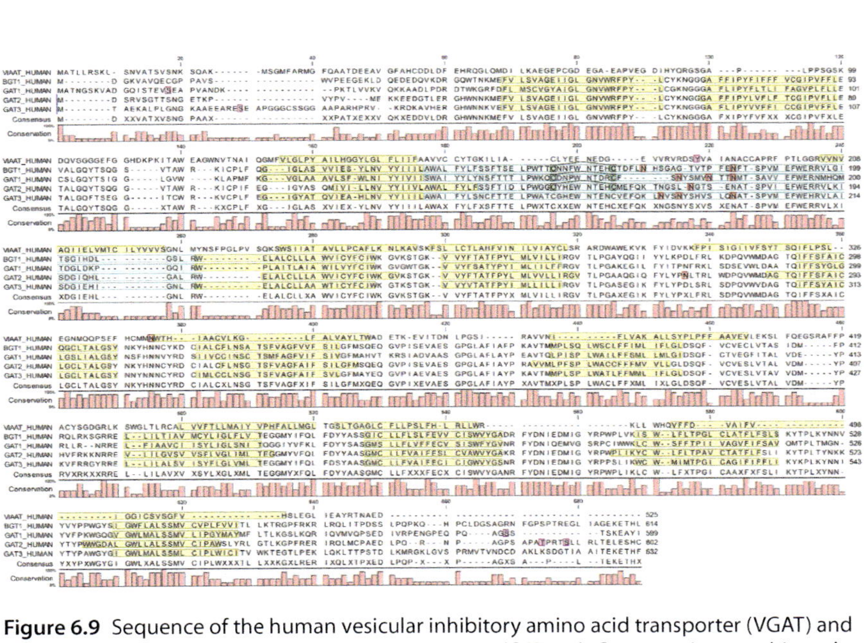

Figure 6.9 Sequence of the human vesicular inhibitory amino acid transporter (VGAT) and sodium- and chloride-dependent GABA transporters (GAT1–3). Conservation graphic and consensus sequence are shown. Disulfide bonds (green), N-glycosylation sites (red), phosphorylation (magenta) sites, metal-binding sites (gray), extracellular domains (light blue), intracellular domains (white), and transmembrane domains (yellow) are indicated on sequences.

nucleus of Meynert, thalamus, globus pallidus, and substantia nigra. Moderate levels were observed in the subthalamic nucleus, periaqueductal gray, spinal trigeminal, and vestibular nuclei [101]. Minor levels of expression in the brain (meninges) have been found for GAT2 and BGT1 [102].

In addition to EGABA, BGT1 transports betaine and GAT2 also transports taurine. Notably, GAT1 is inhibited by acetylcholinesterase (ACHC) but not by β-alanine, while GAT-2 and GAT-3 exhibit the opposite sensitivity to these compounds. GABA transporters, particularly those located on astrocytes, can affect the balance between excitatory and inhibitory pathways. In fact, GABA transporters play a dynamic role in the control of brain excitability [103] and mechanisms of epilepsy [104], which make them a drug target for GABA-related CNS disorders [105] including epilepsy [106]. The functional and pharmacological properties of GABA transporters have been reviewed in several reviews [98, 103, 105, 107] and book chapters [108, 109].

6.3.1 Vesicular GABA Transporter (VGAT)

The VGAT does not have any selectivity by GABA [110]. Contrary to other neurotransmitter transporters, the VGAT has only 10 transmembrane-spanning segments. The VGAT can be glycosylated at Asn341 and nitrated at Tyr186. A natural variant Ser423Gly has been described. Similar to the monoamine vesicular transporters, the VGAT appears to be driven by an H^+ electrochemical gradient generated by an ATP-dependent H^+ pump in the vesicle plasma membrane. The VGAT has no sequence similarity to vesicular monoamine transporter (VMATs) or vesicular acetylcholine transporter (VAChT). It mediates uptake of GABA but also recognizes glycine as a substrate, suggesting that these two inhibitory neurotransmitters can be stored in and released from the same vesicle.

6.3.2 Betaine Transporter (BGT1)

BGT1 transports GABA and betaine. It has a role in regulation of GABAergic transmission in the brain through the reuptake of GABA into presynaptic terminals, as well as in osmotic regulation. BGT1 can be glycosylated at Asn171 and Asn183. A natural variant (Cys10Arg) has been reported [111]. BGT1 is regulated by Janus-activated kinease-2 (JAK2) presumably by enhancing the insertion of carrier protein into the cell membrane [112]. A molecular model of human BGT1 has been constructed based on the structure of LeuT (Section 6.3.3) [113].

6.3.3 Sodium- and Chloride-Dependent GABA Transporters (GAT1–3)

GAT1–3 are responsible for removing GABA from the synaptic cleft [114]. They have 12 transmembrane domains TMD with the C- and N-termini located intracellularly. GAT1 can be *N*-glycosylated at Asn176, Asn181, and Asn184, in the loop between transmembrane domains 3 and 4 [115]. Phosphorylation takes place at Ser18 and Ser591, near the C- and N-termini. Residues Gly59, Ala61,

Ile62, Asn66, Ser295, Asn327, Leu392, Asp395, and Ser396 are involved in sodium binding. Four natural variants (Arg44Gln, Ala288Val, Gly297Arg, and Ala334Pro) have been identified and are related to epilepsy [116]. GAT1 has been identified as a crucial determinant of $GABA_B$ receptor activation in cortical circuits [117]. Studies using mutated and truncated GAT1 demonstrated that C- and N-terminal parts of the transporter are not essential for GABA uptake [118, 119]. On the other hand, Arg69, in TM1, is essential for the activity [120]. Gln291 is essential for the interaction of chloride with the transporter [121]. Mutations Trp68Ser/Leu and Trp222Ser/Leu in TM1 and TM4, respectively, led to a complete loss of the GABA transport activity, demonstrating the necessity of aromatic residues [122]. Also, Glu101Asp abolishes the transport capacity [123], and mutations of Tyr140 by Phe or Trp demonstrated that such a residue is required for GABA binding and transport [124]. Interestingly, deletion of Gly457 in TM10 led to active mutants such as those obtained by deleting the adjacent positions, but replacement mutants such as Gly457Ala were devoid of transport activity. Structural modeling of the transport suggests that Gly457 is likely to form a π-helix required for the activity [125].

GAT2 can be *N*-glycosylated at Asn173 and Asn269. Phosphorylation takes place at Thr587 and Ser591. Mutagenesis experiments with GAT2 showed that Glu48Ala, Glu48Leu reduced GABA uptake and Gly51Ala and Val132Ile blocked it completely [126]. Two alternative isoforms have been described: isoform 2 in which residues 68–159 are missing; isoform 3 differing from canonical isoforms in residues 69–97 and missing residues 98–602. A natural variant (Val426Ile) can also be found. It has been demonstrated that the C-terminal cytoplasmic 22 residues of GAT-2 are necessary for its basolateral distribution and can function as an autonomous signal, independently of the rest of the transporter [127]. Models of GAT2 have been constructed and validated by mutagenesis experiments [126]. These studies conclude that GAT2 is a high selectivity/low affinity transporter that is resistant to inhibition by typical GABAergic inhibitors.

GAT3 can be *N*-glycosylated at Asn187, Asn190, and Asn198. Phosphorylation takes place at Ser21. An alternative isoform with residues 209–632 has been described.

By using a bacterial homologue leucine transporter (PDB IDs: 3TT3, 2A65, 3F3A), 3D models of human BGT1, GAT1, and GAT3 have been constructed. The binding mode for GAT1 shows high similarity for the binding of the carboxylic end of GABA compared to that of leucine in the original structure (Figure 6.10a,b) [128]. The carboxylic end of GABA is bound to one of the sodium atoms and fixed in a network of hydrogen bonds composed of Tyr140 and the backbone-NH of Leu64 and Gly65. The GABA amino group also forms a different set of hydrogen bonds with Thr400, Ser396, and, to a lesser extent, Tyr60 and the backbone-O of Tyr60 and Ser396.

A molecular modeling study of GAT3 based on the above-mentioned leucine transporter allowed the identification of a very selective inhibitor of this transporter [113]. The model suggests potential inhibitory binding sites that were in accordance with experimental SAR studies (Figure 6.10c). A hydrophobic pocket capable of accommodating compounds with aromatic substituents was found. Molecular dynamics studies further corroborated this hypothesis.

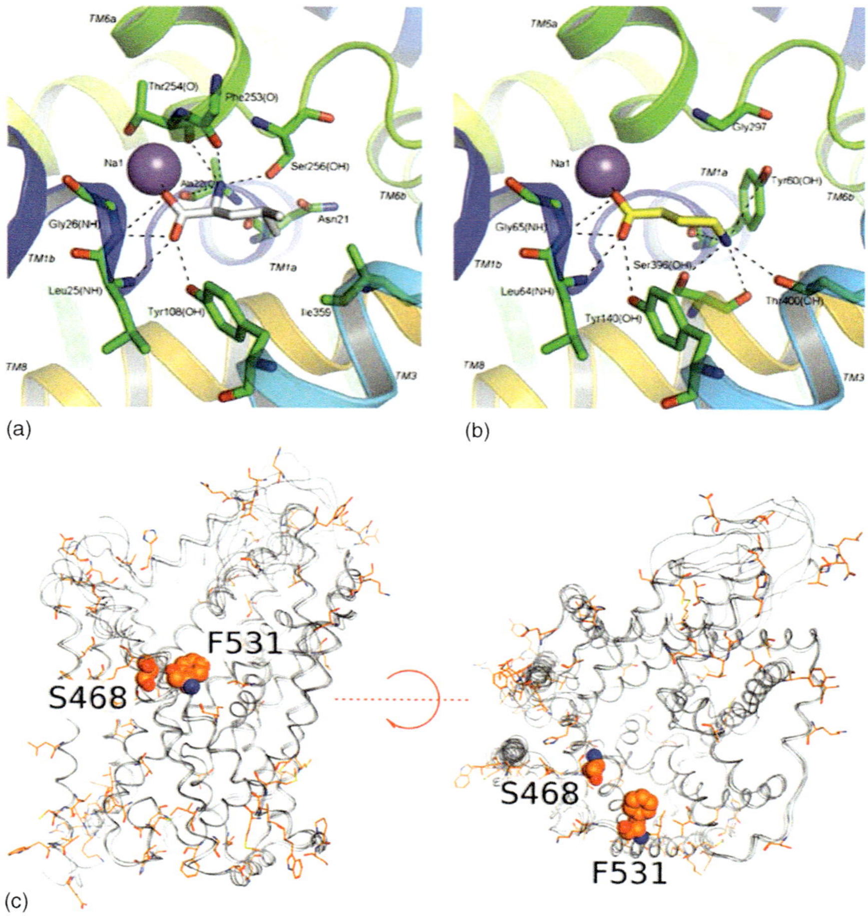

Figure 6.10 (a) Polar interactions of leucine and LeuTAa. Transmembrane helices TM10, TM11, TM12, and side chain of Phe253 are not shown for clarity. Numbering is for LeuTAa. (b) Polar interactions of GABA and GAT-1. Transmembrane helices TM10, TM11, and TM12 are not shown for clarity. Numbering is for hGAT-1. Source: Wein and Wanner 2010 [128]. Reprinted with permission of Springer Nature. (c) Localization of GAT3-selective residues (orange). Amino acids 10.49 (Ser468) and 11.60 (Phe531) located in the putative pocket are depicted in van der Waals surface. Source: Damgaard et al. 2015 [113]. Reprinted with permission of ACS.

References

1 Chronister, R.B. and DeFrance, J.F. (1979). Organization of projection neurons of the hippocampus. *Exp Neurol* 66 (3): 509–523.

2 Alonso, A. and Kohler, C. (1982). Evidence for separate projections of hippocampal pyramidal and non-pyramidal neurons to different parts of the septum in the rat brain. *Neurosci Lett* 31 (3): 209–214.

3 Freund, T.F. and Antal, M. (1988). GABA-containing neurons in the septum control inhibitory interneurons in the hippocampus. *Nature* 336 (6195): 170–173.

4 Jinno, S. and Kosaka, T. (2002). Immunocytochemical characterization of hippocamposeptal projecting GABAergic nonprincipal neurons in the mouse brain: a retrograde labeling study. *Brain Res* 945 (2): 219–231.
5 Melzer, S., Michael, M., Caputi, A. et al. (2012). Long-range-projecting GABAergic neurons modulate inhibition in hippocampus and entorhinal cortex. *Science* 335 (6075): 1506–1510.
6 Muller, M., Faber-Zuschratter, H., Yanagawa, Y. et al. (2012). Synaptology of ventral CA1 and subiculum projections to the basomedial nucleus of the amygdala in the mouse: relation to GABAergic interneurons. *Brain Struct Funct* 217 (1): 5–17.
7 Lübkemann, R., Eberhardt, J., Röhl, F.-W. et al. (2015). Identification and Characterization of GABAergic Projection Neurons from Ventral Hippocampus to Amygdala. *Brain Sciences* 5 (3): 299–317.
8 Schwerdtfeger, W.K. and Buhl, E. (1986). Various types of non-pyramidal hippocampal neurons project to the septum and contralateral hippocampus. *Brain Res* 386 (1-2): 146–154.
9 Jinno, S. (2009). Structural Organization of Long-Range GABAergic Projection System of the Hippocampus. *Front Neuroanat* 3: 13.
10 Higo, S., Akashi, K., Sakimura, K., and Tamamaki, N. (2009). Subtypes of GABAergic neurons project axons in the neocortex. *Front Neuroanat* 3: 25.
11 Tamamaki, N. and Tomioka, R. (2010). Long-Range GABAergic Connections Distributed throughout the Neocortex and their Possible Function. *Front Neurosci* 4: 202.
12 Brown, M.T., Tan, K.R., O'Connor, E.C. et al. (2012). Ventral tegmental area GABA projections pause accumbal cholinergic interneurons to enhance associative learning. *Nature* 492 (7429): 452–456.
13 Creed, M.C., Ntamati, N.R., and Tan, K.R. (2014). VTA GABA neurons modulate specific learning behaviors through the control of dopamine and cholinergic systems. *Front Behav Neurosci* 8: 8.
14 Sgadò, P., Dunleavy, M., Genovesi, S., Provenzano, G., Bozzi, Y. (2011). The role of GABAergic system in neurodevelopmental disorders: a focus on autism and epilepsy. *Int. J. Physiol., Pathophysiol. Pharmacol.,* 3 (3), 223-235
15 Govindpani, K., Calvo-Flores Guzmán, B., Vinnakota, C. et al. (2017). Towards a Better Understanding of GABAergic Remodeling in Alzheimer's Disease. *International Journal of Molecular Sciences* 18 (8): 1813.
16 Prager, E.M., Bergstrom, H.C., Wynn, G.H., and Braga, M.F.M. (2016). The Basolateral Amygdala GABAergic System in Health and Disease. *J. Neurosci. Res.* 94 (6): 548–567.
17 Coghlan, S., Horder, J., Inkster, B. et al. (2012). GABA System Dysfunction in Autism and Related Disorders: From Synapse to Symptoms. *Neurosci Biobehav Rev* 36 (9): 2044–2055.
18 Gatto, C.L., Pereira, D., and Broadie, K. (2014). GABAergic circuit dysfunction in the Drosophila Fragile X syndrome model. *Neurobiol. Dis.* 65: 142–159.
19 Schmidt, M.J. and Mirnics, K. (2014). Neurodevelopment, GABA System Dysfunction, and Schizophrenia. *Neuropsychopharmacology* 40: 190.
20 McIntire, S.L., Reimer, R., and Edwards, R.H. (1997). Isolation and functional expression of a vesicular GABA transporter. *Faseb J.* 11 (9).

21 Grifa, A., Totaro, A., Rommens, J.M. et al. (1998). GABA (gamma-amino-butyric acid) neurotransmission: identification and fine mapping of the human GABAB receptor gene. *Biochem Biophys Res Commun* 250 (2): 240–245.
22 Chebib, M. and Johnston, G.A. (1999). The 'ABC' of GABA receptors: a brief review. *Clin Exp Pharmacol Physiol* 26 (11): 937–940.
23 Korpi, E.R., Grunder, G., and Luddens, H. (2002). Drug interactions at GABA(A) receptors. *Prog Neurobiol* 67 (2): 113–159.
24 Halonen, L.M., Sinkkonen, S.T., Chandra, D. et al. (2009). Brain regional distribution of GABA(A) receptors exhibiting atypical GABA agonism: roles of receptor subunits. *Neurochem. Int.* 55 (6): 389–396.
25 Pizzarelli, R. and Cherubini, E. (2011). Alterations of GABAergic Signaling in Autism Spectrum Disorders. *Neural Plast* 2011.
26 Rudolph, U. and Mohler, H. (2014). GABAA receptor subtypes: therapeutic potential in down syndrome, affective disorders, schizophrenia, and autism. *Annu. Rev. Pharmacol. Toxicol.* 54: 483–507.
27 Watanabe, M., Maemura, K., Kanbara, K., Tamayama, T., Hayasaki, H. (2002). GABA and GABA Receptors in the Central Nervous System and Other Organs, in *International Review of Cytology*, (eds. K. W. Jeon), Academic Press, pp. 1-47
28 Pirker, S., Schwarzer, C., Wieselthaler, A. et al. (2000). GABA(A) receptors: immunocytochemical distribution of 13 subunits in the adult rat brain. *Neuroscience* 101 (4): 815–850.
29 Martin, S.C., Russek, S.J., and Farb, D.H. (1999). Molecular identification of the human GABABR2: cell surface expression and coupling to adenylyl cyclase in the absence of GABABR1. *Mol Cell Neurosci* 13 (3): 180–191.
30 Ng, G.Y., McDonald, T., Bonnert, T. et al. (1999). Cloning of a novel G-protein-coupled receptor GPR 51 resembling GABAB receptors expressed predominantly in nervous tissues and mapped proximal to the hereditary sensory neuropathy type 1 locus on chromosome 9. *Genomics* 56 (3): 288–295.
31 Clark, J.A., Mezey, E., Lam, A.S., and Bonner, T.I. (2000). Distribution of the GABA(B) receptor subunit gb2 in rat CNS. *Brain Res* 860 (1-2): 41–52.
32 Li, Y., Sun, H., Chen, Z. et al. (2016). Implications of GABAergic Neurotransmission in Alzheimer's Disease. *Frontiers in Aging Neuroscience* 8: 31.
33 Błaszczyk, J.W. (2016). Parkinson's Disease and Neurodegeneration: GABA-Collapse Hypothesis. *Frontiers in Neuroscience* 10: 269.
34 Rissman, R.A. and Mobley, W.C. (2011). Implication for treatment: GABA(A) receptors in aging, Down syndrome and Alzheimer's disease. *J. Neurochem.* 117 (4): 613–622.
35 Smith, T.A. (2001). Type A gamma-aminobutyric acid (GABAA) receptor subunits and benzodiazepine binding: significance to clinical syndromes and their treatment. *Br J Biomed Sci* 58 (2): 111–121.
36 Whiting, P. J. (2006). GABA-A receptors: A viable target for novel anxiolytics? *Curr. Opin. Pharmacol.*, 6 (1 SPEC. ISS.), 24-29
37 Tyagi, R.K., Bisht, R., Pant, J. et al. (2015). Possible role of GABA-B receptor modulation in MPTP induced Parkinson's disease in rats. *Exp Toxicol Pathol* 67 (2): 211–217.
38 Nambu, A. (2012). GABA-B receptor: Possible target for Parkinson's disease therapy. *Exp. Neurol.* 233 (1): 121–122.

39 Hillman, R., Sinani, J., and Pendleton, R. (2012). The role of the GABA(B) receptor and calcium channels in a Drosophila model of Parkinson's Disease. *Neurosci Lett* 516 (2): 167–170.

40 Nambu, A. (2012). GABA-B receptor: possible target for Parkinson's disease therapy. *Exp Neurol* 233 (1): 121–122.

41 Gazulla, J., Ruiz-Gazulla, C., and Tintore, M. (2015). GABAergic Pharmacotherapy in the Treatment of Motor Disorders of the Central Nervous System. *Curr. Pharm. Des.* 21 (34): 4989–4995.

42 Shetty, A. K., Bates, A. (2016). Potential of GABA-ergic cell therapy for schizophrenia, neuropathic pain, and Alzheimer's and Parkinson's diseases. *Brain Res.,* 1638 (Part_A), 74-87

43 De Stefano, N. and Giorgio, A. (2015). GABA: a new imaging biomarker of neurodegeneration in multiple sclerosis? *Brain* 138 (9): 2467–2468.

44 Farrar, S.J., Whiting, P.J., Bonnert, T.P., and McKernan, R.M. (1999). Stoichiometry of a ligand-gated ion channel determined by fluorescence energy transfer. *J Biol Chem* 274 (15): 10100–10104.

45 MacDonald, R.L. (2009). *Botzolakis*, 257–282. E. J.: Elsevier.

46 Olsen, R.W. (2014). Analysis of γ-Aminobutyric Acid (GABA) Type A Receptor Subtypes Using Isosteric and Allosteric Ligands. *Neurochem. Res.* 39 (10): 1924–1941.

47 Smart, T.G. (2015). *CRC Press* 345–359.

48 Sigel, E. and Steinmann, M.E. (2012). Structure, Function, and Modulation of GABAA Receptors. *J. Biol. Chem.* 287 (48): 40224–40231.

49 Michels, G. and Moss, S.J. (2007). GABAA receptors: properties and trafficking. *Crit. Rev. Biochem. Mol. Biol.* 42 (1): 3–14.

50 Perfilova, V.N. and Tyurenkov, I.N. (2011). GABAC receptors: structure and functions. *Eksp. Klin. Farmakol.* 74 (1): 45–49.

51 Sieghart, W. (2006). Structure, pharmacology, and function of GABAA receptor subtypes. *Adv. Pharmacol. (San Diego, CA, U. S.),* 54 (GABA), 231-263

52 Carter, C.R.J., Kozuska, J.L., and Dunn, S.M.J. (2010). Insights into the structure and pharmacology of GABAA receptors. *Future Med. Chem.* 2 (5): 859–876.

53 Leinekugel, X., Khalilov, I., McLean, H. et al. (1999). GABA is the principal fast-acting excitatory transmitter in the neonatal brain. *Adv Neurol* 79: 189–201.

54 Nakamura, Y., Darnieder, L.M., Deeb, T.Z., and Moss, S.J. (2015). Regulation of GABAARs by phosphorylation. *Adv Pharmacol* 72: 97–146.

55 Carvill, G.L., Weckhuysen, S., McMahon, J.M. et al. (2014). GABRA1 and STXBP1: novel genetic causes of Dravet syndrome. *Neurology* 82 (14): 1245–1253.

56 Parrini, E., Marini, C., Mei, D. et al. (2017). Diagnostic Targeted Resequencing in 349 Patients with Drug-Resistant Pediatric Epilepsies Identifies Causative Mutations in 30 Different Genes. *Hum Mutat* 38 (2): 216–225.

57 Lachance-Touchette, P., Brown, P., Meloche, C. et al. (2011). Novel alpha1 and gamma2 GABAA receptor subunit mutations in families with idiopathic generalized epilepsy. *Eur J Neurosci* 34 (2): 237–249.

58 Cossette, P., Liu, L., Brisebois, K. et al. (2002). Mutation of GABRA1 in an autosomal dominant form of juvenile myoclonic epilepsy. *Nat Genet* 31 (2): 184–189.

59 Sjoblom, T., Jones, S., Wood, L.D. et al. (2006). The consensus coding sequences of human breast and colorectal cancers. *Science* 314 (5797): 268–274.
60 Lien, E., Vatevik, A.K., Ostern, R. et al. (2016). A second patient with a De Novo GABRB1 mutation and epileptic encephalopathy. *Ann Neurol* 80 (2): 311–312.
61 Allen, A.S., Berkovic, S.F., Cossette, P. et al. (2013). De novo mutations in epileptic encephalopathies. *Nature* 501 (7466): 217–221.
62 Janve, V.S., Hernandez, C.C., Verdier, K.M. et al. (2016). Epileptic encephalopathy de novo GABRB mutations impair GABAA receptor function. *Ann Neurol* 79 (5): 806–825.
63 Coon, H., Sobell, J., Heston, L. et al. (1994). Search for mutations in the beta 1 GABAA receptor subunit gene in patients with schizophrenia. *Am J Med Genet* 54 (1): 12–20.
64 Zhao, C., Xu, Z., Wang, F. et al. (2009). Alternative-splicing in the exon-10 region of GABA(A) receptor beta(2) subunit gene: relationships between novel isoforms and psychotic disorders. *PLoS One* 4 (9): e6977.
65 McKinley, D.D., Lennon, D.J., and Carter, D.B. (1995). Cloning, sequence analysis and expression of two forms of mRNA coding for the human beta 2 subunit of the GABAA receptor. *Brain Res Mol Brain Res* 28 (1): 175–179.
66 Gurba, K.N., Hernandez, C.C., Hu, N., and Macdonald, R.L. (2012). GABRB3 mutation, G32R, associated with childhood absence epilepsy alters alpha1beta3gamma2L gamma-aminobutyric acid type A (GABAA) receptor expression and channel gating. *J Biol Chem* 287 (15): 12083–12097.
67 Allen, A.S., Berkovic, S.F., Cossette, P. et al. (2016). De Novo Mutations in SLC1A2 and CACNA1A Are Important Causes of Epileptic Encephalopathies. *Am J Hum Genet* 99 (2): 287–298.
68 Tanaka, M., Olsen, R.W., Medina, M.T. et al. (2008). Hyperglycosylation and reduced GABA currents of mutated GABRB3 polypeptide in remitting childhood absence epilepsy. *Am J Hum Genet* 82 (6): 1249–1261.
69 Miller, P.S. and Aricescu, A.R. (2014). Crystal structure of a human GABAA receptor. *Nature* 512 (7514): 270–275.
70 Miller, P.S., Scott, S., Masiulis, S. et al. (2017). Structural basis for GABAA receptor potentiation by neurosteroids. *Nat Struct Mol Biol* 24 (11): 986–992.
71 Baulac, S., Huberfeld, G., Gourfinkel-An, I. et al. (2001). First genetic evidence of GABA(A) receptor dysfunction in epilepsy: a mutation in the gamma2-subunit gene. *Nat Genet* 28 (1): 46–48.
72 Unwin, N. and Fujiyoshi, Y. (2012). Gating movement of acetylcholine receptor caught by plunge-freezing. *J Mol Biol* 422 (5): 617–634.
73 Shi, X., Huang, M.C., Ishii, A. et al. (2010). Mutational analysis of GABRG2 in a Japanese cohort with childhood epilepsies. *J Hum Genet* 55 (6): 375–378.
74 Wallace, R.H., Marini, C., Petrou, S. et al. (2001). Mutant GABA(A) receptor gamma2-subunit in childhood absence epilepsy and febrile seizures. *Nat Genet* 28 (1): 49–52.
75 Carvill, G.L., Heavin, S.B., Yendle, S.C. et al. (2013). Targeted resequencing in epileptic encephalopathies identifies de novo mutations in CHD2 and SYNGAP1. *Nat Genet* 45 (7): 825–830.
76 Audenaert, D., Schwartz, E., Claeys, K.G. et al. (2006). A novel GABRG2 mutation associated with febrile seizures. *Neurology* 67 (4): 687–690.

77 Rafiullah, R., Aslamkhan, M., Paramasivam, N. et al. (2016). Homozygous missense mutation in the LMAN2L gene segregates with intellectual disability in a large consanguineous Pakistani family. *J Med Genet* 53 (2): 138–144.
78 Bierzynska, A., Soderquest, K., Dean, P. et al. (2017). MAGI2 Mutations Cause Congenital Nephrotic Syndrome. *J Am Soc Nephrol* 28 (5): 1614–1621.
79 Dibbens, L.M., Feng, H.J., Richards, M.C. et al. (2004). GABRD encoding a protein for extra- or peri-synaptic GABAA receptors is a susceptibility locus for generalized epilepsies. *Hum Mol Genet* 13 (13): 1315–1319.
80 Garret, M., Bascles, L., Boue-Grabot, E. et al. (1997). An mRNA encoding a putative GABA-gated chloride channel is expressed in the human cardiac conduction system. *J Neurochem* 68 (4): 1382–1389.
81 Davies, P.A., Hanna, M.C., Hales, T.G., and Kirkness, E.F. (1997). Insensitivity to anaesthetic agents conferred by a class of GABA(A) receptor subunit. *Nature* 385 (6619): 820–823.
82 Sinkkonen, S.T., Hanna, M.C., Kirkness, E.F., and Korpi, E.R. (2000). GABA(A) receptor epsilon and theta subunits display unusual structural variation between species and are enriched in the rat locus coeruleus. *J Neurosci* 20 (10): 3588–3595.
83 Mott, D. (2015). Chapter 11 - The metabotropic GABAB receptors. In: *Cellular and Molecular Neurophysiology (Fourth Edition)* (ed. C. Hammond), 245–267. Boston: Academic Press.
84 Sullivan, R., Chateauneuf, A., Coulombe, N. et al. (2000). Coexpression of full-length gamma-aminobutyric acid(B) (GABA(B)) receptors with truncated receptors and metabotropic glutamate receptor 4 supports the GABA(B) heterodimer as the functional receptor. *J Pharmacol Exp Ther* 293 (2): 460–467.
85 Villemure, J.F., Adam, L., Bevan, N.J. et al. (2005). Subcellular distribution of GABA(B) receptor homo- and hetero-dimers. *Biochem J* 388 (Pt 1): 47–55.
86 White, J.H., Wise, A., Main, M.J. et al. (1998). Heterodimerization is required for the formation of a functional GABA(B) receptor. *Nature* 396 (6712): 679–682.
87 Kuner, R., Kohr, G., Grunewald, S. et al. (1999). Role of heteromer formation in GABAB receptor function. *Science* 283 (5398): 74–77.
88 Nomura, R., Suzuki, Y., Kakizuka, A., and Jingami, H. (2008). Direct detection of the interaction between recombinant soluble extracellular regions in the heterodimeric metabotropic gamma-aminobutyric acid receptor. *J Biol Chem* 283 (8): 4665–4673.
89 Geng, Y., Bush, M., Mosyak, L. et al. (2013). Structural mechanism of ligand activation in human GABA(B) receptor. *Nature* 504 (7479): 254–259.
90 Schwarz, D.A., Barry, G., Eliasof, S.D. et al. (2000). Characterization of gamma-aminobutyric acid receptor GABAB(1e), a GABAB(1) splice variant encoding a truncated receptor. *J Biol Chem* 275 (41): 32174–32181.
91 Peters, H.C., Kammer, G., Volz, A. et al. (1998). Mapping, genomic structure, and polymorphisms of the human GABABR1 receptor gene: evaluation of its involvement in idiopathic generalized epilepsy. *Neurogenetics* 2 (1): 47–54.
92 Sander, T., Peters, C., Kammer, G. et al. (1999). Association analysis of exonic variants of the gene encoding the GABAB receptor and idiopathic generalized epilepsy. *Am J Med Genet* 88 (4): 305–310.

93 Geng, Y., Xiong, D., Mosyak, L. et al. (2012). Structure and functional interaction of the extracellular domain of human GABA(B) receptor GBR2. *Nat Neurosci* 15 (7): 970–978.

94 Lopes, F., Barbosa, M., Ameur, A. et al. (2016). Identification of novel genetic causes of Rett syndrome-like phenotypes. *J Med Genet* 53 (3): 190–199.

95 Borden, L.A., Smith, K.E., Hartig, P.R. et al. (1992). Molecular heterogeneity of the γ-aminobutyric acid (GABA) transport system. Cloning of two novel high affinity GABA transporters from rat brain. *J. Biol. Chem.* 267 (29): 21098–21104.

96 Borden, L.A. (1996). GABA transporter heterogeneity: pharmacology and cellular localization. *Neurochem Int* 29 (4): 335–356.

97 Soudijn, W. and van Wijngaarden, I. (2000). The GABA transporter and its inhibitors. *Curr Med Chem* 7 (10): 1063–1079.

98 Jin, X.-T., Galvan, A., Wichmann, T., and Smith, Y. (2011). Localization and function of GABA transporters GAT-1 and GAT-3 in the basal ganglia. *Front. Syst. Neurosci.* 5 (July): 63.

99 Zhou, Y. and Danbolt, N.C. (2013). GABA and Glutamate Transporters in Brain. *Front Endocrinol* 4: 165.

100 Itouji, A., Sakai, N., Tanaka, C., and Saito, N. (1996). Neuronal and glial localization of two GABA transporters (GAT1 and GAT3) in the rat cerebellum. *Mol. Brain Res.* 37 (1-2): 309–316.

101 Ng, C.H., Wang, X.S., and Ong, W.Y. (2000). A light and electron microscopic study of the GABA transporter GAT-3 in the monkey basal ganglia and brainstem. *Journal of Neurocytology* 29 (8): 595–603.

102 Zhou, Y., Holmseth, S., Hua, R. et al. (2012). The betaine-GABA transporter (BGT1, slc6a12) is predominantly expressed in the liver and at lower levels in the kidneys and at the brain surface. *American Journal of Physiology - Renal Physiology* 302 (3): F316–F328.

103 Schousboe, A., Sarup, A., Larsson, O.M., and White, H.S. (2004). GABA transporters as drug targets for modulation of GABAergic activity. *Biochem. Pharmacol.* 68 (8): 1557–1563.

104 Richerson, G.B. and Wu, Y. (2004). in *Adv. Exp. Med. Biol.* 548: 76–91.

105 Sarup, A., Larsson, O.M., and Schousboe, A. (2003). GABA transporters and GABA-transaminase as drug targets. *Current drug targets. CNS and neurological disorders* 2 (4): 269–277.

106 Madsen, K.K., White, H.S., and Schousboe, A. (2010). Neuronal and non-neuronal GABA transporters as targets for antiepileptic drugs. *Pharmacology and Therapeutics* 125 (3): 394–401.

107 Scimemi, A. (2014). Structure, function, and plasticity of GABA transporters. *Front Cell Neurosci* 8 (161).

108 Kanner, B. I. (2001). Structure and function of GABA transporters, in *Handb. Exp. Pharmacol.*, (eds. H. Möheler), Springer, New York, pp. 345-354

109 Schousboe, A. and Kanner, B. (2002). GABA transporters: Functional and pharmacological properties. In: *Glutamate and GABA receptors and transporters* (ed. J. Egebjerg, A. Schousboe and P. Krosgaard-Larsen), 337–349. London: Taylor Francis.

110 Soudijn, W. and Van Wijngaarden, I. (2000). The GABA transporter and its inhibitors. *Curr. Med. Chem.* 7 (10): 1063–1079.

111 Rasola, A., Galietta, L.J., Barone, V. et al. (1995). Molecular cloning and functional characterization of a GABA/betaine transporter from human kidney. *FEBS Lett* 373 (3): 229–233.

112 Hosseinzadeh, Z., Shojaiefard, M., Bhavsar, S.K., and Lang, F. (2012). Up-regulation of the betaine/GABA transporter BGT1 by JAK2. *Biochem. Biophys. Res. Commun.* 420 (1): 172–177.

113 Damgaard, M., Al-Khawaja, A., Vogensen, S.B. et al. (2015). Identification of the First Highly Subtype-Selective Inhibitor of Human GABA Transporter GAT3. *ACS Chem. Neurosci.* 6 (9): 1591–1599.

114 Hirunsatit, R., George, E.D., Lipska, B.K. et al. (2009). Twenty-one-base-pair insertion polymorphism creates an enhancer element and potentiates SLC6A1 GABA transporter promoter activity. *Pharmacogenet Genomics* 19 (1): 53–65.

115 Hu, J., Reutter, W., and Fan, H. (2011). Significance of N-glycosylation and sialylation of GABA transporter 1. *Journal of Carbohydrate Chemistry* 30 (4-6): 206–217.

116 Carvill, G.L., McMahon, J.M., Schneider, A. et al. (2015). Mutations in the GABA Transporter SLC6A1 Cause Epilepsy with Myoclonic-Atonic Seizures. *Am J Hum Genet* 96 (5): 808–815.

117 Gonzalez-Burgos, G. (2010). in. *Advances in Pharmacology* 58: 175–204.

118 Bendahan, A. and Kanner, B.I. (1993). Identification of domains of a cloned rat brain GABA transporter which are not required for its functional expression. *FEBS Letters* 318 (1): 41–44.

119 Mabjeesh, N.J. and Kanner, B.I. (1993). The substrates of a sodium- and chloride-coupled gamma-aminobutyric acid transporter protect multiple sites throughout the protein against proteolytic cleavage. *Biochemistry* 32 (33): 8540–8546.

120 Pantanowitz, S., Bendahan, A., and Kanner, B.I. (1993). Only one of the charged amino acids located in the transmembrane alpha-helices of the gamma-aminobutyric acid transporter (subtype A) is essential for its activity. *J Biol Chem* 268 (5): 3222–3225.

121 Ben-Yona, A., Bendahan, A., and Kanner, B.I. (2011). A glutamine residue conserved in the neurotransmitter: Sodium:symporters is essential for the interaction of chloride with the GABA transporter GAT-1. *J. Biol. Chem.* 286 (4): 2826–2833.

122 Kleinberger-Doron, N. and Kanner, B.I. (1994). Identification of tryptophan residues critical for the function and targeting of the gamma-aminobutyric acid transporter (subtype A). *J Biol Chem* 269 (4): 3063–3067.

123 Keshet, G.I., Bendahan, A., Su, H. et al. (1995). Glutamate-101 is critical for the function of the sodium and chloride-coupled GABA transporter GAT-1. *FEBS Letters* 371 (1): 39–42.

124 Bismuth, Y., Kavanaugh, M.P., and Kanner, B.I. (1997). Tyrosine 140 of the gamma-aminobutyric acid transporter GAT-1 plays a critical role in neurotransmitter recognition. *J Biol Chem* 272 (26): 16096–16102.

125 Dayan, O., Nagarajan, A., Shah, R. et al. (2017). An extra amino acid residue in transmembrane domain 10 of the γ-aminobutyric acid (GABA) transporter GAT-1 is required for efficient ion-coupled transport. *J. Biol. Chem.* 292 (13): 5418–5428.

126 Schlessinger, A., Wittwer, M.B., Dahlin, A. et al. (2012). High selectivity of the gamma-aminobutyric acid transporter 2 (GAT-2, SLC6A13) revealed by structure-based approach. *J Biol Chem* 287 (45): 37745–37756.

127 Brown, A., Muth, T., Caplan, M. (2004). The COOH-terminal tail of the GAT-2 GABA transporter contains a novel motif that plays a role in basolateral targeting. *American Journal of Physiology - Cell Physiology,* 286 (5 55-5), C1071-C1077

128 Wein, T. and Wanner, K.T. (2010). Generation of a 3D model for human GABA transporter hGAT-1 using molecular modeling and investigation of the binding of GABA. *J Mol Model* 16 (1): 155–161.

Part II

Metabolism and Proteins

7

Metabolism of Neurotransmitters

7.1 Introduction

Neurotransmitters are chemicals that are used in the brain to relay, amplify, and modulate electrical signals between brain cells. Moreover, the ability to modulate the rate of synthesis of neurotransmitters depending on the activity of neuronal systems is one of the processes contributing to neuronal plasticity. Consequently, metabolism of neurotransmitters is crucial to the correct performance of any biological function controlled by the brain, which strongly depends on the correct balance between neurotransmitters in the synapse. The release of neurotransmitters into the synaptic cleft is carried out from vesicles. The release of monoamine neurotransmitters (dopamine [DA], norepinephrine [NE], epinephrine [EPI], and serotonin [5-HT]) packaged into vesicles is carried out by relatively nonspecific transporters (VMAT1 and VMAT2), and the neuronal selectivity for a particular monoamine is determined by the corresponding biosynthetic enzymes expressed in the neurons; for instance, catecholamines are synthesized in the neurons at their sites. The removal of neurotransmitters from the synaptic cleft is mediated by very selective transporters, which uptake the neurotransmitters into the neurons to reuse or catabolize, depending on the requirements in each moment. Such a catabolism is performed by the corresponding enzymes located inside the neurons. The action of neurotransmitters can also be terminated through their catabolism by enzymes external to the neurons, as in the case of acetylcholine. Any change in the metabolic balance of neurotransmitters leads to the failure of neural control of the communication between brain cells and neurological disorders appear [1]. In some cases, metabolism is affected by age, and neurodegenerative disorders may occur.

7.2 Metabolism of Catecholamines

7.2.1 Dopamine

Dopamine is a catecholamine that had been known since the early twentieth century; however, its role as an independent neurotransmitter was documented much later [2]. Dopamine is biosynthesized in the brain as well in the medulla, adrenal glands, and gastrointestinal tract [3]. Herein, we restrict the study to

Chemical Biology of Neurodegeneration: A Molecular Approach, First Edition. Pedro Merino.

dopamine produced by neurons. In this context, Hiller and coworkers reported a review on dopamine metabolism related to neurodegeneration including clinical aspects regarding Parkinson's disease (PD) [4].

Dopamine **1** is produced in the brain through hydroxylation of L-tyrosine **2** by tyrosine hydroxylase (or, alternatively, by tyrosinase) to yield dihydroxyphenylalanine (DOPA) **5**. The process consumes oxygen, which is transformed into water, with the hydrogen atoms being taken from tetrahydrobiopterin **4**, which is converted into dihydrobiopterin **3** (Scheme 7.1). Cofactor **3** is synthesized from guanosine triphosphate (GTP) by GTP cyclohydroxylase I (GTPCH), 6-pyruvoyl-tetrahydropterin synthase (PTPS), and sepiapterin reductase (SR) in nine steps. The regeneration of **3** from **4** requires pterin-4a-carbinolamine dehydratase (PCD) and dihydropteridine reductase (DHPR) [5]. It was suggested that PD could be a consequence of restricted availability of **3** [6] although oral treatment of PD patients with **3** did not show significant results [7]. On the other hand, more recent studies have shown neurotoxicity of **3**, which induces degeneration of the nigrostriatal pathway by causing a loss of tyrosine hydroxylase [8] and mitochondrial dysfunction [9]. Further decarboxylation of DOPA by aromatic amino acid decarboxylase, AADC (also known as DOPA decarboxylase), furnishes carbon dioxide and dopamine **1**. There is an alternative biosynthetic pathway in which decarboxylation precedes hydroxylation and tyramine **6** is formed from tyrosine. Then, amine **6** is oxidized by cytochrome P450 2D, Cyp2D to dopamine **1** [10].

Neurological disorders related to dopamine deficiency can only be treated with DOPA **5** since it is the only compound, from those illustrated in Scheme 7.1, capable of crossing the blood–brain barrier.

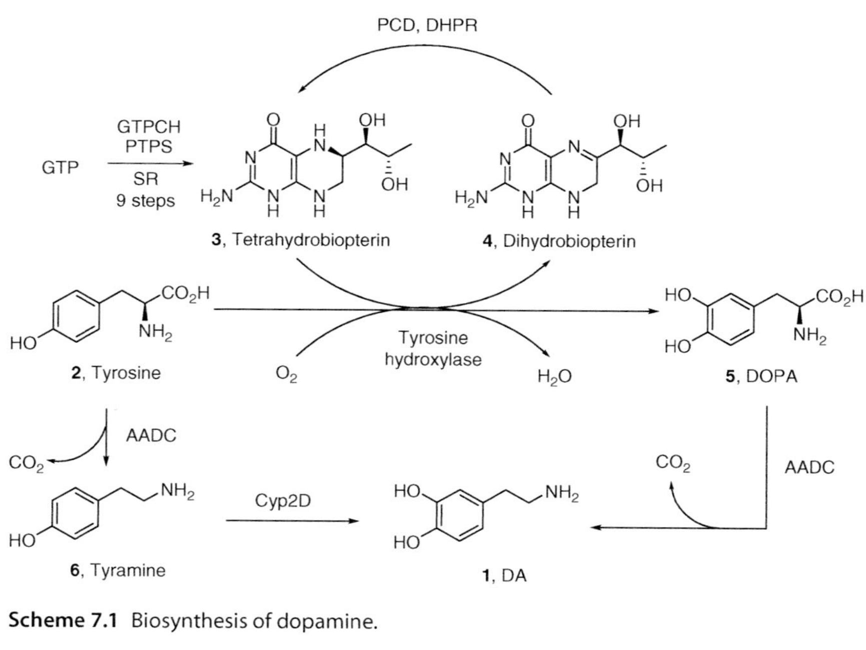

Scheme 7.1 Biosynthesis of dopamine.

Degradation of DA by monoamine oxidase (MAO) when extracellular dopamine is recovered by vesicular monoamine transporter (VMAT) yields 3,4-dihydroxiphenylacetaldehyde (DOPAL) **7** and hydrogen peroxide, one of the main reasons of oxidative stress causing neurodegeneration [11]. Compound **7** is oxidized by aldehyde dehydrogenase to 3,4-dihydroxyphenylacetic acid (DOPAC) **8** and, to a lesser extent, by alcohol dehydrogenase (ADH) to 3,4-dihydroxiphenylethanol (DOPET) **9**. Extracellular dopamine is also retired by glial cells, which, in addition to MAO, possess catechol-*O*-methyltransferase (COMT) that transforms DA into **10** through the transfer of a methyl group from *S*-adenosylmethionine (SAM). Further transformation of **10** into **11** by MAO and oxidation with ADH furnish homovanillic acid (HVA) **12,** the main degradation product of dopamine, which can also be obtained from **8** by the action of COMT (Scheme 7.2). HVA level in the cerebrospinal fluid can be used as a biomarker of the PD stage [12] and complemented with measurements of other metabolites and brain-specific proteins can also be used to distinguish Alzheimer's disease (AD) from dementia with Lewy bodies (DLB) [13].

Before excretion, DA undergoes conjugation reactions that take place in the central nervous system (CNS) and peripheral nervous system. Uridine diphosphoglucuronosyltransferases (UGTs) [14] transfer glucuronic acid from uridine diphosphate (UDP)-glucuronic acid to DA to form equimolar mixtures of 3-*O*-glucuronide **13** and 4-*O*-glucuronide **14** [15]. Phenolsulfotransferases (PSTs) transfer sulfate from 3'- phosphoadenosine-5'-phosphosulfate (PAPS) to phenolic hydroxyls [16], the 3-sulfates **15** being the predominant one (Scheme 7.3).

DA and its precursors tyrosine and DOPA are easily oxidized at the electron-rich catechol moieties by enzymes with peroxidase activity such as prostaglandin H-synthase [17] by iron(III) or even spontaneously [18]. The oxidation results in

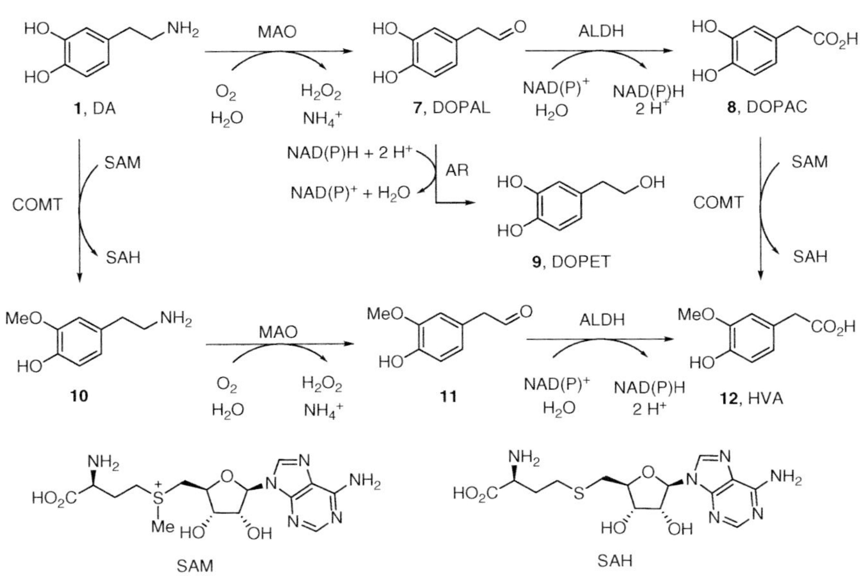

Scheme 7.2 Transformation and degradation of dopamine.

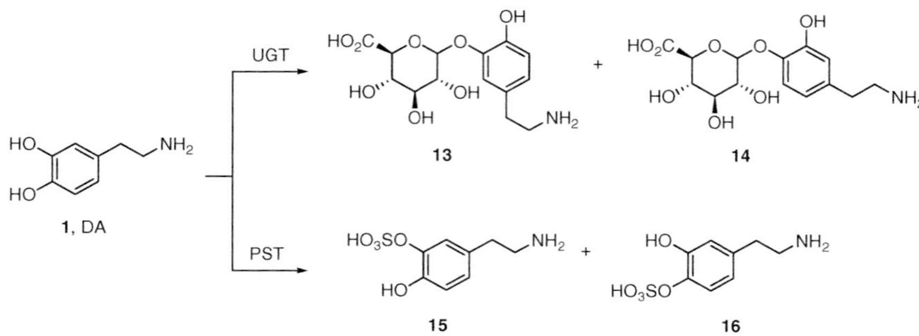

Scheme 7.3 Conjugation reactions of dopamine.

dopamine *o*-semiquinone radical **17**. Compound **17** can disproportionate to one molecule of dopamine and one molecule of electron-poor orthoquinone **18**, which can also be formed directly from **17**. Cyclization of **18** yields cyclic **19**, which is oxidized to aminochrome **20** (Scheme 7.4). Dopamine *o*-quinones have a key role in PD [19]. Since compound **20** is directly involved in mitochondrial and protein degradation dysfunctions, aggregation of α-synuclein, and oxidative stress, it has been proposed as a PD preclinical model [20]. Similar reactions with tyrosine and DOPA are also possible. The oxidation reactions take place with the formation of superoxide radicals, which are responsible for oxidative stress (by reaction with molecular oxygen) and neurodegenerative processes at the dopaminergic neurons of the nigrostriatal system in PD [21].

Dopamine quinone **18** reacts with hydrogen peroxide to form 6-hydroxydopaminequinone **21** or with acetaldehyde to yield salsolinol **23**. Dopamine quinone **20** can form the neurotoxin 6-hydroxydopamine **22**, and salsolinol **23** is a neurotoxin that causes mitochondrial damage [22] and oxidative stress [23] and

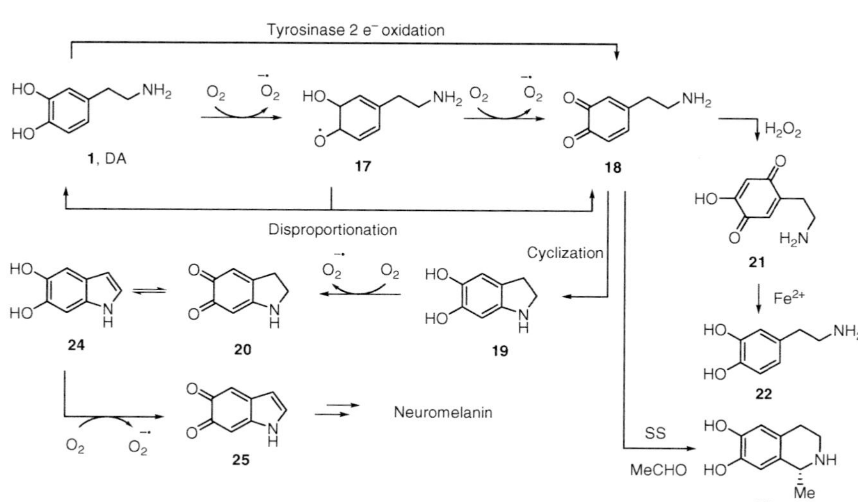

Scheme 7.4 Oxidation reactions of dopamine to *o*-quinones.

inhibits tyrosine hydroxylase (TH), dopamine-β-hydroxylase (DBH), COMT, and MAO [24]. Aminochrome **20** tautomerizes to 5,6-dihydroxyindole **24**, which is oxidized to 5,6-indolequinone **25**, a precursor of the neuronal complex pigment neuromelanin, which increases with age in the substantia nigra and mediates intracellular oxidative mechanisms as supported by recent studies suggesting that such a function may be compromised in PD [25]. Neuromelanin magnetic resonance imaging can be used to distinguish patients with multiple system atrophy (MSA) from normal patients and patients with PD [26].

Conjugation of aminochrome **20** with glutathione by glutathione *S*-transferase M2-2 (GST M2-2) yields 4-*S*-glutathionyl-5,6-dihydroxyindoline **26** [27], which is stable in the presence of oxygen, hydrogen peroxide, and superoxide radicals [28]. Dopamine *o*-quinone **18** also serves as a substrate for GST M2-2, yielding 5-glutathionyl-dopamine **27** that prevents the formation of neurotoxic **20** [29]. Glutathione conjugates undergo degradation to give cysteine conjugates. In the case of 5-glutathionyl-dopamine **27**, the resulting 5-cysteinyl dopamine **28** has been found in the cerebrospinal fluid of PD patients, dopamine-rich regions of the brain [30], and neuromelanin [31]. Similarly, DOPA *o*-quinone **29** leads to 5-glutathionyl DOPA **30**, which generates 5-cysteinyl DOPA **31** (Scheme 7.5) [32]. All these conjugation processes with glutathione can be considered neuroprotective as they remove aminochrome toxicity.

The direct one-electron oxidation of aminochrome **20** catalyzed by flavoenzymes that use nicotinamide adenine dinucleotide H (the H refers to the reduced form) (NADH) as an electron donor generates leukoaminochrome *o*-semiquinone radical **32**, which also reacts with oxygen to regenerate **20** forming superoxide radicals (Scheme 7.6). Radical **32** can also be formed from **19** in the presence of superoxide radicals. This auto-oxidation of **32** is prevented by superoxide dismutase that yields **20** under the same conditions [33]. Then a neuroprotective pathway is the two-electron reduction of **20** catalyzed by DT-diaphorase (alternative name for NAD(P)H dehydrogenase [quinone] 1) [34], which prevents the participation of **20** in the above-mentioned neurotoxic reactions involving

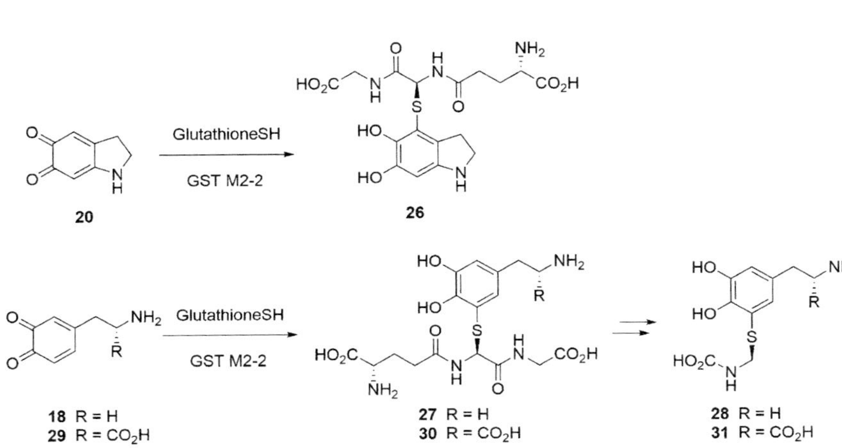

Scheme 7.5 Conjugation reactions of dopamine *o*-quinones.

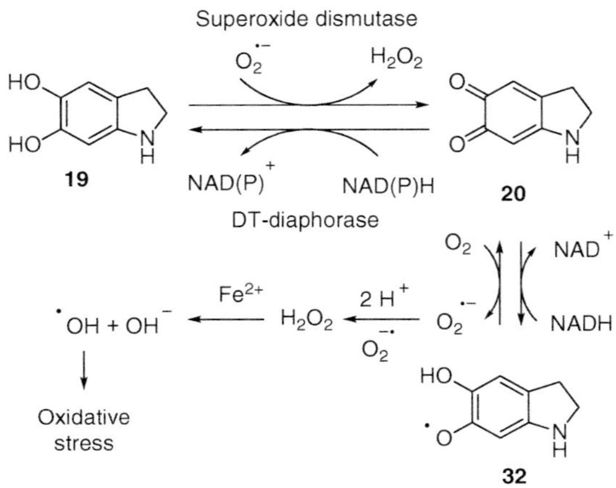

Scheme 7.6 Aminochrome-related redox reactions.

radicals [21]. In general, oxidation of DA **1** to aminochrome **20** is prevented by dopamine degradation catalyzed by MAO and COMT.

Oxidation products of dopamine, including cyclized quinones, inhibit proteasome rather than oxygen radicals. This provides a possible rationale for the observed selectivity of dopaminergic neuron damage in PD [35].

7.2.2 Norepinephrine (Noradrenaline) and Epinephrine (Adrenaline)

DA **1** is also the precursor of norepinephrine (NE) **33** and epinephrine (EPI) **34** (Scheme 7.7) [36]. NE **33** is formed by DBH in a reaction that consumes oxygen and requires ascorbic acid **35**, which is converted into dehydroascorbic acid **36**. Epinephrine **34** is formed by N-methylation of **33** catalyzed by phenylethanolamine-N-methyltransferase (PEMT) (Scheme 7.7). The methylating agent is the ubiquitous SAM that is converted into *S*-adenosylhomocysteine (SAH). NE forms in the CNS, whereas the biosynthesis of EPI occurs in the periphery [37].

Degradation of norepinephrine **33** and epinephrine **34** is carried out by enzymes MAO and COMT after reuptake from the synaptic cleft. By using **33** as a substrate, MAO yields 3,4-dihydroxyphenylglycolaldehyde (DOPEGAL) **37**, which can be either reduced by aldehyde reductase (AR) to 3,4-dihydroxyphenylethyleneglycol (DOPEG) **38** or oxidized by ADH to 3,4-dihydroxymandelic acid (DOMA) **40** (Scheme 7.8). Both **38** and **40** are responsible for the structure of neuromelanin produced at the locus coeruleus [38]. The action of COMT on **38** and **40** yields 3-methoxy-4-hydroxyphenylehtyleneglycol (MHPG) **39** and vanillylmandelic acid (VMA) **41**, respectively, which are excreted in the urine. Compound **41** is the major urinary metabolite of **33** and **34** in humans. The analysis of these metabolites suggested that noradrenergic dysregulation plays an important role in depression [39].

Degradation of **33** by COMT leads to normetanephrine **42**, which is converted by MAO into 3-methoxy-4-hydroxyphenylglycolaldehyde (MOPEGAL) **43**.

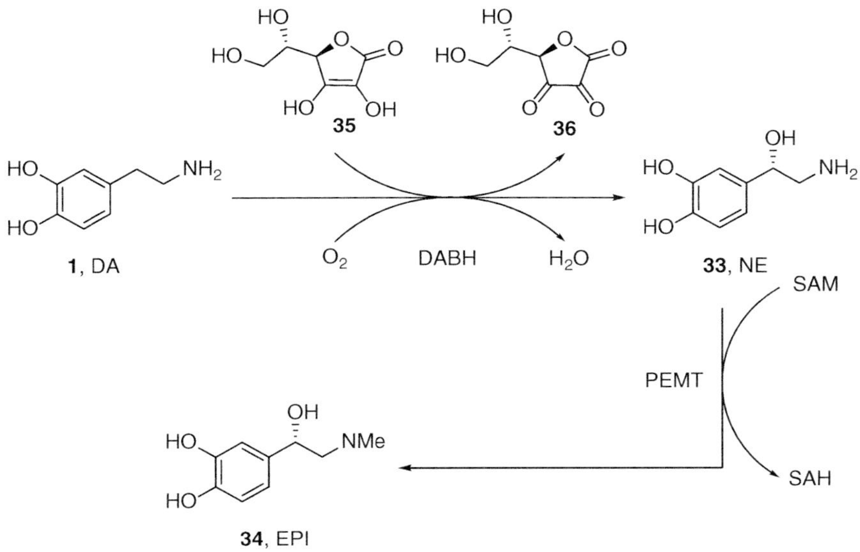

Scheme 7.7 Biosynthesis of norepinephrine and epinephrine from dopamine.

Oxidation of compound **43** with ADH yields VMA **41**, and its reduction by AR produces MHPG **39**. With epinephrine **34** as a substrate, MAO and COMT form **37** and metanephrine **44**, respectively. The latter is converted by MAO into MOPEGAL **43**. Catabolism of norepinephrine and epinephrine is similar to that of dopamine, but the presence of the benzylic hydroxyl group favors the pathway via AR.

In vitro [40] and *in vivo* [41] studies have demonstrated that neurotoxicity of DOPEGAL **37** is high. The intraneuronal production of **37** has been considered to play a role in neurodegenerative diseases affecting the central and peripheral systems. Although the precise mode of action remains unknown, it has been demonstrated that it is capable of inducing apoptosis by generating free radicals [42]. Among the neurodegenerative diseases, PD involves more extensive abnormalities in the metabolism of catecholamines. Implications of catecholamine metabolism for physiology and medicine have been reviewed [43].

7.3 Metabolism of Serotonin

The precursor of serotonin **47** is the essential amino acid L-tryptophan **45**. After being transported from the blood to the brain, the amino acid is hydroxylated by tryptophan hydroxylase to give **46**. In the reaction, the cofactor 6R-BH4 **3** is converted into unstable dihydrobiopterin **4**, which is reconverted into **3** by quinoid DHPR in the presence of NADH (or NADPH). Decarboxylation of **46** by AADC yields the neurotransmitter serotonin **46**, which is placed into storage vesicles ready for being released into the synaptic cleft with depolarization of the presynaptic neuron (Scheme 7.9). The synthesis of serotonin depends on the

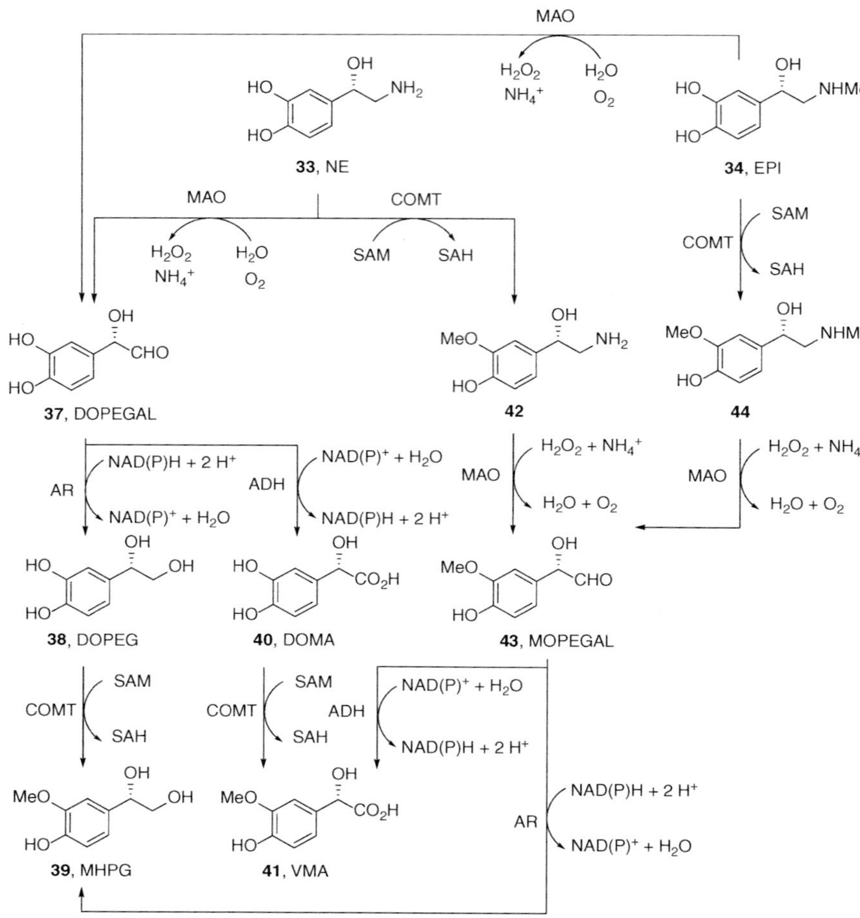

Scheme 7.8 Degradation of norepinephrine and epinephrine.

conversion of L-tryptophan into 5-HTP, the rate-limiting step, which is dependent on the extracellular calcium [44]. Serotonin can form conjugate **48** with L-glutamic [45]. In the brain, serotonin can also be converted into the corresponding sulfate **49** by the action of sulfotransferase, which represents a pathway for inactivation of **47** [46]. The highest activity of serotonin sulfotransferase is found in the hypothalamus and the lowest in the thalamus [47].

Serotonin is removed from the synaptic cleft and then it is either repackaged into storage vesicles or catabolized by MAO (preferentially MAO-A) at the mitochondrial membrane, to form 5-HIAL **51**, which can be either oxidized by ADH to 5-HIAA **53** or reduced to 5-HTOL **54** (Scheme 7.10). The analysis of these metabolites is mainly focused on the diagnosis of carcinoid tumors [48, 49]. 5-HIAA **53** is the major metabolite of serotonin.

An additional important feature of serotonin is that it is the precursor of the hormone melatonin **52**, which is formed after acetylation of **47** to give

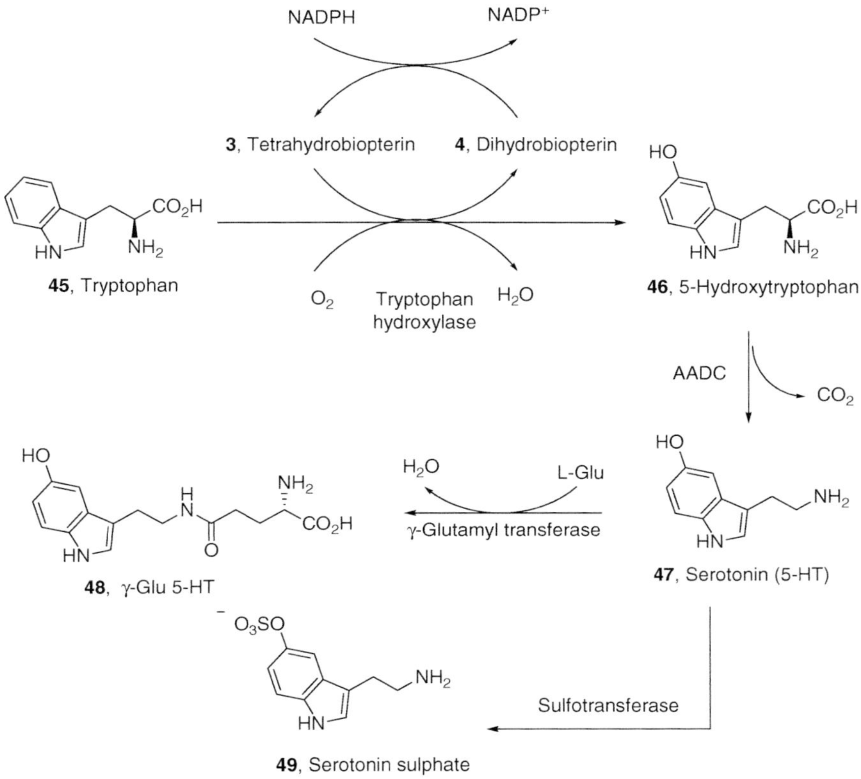

Scheme 7.9 Metabolism of serotonin.

N-acetylserotonin (NAS) **50** and further methylation by the action of hydroxyindole-*O*-methyltransferase. Consequently, there is a close interaction between catabolic pathways of serotonin and melatonergic pathways [50]. Melatonin is produced in the pineal gland outside the blood–brain barrier, but their levels are controlled by numerous types of neurotransmitters and neuropeptides [51]. Melatonin has a significant role in circadian rhythm entrainment and provides seasonal timing cues through the activation of G-protein-coupled receptors [52]. As circadian dysregulation is a clinical aspect in various neurodisorders, the role of melatonin in sleep [53, 54], Alzheimer's disease [55–57], bipolar disorder [58], depression [59], and affective disorders [60] has been investigated.

A variety of clinical disorders involving motor dysfunction/rigidity seems to be associated with an excess of serotonin, but they are also related to dopaminergic deficiency [61], indicating that those disorders may, actually, result from an imbalance between serotonergic and dopaminergic systems [62, 63]. Molecular imaging techniques provide evidences of progressive degeneration of 5-HT terminals in Parkinson's disease, resulting in a lower biosynthesis of serotonin [64]. The brain capacity for synthesizing serotonin plays also a crucial role in obsessive-compulsive disorder (OCD) [65].

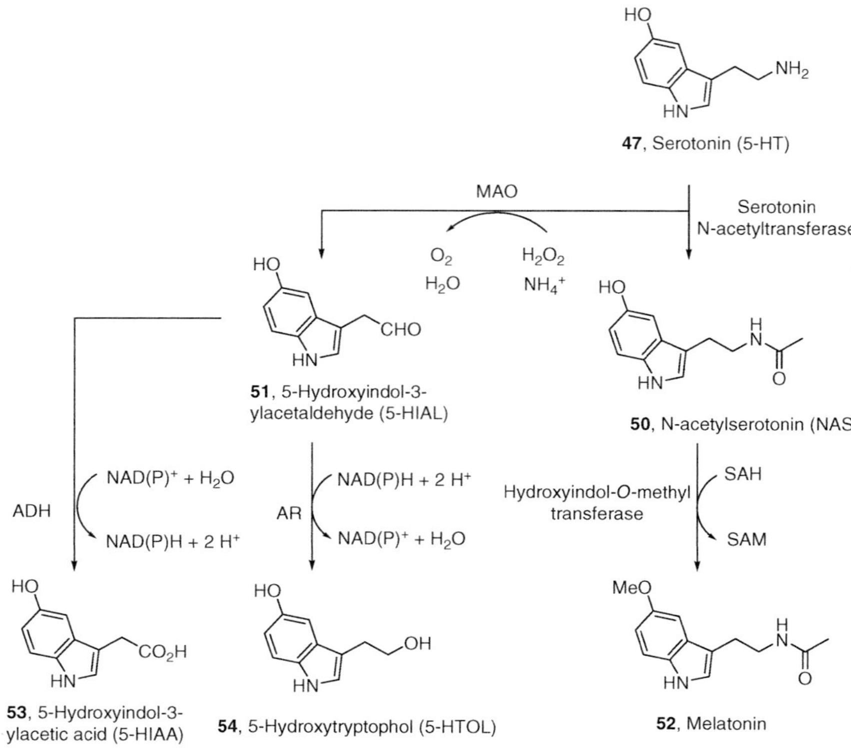

Scheme 7.10 Catabolism of serotonin.

7.4 Metabolism of Acetylcholine

The metabolism of acetylcholine in the CNS has been studied since more than 80 years ago [66–70].

Acetylcholine **54** is synthesized in the synaptoplasmic compartment of cholinergic neurons from choline **53** and acetyl coenzyme A by the enzyme choline acetyltransferase (ChAT). The acetyl coenzyme A is released from the mitochondria through calcium-dependent high-permeability anion channels; alternatively, it can be formed from citrate by the action of adenosine triphosphate (ATP) citrate lyase [71]. Choline is formed from glycine through several steps, including the formation of serine, ethanolamine, phosphatidyl-ethanolamine, and phosphatidyl-choline, which is the immediate precursor of choline. Choline can also have an exogenous origin from food.

The formed acetylcholine is transported in vesicles and released in the synaptic cleft. The action of acetylcholine is terminated by hydrolysis rather than by transport. The hydrolysis is carried out by the enzyme acetylcholine esterase (AChE), which hydrolyzes acetylcholine into its components choline and acetic acid (Scheme 7.11).

The formed acetic acid returns into the Krebs cycle, and choline is reabsorbed by its uptake transporters. The reuptake of choline is the rate-limiting step in the biosynthesis of acetylcholine.

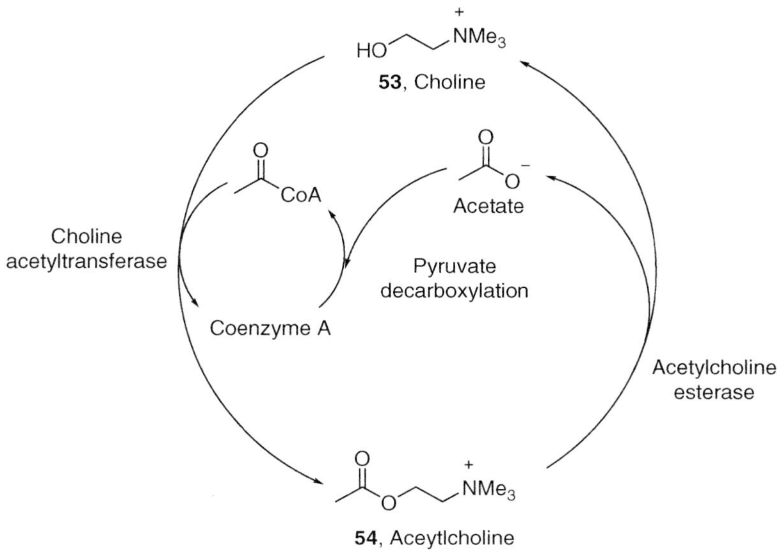

Scheme 7.11 Metabolism of acetylcholine.

7.5 Metabolism of Glutamic Acid and GABA

7.5.1 Glutamic Acid

Glutamic acid **55** can be synthesized in the brain through two pathways: (i) from α-ketoglutaric acid **56**, an intermediate of the Krebs cycle, by transamination with L-aspartic, catalyzed by glutamate oxaloacetate transaminase (GOT), and (ii) from L-glutamine **57**, through hydrolysis catalyzed by glutaminase (Gase) (Scheme 7.12).

The biosynthesis of L-glutamic acid from **56** is considered a minor pathway of glutamate production in the brain due to the high affinity of glutamate dehydrogenase (GDH) for ammonia [72]. Mitochondria of glutamatergic neurons show enhanced production of L-glutamic acid from L-glutamine [73].

In the brain, glutamic acid can be transformed through several routes [74], including those inverse to the above-mentioned one. Glutamic acid can be transformed into α-ketoglutaric acid **56** by the action of GDH [75]. This oxidation for energy proceeds via the Krebs cycle and requires the entry of glutamate into the mitochondria. In neurons, it takes place via the aspartate/glutamate carrier [76], while in astrocytes the glutamate/hydroxyl carrier is responsible for carrying L-glutamate across the mitochondrial membrane [77]. Transformation of L-glutamic acid **55** into **56** in astrocytes is important to sustain energy metabolism of neurons and glutamate homeostasis [78]. Glutamic acid can also lead to L-glutamine by incorporation of ammonia mediated by ATP and catalyzed by glutamine synthetase (GS), a reaction that takes place in astrocytes [74].

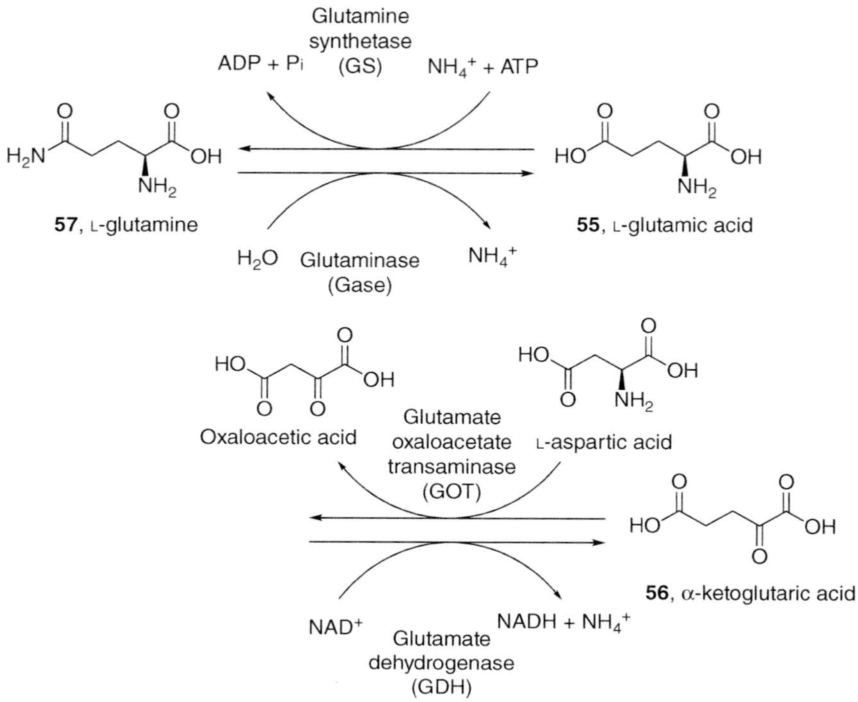

Scheme 7.12 Metabolism of glutamic acid.

7.5.2 γ-Aminobutyric Acid (GABA)

The metabolism of glutamic acid and γ-Aminobutyric Acid (GABA) is closely related. In fact, it has been demonstrated by NMR spectroscopy [79] that L-glutamine **57** formed in astrocytes is a precursor for the synthesis of GABA **58** in neurons due to decarboxylation of L-glutamic **55** catalyzed by glutamate decarboxylase (GD) (Scheme 7.13). In several subclasses of GABAergic neurons and glial cells, GABA is the precursor of homocarnosine **59**, an inhibitory neuromodulator. Compound **59** is synthesized from GABA and L-histidine in an ATP-dependent reaction catalyzed by homocarnosine synthase [80].

GABA is degraded in the mitochondria by GABA transaminase, which converts GABA into succinic semialdehyde **60** [81]. This compound is the precursor of GHB **62**. SSA **60** can be incorporated into the Krebs cycle through its conversion into succinic acid [82].

7.5.3 Neurotransmission Mediated by Glutamic Acid and GABA Metabolisms

L-Glutamic acid and GABA participate in several processes related to neurotransmission, which include the biosynthesis of ornithine, peptides, and proteins taking place in the brain as a whole. There is no single type of cell capable of performing all those processes [72]; thus, GD is mainly expressed in GABAergic

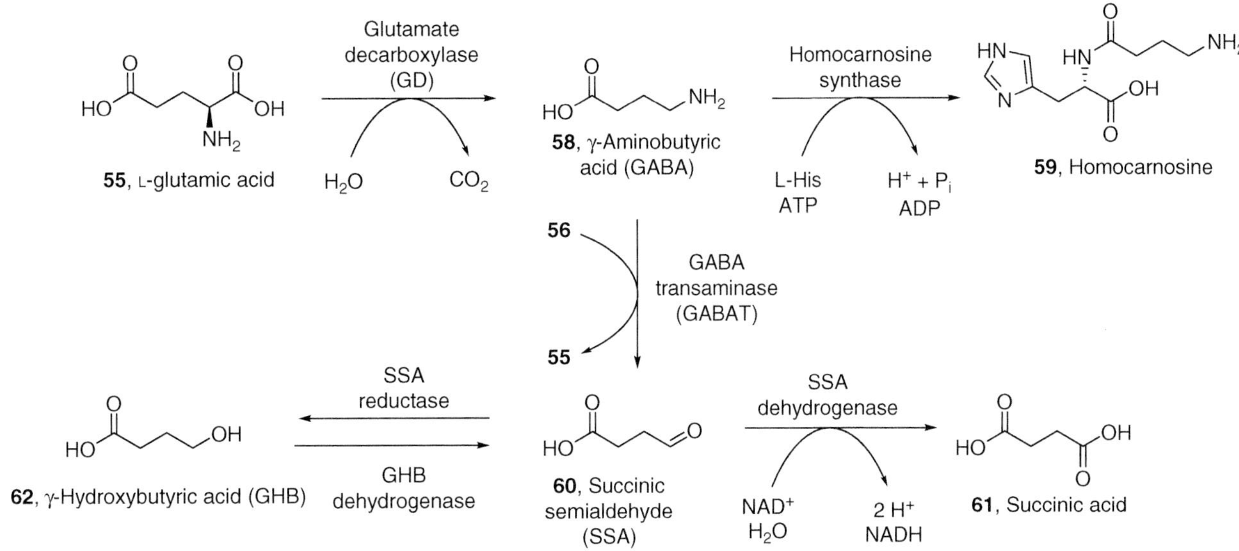

Scheme 7.13 Metabolism of GABA.

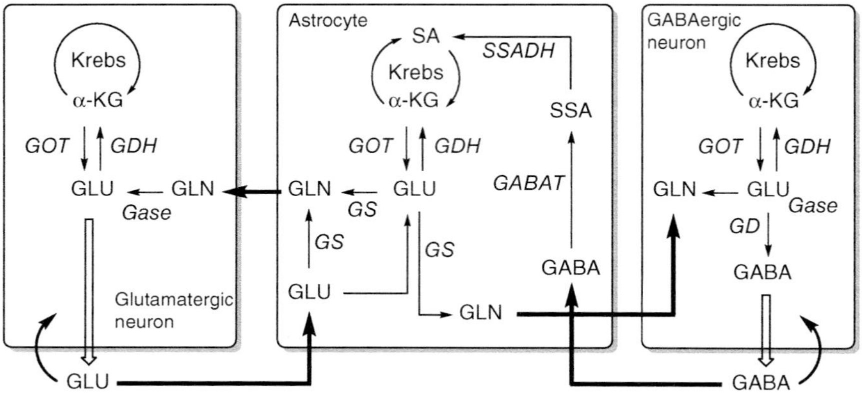

Scheme 7.14 Metabolism of L-glutamic acid and GABA in neurons and astrocytes. GLU: L-glutamic acid; GLN: L-glutamine; α-KG: α-ketoglutaric acid; SSA: succinic semialdehyde.

neurons, while GS is localized in astrocytes. Gase is more active in neurons than in astrocytes [83], but in the latter formation of **56** is also observed [84] .

The Krebs cycle operates in both neurons and astrocytes to produce α-ketoglutaric acid **56** that can be transformed into L-glutamic acid **55**. In glutamatergic neurons, the synthesized L-glutamic acid, once released in the synaptic cleft, is mostly uptaken into astrocytes and converted into L-glutamine, which is then transferred back to the glutamatergic neuron. L-Glutamine is also transferred from astrocytes to GABAergic neurons, in which L-glutamic acid is transformed into GABA and released into the synaptic cleft. After uptake into astrocytes, GABA reenters into the Krebs cycle through succinic semialdehyde (Scheme 7.14).

References

1 Hyland, K. (2003). Disorders of neurotransmitter metabolism. In: *Physician's Guide to the Laboratory Diagnosis of Metabolic Diseases* (ed. N. Blau, M. Duran, M.E. Blaskovics and K.M. Gibson), 107–122. Berlin, Heidelberg: Springer-Verlag.
2 Carlsson, A. (2001). A half-century of neurotransmitter research: impact on neurology and psychiatry. Nobel lectures. *Biosci. Rep.* 21: 691–710.
3 Eisenhofer, G., Aneman, A., Friberg, P. et al. (1997). Substantial production of dopamine in the human gastrointestinal tract. *J. Clin. Endocrinol. Metab.* 82 (11): 3864–3871.
4 Meiser, J., Weindl, D., and Hiller, K. (2013). Complexity of dopamine metabolism. *Cell Commun. Signal.* 11: 34.
5 Thöny, B., Auerbach, G., and Blau, N. (2000). Tetrahydrobiopterin biosynthesis, regeneration and functions. *Biochem. J.* 347 (Pt. 1): 1–16.
6 Curtius, H.C., Niederwieser, A., Levine, R., and Muldner, H. (1984). Therapeutic efficacy of tetrahydrobiopterin in Parkinson's disease. *Adv. Neurol.* 40: 463–466.
7 Moore, A.P., Behan, P.O., Jacobson, W., and Armarego, W.L. (1987). Biopterin in Parkinson's disease. *J. Neurol. Neurosurg. Psychiatry* 50 (1): 85–87.

8 Kim, S.W., Jang, Y.J., Chang, J.W., and Hwang, O. (2003). Degeneration of the nigrostriatal pathway and induction of motor deficit by tetrahydrobiopterin: an in vivo model relevant to Parkinson's disease. *Neurobiol. Dis.* 13 (2): 167–176.
9 Choi, H.J., Lee, S.Y., Cho, Y. et al. (2006). Tetrahydrobiopterin causes mitochondrial dysfunction in dopaminergic cells: implications for Parkinson's disease. *Neurochem. Int.* 48 (4): 255–262.
10 Anna Haduch, A., Bromek, E., and Daniel, W.A. (2013). Role of brain cytochrome P450 (CYP2D) in the metabolism of monoaminergic neurotransmitters. *Pharmacol. Rep.* 65 (6): 1519–1528.
11 Halliwell, B. (2006). Oxidative stress and neurodegeneration: where are we now? *J. Neurochem.* 97 (6): 1634–1658.
12 Stefani, A., Pierantozzi, M., Olivola, E. et al. (2017). Homovanillic acid in CSF of mild stage Parkinson's disease patients correlates with motor impairment. *Neurochem. Int.* 105: 58–63.
13 Morimoto, S., Takao, M., Hatsuta, H. et al. (2017). Homovanillic acid and 5-hydroxyindole acetic acid as biomarkers for dementia with Lewy bodies and coincident Alzheimer's disease: an autopsy-confirmed study. *PLoS One* 12 (2): e0171524.
14 Ouzzine, M., Gulberti, S., Ramalanjaona, N. et al. (2014). The UDP-glucuronosyltransferases of the blood-brain barrier: their role in drug metabolism and detoxication. *Front. Cell Neurosci.* 8: https://doi.org/10.3389/fncel.2014.00349.
15 Wang, P.C., Kuchel, O., Buu, N.T., and Genest, J. (1983). Catecholamine glucuronidation: an important metabolic pathway for dopamine in the rat. *J. Neurochem.* 40 (5): 1435–1440.
16 Renskers, K.J., Feor, K.D., and Roth, J.A. (1980). Sulfation of dopamine and other biogenic amines by human brain phenol sulfotransferase. *J. Neurochem.* 34 (6): 1362–1368.
17 Hastings, T.G. (1995). Enzymatic oxidation of dopamine: the role of prostaglandin H synthase. *J. Neurochem.* 64 (2): 919–924.
18 Munoz, P., Huenchuguala, S., Paris, I., and Segura-Aguilar, J. (2012). Dopamine oxidation and autophagy. *Parkinsons Dis.* 2012: https://doi.org/10.1155/2012/920953.
19 Muñoz, P., Melendez, C., Paris, I., and Segura-Aguilar, J. (2015). Molecular and neurochemical mechanisms dopamine oxidation to O-quinones in Parkinson's disease pathogenesis. In: *Toxicity and Autophagy in Neurodegenerative Disorders*, Current Topics in Neurotoxicity, vol. 9 (ed. J.M. Fuentes), 205–224. Cham: Springer.
20 Segura-Aguilar, J., Munoz, P., and Paris, I. (2016). Aminochrome as new preclinical model to find new pharmacological treatment that stop the development of Parkinson's disease. *Curr. Med. Chem.* 23 (4): 346–359.
21 Graumann, R., Paris, I., Martinez-Alvarado, P. et al. (2002). Oxidation of dopamine to aminochrome as a mechanism for neurodegeneration of dopaminergic systems in Parkinson's disease. Possible neuroprotective role of DT-diaphorase. *Pol. J. Pharmacol.* 54 (6): 573–579.
22 Mravec, B. (2006). Salsolinol, a derivate of dopamine, is a possible modulator of catecholaminergic transmission: a review of recent developments. *Physiol. Res.* 55 (4): 353–364.

23 Napolitano, A., Pezzella, A., and Prota, G. (1999). New reaction pathways of dopamine under oxidative stress conditions: nonenzymatic iron-assisted conversion to norepinephrine and the neurotoxins 6-hydroxydopamine and 6, 7-dihydroxytetrahydroisoquinoline. *Chem. Res. Toxicol.* 12 (11): 1090–1097.

24 Napolitano, A., Manini, P., and D'Ischia, M. (2011). Oxidation chemistry of catecholamines and neuronal degeneration: an update. *Curr. Med. Chem.* 18 (12): 1832–1845.

25 Fedorow, H., Tribl, F., Halliday, G. et al. (2005). Neuromelanin in human dopamine neurons: comparison with peripheral melanins and relevance to Parkinson's disease. *Prog. Neurobiol.* 75 (2): 109–124.

26 Matsuura, K., Maeda, M., Yata, K. et al. (2013). Neuromelanin magnetic resonance imaging in Parkinson's disease and multiple system atrophy. *Eur. Neurol.* 70 (1–2): 70–77.

27 Baez, S., Segura-Aguilar, J., Widersten, M. et al. (1997). Glutathione transferases catalyze the detoxification of oxidized metabolites (*o*-quinones) of catecholamines and may serve as an antioxidant system preventing degenerative cellular processes. *Biochem. J.* 324 (1): 25–28.

28 Segura-Aguilar, J., Baez, S., Widersten, M. et al. (1997). Human class Mu glutathione transferases, in particular isoenzyme M2-2, catalyze detoxication of the dopamine metabolite aminochrome. *J. Biol. Chem.* 272 (9): 5727–5731.

29 Shen, X.-M., Xia, B., Wrona, M.Z., and Dryhurst, G. (1996). Synthesis, redox properties, in vivo formation, and neurobehavioral effects of N-acetylcysteinyl conjugates of dopamine: possible metabolites of relevance to Parkinson's disease. *Chem. Res. Toxicol.* 9 (7): 1117–1126.

30 Rosengren, E., Linder-Eliasson, E., and Carlsson, A. (1985). Detection of 5-*S*-cysteinyldopamine in human brain. *J. Neural Transm.* 63 (3–4): 247–253.

31 Carstam, R., Brinck, C., Hindemith-Augustsson, A. et al. (1991). The neuromelanin of the human substantia nigra. *Biochim. Biophys. Acta, Mol. Basis Dis.* 1097 (2): 152–160.

32 Dagnino-Subiabre, A., Cassels, B.K., Baez, S. et al. (2000). Glutathione transferase M2-2 catalyzes conjugation of dopamine and dopa *o*-quinones. *Biochem. Biophys. Res. Commun.* 274 (1): 32–36.

33 Baez, S., Linderson, Y., and Segura-Aguilar, J. (1995). Superoxide dismutase and catalase enhance autoxidation during one-electron reduction of aminochrome by NADPH-cytochrome P-450 reductase. *Biochem. Mol. Med.* 54 (1): 12–18.

34 Segura-Aguilar, J. and Lind, C. (1989). On the mechanism of the Mn^{3+}-induced neurotoxicity of dopamine: prevention of quinone-derived oxygen toxicity by DT diaphorase and superoxide dismutase. *Chem. Biol. Interact.* 72 (3): 309–324.

35 Zafar, K.S., Siegel, D., and Ross, D. (2006). A potential role for cyclized quinones derived from dopamine, DOPA, and 3,4-dihydroxyphenylacetic acid in proteasomal inhibition. *Mol. Pharmacol.* 70 (3): 1079–1086.

36 Kirshner, N. (1959). Biosynthesis of adrenaline and noradrenaline. *Pharmacol. Rev.* 11 (2 Part 2): 350–357.

37 Gitlow, S.E., Mendlowitz, M., Bertani, L.M. et al. (1971). Human norepinephrine metabolism: its evaluation by administration of tritiated norepinephrine. *J. Clin. Invest.* 50 (4): 859–865.

38 Wakamatsu, K., Tabuchi, K., Ojika, M. et al. (2015). Norepinephrine and its metabolites are involved in the synthesis of neuromelanin derived from the locus coeruleus. *J. Neurochem.* 135 (4): 768–776.
39 Roy, A., Pickar, D., De Jong, J. et al. (1988). Norepinephrine and its metabolites in cerebrospinal fluid, plasma, and urine. Relationship to hypothalamic-pituitary-adrenal axis function in depression. *Arch. Gen. Psychiatry* 45 (9): 849–857.
40 Li, S.W., Lin, T.S., Minteer, S., and Burke, W.J. (2001). 3,4-Dihydroxyphenylacetaldehyde and hydrogen peroxide generate a hydroxyl radical: possible role in Parkinson's disease pathogenesis. *Brain Res. Mol. Brain Res.* 93 (1): 1–7.
41 Burke, W.J., Schmitt, C.A., Gillespie, K.N., and Li, S.W. (1996). Norepinephrine transmitter metabolite is a selective cell death messenger in differentiated rat pheochromocytoma cells. *Brain Res.* 722 (1–2): 232–235.
42 Burke, W.J., Kristal, B.S., Yu, B.P. et al. (1998). Norepinephrine transmitter metabolite generates free radicals and activates mitochondrial permeability transition: a mechanism for DOPEGAL-induced apoptosis. *Brain Res.* 787 (2): 328–332.
43 Eisenhofer, G., Kopin, I.J., and Goldstein, D.S. (2004). Catecholamine metabolism: a contemporary view with implications for physiology and medicine. *Pharmacol. Rev.* 56 (3): 331–349.
44 Boadle-Biber, M.C. (1993). Regulation of serotonin synthesis. *Prog. Biophys. Mol. Biol.* 60 (1): 1–15.
45 Stuart, J.N., Zhang, X., Jakubowski, J.A. et al. (2003). Serotonin catabolism depends upon location of release: characterization of sulfated and gamma-glutamylated serotonin metabolites in *Aplysia californica*. *J. Neurochem.* 84 (6): 1358–1366.
46 Gal, E.M. (1972). 5-Hydroxytryptamine-*O*-sulfate: an alternate route of serotonin inactivation in brain. *Brain Res.* 44 (1): 309–312.
47 Hidaka, H. and Austin, J. (1972). Occurrence and distribution of serotonin-*O*-sulfotransferase in human brain; a new radioisotopic assay. *Biochim. Biophys. Acta, Enzymol.* 268 (1): 132–137.
48 Kema, I.P., de Vries, E.G., and Muskiet, F.A. (2000). Clinical chemistry of serotonin and metabolites. *J. Chromatogr. B Biomed. Sci. Appl.* 747 (1–2): 33–48.
49 Zandee, W.T., Kamp, K., van Adrichem, R.C. et al. (2016). Limited value for urinary 5-HIAA excretion as prognostic marker in gastrointestinal neuroendocrine tumours. *Eur. J. Endocrinol.* 175 (5): 361–366.
50 Anderson, G. and Maes, M. (2017). Interactions of tryptophan and its catabolites with melatonin and the alpha 7 nicotinic receptor in central nervous system and psychiatric disorders: role of the aryl hydrocarbon receptor and direct mitochondria regulation. *Int. J. Tryptophan Res.* 10 (1): https://doi.org/10.1177/1178646917691738.
51 Govitrapong, P., Chetsawang, B., Mukda, S., and Phansuwan-Pujito, P. (2007). *Neural Regulation of Melatonin Synthesis*, 81–115. Nova Science Publishers, Inc.
52 Prada, C. (2005). Melatonin receptors as modulators of GABA(C) and metabotropic glutamate receptors in the optic tectum of Xenopus laevis.

PhD dissertation. 146 pp. University at Buffalo. The State University of New York. CAPLUS AN 2006:89783. https://ubir.buffalo.edu/xmlui/handle/10477/45074.
53 Bersani, G. and Garavini, A. (2000). Melatonin add-on in manic patients with treatment resistant insomnia. *Prog. Neuro-Psychopharmacol. Biol. Psychiatry* 24 (2): 185–191.
54 Hardeland, R. (2009). New approaches in the management of insomnia: weighing the advantages of prolonged-release melatonin and synthetic melatoninergic agonists. *Neuropsychiatr. Dis. Treat.* 5: 341–354.
55 Srinivasan, V., Pandi-Perumal, S., Cardinali, D. et al. (2006). Melatonin in Alzheimer's disease and other neurodegenerative disorders. *Behav. Brain Funct.* 2 (1): 15.
56 Daniel, P.C., Analia, M.F., and Luis, I.B. (2010). Clinical aspects of melatonin intervention in Alzheimer's disease progression. *Curr. Neuropharmacol.* 8 (3): 218–227.
57 Lin, L., Huang, Q.-X., Yang, S.-S. et al. (2013). Melatonin in Alzheimer's disease. *Int. J. Mol. Sci.* 14 (7): 14575–14593.
58 Geoffroy, P.A., Etain, B., Franchi, J.A. et al. (2015). Melatonin and melatonin agonists as adjunctive treatments in bipolar disorders. *Curr. Pharm. Des.* 21 (23): 3352–3358.
59 Hansen, M.V., Madsen, M.T., Hageman, I. et al. (2012). The effect of MELatOnin on depression, anxietY, cognitive function and sleep disturbances in patients with breast cancer. The MELODY trial: protocol for a randomised, placebo-controlled, double-blinded trial. *BMJ Open* 2 (1): e000647.
60 Stefanovic, B., Spasojevic, N., Jovanovic, P. et al. (2016). Melatonin mediated antidepressant-like effect in the hippocampus of chronic stress-induced depression rats: regulating vesicular monoamine transporter 2 and monoamine oxidase A levels. *Eur. Neuropsychopharmacol.* 26 (10): 1629–1637.
61 Ener, R.A., Meglathery, S.B., Van Decker, W.A., and Gallagher, R.M. (2003). Serotonin syndrome and other serotonergic disorders. *Pain Med.* 4 (1): 63–74.
62 Esposito, E., Di Matteo, V., and Di Giovanni, G. (2008). Serotonin-dopamine interaction: an overview. *Prog. Brain Res.* 172: 3–6 (Serotonin-Dopamine Interaction).
63 Dobryakova, E., Genova, H.M., DeLuca, J., and Wylie, G.R. (2015). The dopamine imbalance hypothesis of fatigue in multiple sclerosis and other neurological disorders. *Front. Neurol.* 6: 52.
64 Politis, M. and Niccolini, F. (2015). Serotonin in Parkinson's disease. *Behav. Brain Res.* 277: 136–145.
65 Lissemore, J.I., Sookman, D., Gravel, P. et al. (2018). Brain serotonin synthesis capacity in obsessive-compulsive disorder: effects of cognitive behavioral therapy and sertraline. *Translat. Psychiatry* 8 (1): 82.
66 Mann, P.J.G., Tennenbaum, M., and Quastel, J.H. (1939). Acetylcholine metabolism in the central nervous system. *Biochem. J.* 33 (9): 1506–1518.
67 Eisenstadt, M.L., Treistman, S.N., and Schwartz, J.H. (1975). Metabolism of acetylcholine in the nervous system of Aplysia californica. II. Regional localization and characterization of choline uptake. *J. Gen. Physiol.* 65 (3): 275–291.
68 Giacobini, E., Becker, R., Elble, R. et al. (1986). Acetylcholine metabolism in brain: is it reflected by CSF changes? In: *Alzheimer's and Parkinson's Disease:*

Strategies for Research and Development (ed. A. Fisher, I. Hanin and C. Lachman), 309–316. Boston, MA: Springer US.

69 Scremin, O.U. and Jenden, D.J. (1993). Acetylcholine turnover and release: the influence of energy metabolism and systemic choline availability. In: , Progress in Brain Research, Chapter 22 (ed. A.C. Cuello), 191–195. Elsevier.

70 Karczmar, A.G. (2007). Metabolism of acetylcholine: synthesis and turnover. In: *Exploring the Vertebrate Central Cholinergic Nervous System*, 81–149. Boston, MA: Springer US.

71 Tomaszewicz, M., Rossner, S., Schliebs, R. et al. (2003). Changes in cortical acetyl-CoA metabolism after selective basal forebrain cholinergic degeneration by 192IgG-saporin. *J. Neurochem.* 87 (2): 318–324.

72 Yudkoff, M., Nissim, I., Daikhin, Y. et al. (1993). Brain glutamate metabolism: neuronal-astroglial relationships. *Dev. Neurosci.* 15 (3–5): 343–350.

73 Westergaard, N., Drejer, J., Schousboe, A., and Sonnewald, U. (1996). Evaluation of the importance of transamination versus deamination in astrocytic metabolism of [U-13C]glutamate. *Glia* 17 (2): 160–168.

74 McKenna, M.C. (2007). The glutamate-glutamine cycle is not stoichiometric: fates of glutamate in brain. *J. Neurosci. Res.* 85 (15): 3347–3358.

75 Sonnewald, U., Westergaard, N., Petersen, S.B. et al. (1993). Metabolism of [U-^{13}C]glutamate in astrocytes studied by ^{13}C NMR spectroscopy: incorporation of more label into lactate than into glutamine demonstrates the importance of the tricarboxylic acid cycle. *J. Neurochem.* 61 (3): 1179–1182.

76 Satrustegui, J., Contreras, L., Ramos, M. et al. (2007). Role of aralar, the mitochondrial transporter of aspartate-glutamate, in brain N-acetylaspartate formation and Ca^{2+} signaling in neuronal mitochondria. *J. Neurosci. Res.* 85 (15): 3359–3366.

77 Berkich, D.A., Ola, M.S., Cole, J. et al. (2007). Mitochondrial transport proteins of the brain. *J. Neurosci. Res.* 85 (15): 3367–3377.

78 McKenna, M.C., Sonnewald, U., Huang, X. et al. (1996). Exogenous glutamate concentration regulates the metabolic fate of glutamate in astrocytes. *J. Neurochem.* 66 (1): 386–393.

79 Sonnewald, U., Westergaard, N., Schousboe, A. et al. (1993). Direct demonstration by [^{13}C]NMR spectroscopy that glutamine from astrocytes is a precursor for GABA synthesis in neurons. *Neurochem. Int.* 22 (1): 19–29.

80 Drozak, J., Veiga-da-Cunha, M., Vertommen, D. et al. (2010). Molecular identification of carnosine synthase as ATP-grasp domain-containing protein 1 (ATPGD1). *J. Biol. Chem.* 285 (13): 9346–9356.

81 Yoon, C.S., Kim, D.W., Jang, S.H. et al. (2004). Cysteine-321 of human brain GABA transaminase is involved in intersubunit cross-linking. *Mol. Cells* 18 (2): 214–219.

82 Blasi, P., Boyl, P.P., Ledda, M. et al. (2002). Structure of human succinic semialdehyde dehydrogenase gene: identification of promoter region and alternatively processed isoforms. *Mol. Genet. Metabol.* 76 (4): 348–362.

83 Kvamme, E., Torgner, I.A., and Roberg, B. (2001). Kinetics and localization of brain phosphate activated glutaminase. *J. Neurosci. Res.* 66 (5): 951–958.

84 Hertz, L., Peng, L., and Dienel, G.A. (2007). Energy metabolism in astrocytes: high rate of oxidative metabolism and spatiotemporal dependence on glycolysis/glycogenolysis. *J. Cereb. Blood Flow Metabol.* 27 (2): 219–249.

8

Enzymes Processing Neurotransmitters

8.1 Tyrosine Hydroxylase (TH)

Tyrosine hydroxylase (TH) (also known as tyrosine 3-monooxygenase; EC 1.14.16.2) is the most important enzyme in the biosynthetic route of catecholamines. Indeed, it is the rate-limiting enzyme in the biosynthesis of dopamine, norepinephrine, and epinephrine. In consequence, inhibition of TH will result in a marked depletion of catecholamines in the brain. TH catalyzes the conversion of tyrosine into L-DOPA (dihydroxyphenylalanine) using molecular oxygen and two cofactors: Fe^{2+} and (6*R*)-L-erythro-tetrahydrobiopterin (6R-BH4). Ferrous iron is bound directly to TH and is oxidized to Fe^{3+} by O_2; 6R-BH4 acts as an electron donor. TH is a homomeric enzyme, consisting of four identical 60 kDa subunits, each of which binds Fe^{2+} and BH_4. The enzyme is both soluble and membrane bound. There are 6 isoforms of tryptophan hydroxylase, isoform 3 being the canonical sequence with 528 amino acids (Figure 8.1).

All the alternative isoforms differ from the canonical sequence by having missing residues. Isoform 1 (TH type 3) is missing residues 31–34; isoform 2 (TH type 1) and isoform 4 (TH type 2) are missing residues 31–61 and 35–61, respectively. Isoform 5 is missing residues 35–61 and 264–457 and isoform 6 residues 31–61 and 264–357. Notably, isoforms 5 and 6 lack catalytic activity.

TH has numerous natural variants, some of which related to severe neurodisorders. A complete loss of TH activity is observed for variants Asp227Gly, Glu259Gly, Phe309Ser, Arg319Phe, Arg328Trp, Ile394Thr, Arg441Phe, Ser467Gly, and Leu510Gln [1]. Variants Cys207Tyr, Arg233His, Ala241Thr, Phe301Ala, Cys359Phe, Phe375Leu, Ala376Val, Thr399Met, Gln412Lys, Ser467Gly, and Asp498Gly cause a loss of over 80% of TH activity [2]. Natural variants Leu236Phe, Thr276Phe, Thr314Met, Arg337His, and Thr494Met are closely related to Segawa syndrome, an autosomal recessive disorder – a form of dystonia that causes parkinsonian symptoms in early infancy [3–5]. The TH deficiency can be successfully treated with L-DOPA [6]. It has been suggested that TH may play a role in the pathogenesis of Parkinson's disease (PD) [7]. On the other hand, some studies on Val81Met and Leu205Pro functional polymorphisms have found no evidence in the pathogenesis of schizophrenia or Parkinson's disease [8]. Since TH is the rate-limiting enzyme in the production of dopamine and it has been demonstrated that free dopamine is neurotoxic due to

Chemical Biology of Neurodegeneration: A Molecular Approach, First Edition. Pedro Merino.

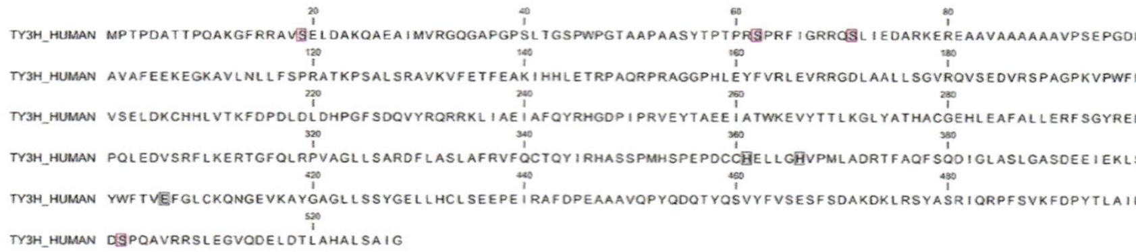

Figure 8.1 Sequence of human tyrosine hydroxylase. Phosphorylation (magenta) sites and metal-binding sites (gray) are indicated on sequence.

the production of reactive oxygen species, a heightened activity of TH can produce oxidative stress and might contribute to neurotoxicity-specific dopaminergic neurons, whose degeneration leads to Parkinson's disease [9].

TH is composed of a largely unstructured N-terminal region (residues 1–71) and a well-folded C-terminal portion (residues 72–159). The region comprising residues 188–456 of TH defines the catalytic domain of the enzyme. Any truncation within these residues results in no detectable activity. The loss of TH activity associated with mutations responsible for parkinsonism is attributed to destabilization of the protein structure. The crystal structure of tyrosine hydroxylase C-term forms a tetramer with two different oligomerization interfaces (Figure 8.2a; PDB ID: 1TOH). The structure revealed an α-helical basket in which the catalytic iron is placed interacting with His331, His336, and Glu376 (Figure 8.2b) [10].

When crystallized with a bound cofactor analogue and iron (PDB ID: 2TOH), self-hydroxylation of Phe300 and the pterin-binding site is evidenced [12]. The pterin binds on one face of the large active-site cleft, forming an aromatic π-stacking interaction with Phe300 (Figure 8.2c). The structure also demonstrates a close interaction between pterin and iron in the active site.

Regulation of TH is complex [13]. On short-term requirements, TH is regulated by end-product inhibition that can be either reversible or irreversible. The reversible process involves binding of catechols to the same enzyme intermedi-

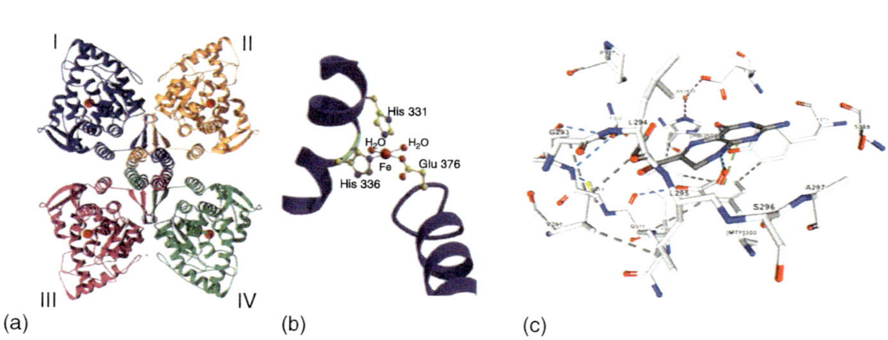

Figure 8.2 (a) The crystal structure of tyrosine hydroxylase C-term in its tetrameric form. The iron atom in the active site is highlighted in red. (b) Close-up view of the iron-binding site. Source: (a, b) Goodwill et al. 1997 [10]. Reprinted with permission of Springer Nature. (c) Interactions at the pterin-binding site.. Source: (c) Image created with the PDB ID: 2TOH, NGL Viewer [11].

ate as does the cofactor, BH4, thus preventing the binding of BH4 in a competitive way and decreasing the enzyme activity. Levels of BH4 are limited, so affinity for this cofactor regulates the enzyme activity. Kinetic analyses and surface plasmon resonance detection demonstrate that TH binds BH4 with negative cooperativity, suggesting that binding of the cofactor at the active site induces a destabilizing conformational change [14].

On the other hand, the levels of tyrosine in the brain are greater than the micromolar affinity of TH for tyrosine, so the levels of tyrosine are unlikely to significantly regulate enzyme activity. The irreversible inhibition involves binding of catecholamines to the Fe^{3+} ion at the catalytic site. This binding induces a more stable but less active form of the enzyme.

TH is modulated by the phosphorylation status of N-terminal serines, which alters the conformation of the enzyme, rendering the catechol binding more dissociable and thus increasing BH4 binding [15]. Indeed, phosphorylation of the enzyme leads to an increase in the catalytic activity, and it can take place at Ser19, Ser40, Ser62, Ser71, and Ser502 [16]. Phosphorylation of Ser40 abolishes the feedback inhibition by catecholamines to a high-affinity dopamine-binding site ($K_D < 4$ nM), thus increasing the enzyme activity. A second low-affinity dopamine-binding site ($K_D = 90$ nM) is also present in both nonphosphorylated and Ser40-phosphorylated forms. Binding of dopamine to these sites increases the K_m for tetrahydrobiopterin contributing to TH regulation [17].

Phosphorylation at Ser19 by p38-regulated/activated kinase and by calcium/calmodulin-dependent protein kinase II activates and stabilizes TH through a mechanism that requires binding to 14–3–3 proteins, a family of proteins present exclusively in the brain [18] acting as a modulator of TH [19]. The first part of the regulatory domain (residues 1–43) interacts with negatively charged membranes. By means of X-ray crystallography and molecular dynamics simulations, the conformations of free N-terminal peptide, unphosphorylated and Ser19-phosphorylated were investigated in three different states: bound to 14-3-3γ, unbound in solution, and bound to negatively charged bilayers [20]. These studies indicated that cationic residues Arg15 and Arg16, and recognition sites of the kinases phosphorylating at Ser19, also contribute to the interaction with the membrane. The X-ray structure of 14-3-3γ bound to TH (residues 1–43) includes four chains in the asymmetric unit and indicates that the region surrounding Ser19 adopted a more extended conformation when bound to 14-3-3γ than was in the case of water (Figure 8.3a). Molecular dynamics calculations confirmed that the number of hydrogen bonds formed between amino acids separated by two, three, or four residues is higher than in the phosphorylated TH (Figure 8.3b). Thus, phosphorylation at Ser19 induces a more stable, compact, and less flexible conformation with high secondary structure content. The X-ray structure of the catalytic domain of human TH has been resolved at 2.68 Å (Figure 8.3c) [21]. By nuclear magnetic resonance (NMR) spectroscopy, the solution structure of the isolated regulatory domain of rat was analyzed [22], and the presence of an ACT domain containing residues 65–159 was established.

The NMR solution structure of the regulatory domain of TH, regulatory domain of tyrosine hydroxylase (RD-TH), shows a largely unfolded N-terminal region (residues 1–71) and a well-folded C-terminal portion (residues 72–159).

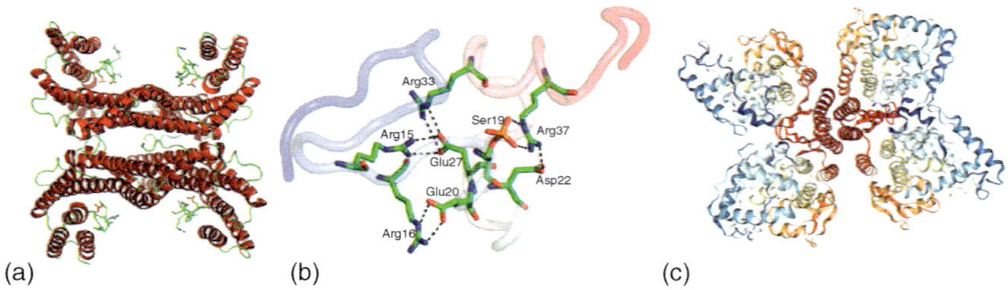

Figure 8.3 (a) The crystal structure of 14–3–3γ bound to the THp fragments. The two 14–3–3γ dimers (chains A–D) in the asymmetric unit (PDB ID: 4J6S) are represented as ribbons colored by secondary structure, complexed to the TH peptides represented by sticks. (b) Representative structure of THp-(1–43) showing the intrapeptide side chain–side chain hydrogen bonds with occupancies >25% (broken lines) in the aqueous simulation. The peptide is colored as a continuous gradient going from blue (N-terminus) to red (C-terminus). The last 100 ns of each simulation was used for the analyses. Source: (a, b) Skjevik et al. 2014 [20]. Reprinted with permission of Elsevier. (c) Catalytic domain of tyrosine hydroxylase. Source: (c) Image created with the PDB ID: 2XSN, NGL Viewer [11].

The entire RD-TH is dimeric at physiological pH, but variants lacking the first 64 residues form monomeric species that present a low conformational stability [23]. Recent experiments suggest that dopamine transporter expression also affects TH expression and phosphorylation in dopamine terminal field compartments, confirming a dichotomic regulation of TH between somatodendritic and terminal field compartments of the nigrostriatal and mesoaccumbens pathways [24].

Other iron-containing, pteridine-binding enzymes related to tyrosine hydroxylase are tryptophan hydroxylase (see next section) and phenylalanine hydroxylase (PAH), an enzyme that converts phenylalanine into tyrosine, all of which require tetrahydrobiopterin (BH4) as a cofactor. There are inherited defects that reduce the concentration of BH4, leading to phenylketonuria and deficiencies of dopamine and serotonin, causing a variety of disorders related to those neurotransmitters [25].

8.2 Tryptophan Hydroxylase (Tryptophan-5-monoxygenase, TPH)

Tryptophan hydroxylase (also known as tryptophan 5-monooxygenase; EC 1.14.16.4) catalyzes the conversion of tryptophan into 5-HTP (hydroxytryptophan) using molecular oxygen and the reduced pteridine cofactor L-erythro-5,6,7,8-tetrahydrobiopterin (6R-BH4), which serves as an electron donor. There are two isoforms of tryptophan hydroxylase [26], referred to as TPH1, expressed in the periphery, and TPH2, expressed in neuronal cells located in the dorsal raphe nucleus of the brain and the myenteric plexus of the gut [27]. It is synthesized in serotonergic cell bodies of the raphe nuclei and is found only in cells that synthesize serotonin. The two isoforms share an overall sequence identity of c. 80% (Figure 8.4) and are homologous to the other two aromatic

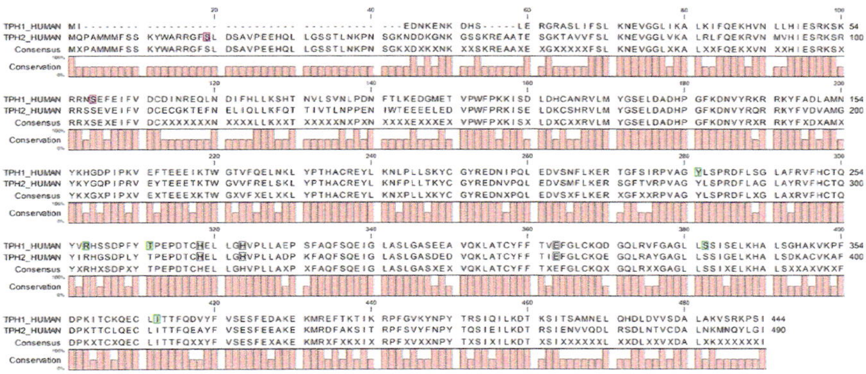

Figure 8.4 Sequences of the two isoforms of human tryptophan hydroxylase. Conservation graphic and consensus sequence are shown. Phosphorylation (magenta), binding (green), and metal-binding (gray) sites are indicated on sequences.

amino acid hydroxylases, PAH and TH, with approximately 50% sequence identity.

The isoform expressed in the brain, TPH2, can be phosphorylated at Ser19. An isoform in which Glu147 is replaced by Gly-Lys-Glu sequence has been identified. TPH2 presents several natural variants including Ala328Vaal, Arg433Gly, and Gln468Arg. A moderate loss-of-function has been observed for the variants Ala328Val and Asp479Glu. The properties of natural variants Leu36Pro, Leu36Val, Ser41Tyr, and Arg55Cys are indistinguishable from those of wild type [28]. The variants Arg303Trp and Arg441His are identified as a loss-of-function mutation that may represent an important risk factor for major depressive disorder [29] and attention deficit-hyperactivity disorder (ADHD) [30]. Examination of the functional effects of the variant Pro206Ser provided evidence for a reduced thermal stability and solubility of the mutated enzyme, suggesting reduced 5-HT production in the brain as a pathophysiological mechanism in bipolar affective disorder [31].

As TH for catecholamines, TPH is the rate-limiting enzyme in the biosynthesis of serotonin, and, consequently, inhibition of this enzyme will result in an evident depletion of serotonin in the brain. V_{max} is 833 and K_m of TPH2 for tryptophan is c. 41 μM, a concentration comparable to that of tryptophan in the brain, so the enzyme would not be saturated with the substrate [31]. Production of serotonin can be stimulated by administration of pure tryptophan (but not of aliments rich in that amino acid), which in fact has an antidepressant effect. TPH2 plays an active role in stress response; consequently, the enzyme is a therapeutic target for stress disorders [32] and other disorders related to cognitive control and emotion regulation [33].

The X-ray structure (PDB ID: 1MLW) of a truncated functional form of TPH1 bound to 7,8-dihydro-L-biopterin provided molecular information of the catalytic domain (Figure 8.5a) [34]. The active site of human THP is a cavity of 9 × 10 Å. The enzyme contains a Fe^{3+} ion c. 13 Å below the surface, on the floor of the active site, and coordinated to His272, His277, and Glu317; the octahedral environment of the iron atom is completed by three water molecules (Figure 8.5b).

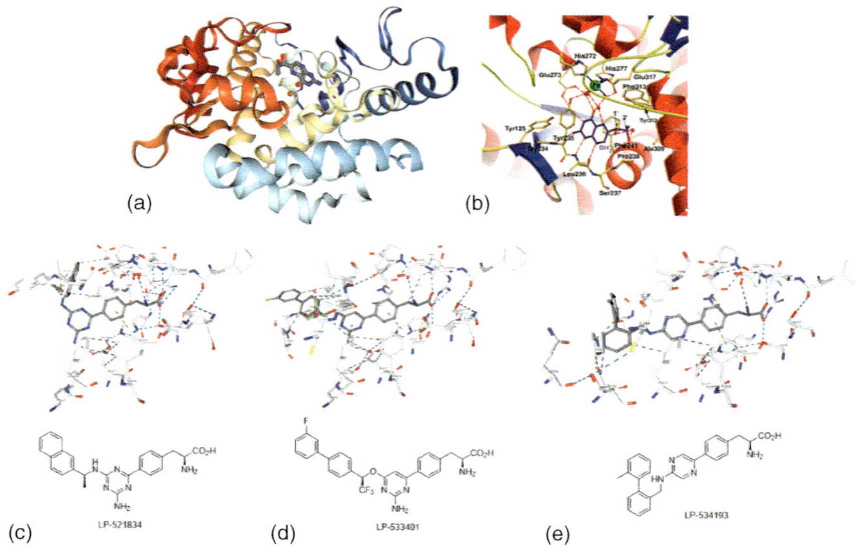

Figure 8.5 (a) Crystal structure (PDB ID: 1MLW) of human tryptophan hydroxylase with bound 7,8-dihydro-L-biopterin cofactor (ball and sticks) and Fe^{3+} (orange). (b) Active site of the iron- and cofactor (7,8-dihydro-L-biopterin)-bound form of human TPH. The region mainly responsible for cofactor binding (Gly234-Pro238) is shown in ball and stick mode for clarification of hydrogen bonding interactions with the pterin. The R-helices of the catalytic domain are colored red, the â-strands blue, and the coiled regions yellow. The active-site iron is shown as a green sphere, and the bound 7,8-BH2 cofactor is shown as a ball-and-stick model with carbon atoms in purple, oxygen atoms in red, and nitrogen atoms in blue. Hydrogen bonds are represented by dashed red lines between atoms, and iron coordination is represented by black solid lines. Source: (b) Reprinted with permission from Wang et al. [34]. Copyright 2008, American Chemical Society. (c–e) Interactions at the active site of human THP1 with ligands LP-521834 (PDB ID: 3HF6; c), LP-533401 (PDB ID: 3HF8; d), and LP-534193 (PDB ID: 3HFB; e). Source: (a, c–e) Images created with the indicated PDB IDs, NGL Viewer [11]. Reprinted with permission of OUP.

The cofactor forms π-interactions with Phe241 and Tyr235 and H-bond interactions with Gly234 and Leu236. Additional hydrogen bonds are formed with water molecules. Cocrystal structures of the catalytic domain of human TPH1 with inhibitors LP-521834 (PDB ID: 3HF6), LP-533401 (PDB ID: 3HF8), and LP-534193 (PDB ID: 3HFB) have been resolved in complex with one iron atom and without pterin [35]. The three inhibitors form similar interactions with TPH1. Indeed, the common substructure across the ligands, 2-amino-3-phenylpropionic acid, has the same interactions in the three structures (Figure 8.5c–e).

On the other hand, the pattern of the iron ion coordination is different among the three complexes. While structures complexed with LP-521834 and LP-533401 show a monodentate coordination by Glu317, identical to the pattern in TPH1-BH2-Fe(III), the structure of the complex with LP-533401 shows a bidentate interaction with Glu317, causing a marked distortion in the geometry coordination. Under such a disposition, iron ion is only coordinated to five sites, allowing one open site for oxygen [36]. A major difference among the three structures is the conformation of Tyr235, conserved in all known TPH1 and TPH2 sequences.

Mutagenesis experiments demonstrate the role of that residue in both pterin and Trp binding [36]. Indeed, Tyr235 was found to interact with both tryptophan and pterin in both TPH1-Trp-Fe^{3+} and TPH1-BH2-Fe^{3+} structures [34, 36]. In the complex with LP-521834, Tyr235 has a conformation similar to that in Trp- or BH2-bound structure. However, in the structures of complexes LP-533401 and LP-534193, it is rotated inward by c. 90° forming a new hydrogen bond with Pro268. Such a conformational freedom of Tyr235 might be critical to the reaction mechanism of TPH.

The structure of chicken THP1 in complex with the tryptophan substrate and an iron-bound imidazole has also been resolved (PDB ID: 3E2T) [36]. The binding of ligands caused structural changes in the catalytic domain compared to the structure of the human TPH discussed earlier. The structure of chicken TPH1 is more compact, and the loops of residues Leu124-Asp139 and Ile367-Thr369 close around the active site. This sort of structural changes in the catalytic domain has also been seen in other aromatic amino-acid hydroxylases upon binding of substrate analogues.

A dimeric form has been found in the active conformation of THP1 in complex with acyl guanidine-derived inhibitors [37] that served as cocrystal guided design for the development of spirocyclic proline-derived inhibitors [38]. The structure of the catalytic domain of TPH2 has been resolved [39] although it has not been published (PDB ID: 4V06).

8.3 Aromatic Amino Acid Decarboxylase (AADC)

Aromatic amino acid decarboxylase (AADC; (EC 4.1.1.28), also known as DOPA decarboxylase, converts L-DOPA and 5-HTP into dopamine and serotonin, respectively [40, 41]. Interestingly, *in vitro* experiments showed that different pH optima or concentrations of substrate or cofactor are required for optimum activity of the enzyme when using either 5-HTP or DOPA as the substrate. The enzyme uses pyridoxal-5′-phosphate (PLP) as a cofactor. AADC is a dimer of 380 amino acids, corresponding to a molecular weight of c. 54 000 Da (Figure 8.6). It requires free sulfhydryl groups for activity and is readily inhibited by heavy metals. AADC is present in both dopaminergic and serotonergic neurons.

Posttranslational modifications of AADC include *N*-AcMet1 and *N*6-(pyridoxal phosphate)Lys303. AADC presents several natural variants including Met17Val, Pro47His, Glu61Asp, Ala91Val, Gly102Ser, Ser147Arg, Pro210Leu, Met217Val, Met239Ile, Met239Leu, Ser250Phe, Ala275Thr, Phe309Leu,

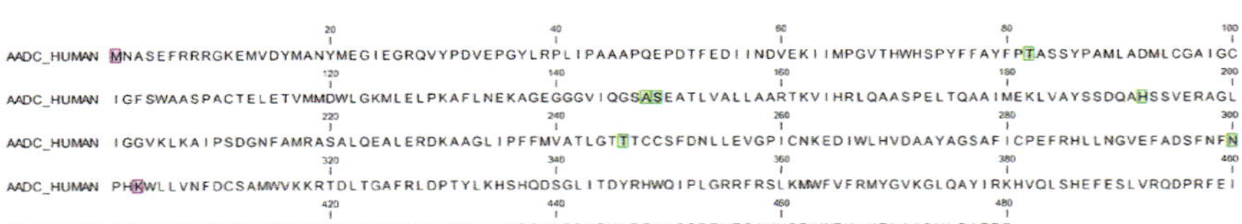

Figure 8.6 Sequence of human aromatic amino acid decarboxylase. Modified (magenta) and metal-binding sites (gray) are indicated on sequence.

Arg347Gln, Leu408Ile, and Arg462Gln [42]. Some of these variants are involved in AADC deficiency, an inborn error in neurotransmitter metabolism that causes combined serotonin and catecholamine deficiency [43, 44]. There are three alternative isoforms; isoform 2 differs from the canonical sequence in residues 316–338 and missing residues 339–480. Isoforms 3 and 4 are missing residues 68–145 and 146–238, respectively. The AADC deficiency has been identified as an autosomal recessive neurotransmitter defect resulting in a combined central deficiency of monoamines [45].

AADC is directly implicated in Parkinson's disease. Since dopamine cannot cross the brain–blood barrier, it cannot be used directly to treat PD. On the other hand, L-DOPA is capable of crossing the brain–blood barrier, but peripheral AADC rapidly converts L-DOPA into dopamine and less than 10% of the administered L-DOPA reaches the brain. Consequently, inhibition of AADC is the primary target in PD [46]. K_m of AADC for L-DOPA is 165 μM, and the enzyme is inhibited by carbidopa (used to treat PD). The crystal structure of free AADC from drosophila melanogaster (PDB ID: 3K40) and pig (PDB ID: 1JS6) has been resolved. Pig liver AADC bound with carbidopa (PDB ID: 1JS3) has also been reported (Figure 8.7a) [46]. The active site of AADC is located near the monomer–monomer interface although it is composed mainly of residues from one monomer. The cofactor PLP binds to Lys303 through an imine bond. The salt bridge between the carboxylate group of Asp271 and the protonated pyridine nitrogen of the cofactor contributes to stabilize the presumed carbanionic inter-

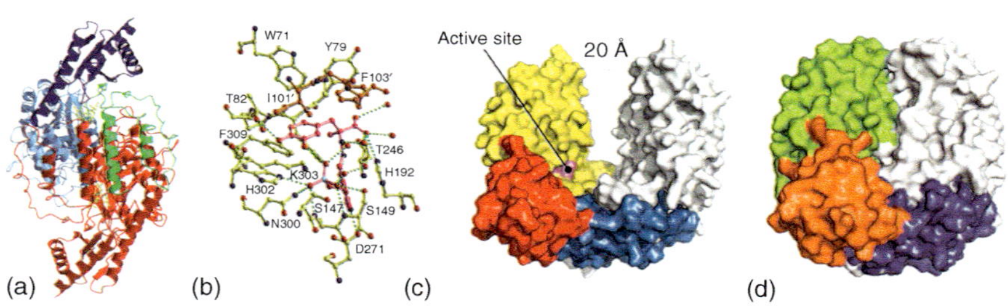

Figure 8.7 (a) Crystal structure of AADC in complex with cofactor PLP and inhibitor carbiDOPA (PDB ID: 1JS3. One monomer is completely red, whereas the other one is green (N-terminal domain), cyan (large domain), and blue (small domain). The cofactors (PLP) and the inhibitors (carbiDOPA) are shown in ball-and-stick representation in yellow. The N-terminal domain of one monomer packs on top of the other monomer, resulting in an extended dimer interface. (b) Detailed view of the hydrogen bond interactions, including all structural water molecules at the active-site cleft. Nitrogen, phosphate, and oxygen atoms are marked blue, cyan, and red, respectively. Carbon atoms are colored in yellow for the enzyme, in magenta for the PLP–carbiDOPA complex, and in orange for the residues of the other monomer. Hydrogen bonds are indicated in green dotted lines. Source: Reproduced with permission from ref. [46]. (c) Apo human AADC in the open conformation (PDB ID: 3RBL). The domain organization of one monomer is highlighted using different colors (blue – N-domain; yellow – large domain; red – C-domain), while the other monomer is shown in white. (d) Holo pig kidney AADC (PDB ID: 1JS6) in the closed conformation. The structure is shown in the same orientation for comparison; the three domains are colored differently for clarity (purple – N-domain; green – large domain; orange – C-domain). Source: Giardina et al. 2011 [47]. Reproduced with permission of PNAS.

mediates. The catechol ring of the inhibitor carbiDOPA forms London interactions with Ile1012 and Phe1032 (Figure 8.7b). The inhibitor carbiDOPA binds to the enzyme by forming a hydrazone bond with the PLP cofactor through its hydrazine functionality, mimicking the aldimine enzyme–substrate intermediate. The 4′-catechol hydroxyl group of the inhibitor forms a hydrogen bond with the hydroxyl group of Thr82, located at the very bottom of the active-site cleft. The phosphate group of the cofactor is also involved in substrate binding, since it forms a hydrogen bond with the 3′-catechol hydroxyl group of carbiDOPA. Further crystallographic studies allowed trapping the open conformation of the enzyme (Figure 8.7c; PDB ID: 3RBL), which could be compared with the previously resolved closed conformation (Figure 8.7d; PDB ID: 1JS6). The structure of pig kidney AADC in complex with PLP and vitamin B6 phosphate (PDB ID: 3RCH) has also been resolved [47].

By tuning the PLP concentration, it was possible to obtain two crystal structures with different conformations of the active site. The analysis of these structures in parallel with kinetic studies allowed to propose a regulatory model for AADC levels in the brain. According to this model, the active holoform is responsible for the synthesis of biogenic amines, while the apo open conformation is recognized by ubiquitin ligases and preferentially degraded by the ubiquitin proteasome system [47].

K_m of AADC for 5-HTP is 10 μM, higher than the concentration of 5-HTP in the brain, so the enzyme is not saturated with the substrate. In consequence, AADC activity appears to be determined by the concentration of substrate, 5-HTP. Furthermore, the affinity for 5-HTP is c. 3 orders of magnitude higher than that for L-tryptophan, ensuring that 5-HT rather than tryptamine is synthesized in serotonergic neurons.

The regulation of AADC at a posttranslational level is exerted by protein kinases. The enzyme is responsible for the synthesis of other amines, such as 2-phenetylamine, tryptamine, or tyramine that can act as endogenous modulators of central transmission by activating G-protein-coupled receptors.

AADC plays an important role in Parkinson's disease since it is responsible for converting L-DOPA administered therapeutically into dopamine. In Parkinson's disease, it is critical to inhibit peripheral AADC with carbidopa to allow L-DOPA crossing the blood–brain barrier and increasing its bioavailability in the brain.

8.4 Dopamine β-Hydroxylase (DBH)

Dopamine β-hydroxylase (DBH) (also called dopamine β-monooxygenase; EC 1.14.17.1) catalyzes the synthesis of norepinephrine from dopamine in the vesicles of noradrenergic and adrenergic neurons [48, 49]. DBH is also found in chromaffin cells [50] and in both membrane-bound and soluble forms [51]. K_m for dopamine is c. 5 mM, but the concentration of catecholamine in the synaptic vesicles or chromaffin cells is high. In addition to dopamine, phenylethylamines such as tyramine (K_m = 1.8 mM) [49] can serve as substrates.

The enzyme uses molecular oxygen as the source of hydroxyl oxygen and ascorbic acid as the electron donor. Actually, ascorbic acid is oxidized to semihydroascorbate, which is then reduced back to ascorbate by the action of cytochrome b-561. The enzyme uses copper(II) as a cofactor, each subunit binding two copper ions (Cu_H and Cu_M), which are essential for the enzymatic activity [52]. Residues His262, His263, and His333 are involved in binding of one copper ion; residues His412, His414, and Met487 are involved in binding of the second copper atom. Cu_H is the site of electron transfer.

N-Glycosylation of the enzyme takes place at Asn64, Asn184, and Asn566 (Figure 8.8). Several disulfide bonds are present (Cys154/Cys196; Cys232/Cys283; Cys269/Cys285; Cys390/503; Cys394/Cys565; and Cys466/Cys488); some of them are between different chains (A-Cys528/Cys530; A-Cys530/C-Cys528) [52].

DBH is involved in several neurodisorders including depression, anxiety, Parkinson's disease, Tourette syndrome, schizophrenia, and ADHD. DBH deficiency, causing minor central nervous system (CNS) dysfunctions (but profound deficits in autonomic and cardiovascular function), is related to three natural variants Val101Met, Asp114Glu, and Asp345Asn [53]. DBH has been suggested to be involved in schizophrenia although no evidences for allelic association between the disease and functional variant Ala211Thr have been found [54]. Other natural variants include Gly12Ser, Glu181Gln, Lys239Asn, Glu250Gln, Asp290Asn, Leu317Pro, and Arg549Cys [55].

The X-ray structure of dopamine β-hydroxylase N-terminal (DOMON) domain has been resolved at 2.9 Å resolution as a dimer (PDB ID: 4ZEL) [52]. Each chain of the dimer is composed of four domains: the DOMON domain, the catalytic Cu_H and Cu_M domains, and the C-terminal dimerization domain. The crystallized dimer is asymmetric, whereas in the A chain the two catalytic Cu_H and Cu_M domains adopt a closed conformation, in the B chain they are in an open conformation (Figure 8.9a,b).

The different conformations imply different distances between the copper atoms. The catalytic site is formed by two domains: an N-terminal domain where Cu_H binds and a C-terminal domain where Cu_M binds. The Cu_H domain is positioned significantly different in the A and B chains; the Cu_H domain in chain A is being moved away from the DOMON domain and closer to the Cu_M domain (Figure 8.9c,d). The dimerization domain consists of two antiparallel α helices from each chain and is followed by a long C-terminal extension with two β-strands (Figure 8.9e,f). The two different conformations of the DBH catalytic site

```
                        20                  40                  60                  80                 100
DBH_HUMAN  MPALSRWASLPGPSMREAAFMYSTAVAIFLVILVAALQGSAPRESPLPYHIPLDPEGSLELSWNVSYTQEAIHFQLLVRRLKAGVLFGMSDRGELENADL
                       120                 140                 160                 180                 200
DBH_HUMAN  VVLWTDGDTAYFADAWSDQKGQIHLDPQQDYQLLQVQRTPEGLTLLFKRPFGTCDPKDYLIEDGTVHLVYGILEEPFRSLEAINGSGLQMGLQRVQLLKP
                       220                 240                 260                 280                 300
DBH_HUMAN  NIPEPELPSDACTMEVQAPNIQIPSQETTYWCYIKELPKGFSRHHIIKYEPIVTKGNEALVHHMEVFQCAPEMDSVPHFSGPCDSKMKPDRLNYCRHVLA
                       320                 340                 360                 380                 400
DBH_HUMAN  AWALGAKAFYYPEEAGLAFGGPGSSRYLRLEVHYHNPLVIEGRNDSSGIRLYYTAKLRRFNAGIMELGLVYTPVMAIPPRETAFILTGYCTDKCTQLALP
                       420                 440                 460                 480                 500
DBH_HUMAN  PSGIHIFASQLHTHLTGRKVVTVLVRDGREWEIVNQDNHYSPHFQEIRMLKKVVSVHPGDVLITSCTYNTEDRELATVGGFGILEEMCVNYVHYYPQTQL
                       520                 540                 560                 580                 600
DBH_HUMAN  ELCKSAVDAGFLQKYFHLINRFNNEDVCTCPQASVSQQFTSVPWNSFNRDVLKALYSFAPISMHCNKSSAVRFQGEWNLQPLPKVISTLEEPTPQCPTSQ

DBH_HUMAN  GRSPAGPTVVSIGGGKG
```

Figure 8.8 Sequence of human dopamine-β-hydroxylase. Disulfide bonds (green), N-glycosylation (red), and metal-binding (gray) sites as well as cytoplasmic (light blue), extracellular (white), and transmembrane domains (yellow) are indicated on sequences.

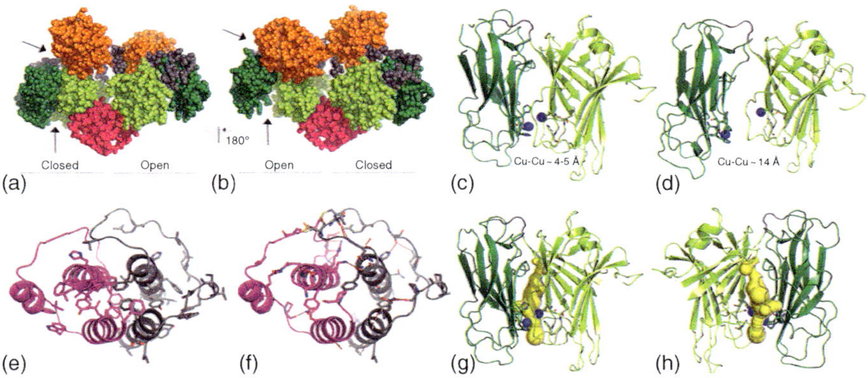

Figure 8.9 Structure of the human DBH dimer. The two conformations of the DBH catalytic core showing a closed and an open active site, with chain A to the left (a) and view from the back with chain A to the right (b). The DOMON domain is displayed in orange, the Cu_H domain in dark green, the Cu_M domain in light green, and the dimerization domain in magenta. The interdomain regions are in gray. (c) Closed conformation of the catalytic domain as seen in chain A. (d) Open conformation of the catalytic domain as seen in chain B. (e, f) Domain interactions in the dimerization domain. Chain A is shown in magenta and chain B in gray. (e) Residues involved in hydrophobic interactions are shown as sticks. (f) Residues involved in hydrophilic interactions are shown as sticks, and the disulfide bridges are shown in yellow. (g, h) Binding pockets and channels in the vicinity of the closed active site seen in chain A. CAVER identified binding pocket and channel (yellow) in the closed catalytic core (chain A). Two different orientations are shown. (g) Same orientation as in (c), with the CuH domain to the left. (h) Viewed from the back, with the CuH domain to the right. Source: Vendelboe et al. 2016 [52]. https://creativecommons.org/licenses/by/4.0/

provide possible catalytic mechanisms, one hypothesis being that the closed conformation is an artifact and inactive and the open conformation is the active form of the enzyme, with coppers 11–14 Å apart. This hypothesis is supported by a number of structures as well as spectroscopic and kinetic data [56, 57]. However, a CAVER search suggests that there is room for the binding of a substrate in the closed conformation (Figure 8.9g,h), and according to the resolved structure, an alternative mode of action could be that the closed conformation represents the catalytically active form and that the open conformation is inactive although it allows for loading of substrate, release of product, and recycling of the copper redox states.

8.5 Monoamine Oxidase (MAO)

Monoamine oxidase (MAO; EC. 1.4.3.4) is a flavin-dependent enzyme that catalyzes the oxidative deamination of several monoamines, including catecholamines, serotonin, and trace amines. The substrate amino group is oxidized through the enzyme-bound flavin adenine dinucleotide (FAD) cofactor, which is then reoxidized by molecular oxygen with the release of hydrogen peroxide. In particular, the enzyme converts dopamine into dihydroxyphenylacetaldehyde (DOPAL), norepinephrine and epinephrine into dihydroxyphenylglycolaldehyde

(DOPEGAL), and serotonin into 5-hydroxyindoleacetaldehyde (5-HIAL). There are two MAO isoenzymes (A and B) that are integral flavoproteins of outer mitochondrial membranes in neurons, glia, and other cells. MAO-A appears to be a specific enzyme for serotonin and norepinephrine. MAO-B is a more broadly active enzyme that acts on phenylethylamines. Tyramine and dopamine are equally good substrates for both MAO-A and MAO-B. MAO-A is selectively inhibited by clorgyline, whereas MAO-B is highly inhibited by deprenyl. There is an anatomic distribution of the isoenzymes. In the brain, MAO-A predominates in noradrenergic cell bodies in the locus coeruleus, while MAO-B is the major form in serotonergic cell bodies in the raphe nucleus; this isoform is not located in the dopamine nerve endings in the basal ganglia, but instead is located in the glia. In general, MAO-A is most abundant in catecholaminergic neurons, whereas MAO-B is found primarily in serotonergic neurons and glial cells.

The amino acid sequences of MAO-A and MAO-B (Figure 8.10) show c. 70% homology with molecular weights of 59 700 and 58 800 Da, respectively. MAO-A and MAO-B have 527 and 520 residues, respectively, with transmembrane helical domains (residues 498–518 for MAO-A and residues 490–516 for MAO-B) that are the anchor for type 4 membrane protein and mitochondrial intermembrane topological domains (residues 519–527 for MAO-A and residues 517–520 for MAO-B) [58]. Posttranslational modifications include acetylation (at Met1 for MAO-A [59] and Ser2 and Lys52 for MAO-B [60]) and phosphorylation at Ser383 in the case of MAO-A. Incorporation of 8-α-FAD takes place at Cys406 in MAO-A and Cys397 in MAO-B. The role of cysteines in MAO-A and MAO-B has been investigated by mutagenesis [61]. While catalytic activities were retained in seven MAO-A cysteine to serine mutants (at residues 165, 210, 266, 306, 321, 323, and 398) and in six MAO-B cysteine to serine mutants (at residues 5, 172, 192, 297, 312, and 389), mutations at MAO-A Cys-374 and -406 and MAO-B Cys-156, -365, and -397 resulted in a complete loss of catalytic activity. For MAO-A, natural variants Asp15Glu, Glu188Lys, Cys266Phe, Phe314Val, and Lys520Arg have been found [62]. An alternative isoform of MAO-A is missing

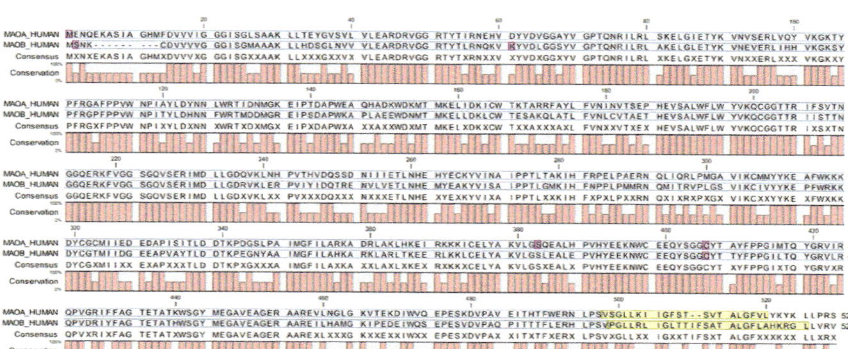

Figure 8.10 Sequences of the human monoamine oxidases A and B. Conservation graphic and consensus sequence are shown. Modified sites (magenta) and cytoplasmic (light blue), extracellular (white), and transmembrane domains (yellow) are indicated on sequences.

residues 1–133. In the case of MAO-B, the alternative isoform is missing residues 1–16 and 428–520 and presents a different sequence for residues 380–427.

MAO has a dramatic effect on the concentration of amines. The oxidative deamination catalyzed by MAO yields hydrogen peroxide that can generate free radicals [63]. For this reason, its inhibition is a therapeutic target [64] for disorders caused by a limited availability of monoamines such as dopamine in Parkinson's disease (PD) as well as to provide neuroprotective effects by preventing or delaying neurodegeneration itself [65]. Both isoenzymes play a decisive role in some neurological disorders such as depression and PD. A deficiency of MAO-A results in aggressive behavior [66].

The first X-ray structure of MAO was determined for MAO-B [67] and showed the main structural features of the enzyme (PDB ID: 1GOS). The enzyme crystallized as a dimer although it was suggested that *in vivo* it forms a quaternary assembly. MAO-B is bound to the outer mitochondrial membrane, the region responsible for the attachment being formed by the C-terminal amino acids 461–520; indeed, the crystal structure reveals that residues 461–488 form an extended polypeptide chain followed by an α-helix that starts at Val489 forming the transmembrane helical domain (Figure 8.11a–c). More particularly, residues 481–488 have hydrophobic side chains (Phe481, Leu482, Leu486, and Pro487) pointing toward the membrane. The presence of these side chains further suggests that membrane attachment also involves additional hydrophobic interactions of the protein surface. The inhibitor pargyline binds covalently to the N5 atom of the flavine. The active site consists of a hydrophobic substrate cavity of 420 $Å^3$ interconnected to an entrance cavity of 290 $Å^3$. The recognition site is formed by a hydrophobic site composed of Tyr398 and Tyr435 (Figure 8.11d), which are oriented perpendicular to the flavin ring, playing a crucial role in the catalysis as demonstrated by mutating Tyr435 with Phe, His, Leu, and Trp [68]. The structural studies also demonstrate that the entrance cavity connects the protein surface with the substrate cavity.

Further insights into the mode of inhibition of human MAO-B have been obtained from crystal structures in complex with inhibitors including isatin (PDB ID: 1OJA), tranylcypromine (PDB IDs: 1OJB and 2XFU), *N*-(2-aminoethyl)-*p*-chlorobenzamide (PDB ID: 1OJC), and 1,4-diphenyl-2-butene (PDB ID: 1OJ9) [69]. Complexes of MAO-B with inhibitors of the *N*-propargylaminoindan class such as rasagiline (PDB ID: 1S2Q), *N*-propargyl-1(*S*)-aminoindan (PDB ID: 1S2Y), 6-hydroxy-*N*-propargyl-1(*R*)-aminoindan (PDB ID: 1S3E), and *N*-methyl-*N*-propargyl-1(*R*)-aminoindan (PDB ID: 1S3B) showed similar conformations with extended propargyl chains covalently bound to the flavin and the indan ring located in the rear of the substrate cavity [70]. Addition of different substituents to the rasagiline scaffold modified the binding affinity depending on the position on the aminoindan ring and on the size of the substituent. In fact, the crystal structure of MAO-B in complex with *N*-methyl-1(*R*)-aminoindan shows that its aminoindan ring adopts a different orientation from that of rasagiline [71].

The crystal structures of human MAO-B in complex with 1,4-diphenyl-2-butene (PDB ID: 1OJ9) or with *trans,trans*-farnesol (PDB ID: 2BK3) show that these inhibitors span the substrate and entrance cavities with the side chain of Ile199 rotated out of its normal conformation, suggesting that Ile199 is gating the

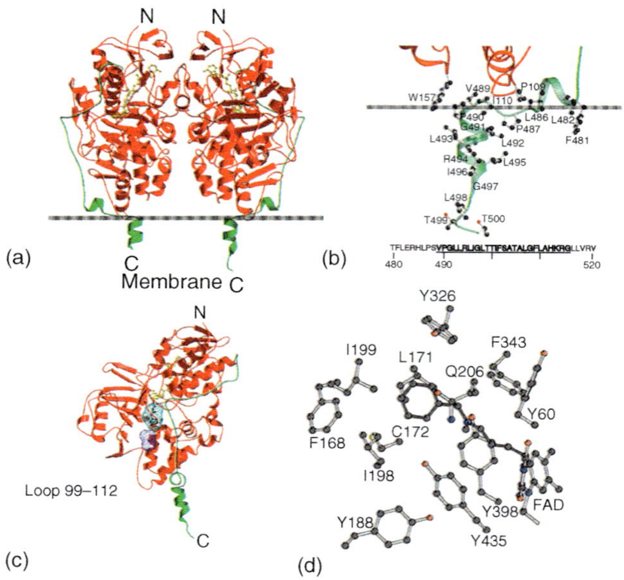

Figure 8.11 Structure of human MAO-B. (a) Ribbon diagram of the MAO-B dimer with the twofold axis vertical in the plane of the paper. Monomer A is on the right and monomer B is on the left. Residues 4–460 are in red, and the C-terminal tail (residues 461–500) is in green. The FAD is shown in ball-and-stick representation and colored in yellow. The membrane region boundary is indicated by a thick dashed line for illustration purposes and is not intended to define the exact depth of protein insertion into the membrane. (b) Close-up view of the membrane-binding region in monomer A. Carbons are in black, nitrogens in blue, and oxygens in red. The amino-acid sequence of the C-terminal residues (480–520) is indicated. Residues 489–515 forming the transmembrane helix are underlined. (c) Cavities constituting the substrate path from the protein surface to the flavin in the MAO-B monomer. Adjacent to the active-site cavity (cyan) is the "entrance" cavity (blue), which is underneath loop 99–112. The FAD is in yellow and the covalent pargyline inhibitor is in black. (d) View of the pargyline inhibitor and the residues lining the binding site at the re side of the flavin. Carbons are in black, nitrogens in blue, oxygens in red and sulfurs in yellow. The inhibitor is outlined by shaded bonds. Source: Binda et al. 2002 [67]. Reproduced with permission of Springer Nature.

substrate cavity [72]. Both Ile199 and Tyr326 function in substrate and inhibitor recognition [73].

Structural studies with the antiepileptic drug zonisamide (PDB ID: 3PO7) [74] demonstrate that this inhibitor binds within the substrate cavity. Similarly, chromone derivatives dimethylphenyl- (PDB ID: 6FVZ), chlorophenyl- (PDB ID: 6FW0), and fluorophenyl- (PDB ID: 6FWC) chromone carboxamides all bind in the active-site cavity of the enzyme with the chromone moiety located in front of the FAD cofactor. Two hydrogen bonds with Tyr435 and Cys172 are formed, and the inhibitors perfectly fit the hydrophobic flat active site of human MAO-B [75]. On the other hand, crystal structures of MAO-B in complex with selective noncovalent inhibitors demonstrated that, in these cases, both the entrance and substrate cavities are occupied [76]. The same inhibition mode was found for glitazone antidiabetic drugs such as pioglitazone (PDB ID: 4A79) and rosiglitazone (PDB ID: 4A7A) [77].

The crystal structures of MAO-B in complex with phenylethylhydrazine (PDB ID: 2VRM) and benzylhydrazine (PDB ID: 2VRL) confirmed that the mechanism of inhibition takes place through alkylation at the N5 position on the *re* face of the covalent flavin with the loss of the hydrazyl nitrogens [78]. The same mechanism was confirmed for the irreversible inhibition of human MAO-B by the inhibitors mofegiline (PDB ID: 2VZ2) [79], *N*-(furan-2yl-methyl)-*N*-methylprop-2yl-1-amine (F2MPA) (PDB ID: 5MRL) [80], and *N*-((5-(3-(1-benzylpiperidin-4-yl)propoxy)-1-methyl-1*H*-indol-2-yl)methyl)-*N*-methylprop-2-yn-1-amine (ASS234) PDB ID: 4CRT) [81] through the corresponding crystal structures of the complexes.

Crystal structures of MAO-B in complex with 2-(2-benzofuranyl)-2-imidazoline (2-BFI) (PDB ID: 2XCG) showed that the imidazoline ligand bound in a different site from the substrate-binding cavity. Conformational changes in Gln206, Ile199, and Ile316 giving rise to favorable London interactions on each face of the benzofuran ring of 2-BFI are responsible for binding. Further structural studies with Ile199Ala mutant (PDB ID: 2XFO) support an allosteric mechanism for the inhibition of MAO-B by 2-BFI [82].

Two crystal forms of human MAO-A in complex with the inhibitor clorgyline (PDB IDs: 2BXR and 2BXS) revealed the three-dimensional structure of the enzyme as a monomer (Figure 8.12a) [83] in contrast to MAO-B, identified as a dimer in its active form. Interestingly, although the amino acid sequences in the actives sites of human MAO-A and rat MAO-A are identical, only the latter is identical to that of human MAO-B. The origin of the observed difference for human MAO-A resides in the loop conformation of residues 210–216, which is different for human MAO-B and rat MAO-B. This structural alteration is suggested to be due to long-range interactions in the monomeric form of the enzyme. The structure of the active site of MAO-A resembles that of MAO-B; indeed, the complex with the inhibitor is formed as shown previously for the pargyline [69] and rasagiline [70] adducts of human MAO-B and for the clorgyline adduct of rat MAO-A [84], the inhibitor-binding site being formed by a single cavity that extends from the flavin ring to the cavity-shaping loop formed by residues 210–216. A structural comparison between human MAO-A and MAO-B showed clear similarities as well as differences (Figure 8.12b–d). The structures of the covalent FAD coenzymes and two tyrosines comprising an aromatic pocket in the active sites are identical, but there are major differences in the area of the active sites opposite to the flavin, which governs substrate recognition.

Further structural studies of MAO-A (PDB IDs: 2Z5X and 2Z5Y) illustrate the structure–function relationship between a transmembrane helical anchor and an extramembrane domain [58]. A series of positively charged residues (Arg129, His148, Lys151, Lys163, Arg493, Lys503, Lys520, and Lys522) presumed to interact with the phospholipid hydrophilic head groups at the membrane surface are arranged in a horizontal disposition, thus determining the location of the outer-membrane surface and providing a model of the orientation of the enzyme (Figure 8.12e). Superposition studies of the residues surrounding the substrate/inhibitor cavity of human MAO-A, rat MAO-A, and human MAO-B (Figure 8.12f,g) highlight the structural differences between the enzymes and provide a rationale for the observed selectivity of reversible inhibitors, caused by

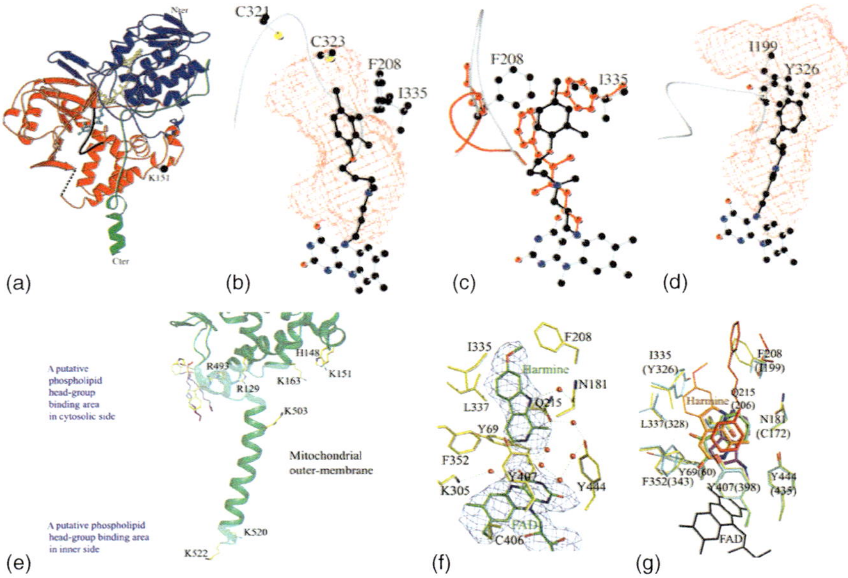

Figure 8.12 (a) Ribbon representation of the monomer. The FAD-binding domain (residues 13–88, 220–294, and 400–462) is in blue; the substrate-binding domain (89–219 and 295–399) is in red; and the C-terminal membrane region (463–506) is in green. A dashed line connects residues 110–116. FAD and clorgyline are depicted in yellow and cyan ball-and-stick representation, respectively. The active-site cavity-shaping loop 210–216 is depicted as black coil. (b) Site cavity in *h*MAO-A is shown in red wire representation. Clorgyline is depicted in black. (c) Active-site comparison of *h*MAO-A and *h*MAO-B with the crucial Phe208 and Ile335 residues of *h*MAO-A superimposed to the corresponding Ile199 and Tyr326 residues of *h*MAO B. The protein and inhibitor atoms of *h*MAO-B are in red. With respect to (a), the model has been rotated by 90° around the vertical axis in the plane of the drawing. (d) The active-site cavity (red surface) of *h*MAO-B in complex with deprenyl (black) is depicted in the same orientation as in (a). (e) Binding model of MAOA into the mitochondrial outer membrane. The positively charged residues Arg129, His148, Lys151, Lys163, Arg493, Lys503, Lys520, and Lys522 are shown. The upper area represents the cytosolic side. (f) Substrate/inhibitor-binding site. Amino acid residues are shown in yellow, and FAD and harmine are shown in green. Dotted lines indicate hydrogen bonds. (g) The structure of the substrate/inhibitor-binding sites in human MAO-A, and the numbers in parentheses are for human MAO-B. MAO-A and MAO-B residues are shown in yellow and light blue, respectively. Inhibitors are colored as follows: orange, harmine; green, isatin (PDB ID: 1OJA); purple, rasagiline analogue (PDB ID: 2C67); and red, 1,4-diphenyl-2-butene (PDB ID: 1OJ9). FAD is shown in black. Nitrogen and oxygen atoms are shown in blue and red, respectively. The rotamer of this residue, in MAO-B with 1,4-diphenyl-2-butene, is shown in red. The residues Q215 and Y407 that form important hydrophobic interactions to the inhibitors are shown as thick stick models. Source: Reproduced with permission from Son et al. 2008 [58] and De Colibus et al. 2005 [83]. Copyright 2008 and 2005, National Academy of Sciences, USA.

the different size and shape of the substrate/inhibitor cavity. These differences are mainly attributed to the different conformations of Ile335. These structural studies also suggest that the discrepancy between the motions of the extramembrane and transmembrane domain gives a flexibility that might cause conformational changes in or around the entrance loops near the active site, which in turn opens the entrance for the substrate/inhibitor. Such a flexibility of loop 108–118 is therefore essential for controlling substrate access to the active site.

8.6 Catecholamine-O-methyl Transferase (COMT)

Catecholamine-*O*-methyl transferase (COMT, EC 2.1.1.6) catalyzes the transfer of the methyl group of *S*-adenosyl-L-methionine (SAM) to the 3-hydroxyl group of catecholamines, their hydroxylated metabolites, and L-DOPA. The corresponding neurotransmitter is thus inactivated, and SAM is converted into *S*-adenosyl-L-homocysteine (SAH). There are an additional number of compounds that can be substrates for COMT. In general, COMT eliminates biologically active and toxic catechols and other hydroxylated metabolites. The enzyme is dependent on Mg^{2+} and inhibited by Ca^{2+}.

COMT has 271 residues that are composed of cytoplasmic (residues 1–6) and extracellular (residues 27–271) topological domains (Figure 8.13). Residues 7–26 form a transmembrane helical domain that serves as a signal anchor for type 2 membrane proteins. The enzyme is phosphorylated at Ser267. Residues 167–170 are involved in SAM binding.

The enzyme has two isoforms derived from a single gene. COMT-S is the soluble form, while MB-COMT is the membrane-bound form, which contains 50 hydrophobic residues not present in COMT-S. In the brain, the predominant form is MB-COMT, and it is generally referred to as COMT. There is some controversy over the location of COMT in the brain. At a cellular level, COMT appears to be on postsynaptic neurons, but low levels may also exist on glia. At a subcellular level, COMT has been assigned to the plasmalemmal membrane and the outer membrane of mitochondria, but more recent evidence suggests its presence on the rough endoplasmic reticulum [85].

K_m values for MB-COMT for dopamine (3 μM) and norepinephrine (24 μM) are significantly less than those of S-COMT (280 μM for dopamine and 369 μM for norepinephrine). Since MB-COMT is the form present in the brain, the enzyme plays an important role in the metabolism of synaptic catecholamines. Variants affecting the expression of COMT are closely related to schizophrenia, in particular Ala72Ser polymorphism [86]. Other natural variants include Cys34Ser, Ala102Thr, Ala146Val, and Val158Met [87, 88]. Since COMT also inactivates L-DOPA, which is used in the treatment of PD, COMT inhibitors are also employed as adjuvants in the L-DOPA therapy for Parkinson's disease. Enhancement of dopamine levels in the prefrontal cortex, caused by COMT

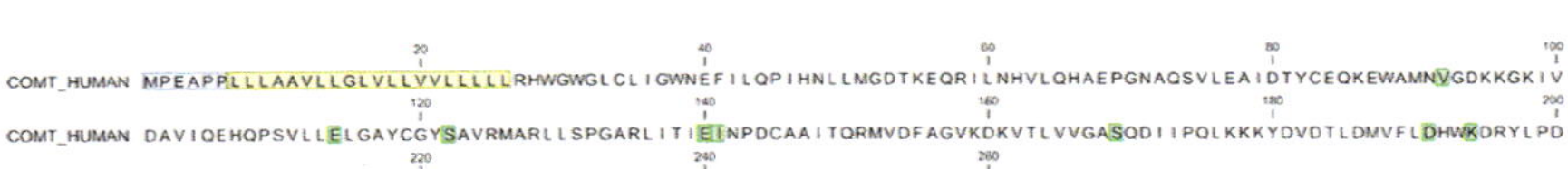

Figure 8.13 Sequence of human catechol-*O*-methyl transferase. Phosphorylation (magenta) and metal-binding (grey) sites as well as cytoplasmic (light blue), extracellular (white), and transmembrane domains (yellow) are indicated on sequence.

inhibition [89, 90], may also have therapeutic benefits for the treatment of schizophrenia [91].

The crystal structure of COMT was first resolved in 1994 for the rat enzyme [92]. The structure (PDB ID: 1VID) consisted of eight α-helices and seven β-strands with a typical α/β fold (Figure 8.14a). The helices α_4–α_8 and the parallel β-strands β_1–β_5 form a typical Rossmann fold. The two antiparallel β-strands β_6 and β_7 form the C-terminal part. The active site of COMT includes both the catalytic site, in the vicinity of magnesium ion, and the coenzyme binding area. This structure in complex with SAM and unreactive dinitrocatechol (DNC) can be considered a surrogate of the Michaelis complex and shows the precise preorganization of one of the magnesium-bound catechol hydroxyl groups with respect to the electrophilic methyl group of SAM, providing an excellent insight into the mechanism of methyl transfer.

These structural studies were further used for designing a bisubstrate inhibitor of COMT (IC_{50} = 2 μM) [94]; the ternary complex of that inhibitor, COMT, and magnesium ion was also characterized by X-ray crystallography (PDB ID: 1JR4) (Figure 8.14b–c) [93]. The structure showed, as expected, that the inhibitor occupies both the SAM- and catechol-binding sites. Some flexibility of the side chains of Met40 and Trp143 was also observed to accommodate the inhibitor. Soluble COMT was crystallized in complex with the inhibitors 1-(3,4-dihydroxy-5-nitrophenyl)-3-(*n*-3′-trifluoromethylphenyl)piperazine-1-propanone dihydrochloride (BIA 3-335)

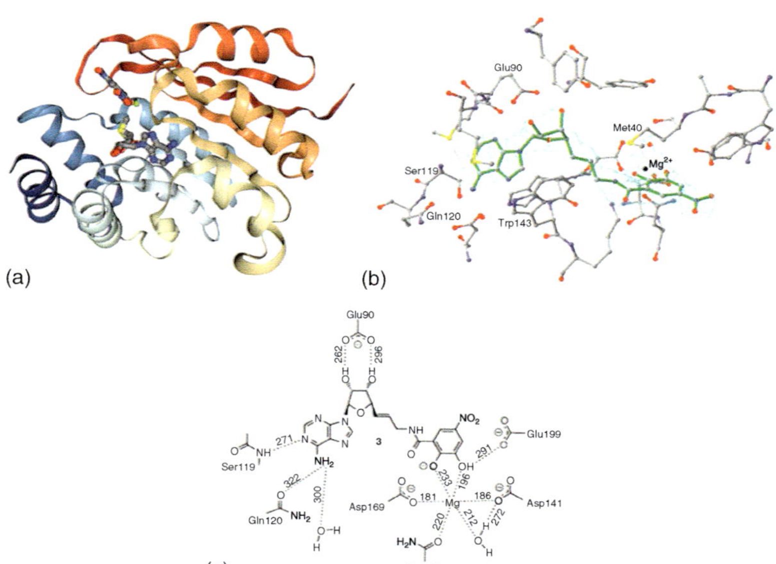

Figure 8.14 (a) X-ray structure of COMT. Source: (a) Image created from PDB ID: 1VID (ref. [91]) using NGL Viewer [11], and RCSB PDB. (b) The bisubstrate inhibitor-binding site in the refined X-ray structure. Coloring: C: gray, inhibitor framework: green, O: red, N: blue, S: yellow; H atoms omitted. (c) Schematic drawing of the H-bonding interactions (dashed lines) in the ternary complex. Distances are given in picometer. Source: (c) Lerner et al. 2001 [93]. Reproduced with permission of John Wiley & Sons.

[95] and 3,4-dihydroxy-2-nitrophenyl)phenyl-methanone (BIA 8-176) [96]. The crystallographic analysis of the former in complex with SAM (PDB ID: 1H1D) is identical to that observed for the same enzyme in complex with SAM and inhibitor 3,5-DNC [92]. BIA 3–334 binds into the catalytic site extending its side chain within a long groove, where interactions with hydrophobic residues are predominant [97]. The crystallographic analysis of BIA 8–176 (PDB ID: 2CL5) showed that the nitrocatechol moiety binds to the catalytic pocket in the *ortho* configuration positioning the *ortho*-hydroxyl oxygen close to the methylsulfonium of SAM, whereas the benzoyl side chain is facing the edge of the catalytic pocket flanked by the flexible Met40 and Trp143 [98].

The crystal structures of Val108 (PDB ID: 3BWM) and Met108 (PDB ID: 3BWY) isoforms of human COMT are superposable although Val108Met mutation decreases the structural stability of the enzyme [88]. The orientations of SAM, magnesium, and 3,5-DNC and their interactions with residues within the SAM- and catechol-binding sites are essentially identical [99]. Only slight differences in the structures are observed at the surroundings in the polymorphic site. The binding mode of nucleotide analogues in which the adenine has been replaced by other heterocycles, including thiopyridine, purine, *N*-methyladenine, and 6-methylpurine, has been studied by X-ray crystallography [100]. Crystal structures of bisubstrate inhibitors in ternary complexes with COMT and magnesium ion (PDB IDs: 3NW9, 3OE4, 3OE5, 3OZR, 3OZS, and 3OZT) identified key interactions in the nucleobase pocket, which can be occupied by a variety of heterocycles, and best results were obtained with those having a nitrogen atom in position 1 due to the hydrogen bond interaction with Ser119 (Figure 8.15). Crystal structures of COMT in complex with ribose-modified bisubstrate inhibitors (PDB IDs: 3NWB, 3NWE, and 3R6T) have also been studied, and subtle alterations of the ribose-binding site were identified [101]. In particular, the 3′-*endo* fluorine and methyl substituents, pointing into a similar hydrophobic pocket, display an enhanced binding affinity.

A series of crystal structures of COMT in several crystal forms and several conformational states (PDB IDs: 4PYK, 4PYQ, 4PYM, 4PYJ, 4P7F, 4PYN, 4PYI, 4P7J, 4P7K, 4PYL, 4PYO, and 4P7G) served for mapping the conformational space accessible to COMT [102]. The study revealed a set of three loop regions that come together when the Michaelis complex is formed. These observations confirm the malleability of the enzyme, which is necessary to accommodate the large volume and flexibility of SAM, but it makes the enzyme a difficult drug target.

The analysis of X-ray cocrystal structures of biaryl 3-hydroxy-4-pyridinones (PDB ID: 4XUC and 4XUD) and 5-hydroxy-4-pyrimidinones (PDB ID: 4XUE), identified as COMT inhibitors, in complex with the enzyme and cofactors SAM and magnesium suggested a different mechanism of inhibition than that previously stated for COMT inhibitors [103]. The absence of an aromatic nitro group promotes the biaryl group to be oriented either meta or para to the chelating hydroxyl group of the inhibitors. The proposed mechanism is based on the different nucleophilicity of the carbonyl oxygen atoms of the heterocyclic rings.

A crucial aspect of the activity of COMT is the observed regioselectivity in the methylation of hydroxyl groups of catechol moieties. The above-mentioned structural studies suggest that such regioselectivity is attributed to a

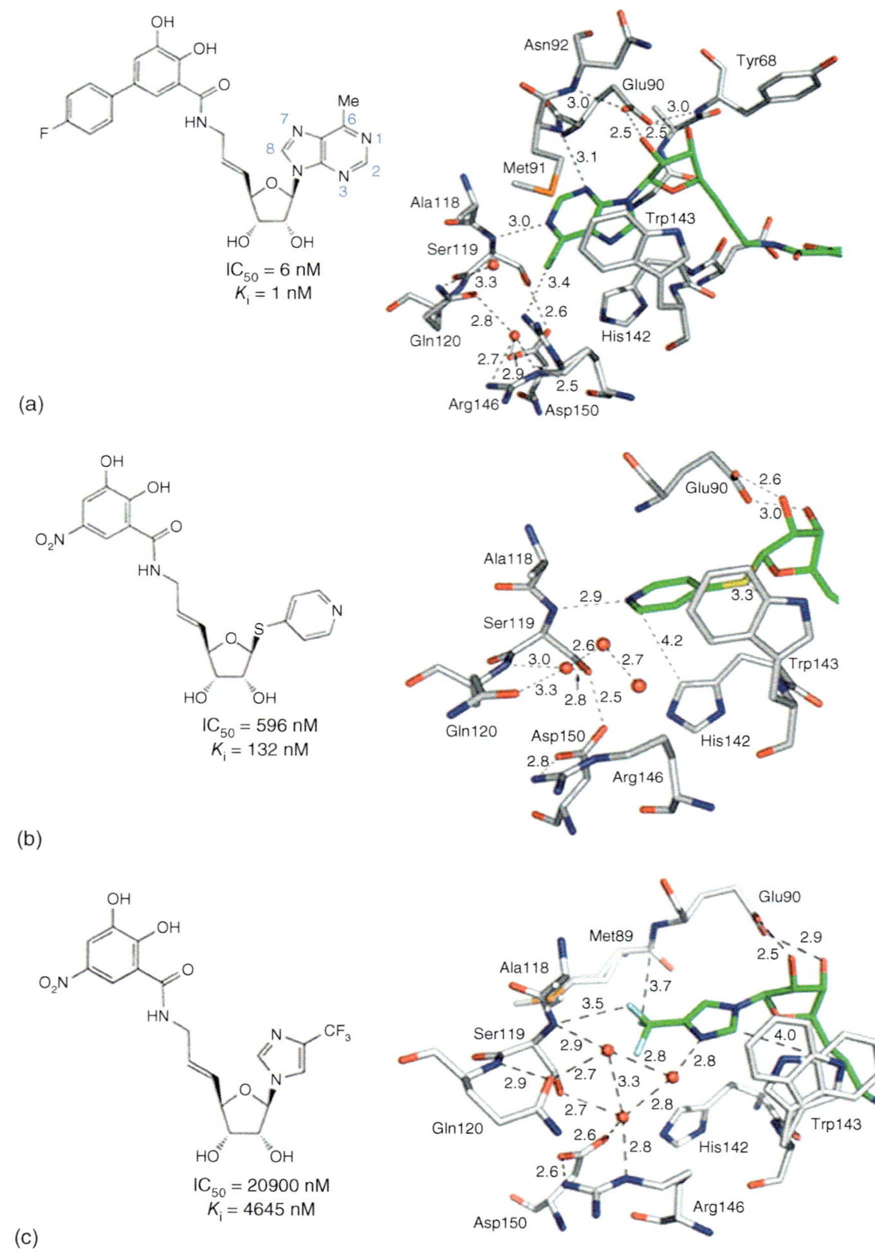

Figure 8.15 X-ray crystal structures of the adenine-binding site of the ternary complexes of COMT and magnesium with bisubstrate inhibitors, (a) PDB ID: 3NW9; (b) PDB code: 3OZT; (c) PDB ID: 3OE4; (d) PDB ID: 3OZS; (e) PDB code: 3OE5; and (f) PDB ID: 3OZR. Source: Ellermann et al. 2011 [100]. Reproduced with permission of John Wiley & Sons.

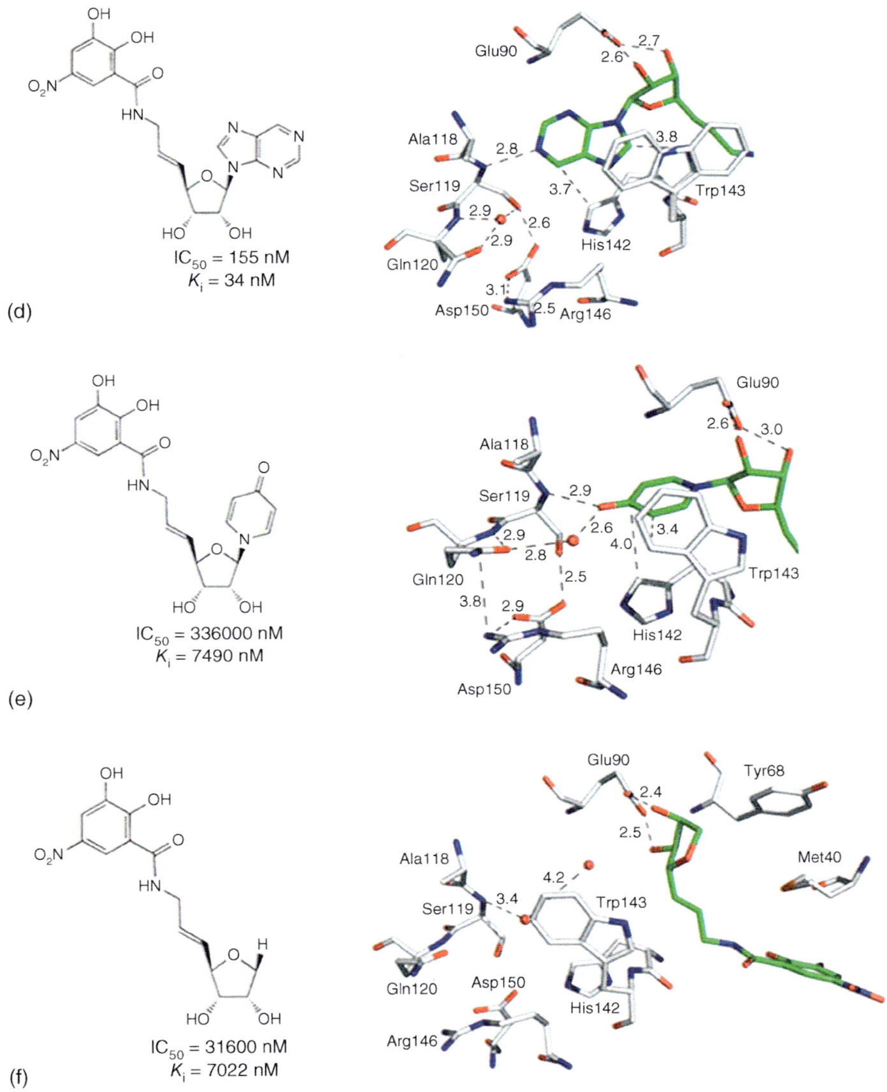

Figure 8.15 *(Continued)*

hydrophobic region found in the active site that prevents methylation in the undesired position [92, 98]. The nature of the substrate is also crucial for good regioselectivity. Polar groups such as those present in the natural substrates of COMT (catecholamines) are preferentially oriented out of the active site, favoring meta-methylation; on the other hand, para-methylation is observed in substrates with neutral or hydrophobic substituents that are preferably oriented toward the hydrophobic region. In fact, an effect of pH on the site of methylation of polar-substituted catecholamines was clearly observed [104]. A recent study demonstrates that the regioselectivity of COMT is governed by both the active-site residues and the quaternary structure of the enzyme [105]. Indeed, active-site modifications cause significant changes in regioselectivity independently of the enzyme oligomeric state.

8.7 Phenylethanolamine-N-methyl Transferase (PNMT)

Phenylethanolamine-*N*-methyl transferase (PNMT, noradrenaline *N*-methyltransferase; EC 2.1.1.28) is a soluble monomeric enzyme of 282 residues (Figure 8.16) and 30 000 Da that catalyzes the conversion of norepinephrine into epinephrine, using SAM as a methyl donor, which is transformed into SAH. Mutagenesis studies suggest that residues Tyr35, Tyr40, Tyr85, Asp101, Asn106, and Ala181 are involved in SAM binding and Glu219 in substrate binding. Two additional regions (residues Gly79-Ser80 and residues Asp158-Val159) are involved in SAM binding. The enzyme is phosphorylated at Ser7. The following natural variants have been reported for PMNT [106]: Asn9Ser, Thr98Ala, Arg112Cys, Ala175Thr, Ser188Cys, Leu211His, Leu217Gln, Arg254His, and Trp276Arg.

K_m for phenylethanolamine is 99 and 3.4 μM for SAM. Mutagenesis experiments [107] showed that replacement of Tyr35 by Phe strongly increases K_m for substrate and SAM, whereas replacement of Glu219 by Ala reduces enzyme activity, decreasing K_m sixfold for substrate and twofold for SAM. Similarly, replacement of Asp267 by Ala strongly reduces enzyme activity, decreasing K_m 200-fold for substrate and threefold for SAM. The enzyme activity is also considerably reduced by replacing Glu185 by Ala but affinity by SAM is increased. When Glu185 is replaced by Asp, the enzyme activity is also reduced and affinity for substrate and SAM decreases threefold.

PNMT activity plays important roles in neurodegenerative diseases such as AD and PD. In fact, PMNT activity is decreased in AD brains [108] due to the reduced amount of the enzyme [109]. More particularly, in early AD there is a decrease in PNMT in axon terminals and in total PNMT in epinephrine cell bodies and terminals. However, in advanced Alzheimer's disease (AD), the diminished transport of PNMT to axon terminals causes the accumulation of PNMT in epinephrine cell bodies that may contribute to the degeneration of these neurons in AD [110]. Immunohistochemical studies demonstrated a selective loss of C1 and C3 PNMT-positive neurons, confirming a damage to these presumed brainstem sympathetic premotor neurons in Parkinson's disease [111]. The formation of toxic 2*N*-methylated β-carbolinium cations by PNMT has been suggested as a possibility for the pathogenesis of idiopathic PD [112]. A reduction of PNMT activity in hypothalamus was also observed in parkinsonism [113]. The

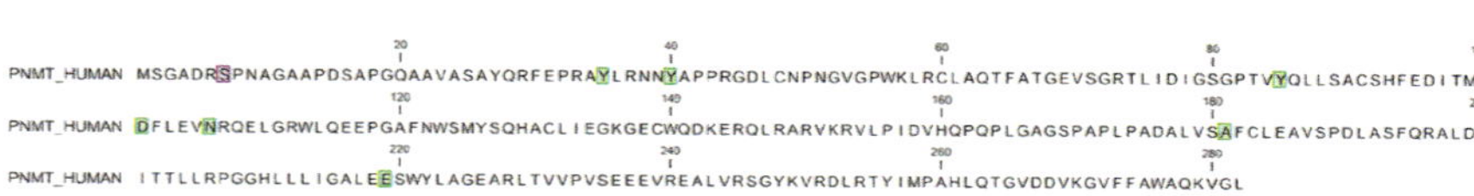

Figure 8.16 Sequence of human phenylethanolamine-*N*-methyl transferase. Phosphorylation (magenta) and binding (green) sites are indicated on sequence.

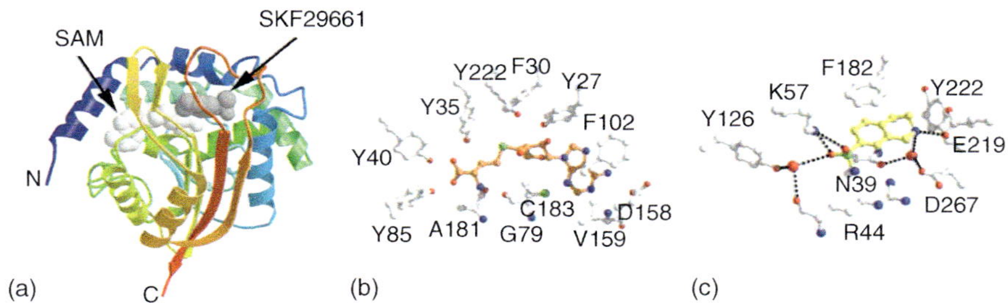

Figure 8.17 (a) A backbone trace of hPNMT shown colored from blue (N terminus) to red (C-terminus), with helices shown as coils and strands shown as arrows. Ligands bound to the enzyme are shown in light gray (SAM) and dark gray (SKF29661). (b) The binding mode of SAM (orange), showing the five tyrosines and two phenylalanines of hPNMT that surround one face of the cofactor product. (c) The binding mode of SKF29661 (yellow) to hPNMT. Source: Martin et al. 2001 [115]. Reproduced with permission of Elsevier.

role of catecholamine-related enzymes including PNMT in the pathophysiology of PD has been reviewed recently [114]. In consequence, PNMT inhibitors are potential therapeutics in those diseases.

The crystal structure of human PNMT was determined (PDB ID: 1HNN) for the first time in complex with the inhibitor 7-aminosulfonyl-1,2,3,4-tetrahydroisoquinoline (SKF29661) and the cofactor product SAH (Figure 8.17a) [115, 116]. Interestingly, the binding site of the cofactor SAM is composed of five tyrosines (Tyr27, Tyr35, Tyr40, Tyr85, and Tyr222) and two phenylalanines (Phe30 and Phe102), forming a hydrophobic environment conserved in the known PNMTs but unusual in other methyl transferases (Figure 8.17b). The inhibitor, having a tetrahydroisoquinoline ring, binds in the norepinephrine-binding site; the aromatic ring presents interactions with the side chains of Phe182 and Asn39. The aliphatic amine forms a direct hydrogen bond to Glu219 and water-mediated interactions with Asp267 and Asn39 (Figure 8.17c). The structural data suggest that the inhibitor does not use the full binding capacity present in the active site, leaving space for additional substituents at the aromatic ring, which might increase binding and thus inhibitor potency.

A comparison of the ternary complex formed with SKF29661 (PDB ID: 1HNN) and the complex formed with 7-iodo-1,2,3,4-tetrahydroisoquinoline (PDB ID: 1N7J) or 8,9-dichloro-2,3,4,5-tetrahydro-1*H*-2-benzazepine (LY134046) (PDB ID: 1N7I) confirmed that the enzyme can accommodate a variety of substituents on the aromatic ring of the inhibitors by adapting the binding pocket and subtle changes in the orientation of the ligand [117]. Further structural studies with a variety of tetrahydroisoquinoline derivatives (PDB IDs: 2G70, 2G71, 2G72, 2OBF, 2ONY, 2ONZ, and 2OPB) confirmed the importance of the enzyme flexibility for inhibitor binding [118]. Crystallographic studies allowed the design of a potent inhibitor (SKF64139) that formed a ternary complex with PNMT and SAM (PDB ID: 1YZ3), which was used to identify the residues in inhibitor binding [119]. Combination of structural data with mutagenesis studies provided kinetic insights into the catalytic mechanism. Moreover, crystalline complexes with PNMT of substrate analogues (PDB IDs: 2AN3, 2AN4, and 2AN5) furnished crucial information for determining the mode of action [107]. Thus, Asp267 is likely to play a key role in positioning the side chain correctly for methylation to occur at the amine and catalysis likely occurs via deprotonation of the amine through the action of Glu185. Mutagenesis experiments confirmed this hypothesis. However, molecular dynamics and free energy calculations enabled the refinement of ligand interactions, suggesting a molecular basis for the observed enantioselectivity at position 3 of the tetrahydroisoquinoline ring [120]. These calculations also identified the role of specific structural waters. According to the calculations, the preferred binding mode of *trans*-(1*S*,2*S*)-2-amino-1-tetralol is similar to that of its epimer *cis*-(1*R*,2*S*)-2-amino-1-tetralol, indicating that the ligand does not adopt the binding mode proposed previously (PDB ID: 2AN5) [107].

3-Fluoromethyl-7-(*N*-substituted aminosulfonyl)-1,2,3,4-tetrahydroisoquinolines have been identified as potent and selective inhibitors of PNMT in which the sulfonamide amino group could form H-bond interactions with Lys57. The crystal structure of human PNMT in complex with 3-hydroxymethyl-7-(*N*-4-chlorophenylaminosulfonyl)-tetrahydroisoquinoline and SAM (PDB ID: 2GN) showed additional interactions with the main chain carbonyl oxygen of Asn39 [121].

The crystal structure of PNMT with its natural substrate *R*-norepinephrine and cofactor SAM (PDB ID: 3HCD) revealed identical binding modes to those found in previously studied models [122]. Additional crystallographic studies with substrate analogues (PDB IDs: 3HCA, 3HCB, 3HCC, 3hCE, and 3HCF) provided a rationale for those substrates that are methylated by the enzyme and those inducing conformational changes or orientations, thus becoming inhibitors. A fragment-based screening using X-ray crystallography (PDB IDs: 3KPJ, 3KPU, 3KPV, and 3KPW), among other techniques, has been used to identify PNMT inhibitors [123]. However, some limitations to these studies should be taken into consideration [124].

The transition structure for human PNMT has been modeled by matching predicted kinetic isotopic effects with experimental observations. The model suggests an early S_N2-type transition state in which methyl transfer is rate limiting [125].

8.8 5-Hydroxyindole-O-methyltransferase (HIOMT)

5-Hydroxyindole-*O*-methyltransferase (HIOMT) (acetylserotonin O-methyltransferase (ASMT); EC 2.1.1.4) catalyzes the transfer of a methyl group from SAM onto *N*-acetylserotonin (NAS), yielding *N*-acetyl-5-methoxyriptamine (melatonin) and SAH. The enzyme has 345 residues (Figure 8.18) and presents

```
HIOMT_HUMAN  MGSSEDQAYRLLNDYANGFMVSQVLFAACELGVFDLLAEAPGPLDVAAVAAGVRASAHGTELLLDICVSLKLLKVETRGGKAFYRNTELSSDYLTTVSPT
HIOMT_HUMAN  SQCSMLKYMGRTSYRCWGHLADAVREGRNQYLETFGVPAEELFTAIYRSEGERLQFMQALQEVWSVNGRSVLTAFDLSVFPLMCDLGGGAGALAKECMSL
HIOMT_HUMAN  YPGCKITVFDIPEVVWTAKQHFSFQEEEQIDFQEGDFFKDPLPEADLYILARVLHDWADGKCSHLLERIYHTCKPGGGILVIESLLDEDRRGPLLTQLYS
HIOMT_HUMAN  LNMLVQTEGQERTPTHYHMLLSSAGFRDFQFKKTGAIYDAILARK
```

Figure 8.18 Sequence of human 5-hydroxyindole-*O*-methyltransferase. Binding sites (green) are indicated on sequence.

two alternative isoforms (with residues 1989-235 missing and with residue 188 replaced by a chain of 27 amino acids) that are inactive.

In the brain, the enzyme is expressed in the pineal gland [126]. Several natural variants have been found for HIOMT. Among them, Glu61Gln, Ile269Met, Gly278Ala, and Val305Met reduced enzyme activity [127] [128]. Asn17Lys, Val171Met, Asp210Gly, Arg291Gln, and Leu298Phe practically abolish the enzyme activity and are suggested to be related to autism [129]. In fact, mutations Arg115Trp, Val166Ile, Val179Gly, and Trp257X were exclusively detected in patients with autism [130]. Tyr170X was detected exclusively in patients with ADHD [131]. Asn13His, Lys81Glu, Arg11Glu, Lys219Arg, and Glu288Asp have no effect on enzyme activity [128]. Other natural variants include Gly151Ser, Val161Ile, Ile211Met, and Thr217Met.

Crystal structures of HIOMT in complex with SAM (PDB ID: 4A6D) and in complex with SAM and NAS (PDB ID: 4A6E) have been resolved [132]. The crystal structure of human HIOMT shows the protein as a homodimer (Figure 8.19a). Residues Tyr147, Trp164, Asp210, and Arg252 are involved in binding of SAM, whereas residues Asp256, Asn302, and Gln306 are involved in substrate binding. His255 is involved in proton donor/acceptor functionality at the active site (Figure 8.19b,c).

Mutagenesis experiments confirmed that Leu11Phe and His318Asp reduce enzyme activity and Thr296Met nearly abolishes enzyme activity. On the other

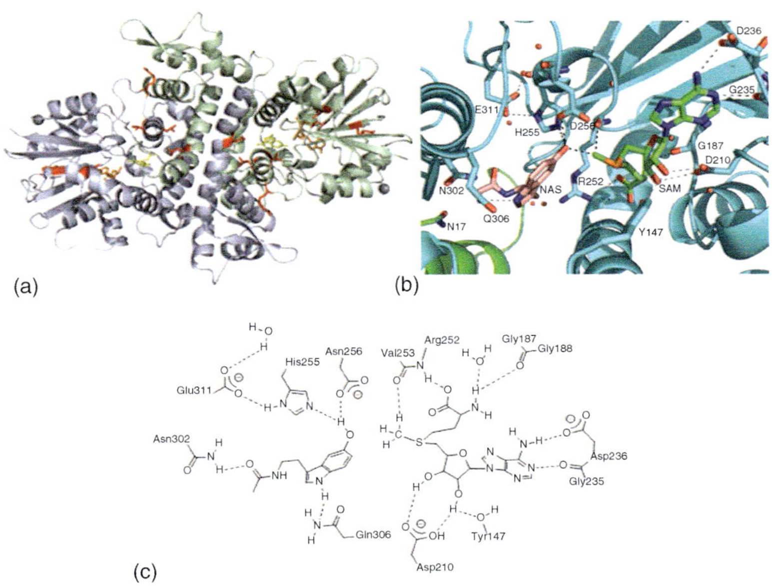

Figure 8.19 (a) Ribbon diagram of the crystal structure of HIOMT bound to NAS and (SAM) (PDB ID: 4A6E). (b) Close-up of the active site of HIOMT (cyan) with NAS (pink) and SAM (green). The carbonyl of Arg252 and the side chains of His255 and Asp256 are situated just above the activated methyl group. (c) Schematic diagram of the active site of HIOMT indicating residues that make H-bonds to the SAM and NAS molecules. Source: Botros et al. 2013 [132]. Reproduced with permission of John Wiley & Sons.

hand, Leu31His has no effect on enzyme activity [132]. The crystal structures provide a rationale for explaining the role of all mutations and natural variants that induce loss of enzymatic activity. Most of the mutations occur at the residues located near the substrate-binding sites or in the dimerization domain of the enzyme.

8.9 Serotonin-N-acetyl Transferase (SNAT)

Serotonin-*N*-acetyl transferase (SNAT, aralkylamine *N*-acetyltransferase [AANAT]; EC 2.3.1.87) catalyzes *N*-acetylation of serotonin into NAS from acetyl-CoA, the penultimate step in the synthesis of melatonin [133]. The enzyme has 207 residues. A posttranslational modification involves cyclic adenosine monophosphate (c-AMP)-dependent phosphorylation at N-terminal Thr31 and C-terminal Ser205 (Figure 8.20). The phosphorylation regulates SNAT activity by promoting an interaction with 14–3–3 proteins. Regions comprising 124–126, 132–137, and 168–170 are involved in acetyl-CoA.

Natural variants found for SNAT include Arc15Cys and Ala129T, the latter being found in individuals with delayed sleep phase syndrome [134]. An alternative isoform of the enzyme has residue 1 replaced by a chain of 40 residues.

K_m for tryptamine is 0.13 mM, 2.6 mM for 5-hydroxytryptamine, 0.55 mM for phenylethylamine, and 10.6 mM for tyramine. His120 and His122 are involved in substrate deprotonation, playing an important role in catalytic mechanism. Leu124 and Met159 are involved in substrate binding via amide nitrogen and carbonyl oxygen, respectively. The enzyme is activated by cAMP; such an activation is lost when Thr31 is replaced by Ala [135].

8.10 Choline Acetyltransferase (ChAT)

Choline acetyltransferase (ChAT, EC 2.3.1.6) catalyzes the transfer of an acetyl group from acetyl-CoA to choline to yield acetylcholine and coenzyme A. The enzyme has 748 residues (83 000 Da), and it can be phosphorylated at least at six residues [136] including Ser125, Thr456, and Ser473 (Figure 8.21) [137]. Phosphorylation at Ser440 by protein kinase C is involved in regulation of the enzyme's catalytic activity and binding to subcellular membranes [138]. There are two alternative isoforms: isoform S, which differs from the canonical isoform M in the sequence of residues 1–95, and isoform R, which is missing

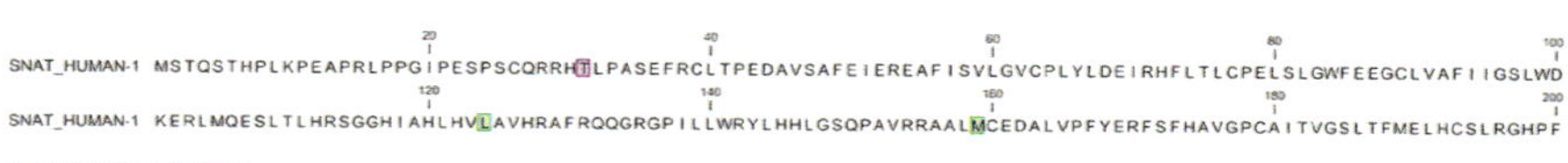

Figure 8.20 Sequence of human serotonin-*N*-acetyl transferase. Phosphorylation (magenta) and binding (green) sites are indicated on sequence.

```
ChAT_HUMAN MGLRTAKKRGLGGGGKWKREEGGGTRGRREVRPACFLQSGGRGDPGDVGGPAGNPGCSPHPRAATRPPPLPAHTPAHTPEWCGAASAEAAEPRRAGPHLC
ChAT_HUMAN IPAPGLTKTPILEKVPRKMAAKTPSSEESGLPKLPVPPLQQTLATYLQCMRHLVSEEQFRKSQAIVQQFGAPGGLGETLQQKLLERQEKTANWVSEYWLN
ChAT_HUMAN DMYLNNRLALPVNSSPAVIFARQHFPGTDDQLRFAASLISGVLSYKALLDSHSIPTDCAKGQLSGQPLCMKQYYGLFSSYRLPGHTQDTLVAQNSSIMPE
ChAT_HUMAN PEHVIVACCNQFFVLDVVINFRRLSEGDLFTQLRKIVKMASNEDERLPPIGLLTSDGRSEWAEARTVLVKDSTNRDSLDMIERCICLVCLDAPGGVELSD
ChAT_HUMAN THRALQLLHGGGYSKNGANRWYDKSLQFVVGRDGTCGVVCEHSPFDGIVLVQCTEHLLKHVTQSSRKLIRADSVSELPAPRRLRWKCSPEIQGHLASSAE
ChAT_HUMAN KLQRIVKNLDFIVYKFDNYGKTFIKKQKCSPDAFIQVALQLAFYRLHRRLVPTYESASIRRFQEGRVDNIRSATPEALAFVRAVTDHKAAVPASEKLLLL
ChAT_HUMAN KDAIRAQTAYTVMAITGMAIDNHLLALRELARAMCKELPEMFMDETYLMSNRFVLSTSQVPTTTEMFCCYGPVVPNGYGACYNPQPETILFCISSFHSCK
ChAT_HUMAN ETSSSKFAKAVEESLIDMRDLCSLLPPTESKPLATKEKATRPSQGHQP
```

Figure 8.21 Sequence of human choline acetyltransferase. Phosphorylation (magenta) and binding (green) sites are indicated on sequence.

residues 1–118. Natural variants include Asp47Glu, Ala120Thr, Leu210Pro, Pro211Ala, Arg222Pro, Leu243Phe, Pro299Leu, Ile305Thr, Ile336Thr, Ala392Gly, Asp400Asn, Arg420Cys, Glu441Lys, Val461Met, Arg482Gly, Ser498Leu, Val506Leu, and Arg560His [139–141].

Historically, acetylcholine was the first neurotransmitter to be considered as such, developing its functions in the cholinergic neurons of the peripheral and CNSs. In the CNS, cholinergic neurons have been involved in the pathophysiology of several neurodegenerative disorders including Alzheimer's disease, schizophrenia, and amyotrophic lateral sclerosis (ALS) [142].

Site-directed mutagenesis studies suggested that His334 in rat ChAT is essential for acting as an acid/base catalyst [143]. Arg452 has been identified as a residue interacting directly with acetyl-CoA [144]. Ser558 and Gln669 have also been suggested to interact with coenzyme A at the binding site. However, although this information has been known for many years, a three-dimensional structure remained elusive. The first crystal structure of rat ChAT (PDB IDs: 1Q6X and 1T1U) revealed that the enzyme is a monomer consisting of two domains [145]. Both domains are composed of six stranded mixed β-sheets (S1, S8, S7, S6, S4, and S5 for domain I and S10, S16, S15, S13, S11, and S12 for domain II). Domains I and II are interconnected on both flanks by 8 and 11 α-helices. For domain II, seven α-helices are from the C-terminal and four from the N-terminal. The catalytic site is located between the interface of domains I and II [146]. The structure of human ChAT was resolved at 2.2 Å resolution (PDB ID: 2FY2) along with binary structures in complex with choline (PDB ID: 2FY3), acetyl-CoA (PDB ID: 2FY4), and a nonhydrolyzable acetyl-CoA analogue, *S*-(2-oxopropyl)-CoA (SOP) (PDB ID: 2FY5) [147]. The two domains of human ChAT are the substrate binding domain that comprises residues 1–89 and 392–615 and the catalytic domain comprising residues 90–391 (Figure 8.22a). Choline-binding site (Figure 8.22b) is made up of residues belonging to the binding domain. Coenzyme A forms five hydrogen bonds. The pyrophosphate moiety sits above the two closely interacting carboxylic acid side chains of Asp414 and Glu437 (Figure 8.22c). These structural studies showed that a P-loop, disordered in the unliganded protein, interacts with the phosphates of CoA being responsible for the low K_m of human ChAT.

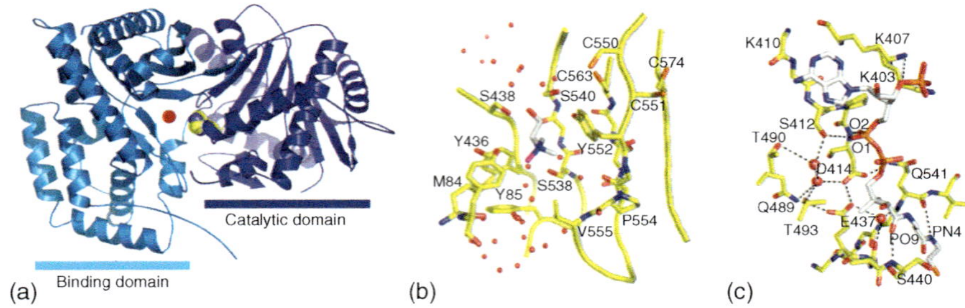

Figure 8.22 (a) X-ray structure of human ChAT. The active-site tunnel, which runs through the enzyme and is located at the domain interface, is indicated by the red circle. The side chain of the catalytic H324 is shown as a space-filling CPK model. (b) Specific residues involved in binding choline. (c) Hydrogen-bonding interactions in the CoA-binding site of ChAT. Source: Reprinted with permission from Kim et al. 2006 [147]. Copyright 2006, American Chemical Society.

8.11 Acetylcholinesterase (AChE)

Acetylcholinesterase (AChE, EC 3.1.1.7) catalyzes the hydrolysis of acetylcholine into choline and acetate. The enzyme has 614 residues, and it can be *N*-glycosylated at Asn296, Asn381, and Asn495 (Figure 8.23). Alternative splicing produces four isoforms. Isoform T has been chosen as the canonical sequence. Isoform H differs in the sequence of 575–614 residues. Isoform R differs in the sequence of 575–603 residues and residues 604–614, which are missing. In isoform 4, residues 357–444 are missing. Natural variants include Arg34Gln, Pro135Ala, Val333Glu, and His353Asn.

AChE is one of the most efficient enzymes with a turnover greater than $10^4\,s^{-1}$. In the CNS, the major function of AChE is the elimination of acetylcholine from the synapses. Also, AChE has a role in neuronal apoptosis [148]. AChE can exist in a variety of molecular forms, the catalytic subunit being common to all forms [149]. The dimeric form can be posttranslationally modified by attaching a glycophosphatidylinositol unit to the C-terminus. The only form expressed in the mammalian brain is isoform T, which can generate monomers, dimers, and tetramers. In addition, collagen-tailed [150, 151] and hydrophobic-tailed [152] heteroligomers can also be formed. The structure of AChE has been characterized by many crystallographic studies. The first studies showed binding sites for

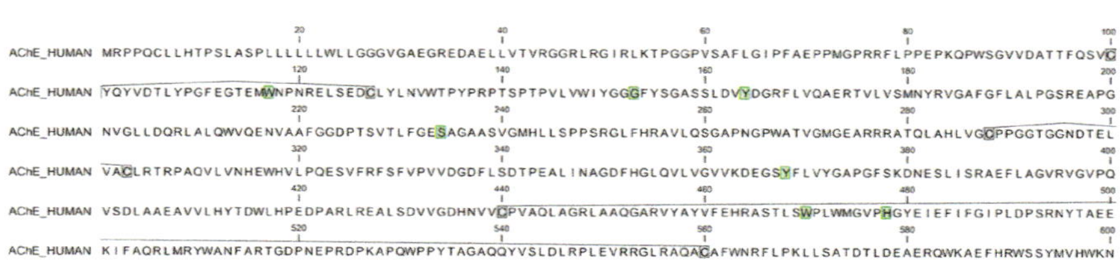

Figure 8.23 Sequence of human acetylcholinesterase. Disulfide bonds (dark green), modified (magenta), binding (green), and metal-binding (gray) sites are indicated on sequence.

quaternary ligands (PDB IDs: 1ACJ and 1ACL) [153] illustrating the important role of aromatic groups as binding sites for such types of ligands. Additional evidences were obtained assigning Trp84 and Phe330 to the "anionic" subsite of the active site and Trp279 to the "peripheral" anionic site. The X-ray structure of *Torpedo californica* AChE in complex with fasciculin-II (FAS), a snake-venom toxin, (PDB ID: 1FSS) [154] demonstrated that conserved aromatic residues in the AChE peripheral anionic site make important contacts with FAS. Also, several basic residues in FAS make important contacts with AChE. Similar observations were made with mouse AChE (PDB ID: 1MAH) [155]. The same structure with human AChE (PDB ID: 1F8U) [156] was very similar. The structure of native AChE from *T. californica* (PDB ID: 2ACE) and its complex with the nootropic alkaloid huperzine (PDB ID: 1VOT) revealed the importance of hydrophobic residues in the active-site gorge of the enzyme [157].

The crystal structure of mouse AChE in the apo form (PDB ID: 1J06) served to characterize the peripheral anionic site, located at the active center gorge entry and suggested to be involved in heterologous protein associations occurring during synaptogenesis or upon neurodegeneration [158]. The crystal structure of mouse AChE has been also resolved more recently (PDB ID: 5DTI), and a ternary complex between diisopropylphosphorofluoridate (DFP)-inhibited enzyme and the reactivator SP-134 has been crystallized (PDB ID: 5DTJ) [159]. Mouse AChE has been crystallized in complex with various ligands such as gallamine (PDB ID: 1N5M) or propidium (PDB ID: 1N5R) among others. Syn- and anticomplexes of mouse AChE with TZ2PA6 (PDB IDs: 1Q83 and 1Q84, respectively) have also been crystallized [160]. The crystal structure of a complex formed by four units of the tryptophan amphiphilic tetramerization sequence at the C-terminus and a proline-rich attachment domain enabled to propose a model for the collagen-linked AChE tetramer [161].

The first structure of wild-type human AChE was resolved in complex with FAS II (PDB ID: 1B41) [156]. The overall structure is very similar to that of the same complex with the *T. californica* enzyme and to that of the mouse AChE (Figure 8.24). In all these structures, the active site of AChE is located at the bottom of a deep gorge lined largely by aromatic residues.

A comparison of the Glu202Gln mutant complex revealed that substitution of the charged group by a neutral isosteric moiety does not affect the architecture of the active center. A key element of this architecture is a hydrogen-bond network formed by residues Glu202, Glu450, Tyr133 and two bridging molecules of water. This observation supports the hypothesis that the main role of that network is proper orientation of the Glu202 carboxylate relative to the catalytic triad, thus defining its functional role in the interaction with ligands. As shown in Figure 8.24, human AChE in its apo-state is very similar to that of the *Torpedo* enzyme; however, the unique crystal packing reveals a novel peptide sequence, which blocks access to the active-site gorge [163].

Organophosphorus (OP) compounds are able to block the catalytic activity of AChE by rapidly phosphorylating the catalytic Ser203 residue. The inhibited enzyme can at least partly be reactivated with nucleophilic reactivators such as oximes. However, the covalently attached OP conjugate may undergo further intramolecular dealkylation or deamidation reactions (aging process), resulting

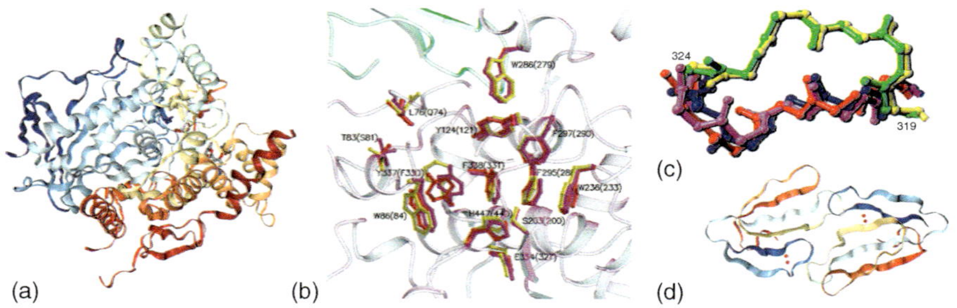

Figure 8.24 (a) Human AChE complexed with fasciculin II (FAS II). Source: (a) Image from the RCSB PDB (www.rcsb.org) of PDB ID: 1B41 [156]. (b) Comparison of the key residues within the active-site gorge in the three complex structures: magenta, *h*AChE-FAS-II; red, *m*AChE-FAS-II; yellow, *Tc*AChE-FAS-II. The gray ribbon denotes the *h*AChE backbone and the green ribbon the FAS-II backbone. (c) Overlay of the loop 319(310)–324(317) in three AChE-FAS-II complexes and two AChE apo structures: magenta, *h*AChE-FAS-II; red, *m*AChE-FAS-II; blue, *m*AChE (chain A); yellow, *Tc*AChE-FAS-II; green, TcAChE. Source: le Du et al. 1996 [156]. Reprinted with permission of IUCR. (d) Fasciculin II. Source: Image from the RCSB PDB (www.rcsb.org) of PDB ID: 1FSC [162].

in an enzyme resistant to reactivation [164]. Several crystallographic studies have been carried out with mouse AChE (PDB IDs: 2C0P, 2C0Q, 2JEZ, 2JF0) to study those processes [164, 165]. The first crystal structure of human AChE inhibited by a nerve agent (Tabun) provided structural insights into the inhibition and aging mechanisms (PDB ID: 2X8B) [166]. In agreement with previous studies carried out with mouse AChE (PDB ID: 3DLT) [167], (−)-Tabun reacts 6.3 times faster than (+)-Tabun with *h*AChE.

Different structures of AChE have also been resolved in different states in complex with OP derivatives and antidotes. Combined diffusion trap cryocrystallography and density functional theory (DFT) calculations have been used to generate and analyze crystallographic data for a ternary complex between AChE, the nerve agent sarin, and the therapeutic antidote HI-6 [168]. Recent crystallographic studies of human AChE inhibited by nerve agent VX (PDB ID: 6QCZ) and its enantiomers (+)-VX and (−)-VX and *rac*-VX (PDB ID: 6CQX and 6CQT, respectively) and the corresponding ternary complexes with the antidote oxime HI-6 (PDB ID: 6CQV, 6CQW, 6CQY) demonstrated that the therapeutic reversal of human AChE inhibition by reactivators is also dependent on the stereochemistry of the inhibitor [169]. Ternary structures of human AChE inhibited by insecticide paraoxon in complex with the oximes HI6 (PDB ID: 5HF9) and 2-PAM (positive allosteric modulator) (PDB ID: 5HFA) reveal unique conformational changes of the acyl loop and the dimer interface [170]. Such perturbations are different from those observed from inhibition by traditional nerve agents. These differences should be considered for the development of new insecticides [171] and specific treatments for exposure to diverse OP derivatives [172].

AChE has been crystallized in complex with the anti-Alzheimer drug Aricept (PDB ID: 1EVE) [173]. This study revealed specific interactions responsible for the high affinity and selectivity of Aricept. The ligand showed a unique orientation along the active-site gorge, extending from the anionic subsite of the active

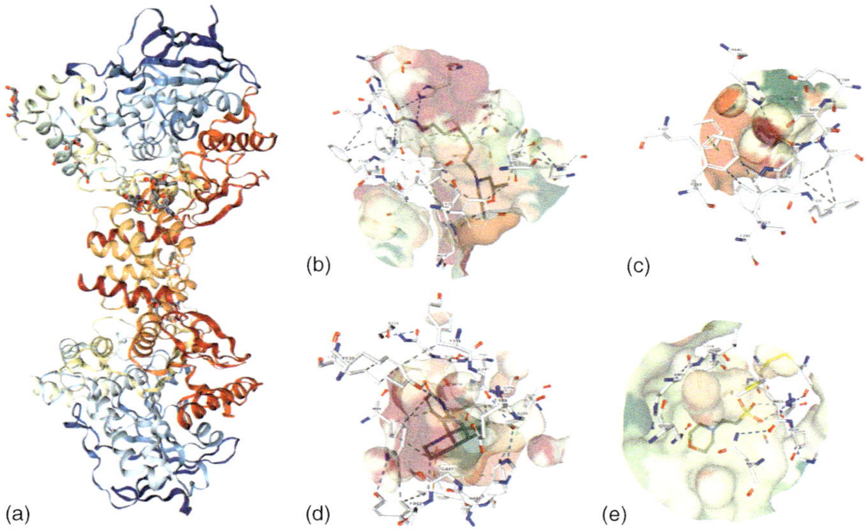

Figure 8.25 (a) Structure of human acetylcholinesterase in the apo form. Protein *N*-glycosylations are shown as sticks with carbon, oxygen, and nitrogen atoms colored green, red, and blue, respectively. (b–e) Ligand interactions in AChE complexed with MF268 (b), SOMAN (c), (+)-huperzine (d), and DFP (e). H-bonds in blue, halogen bonds in turquoise, hydrophobic contacts in gray, π-interactions in orange/green, metal interactions in purple. Ligand pocket is colored according to hydrophobicity. Source: Images from the RCSB PDB (www.rcsb.org) of the corresponding PDB IDs and references.

site, at the bottom, to the peripheral anionic site, at the top, via aromatic stacking interactions with conserved aromatic acid residues. On the other hand, no direct interactions with either the catalytic triad or the "oxyanion hole" were observed (only indirectly via solvent molecules). The complex with other anti-Alzheimer drugs, rivastigmine (PDB ID: 1GQR) [174] and galantamine (nivalin) (PDB ID: 1QTI), showed an unexpected orientation of the ligand within the active site, as well as unusual protein–ligand interactions revealing important implications for the design of new anti-Alzheimer drugs [175]. AChE from *T. californica* has also been crystallized with a variety of inhibitors including MF268 (PDB ID: 1OCE) [176], (−)-huprine X [177], (+)-huperzine B (PDB ID: 1GPN) [177], and SOMAN, a nerve agent (PDB ID: 1SOM), and DFP (PDB ID: 2DFP) (Figure 8.25) [178]. *Torpedo* AChE has also been employed for kinetic and structural studies on the interactions with two rigid analogues (PDB IDs: 5NAP and 5NAU) of the anti-Alzheimer drug donepezil [179]. Structural studies of human AChE alone (PDB ID: 4EY4) and in complex with FAS II (PDB ID: 4EY8) and several Alzheimer's disease drugs, such as (−)-huperzine A (PDB ID: 4EY5), (−)-galantamine (PDB ID: 4EY6), and donepezil (PDB ID: 4EY7), showed that binding in *h*AChE is different from that in *Torpedo* AChE [180]. Consequently, the human enzyme should be considered a better starting model than the *Torpedo* enzyme for the design of new therapeutic drugs.

Additional structural studies with dihydrotanshinone I (PDB ID: 4M0E) and Territrem B (PDB ID: 4M0F), natural inhibitors of AChE, showed peripheral site flexibility [181], revealing important mechanistic details about ligand binding and protein conformational changes that should also be considered very relevant for future drug design. Crystal structures of human AChE in complex with anti-Alzheimer drugs huprine W (PDB ID: 4BDT) and tacrine (PDB ID: 2AOW) confirmed key interactions with Trp82 and Trp86 as well as H-bonding with catalytic His residue. The higher affinity of huprine W is due to additional interactions with a group of aromatic residues including Phe295, Tyr337, and Ph338 [182].

8.12 Glutamic Acid Decarboxylase (GAD)

Glutamic acid decarboxylase (GAD; EC 4.1.1.15) is a PLP-dependent enzyme that catalyzes decarboxylation of L-glutamate onto γ-aminobutyric acid (GABA), yielding carbon dioxide. There is, however, a side reaction that produces pyridoxamine 5′-phosphate (PMP), succinic semialdehyde (SSA), and inactive apo enzyme. In mammals, there are two active isoforms GAD1 and GAD2, the latter being 15 times more efficient than the former in catalyzing the side reaction [183]. Some studies suggested that this behavior actually represents a mechanism of control for GABA production [184]. Indeed, GAD1 is responsible for the production of basal GABA, whereas GAD2 only works when extra GABA is required [185].

The amino acid sequences of GAD1 and GAD2 (Figure 8.26) show c. 70% homology with molecular weights of c. 67 000 and 65 000 Da, respectively. GAD1 and GAD2 have 594 and 585 residues, respectively. Posttranslational modifications of GAD1 include phosphorylation at Ser78 and linking of *N*6-(pyridoxal phosphate) at Lys405. GAD2 is phosphorylated at Ser3, Ser6, Ser10, and Ser13 [186] and palmitoylated at Cys30 and Cys45 [187]. The cofactor in GAD2 is linked at Lys396. An inactive isoform of GAD1 differing in residues 214–224

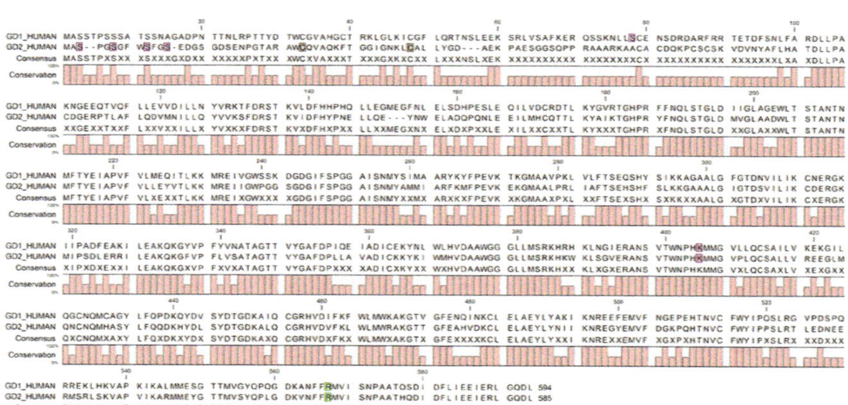

Figure 8.26 Sequences of human glutamic acid decarboxylases 1 and 2. Conservation graphic and consensus sequence are shown. Phosphorylation (magenta) and binding (green) and lipid-binding (brown) sites are indicated on sequences.

and missing residues 225–594 has been found. GAD2 can present several natural variants: Gly12Arg, Lys124Asn, Pro153Gln, Gly232Glu, Lys286Arg, Gly326Ala, and Arg375Gln.

In the CNS, the overall expression of GAD1 is higher, but regions corresponding to the visual and neuroendocrine systems have higher levels of GAD2 [188]. While GAD1 is widely distributed in the cell body of the neurons, GAD2 expression is localized in the terminals of the neurons and is associated with synaptic vesicles [189]. The N-terminal domain of the two proteins appears to be responsible for this differential subcellular distribution [190].

A decrease in GAD1 in the human brain is associated with schizophrenia and related neurological disorders [191]. Mutations affecting GAD1 are involved in cerebral palsy, spastic quadriplegic 1 (CPSQ1), a nonprogressive disorder of movement and/or posture resulting from defects in the developing CNS [192].

The monomeric unit of GAD comprises three domains: *N*-terminal, *C*-terminal, and PLP-binding. The N-terminal domain and the C-terminal domain are the two main functional domains of GAD. The N-terminal domains of the two proteins have only a 23% identity, whereas the C-terminal domains have a 73% identity. In human GAD1, the N-terminal domain is composed of residues 1–101 and the GAD2 is composed of residues 1–95. For both GAD1 and GAD2, the C-terminal domain is the catalytic domain [193]. GAD forms homodimers and heterodimers [194]. GAD1 and GAD2 interact differently with the cofactor, which is thought to be one of the reasons for functional differences between the two isoforms.

The X-ray structure of *N*-terminal truncations of GAD1 (PDB ID: 2OKJ) and GAD2 (PDB ID: 2OKK) showed important differences between the two isoforms [195]. The structure of GAD1 has a tethered loop covering the active site, providing a catalytic environment for GABA production. However, the same catalytic loop is inherently mobile in GAD2, promoting the above-mentioned side reaction that regulates GABA production (Figure 8.27). Further crystallographic studies (PDB ID: 3VP6) showed that the conformation of the catalytic loop is intimately linked to the C-terminal domain, confirming its key role in controlling auto-inactivation of GAD2 [196].

8.13 Glutamine Synthetase (GLUL)

Glutamine synthetase (GLUL, GS; EC 6.3.1.2) catalyzes the conversion of glutamate, ATP, and ammonia to glutamine yielding adenosine diphosphate (ADP) and inorganic phosphate [197]. The enzyme has 373 residues (Figure 8.28) and can present several natural variants including Arg324Cys and Arg341Cys, both of which lead to a reduced enzymatic activity [198]. Ser482Cys variant is associated with profound developmental delay and infantile cataract [199]. Acetylation can take place at Thr2 and phosphorylation at Tyr104 and Ser343. Posttranslational modifications include autopalmitoylation and ubiquitination [197].

In the brain, GLUL is mainly located in astrocytes, which act as neuroprotecting cells against an excess of ammonia by converting glutamate into glutamine [200]. The glutamate–glutamine cycle regulates glutamatergic synaptic transmission in

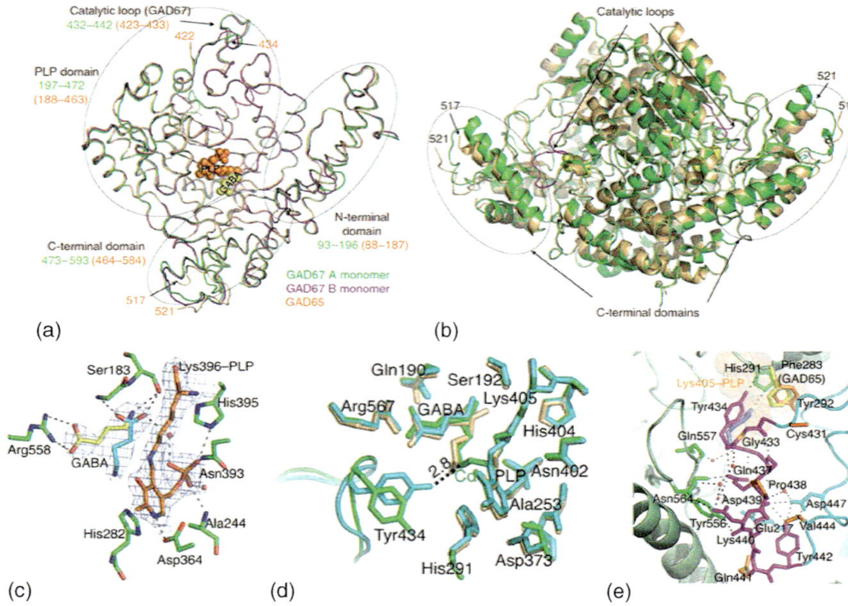

Figure 8.27 (a, b) Structural comparisons between GAD1 and GAD2. (a) Structural superposition between GAD1 monomer A (green), GAD1 monomer B (purple), and GAD2 (light brown). Disordered regions in GAD2 are numbered. (b) Structural superposition of GAD1 dimer (green) and GAD2 dimer (light brown). PLP is shown as orange spheres and GABA moieties bound in active site are shown as yellow spheres. The catalytic loops of GAD1 are colored magenta; this region is not visible in electron density in GAD65. Both figures highlight the structural shifts in the C terminus. (c, e) Comparison of GAD1 and GAD2 active sites. (c) Active site of GAD2. Lys-PLP moiety is shown as orange sticks; noncovalently bound GABA molecules are colored yellow and cyan. (d) Superposition of active-site residues of GAD1 (green, monomer A; cyan, monomer B) and GAD2 (light brown). Both Tyr434 side chain conformations from the catalytic loops of GAD1 are shown. Tyr434 from monomer B (in cyan) enters the active site of monomer A. In this conformation, the hydroxyl group of Tyr434 is 2.8 Å from the Ca atom. (d) Interactions between catalytic loop and adjacent monomer. Residues that are different in GAD2 are colored orange. The alternative conformation of Tyr434 (from monomer B) is shown as mauve bonds. Phe283 of GAD2 is shown as yellow bonds. Residues in GAD2 corresponding to GAD1 residues 432–442 are disordered. The PLP moiety in the active site is shown as orange spheres, hydrogen bonds as dotted lines, and water molecules as red spheres. Source: Reprinted with permission from Langendorf et al. 2013 [196]. Copyright 2013, BioScience Reports.

```
GLUL_HUMAN  MTTSASSHLNKGIKQVYMSLPQGEKVQAMYIWIDGTGEGLRCKTRTLDSEPKCVEELPEWNFDGSSTLQSEGSNSDMYLVPAAMFRDPFRKDPNKLVLCE
GLUL_HUMAN  VFKYNRRPAETNLRHTCKRIMDMVSNQHPWFGMEQEYTLMGTDGHPFGWPSNGFPGPQGPYYCGVGADRAYGRDIVEAHYRACLYAGVKIAGTNAEVMPA
GLUL_HUMAN  QWEFQIGPCEGISMGDHLWVARFILHRVCEDFGVIATFDPKPIPGNWNGAGCHTNFSTKAMREENGLKYIEEAIEKLSKRHQYHIRAYDPKGGLDNARRL
GLUL_HUMAN  TGFHETSNINDFSAGVANRSASIRIPRTVGQEKKGYFEDRRPSANCDPFSVTEALIRTCLLNETGDEPFQYKN
```

Figure 8.28 Sequence of human glutamine synthetase. Modified sites (magenta) are indicated on sequence.

the adult hippocampus, and recent studies demonstrated its implication in the modulation of sexual differentiation [201]. Changes in the expression of GLUL may affect neuronal functions [202]. Genetic defects in ammonia metabolism can produce irreversible damage to the developing CNS; in this context, recent studies demonstrated that systemic hyperammonemia resulted in significant deficits in novelty acquisition and disturbed synaptic plasticity in corticostriatal and hippocampal pathways involved in learning and goal-directed behavior [203]. The induction of GLUL expression in glutamatergic areas protects against neurodegeneration, and, consequently, reduction of GLUL activity has been suggested as a mechanism mediating neurotoxicity in neurodegenerative disorders [204]. Extra-astrocytic expression of GLUL in oligodendrocytes, ependymal cells, and certain neurons have also been suggested, but functional implications are still under discussion [205]. Human GLUL deficiency causes severe brain malformations resulting in multiorgan failure and neonatal death [198]. In Alzheimer's patients, levels of GLUL are higher in the cerebral spinal fluid. Modification of enzyme activity contributes to glutamate dysregulation and impairment of neuroplasticity during the development of Alzheimer's disease [206].

Several X-ray structures of bacterial GLUL have been reported. Among these structures are *Salmonella typhimurium* GLUL alone (PDB ID: 2GLS) [207] and in complex with ADP (PDB ID: 1F52), thallium (PDB ID: 1F1H) and inhibitor phosphinothricin (PDB ID: 1FPY) [208]. Interactions of unadenylated *St*GLUL with nucleotides (PDB ID: 1LGR) [209] and amino acids (PDB ID: 2LGS) [210] have been studied. *Mycobacterium tuberculosis* GLUL has also been crystallized in an inactive form (PDB IDs: 1HTO and 1HTQ) [211] and in complex with L-methionine-*S*-sulfoximine phosphate (MSO-P), a transition-state mimic, Mg^{2+}, and ATP (PDB ID: 2BVC) [212]. Structures from *Bacillus subtilis* (PDB IDs: 4LNO, 4LNI, 4LNF, 4LNN, and 4LNK) revealed some basis for isoenzyme-specific regulation [213]. Other structures include GLUL from *Streptomyces coelicolor* (PDB ID: 4BAX) and *Bacteroides fragilis* (PDB ID: 3O6X) [214]. More recently, the crystal structure of GLUL from *Helicobacter pylori* has been reported (PDB ID: 5ZLI) [215]. These bacterial GLULs are all dodecameric, with two hexameric rings of GLUL subunits stacked onto each other. The first structures of mammalian GLULs corresponded to canine GLUL (PDB ID: 2UU7) and human GLUL that was crystallized in complex with ADP and MSO-P (PDB ID: 2QC8) and with ADP and phosphate (PDB ID: 2OJW) [216]. The structure of *Cf*GLUL consists of three pentamers (Figure 8.29a), two of which constitute the biologically relevant decamer. This decamer has 10 active sites, each located at the interface of two adjacent subunits in the pentameric ring. The structure of *h*GLUL in complex with ADP, manganese ions, and MSO-P showed key interactions in the active site (Figure 8.29b). The adenine ring of ADP fits in a hydrophobic pocket formed by Trp130, Arg262, Tyr336, and Pro208. The side chains of Trp130 and Arg262 give rise to an aromatic–arginine planar stacking interaction, whereas Pro208 adopts a cis conformation. Mammalian and bacterial GLULs share less than 20% amino acid sequence identity. A comparison of the active sites of mammalian and *Mt*GLUL structures is given in Figure 8.29c. Interestingly, all residues binding to MSO-P and Mg^{2+}/Mn^{2+} are conserved, as well as the

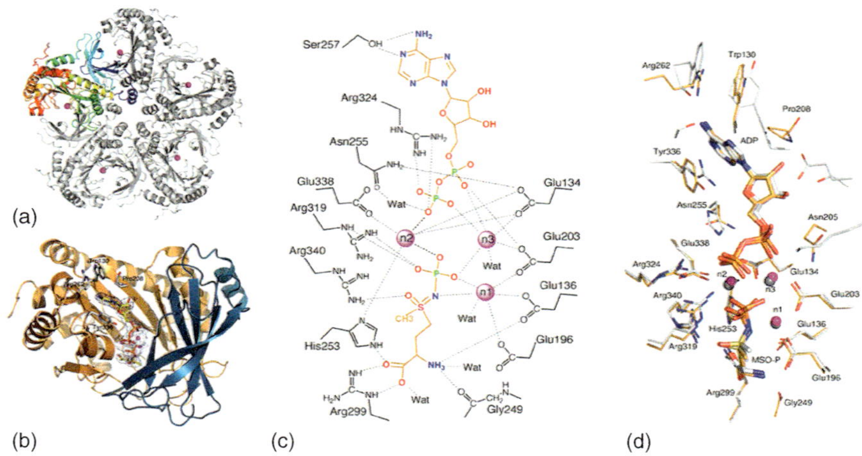

Figure 8.29 (a) Cartoon representation of a *Cf*GS pentamer. One subunit is colored from blue to red beginning from the N-terminus. The blue-cyan unit corresponds to the N-terminal β-grasp domain, whereas the remainder represents the C-terminal catalytic domain. The positions of the n1 magnesium ions (magenta spheres) mark the funnel-shaped active-site cavities that face outward. The biologically relevant decamer is formed via a noncrystallographic or crystallographic twofold axis that places the second pentamer ring behind the first. The active sites in the second pentamer are placed between those in the first and face the opposite direction. (b) *h*GS active site with bound ADP, MSO-P, and manganese ions. The N-terminal β-grasp domain is in blue, whereas the C-terminal catalytic domain of the adjoining subunit is in gold. Conserved hydrophobic residues interacting with the adenine ring of ADP are shown with ball-and-stick representations. (c) Polar interactions with ligands in the *h*GS/MnADP/MSO-P structure. (d) Comparison of active sites of mammalian and mycobacterial GSs. The *Mt*GS/MgADP/MSO-P structure (gray) is superposed on that of *h*GS/MnADP/MSO-P (gold). MSO-P and ADP of *h*GS are shown as ball-and-stick models; the *h*GS manganese ions are shown as magenta spheres, whereas gray spheres represent the magnesium ions of *Mt*GS. For clarity, only residues of *h*GS are labeled.

positions of the three metal ions. On the other hand, all side chains lining the nucleotide-binding site are different; some differences are also found in the positions of main-chain atoms (Figure 8.29d).

8.14 GABA Transaminase (GABA-T)

GABA transaminase (4-aminobutyrate aminotransferase [ABAT], GABA-T; EC 2.6.1.19) is a PLP-dependent enzyme that catalyzes the conversion of 2-oxoglutarate and 4-aminobutanoate into L-glutamate and succinate semialdehyde. The enzyme has 500 residues (Figure 8.30) that can be acetylated at Lys252, Lys279, Lys318, Lys413, Lys452, and Lys470. The enzyme can also be succinylated at Lys231, Lys252, and Lys413. The PLP is linked at Lys357. The natural variant Arg220Lys has been found to induce 25% reduction in activity. The enzyme is a disulfide-linked homodimer; mutagenesis experiments demonstrated the essential role of Cys321 for the catalytic function of γ-aminobutyric acid transporter (GABA-T) [217].

Figure 8.30 Sequence of GABA transaminase. Disulfide bonds (dark green), modified (magenta), binding (green), and metal-binding (gray) sites are indicated on sequence.

The neurotransmitter GABA is involved in many neurologic and psychiatric diseases including AD [218]. Since GABA is catabolized by the action of GABA-T, inhibition of this enzyme causes accumulation of GABA in the brain [219], leading to increased extracellular GABA concentration, which inhibits neuronal activity and causes lethargy [220]. The activity of GABAT in the brain has been correlated to several neuropsychiatric disorders including epilepsy and Alzheimer's disease [221]. Decreased GABA-T activities have been found in the brain of patients with AD [222]. GABA-T deficiency is a rare autosomal recessive disorder [223] resulting in psychomotor retardation, hypotonia, hyperreflexia, lethargy, refractory seizures, and electroencephalogram (EEG) abnormalities [224, 225].

The X-ray structure of pig GABA-T was initially reported by molecular replacement in ornithine aminotransferase [226], but later the enzyme was crystallized and resolved in its native form (PDB ID: 1OHV) and in complex with vigabatrin and 4-vinyl-GABA (PDB ID: 1OHW) and 4-ethynyl-GABA (PDB ID: 1OHY) [227]. The inactivators form a covalent ternary adduct with the active-site Lys329 and PLP cofactor. The X-ray crystal structures of *Escherichia coli* GABA-T unbound (PDB: 1SF2) and bound to the inhibitor aminooxyacetate (PDB ID: 1SFF) showed the absence of iron–sulfur clusters, found in pig GABA-T. Small changes in the active-site structure were observed [228]. Mutants Ile50Gln (PDB ID: 1SZS), Val241Ala (PDB ID: 1SZU), and Glu112Ser (PDB ID: 1SZK) of *Ec*GABA-T were investigated to change the reaction specificity to that of a decarboxylase [229]. The crystal structures of *Mycobacterium marinum* GABA-T (PDB ID: 3R4T) and *Mycobacterium abscessus* (PDB ID: 4FFC) have also been resolved [230].

Further investigations on the inactivation of GABA-T led to the development of potent inhibitors such as (1*S*,3*S*)-3-amino-4-difluoromethylenyl-1-cyclopentanoic acid (CPP-115), which showed 187 times greater potency than that of vigabatrin, an approved drug for the treatment of refractory adult epilepsy. The X-ray structure of GABA-T (Figure 8.31a,b) in complex with CPP-115 (PDB ID: 4Y0H) showed that inactivation takes place via a covalent Michael addition mechanism consisting of a conformational change induced by the formation of an aldimine that disrupts an electrostatic interaction between Glu270 and Arg445 by forming a new electrostatic interaction between Arg445 and the newly formed carboxylate produced by hydrolysis of the difluoromethylene group in CPP-115 (Figure 8.31c) [231]. The same mechanism of inactivation has been found for (*E*)- and (*Z*)-(1*S*,3*S*)-3-amino-4-fluoromethylenyl-1-cyclopentanoic acid (PDB IDs: 4ZSW and 4ZSY, respectively), monofluorinated analogues of CPP-115, which are comparable to vigabatrin in inactivating GABA-T [232]. All

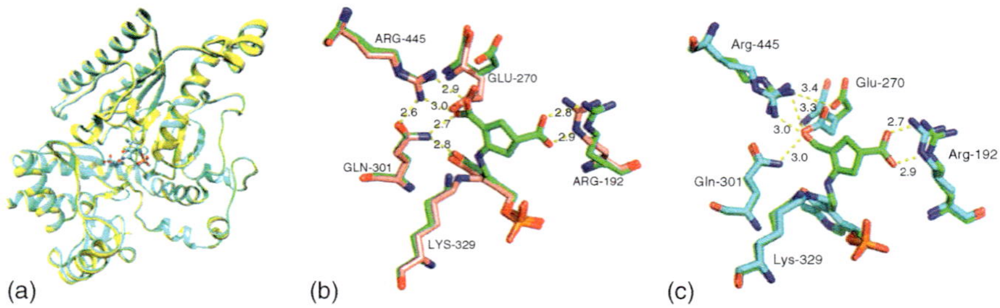

Figure 8.31 (a) Ribbon diagram of the superimposed native GABA-T (yellow) and GABA-T (cyan) bound to CPP-115. (b) Superimposition of the crystal structures of native GABA-T (pink) and CPP-115-inactivated GABA-T (green). Source: (a, b) Lee et al. 2015 [231]. Reproduced with permission of ACS. (c) Superimposition of the crystal structure of inactivated GABA-T by (*E*)-(1*S*,3*S*)-3-amino-4-fluoromethylenyl-1-cyclopentanoic acid (green) and native GABA-T (cyan) monomer. Source: (c) Lee et al. 2015 [232]. Reproduced with permission of ACS.

these findings served to design (*S*)-3-amino-4-(difluoromethylenyl)cyclopent-1-ene-1-carboxylic acid as a new inactivator of GABA-T 10 times more potent than CPP-115. The crystal structure of this compound bound to GABA-T (PDB ID: 6B6G) confirmed the hydrolysis of the difluoromethylene group to a carboxylate; inactivation of the enzyme appeared to be caused by a tight binding rather than a covalent modification [233].

References

1 Haugarvoll, K. and Bindoff, L.A. (2011). A novel compound heterozygous tyrosine hydroxylase mutation (p.R441P) with complex phenotype. *J. Parkinsons Dis.* 1 (1): 119–122.

2 Fossbakk, A., Kleppe, R., Knappskog, P.M. et al. (2014). Functional studies of tyrosine hydroxylase missense variants reveal distinct patterns of molecular defects in Dopa-responsive dystonia. *Hum Mutat.* 35 (7): 880–890.

3 Ludecke, B., Knappskog, P.M., Clayton, P.T. et al. (1996). Recessively inherited l-DOPA-responsive parkinsonism in infancy caused by a point mutation (L205P) in the tyrosine hydroxylase gene. *Hum. Mol. Genet.* 5 (7): 1023–1028.

4 Swaans, R.J., Rondot, P., Renier, W.O. et al. (2000). Four novel mutations in the tyrosine hydroxylase gene in patients with infantile parkinsonism. *Ann. Hum. Genet.* 64 (Pt 1): 25–31.

5 Schiller, A., Wevers, R.A., Steenbergen, G.C. et al. (2004). Long-term course of l-dopa-responsive dystonia caused by tyrosine hydroxylase deficiency. *Neurology* 63 (8): 1524–1526.

6 Willemsen, M.A., Verbeek, M.M., Kamsteeg, E.J. et al. (2010). Tyrosine hydroxylase deficiency: a treatable disorder of brain catecholamine biosynthesis. *Brain* 133 (Pt 6): 1810–1822.

7 Tolleson, C. and Claassen, D. (2012). The function of tyrosine hydroxylase in the normal and Parkinsonian brain. *CNS Neurol. Disord. Drug. Targets* 11 (4): 381–386.

8 Kunugi, H., Kawada, Y., Hattori, M. et al. (1998). Association study of structural mutations of the tyrosine hydroxylase gene with schizophrenia and Parkinson's disease. *Am. J. Med. Genet.* 81 (2): 131–133.

9 Vecchio, L. M., Bermejo, M. K., Dunn, A. R., Milenkovic, M., Urs, N., Ramsey, A., Miller, G. W., Salahpour, A. (2017) Increased activity of tyrosine hydroxylase leads to elevated amphetamine response and markers of oxidative stress in transgenic mice. https://www.biorxiv.org/content/10.1101/188318v1. DOI: https://doi.org/10.1101/188318.

10 Goodwill, K.E., Sabatier, C., Marks, C. et al. (1997). Crystal structure of tyrosine hydroxylase at 2.3 A and its implications for inherited neurodegenerative diseases. *Nat. Struct. Biol.* 4 (7): 578–585.

11 Rose, A.S., Bradley, A.R., Valasatava, Y. et al. (2018). NGL viewer: web-based molecular graphics for large complexes. *Bioinformatics* https://doi.org/10.1093/bioinformatics/bty419.

12 Goodwill, K.E., Sabatier, C., and Stevens, R.C. (1998). Crystal structure of tyrosine hydroxylase with bound cofactor analog and iron at 2.3 Å resolution: self-hydroxylation of Phe300 and the pterin-binding site. *Biochemistry* 37 (39): 13437–13445.

13 Daubner, S.C., Le, T., and Wang, S. (2011). Tyrosine hydroxylase and regulation of dopamine synthesis. *Arch. Biochem. Biophys.* 508 (1): 1–12.

14 Flatmark, T., Almas, B., Knappskog, P.M. et al. (1999). Tyrosine hydroxylase binds tetrahydrobiopterin cofactor with negative cooperativity, as shown by kinetic analyses and surface plasmon resonance detection. *Eur. J. Biochem.* 262 (3): 840–849.

15 Nakashima, A., Hayashi, N., Kaneko, Y.S. et al. (2009). Role of N-terminus of tyrosine hydroxylase in the biosynthesis of catecholamines. *J. Neural. Transm.* 116 (11): 1355–1362.

16 Dunkley, P.R., Bobrovskaya, L., Graham, M.E. et al. (2004). Tyrosine hydroxylase phosphorylation: regulation and consequences. *J. Neurochem.* 91 (5): 1025–1043.

17 Gordon, S.L., Quinsey, N.S., Dunkley, P.R., and Dickson, P.W. (2008). Tyrosine hydroxylase activity is regulated by two distinct dopamine-binding sites. *J. Neurochem.* 106 (4): 1614–1623.

18 Ichimura, T., Isobe, T., Okuyama, T. et al. (1988). Molecular cloning of cDNA coding for brain-specific 14–3–3 protein, a protein kinase-dependent activator of tyrosine and tryptophan hydroxylases. *Proc. Natl. Acad. Sci.* 85 (19): 7084–7088.

19 Hritz, J., Byeon, I.J., Krzysiak, T. et al. (2014). Dissection of binding between a phosphorylated tyrosine hydroxylase peptide and 14–3–3zeta: a complex story elucidated by NMR. *Biophys. J.* 107 (9): 2185–2194.

20 Skjevik, A.A., Mileni, M., Baumann, A. et al. (2014). The N-terminal sequence of tyrosine hydroxylase is a conformationally versatile motif that binds 14–3–3 proteins and membranes. *J. Mol. Biol.* 426 (1): 150–168.

21 Muniz, J.R.C., Cooper, C.D.O., Yue, W.W. et al. (2011). Crystal structure of human tyrosine hydroxylase catalytic domain. *RSC PDB* 2XSN.

22 Zhang, S., Huang, T., Ilangovan, U. et al. (2014). The solution structure of the regulatory domain of tyrosine hydroxylase. *J. Mol. Biol.* 426 (7): 1483–1497.

23 Neira, J.L., Hornos, F., Bacarizo, J. et al. (2016). The monomeric species of the regulatory domain of tyrosine hydroxylase has a low conformational stability. *Biochemistry* 55 (24): 3418–3431.
24 Salvatore, M.F., Calipari, E.S., and Jones, S.R. (2016). Regulation of tyrosine hydroxylase expression and phosphorylation in dopamine transporter-deficient mice. *ACS Chem. Neurosci.* 7 (7): 941–951.
25 Hyland, K. (2007). Inherited disorders affecting dopamine and serotonin: critical neurotransmitters derived from aromatic amino acids. *J. Nutr.* 137 (6 Suppl 1): 1568S–1572S; discussion 1573S–1575S.
26 Gentile, M.T., Nawa, Y., Lunardi, G. et al. (2012). Tryptophan hydroxylase 2 (TPH2) in a neuronal cell line: modulation by cell differentiation and NRSF/rest activity. *J. Neurochem.* 123 (6): 963–970.
27 Walther, D.J., Peter, J.U., Bashammakh, S. et al. (2003). Synthesis of serotonin by a second tryptophan hydroxylase isoform. *Science* 299 (5603): 76.
28 McKinney, J.A., Turel, B., Winge, I. et al. (2009). Functional properties of missense variants of human tryptophan hydroxylase 2. *Hum. Mutat.* 30 (5): 787–794.
29 Zhang, X., Gainetdinov, R.R., Beaulieu, J.M. et al. (2005). Loss-of-function mutation in tryptophan hydroxylase-2 identified in unipolar major depression. *Neuron* 45 (1): 11–16.
30 McKinney, J., Johansson, S., Halmoy, A. et al. (2008). A loss-of-function mutation in tryptophan hydroxylase 2 segregating with attention-deficit/ hyperactivity disorder. *Mol. Psychiatry* 13 (4): 365–367.
31 Cichon, S., Winge, I., Mattheisen, M. et al. (2008). Brain-specific tryptophan hydroxylase 2 (TPH2): a functional Pro206Ser substitution and variation in the 5'-region are associated with bipolar affective disorder. *Hum. Mol. Genet.* 17 (1): 87–97.
32 Chen, G.-L. and Miller, G.M. (2013). Tryptophan hydroxylase-2: an emerging therapeutic target for stress disorders. *Biochem. Pharmacol.* 85 (9): 1227–1233.
33 Waider, J., Araragi, N., Gutknecht, L., and Lesch, K.-P. (2011). Tryptophan hydroxylase-2 (TPH2) in disorders of cognitive control and emotion regulation: a perspective. *Psychoneuroendocrinology* 36 (3): 393–405.
34 Wang, L., Erlandsen, H., Haavik, J. et al. (2002). Three-dimensional structure of human tryptophan hydroxylase and its implications for the biosynthesis of the neurotransmitters serotonin and melatonin. *Biochemistry* 41 (42): 12569–12574.
35 Cianchetta, G., Stouch, T., Yu, W. et al. (2010). Mechanism of inhibition of novel tryptophan hydroxylase inhibitors revealed by co-crystal structures and kinetic analysis. *Curr. Chem. Genomics* 4: 19–26.
36 Windahl, M.S., Petersen, C.R., Christensen, H.E., and Harris, P. (2008). Crystal structure of tryptophan hydroxylase with bound amino acid substrate. *Biochemistry* 47 (46): 12087–12094.
37 Goldberg, D.R., De Lombaert, S., Aiello, R. et al. (2016). Discovery of acyl guanidine tryptophan hydroxylase-1 inhibitors. *Bioorg. Med. Chem. Lett.* 26 (12): 2855–2860.
38 Goldberg, D.R., De Lombaert, S., Aiello, R. et al. (2017). Optimization of spirocyclic proline tryptophan hydroxylase-1 inhibitors. *Bioorg. Med. Chem. Lett.* 27 (3): 413–419.

39 Kopec, J., Oberholzer, A., Fitzpatrick, F. et al. (2014). Crystal structure of human tryptophan hydroxylase 2 (TPH2), catalytic domain. *RCSB PDB* 4V06.

40 Siow, Y.L. and Dakshinamurti, K. (1990). Neuronal dopa decarboxylase. *Ann. N.Y. Acad. Sci.* 585: 173–188.

41 Bertoldi, M. (2014). Mammalian Dopa decarboxylase: structure, catalytic activity and inhibition. *Arch. Biochem. Biophys.* 546: 1–7.

42 Vassilacopoulou, D., Sideris, D.C., Vassiliou, A.G., and Fragoulis, E.G. (2004). Identification and characterization of a novel form of the human l-dopa decarboxylase mRNA. *Neurochem. Res.* 29 (10): 1817–1823.

43 Chang, Y.T., Sharma, R., Marsh, J.L. et al. (2004). Levodopa-responsive aromatic l-amino acid decarboxylase deficiency. *Ann. Neurol.* 55 (3): 435–438.

44 Pons, R., Ford, B., Chiriboga, C.A. et al. (2004). Aromatic l-amino acid decarboxylase deficiency: clinical features, treatment, and prognosis. *Neurology* 62 (7): 1058–1065.

45 Tay, S.K.H., Wang, F.S.J., and Lin, J.B.Y. (2013). Aromatic-amino acid decarboxylase deficiency: perspectives on diagnosis and management. *Pediatr. Health. Med. Ther.* 4: 89–99.

46 Burkhard, P., Dominici, P., Borri-Voltattorni, C. et al. (2001). Structural insight into Parkinson's disease treatment from drug-inhibited DOPA decarboxylase. *Nat. Struct. Biol.* 8 (11): 963–967.

47 Giardina, G., Montioli, R., Gianni, S. et al. (2011). Open conformation of human DOPA decarboxylase reveals the mechanism of PLP addition to Group II decarboxylases. *Proc. Natl. Acad. Sci. USA* 108 (51): 20514–20519.

48 Lamouroux, A., Vigny, A., Faucon Biguet, N. et al. (1987). The primary structure of human dopamine-beta-hydroxylase: insights into the relationship between the soluble and the membrane-bound forms of the enzyme. *EMBO J.* 6 (13): 3931–3937.

49 Li, B., Tsing, S., Kosaka, A.H. et al. (1996). Expression of human dopamine beta-hydroxylase in *Drosophila* Schneider 2 cells. *Biochem. J* 313 (Pt 1): 57–64.

50 Chen, Y., Zhang, K., Wen, G. et al. (2011). Human dopamine beta-hydroxylase promoter variant alters transcription in chromaffin cells, enzyme secretion, and blood pressure. *Am. J. Hypertens.* 24 (1): 24–32.

51 Kobayashi, K., Morita, S., Mizuguchi, T. et al. (1994). Functional and high level expression of human dopamine beta-hydroxylase in transgenic mice. *J. Biol. Chem.* 269 (47): 29725–29731.

52 Vendelboe, T.V., Harris, P., Zhao, Y. et al. (2016). The crystal structure of human dopamine β-hydroxylase at 2.9 Å resolution. *Sci. Adv.* 2 (4): e1500980/1500981–e1500980/1500910.

53 Kim, C.H., Zabetian, C.P., Cubells, J.F. et al. (2002). Mutations in the dopamine beta-hydroxylase gene are associated with human norepinephrine deficiency. *Am. J. Med. Genet.* 108 (2): 140–147.

54 Williams, H.J., Bray, N., Murphy, K.C. et al. (1999). No evidence for allelic association between schizophrenia and a functional variant of the human dopamine beta-hydroxylase gene (DBH). *Am. J. Med. Genet.* 88 (5): 557–559.

55 Cargill, M., Altshuler, D., Ireland, J. et al. (1999). Characterization of single-nucleotide polymorphisms in coding regions of human genes. *Nat. Genet.* 22 (3): 231–238.

56 Osborne, R.L. and Klinman, J.P. (2011). Insights into the proposed copper–oxygen intermediates that regulate the mechanism of reactions catalyzed by dopamine beta-monooxygenase, peptidylglycine alpha-hydroxylating monooxygenase, and tyramine beta-monooxygenase. In: *Copper-Oxygen Chemistry* (ed. K.D. Karlin and S. Itoh), 1–22. Hoboken, NJ: Wiley.
57 Solomon, E.I., Heppner, D.E., Johnston, E.M. et al. (2014). Copper active sites in biology. *Chem. Rev.* 114 (7): 3659–3853.
58 Son, S.-Y., Mat, J., Kondou, Y. et al. (2008). Structure of human monoamine oxidase A at 2.2-Å resolution: The control of opening the entry for substrates/inhibitors. *Proc. Natl. Acad. Sci. U.S.A.* 105 (15): 5739–5744.
59 Li, M., Hubalek, F., Newton-Vinson, P., and Edmondson, D.E. (2002). High-level expression of human liver monoamine oxidase A in Pichia pastoris: comparison with the enzyme expressed in Saccharomyces cerevisiae. *Protein Expression Purif.* 24 (1): 152–162.
60 Newton-Vinson, P., Hubalek, F., and Edmondson, D.E. (2000). High-level expression of human liver monoamine oxidase B in Pichia pastoris. *Protein Expression Purif.* 20 (2): 334–345.
61 Wu, H.F., Chen, K., and Shih, J.C. (1993). Site-directed mutagenesis of monoamine oxidase A and B: role of cysteines. *Mol. Pharmacol.* 43 (6): 888–893.
62 Piton, A., Poquet, H., Redin, C. et al. (2014). 20 ans apres: a second mutation in MAOA identified by targeted high-throughput sequencing in a family with altered behavior and cognition. *Eur. J. Hum. Genet.* 22 (6): 776–783.
63 Youdim, M.B. and Riederer, P. (1993). The relevance of glial monoamine oxidase-B and polyamines to the action of selegiline in Parkinson's disease. *Mov. Disord.* 8 (Suppl 1): S8–S13.
64 Youdim, M.B., Edmondson, D., and Tipton, K.F. (2006). The therapeutic potential of monoamine oxidase inhibitors. *Nat. Rev. Neurosci.* 7 (4): 295–309.
65 Saura, J., Nadal, E., van den Berg, B. et al. (1996). Localization of monoamine oxidases in human peripheral tissues. *Life Sci.* 59 (16): 1341–1349.
66 Caspi, A., McClay, J., Moffitt, T.E. et al. (2002). Role of genotype in the cycle of violence in maltreated children. *Science* 297 (5582): 851–854.
67 Binda, C., Newton-Vinson, P., Hubalek, F. et al. (2002). Structure of human monoamine oxidase B, a drug target for the treatment of neurological disorders. *Nat. Struct. Biol.* 9 (1): 22–26.
68 Li, M., Binda, C., Mattevi, A., and Edmondson, D.E. (2006). Functional role of the "aromatic cage" in human monoamine oxidase B: structures and catalytic properties of Tyr435 mutant proteins. *Biochemistry* 45 (15): 4775–4784.
69 Binda, C., Li, M., Hubalek, F. et al. (2003). Insights into the mode of inhibition of human mitochondrial monoamine oxidase B from high-resolution crystal structures. *Proc. Natl. Acad. Sci. U.S.A.* 100 (17): 9750–9755.
70 Binda, C., Hubalek, F., Li, M. et al. (2004). Crystal structures of monoamine oxidase B in complex with four inhibitors of the N-propargylaminoindan class. *J. Med. Chem.* 47 (7): 1767–1774.
71 Binda, C., Hubalek, F., Li, M. et al. (2005). Binding of rasagiline-related inhibitors to human monoamine oxidases: a kinetic and crystallographic analysis. *J. Med. Chem.* 48 (26): 8148–8154.

72 Hubalek, F., Binda, C., Khalil, A. et al. (2005). Demonstration of isoleucine 199 as a structural determinant for the selective inhibition of human monoamine oxidase B by specific reversible inhibitors. *J. Biol. Chem.* 280 (16): 15761–15766.

73 Milczek, E.M., Binda, C., Rovida, S. et al. (2011). The gating residues Ile199 and Tyr326 in human monoamine oxidase B function in substrate and inhibitor recognition. *FEBS J.* 278 (24): 4860–4869.

74 Binda, C., Aldeco, M., Mattevi, A., and Edmondson, D.E. (2011). Interactions of monoamine oxidases with the antiepileptic drug zonisamide: specificity of inhibition and structure of the human monoamine oxidase B complex. *J. Med. Chem.* 54 (3): 909–912.

75 Reis, J., Manzella, N., Cagide, F. et al. (2018). Tight-binding inhibition of human monoamine oxidase B by chromone analogs: a kinetic, crystallographic, and biological analysis. *J. Med. Chem.* 61 (9): 4203–4212.

76 Binda, C., Wang, J., Pisani, L. et al. (2007). Structures of human monoamine oxidase B complexes with selective noncovalent inhibitors: safinamide and coumarin analogs. *J. Med. Chem.* 50 (23): 5848–5852.

77 Binda, C., Aldeco, M., Geldenhuys, W.J. et al. (2011). Molecular insights into human monoamine oxidase B inhibition by the glitazone anti-diabetes drugs. *ACS Med. Chem. Lett.* 3 (1): 39–42.

78 Binda, C., Wang, J., Li, M. et al. (2008). Structural and mechanistic studies of arylalkylhydrazine inhibition of human monoamine oxidases A and B. *Biochemistry* 47 (20): 5616–5625.

79 Milczek, E.M., Bonivento, D., Binda, C. et al. (2008). Structural and mechanistic studies of mofegiline inhibition of recombinant human monoamine oxidase B. *J. Med. Chem.* 51 (24): 8019–8026.

80 De Deurwaerdere, P., Binda, C., Corne, R. et al. (2017). Comparative analysis of the neurochemical profile and MAO inhibition properties of *N*-(furan-2-ylmethyl)-*N*-methylprop-2-yn-1-amine. *ACS Chem. Neurosci.* 8 (5): 1026–1035.

81 Esteban, G., Allan, J., Samadi, A. et al. (2014). Kinetic and structural analysis of the irreversible inhibition of human monoamine oxidases by ASS234, a multi-target compound designed for use in Alzheimer's disease. *Biochim. Biophys. Acta* 1844 (6): 1104–1110.

82 Bonivento, D., Milczek, E.M., McDonald, G.R. et al. (2010). Potentiation of ligand binding through cooperative effects in monoamine oxidase B. *J. Biol. Chem.* 285 (47): 36849–36856.

83 De Colibus, L., Li, M., Binda, C. et al. (2005). Three-dimensional structure of human monoamine oxidase A (MAO A): relation to the structures of rat MAO A and human MAO B. *Proc. Natl. Acad. Sci. U.S.A.* 102 (36): 12684–12689.

84 Ma, J., Yoshimura, M., Yamashita, E. et al. (2004). Structure of rat monoamine oxidase A and its specific recognitions for substrates and inhibitors. *J. Mol. Biol.* 338 (1): 103–114.

85 Chen, J., Song, J., Yuan, P. et al. (2011). Orientation and cellular distribution of membrane-bound catechol-*O*-methyltransferase in cortical neurons: implications for drug development. *J. Biol. Chem.* 286 (40): 34752–34760.

86 Lee, S.G., Joo, Y., Kim, B. et al. (2005). Association of Ala72Ser polymorphism with COMT enzyme activity and the risk of schizophrenia in Koreans. *Hum. Genet.* 116 (4): 319–328.

87 Lachman, H.M., Papolos, D.F., Saito, T. et al. (1996). Human catechol-*O*-methyltransferase pharmacogenetics: description of a functional polymorphism and its potential application to neuropsychiatric disorders. *Pharmacogenetics* 6 (3): 243–250.
88 Rutherford, K., Alphandery, E., McMillan, A. et al. (2008). The V108M mutation decreases the structural stability of catechol *O*-methyltransferase. *Biochim. Biophys. Acta* 1784 (7–8): 1098–1105.
89 Mannisto, P.T., Ulmanen, I., Lundstrom, K. et al. (1992). Characteristics of catechol *O*-methyl-transferase (COMT) and properties of selective COMT inhibitors. *Prog. Drug Res.* 39: 291–350.
90 Mannisto, P.T. and Kaakkola, S. (1999). Catechol-*O*-methyltransferase (COMT): biochemistry, molecular biology, pharmacology, and clinical efficacy of the new selective COMT inhibitors. *Pharmacol. Rev.* 51 (4): 593–628.
91 McCarthy, J.J. (2001). COMT and schizophrenia: biology meets genetics. *Trends Biotechnol.* 19 (8): 283–284.
92 Vidgren, J., Svensson, L.A., and Liljas, A. (1994). Crystal structure of catechol *O*-methyltransferase. *Nature* 368 (6469): 354–358.
93 Lerner, C., Ruf, A., Gramlich, V. et al. (2001). X-ray crystal structure of a bisubstrate inhibitor bound to the enzyme catechol-*O*-methyltransferase: a dramatic effect of inhibitor preorganization on binding affinity. We thank F. Hoffmann-La Roche for generous support of this work. We are grateful to P. Malherbe for the cloning of COMT, P. Caspers for the expression of COMT, A. Cesura for enzyme purification, B. Wipf for fermentation, and H. W. Lahm for sequencing. *Angew. Chem. Int. Ed.* 40 (21): 4040–4042.
94 Masjost, B., Ballmer, P., Borroni, E. et al. (2000). Structure-based design, synthesis, and in vitro evaluation of bisubstrate inhibitors for catechol *O*-methyltransferase (COMT). *Chem. Eur. J.* 6 (6): 971–982.
95 Rodrigues, M.L., Archer, M., Bonifacio, M.J. et al. (2001). Crystallization and preliminary crystallographic characterization of catechol-O-methyltransferase in complex with its cosubstrate and an inhibitor. *Acta Crystallogr., Sect. D: Biol. Crystallogr.* D57 (6): 906–908.
96 Rodrigues, M.L., Bonifacio, M.J., Soares-da-Silva, P. et al. (2005). Crystallization and preliminary x-ray diffraction studies of a catechol-O-methyltransferase/inhibitor complex. *Acta Crystallogr., Sect. F: Struct. Biol. Cryst. Commun.* 61 (1): 118–120.
97 Bonifacio, M.J., Archer, M., Rodrigues, M.L. et al. (2002). Kinetics and crystal structure of catechol-*O*-methyltransferase complex with co-substrate and a novel inhibitor with potential therapeutic application. *Mol. Pharmacol.* 62 (4): 795–805.
98 Palma, P.N., Rodrigues, M.L., Archer, M. et al. (2006). Comparative study of ortho- and meta-nitrated inhibitors of catechol-*O*-methyltransferase: interactions with the active site and regioselectivity of *O*-methylation. *Mol. Pharmacol.* 70 (1): 143–153.
99 Rutherford, K., Le Trong, I., Stenkamp, R.E., and Parson, W.W. (2008). Crystal structures of human 108V and 108M catechol *O*-methyltransferase. *J. Mol. Biol.* 380 (1): 120–130.

100 Ellermann, M., Paulini, R., Jakob-Roetne, R. et al. (2011). Molecular recognition at the active site of catechol-O-methyltransferase (COMT): adenine replacements in bisubstrate inhibitors. *Chemistry* 17 (23): 6369–6381.

101 Ellermann, M., Lerner, C., Burgy, G. et al. (2012). Catechol-*O*-methyltransferase in complex with substituted 3'-deoxyribose bisubstrate inhibitors. *Acta Crystallogr., Sect. D: Biol. Crystallogr.* 68 (3): 253–260.

102 Ehler, A., Benz, J., Schlatter, D., and Rudolph, M.G. (2014). Mapping the conformational space accessible to catechol-*O*-methyltransferase. *Acta Crystallogr., Sect. D: Biol. Crystallogr.* 70 (Pt 8): 2163–2174.

103 Harrison, S.T., Poslusney, M.S., Mulhearn, J.J. et al. (2015). Synthesis and evaluation of heterocyclic catechol mimics as inhibitors of catechol-*O*-methyltransferase (COMT). *ACS Med. Chem. Lett.* 6 (3): 318–323.

104 Thakker, D.R., Boehlert, C., Kirk, K.L. et al. (1986). Regioselectivity of catechol *O*-methyltransferase. The effect of pH on the site of *O*-methylation of fluorinated norepinephrines. *J. Biol. Chem.* 261 (1): 178–184.

105 Law, B.J.C., Bennett, M.R., Thompson, M.L. et al. (2016). Effects of active-site modification and quaternary structure on the regioselectivity of catechol-*O*-methyltransferase. *Angew. Chem. Int. Ed.* 55 (8): 2683–2687.

106 Ji, Y., Salavaggione, O.E., Wang, L. et al. (2005). Human phenylethanolamine *N*-methyltransferase pharmacogenomics: gene re-sequencing and functional genomics. *J. Neurochem.* 95 (6): 1766–1776.

107 Gee, C.L., Tyndall, J.D., Grunewald, G.L. et al. (2005). Mode of binding of methyl acceptor substrates to the adrenaline-synthesizing enzyme phenylethanolamine *N*-methyltransferase: implications for catalysis. *Biochemistry* 44 (51): 16875–16885.

108 Burke, W.J., Chung, H.D., Nakra, B.R. et al. (1987). Phenylethanolamine *N*-methyltransferase activity is decreased in Alzheimer's disease brains. *Ann. Neurol.* 22 (2): 278–280.

109 Mann, M.B., Wu, S., Rostamkhani, M. et al. (2001). Phenylethanolamine *N*-methyltransferase (PNMT) gene and early-onset Alzheimer disease. *Am. J. Med. Genet.* 105 (4): 312–316.

110 Burke, W.J., Chung, H.D., Marshall, G.L. et al. (1990). Evidence for decreased transport of PNMT protein in advanced Alzheimer's disease. *J. Am. Geriatr. Soc.* 38 (12): 1275–1282.

111 Gai, W.P., Geffen, L.B., Denoroy, L., and Blessing, W.W. (1993). Loss of C_1 and C_3 epinephrine-synthesizing neurons in the medulla oblongata in Parkinson's disease. *Ann. Neurol.* 33 (4): 357–367.

112 Gearhart, D.A., Neafsey, E.J., and Collins, M.A. (2002). Phenylethanolamine *N*-methyltransferase has beta-carboline 2*N*-methyltransferase activity: hypothetical relevance to Parkinson's disease. *Neurochem. Int.* 40 (7): 611–620.

113 Nagatsu, T., Kato, T., Numata, Y. et al. (1977). Phenylethanolamine *N*-methyltransferase and other enzymes of catecholamine metabolism in human brain. *Clin. Chim. Acta* 75 (2): 221–232.

114 Nagatsu, T. and Nagatsu, I. (2016). Tyrosine hydroxylase (TH), its cofactor tetrahydrobiopterin (BH4), other catecholamine-related enzymes, and their human genes in relation to the drug and gene therapies of Parkinson's disease

(PD): historical overview and future prospects. *J. Neural. Transm.* 123 (11): 1255–1278.

115 Martin, J.L., Begun, J., McLeish, M.J. et al. (2001). Getting the adrenaline going. Crystal structure of the adrenaline-synthesizing enzyme PNMT. *Structure* 9 (10): 977–985.

116 Begun, J., McLeish, M.J., Caine, J.M. et al. (2002). Crystallization of PNMT, the adrenaline-synthesizing enzyme, is critically dependent on a high protein concentration. *Acta Crystallogr., Sect. D: Biol. Crystallogr.* D58 (2): 314–315.

117 McMillan, F.M., Archbold, J., McLeish, M.J. et al. (2004). Molecular recognition of sub-micromolar inhibitors by the epinephrine-synthesizing enzyme phenylethanolamine *N*-methyltransferase. *J. Med. Chem.* 47 (1): 37–44.

118 Gee, C.L., Drinkwater, N., Tyndall, J.D.A. et al. (2007). Enzyme adaptation to inhibitor binding: a cryptic binding site in phenylethanolamine *N*-methyltransferase. *J. Med. Chem.* 50 (20): 4845–4853.

119 Wu, Q., Gee, C.L., Lin, F. et al. (2005). Structural, mutagenic, and kinetic analysis of the binding of substrates and inhibitors of human phenylethanolamine *N*-methyltransferase. *J. Med. Chem.* 48 (23): 7243–7252.

120 Nair, P.C., Malde, A.K., and Mark, A.E. (2011). Using theory to reconcile experiment: the structural and thermodynamic basis of ligand recognition by phenylethanolamine *N*-methyltransferase (PNMT). *J. Chem. Theory Comput.* 7 (5): 1458–1468.

121 Grunewald, G.L., Seim, M.R., Regier, R.C. et al. (2006). Comparison of the binding of 3-fluoromethyl-7-sulfonyl-1,2,3,4-tetrahydroisoquinolines with their isosteric sulfonamides to the active site of phenylethanolamine *N*-methyltransferase. *J. Med. Chem.* 49 (18): 5424–5433.

122 Drinkwater, N., Gee, C.L., Puri, M. et al. (2009). Molecular recognition of physiological substrate noradrenaline by the adrenaline-synthesizing enzyme PNMT and factors influencing its methyltransferase activity. *Biochem. J* 422 (3): 463–471.

123 Drinkwater, N., Vu, H., Lovell, K.M. et al. (2010). Fragment-based screening by X-ray crystallography, MS and isothermal titration calorimetry to identify PNMT (phenylethanolamine *N*-methyltransferase) inhibitors. *Biochem. J* 431 (1): 51–61.

124 Nair, P.C., Malde, A.K., Drinkwater, N., and Mark, A.E. (2012). Missing fragments: detecting cooperative binding in fragment-based drug design. *ACS Med. Chem. Lett.* 3 (4): 322–326.

125 Stratton, C.F., Poulin, M.B., Du, Q., and Schramm, V.L. (2017). Kinetic isotope effects and transition state structure for human phenylethanolamine n-methyltransferase. *ACS Chem. Biol.* 12 (2): 342–346.

126 Bernard, M., Donohue, S.J., and Klein, D.C. (1995). Human hydroxyindole-*O*-methyltransferase in pineal gland, retina and Y79 retinoblastoma cells. *Brain Res.* 696 (1–2): 37–48.

127 Pagan, C., Botros, H.G., Poirier, K. et al. (2011). Mutation screening of ASMT, the last enzyme of the melatonin pathway, in a large sample of patients with intellectual disability. *BMC Med. Genet.* 12: 17.

128 Etain, B., Dumaine, A., Bellivier, F. et al. (2012). Genetic and functional abnormalities of the melatonin biosynthesis pathway in patients with bipolar disorder. *Hum. Mol. Genet.* 21 (18): 4030–4037.

129 Melke, J., Goubran Botros, H., Chaste, P. et al. (2008). Abnormal melatonin synthesis in autism spectrum disorders. *Mol. Psychiatry* 13 (1): 90–98.

130 Wang, L., Li, J., Ruan, Y. et al. (2013). Sequencing ASMT identifies rare mutations in Chinese Han patients with autism. *PLoS One* 8 (1): e53727.

131 Chaste, P., Clement, N., Botros, H.G. et al. (2011). Genetic variations of the melatonin pathway in patients with attention-deficit and hyperactivity disorders. *J. Pineal. Res.* 51 (4): 394–399.

132 Botros, H.G., Legrand, P., Pagan, C. et al. (2013). Crystal structure and functional mapping of human ASMT, the last enzyme of the melatonin synthesis pathway. *J. Pineal. Res.* 54 (1): 46–57.

133 Coon, S.L., Weller, J.L., Korf, H.W. et al. (2001). cAmp regulation of arylalkylamine *N*-acetyltransferase (AANAT, EC 2.3.1.87): a new cell line (1E7) provides evidence of intracellular AANAT activation. *J. Biol. Chem.* 276 (26): 24097–24107.

134 Hohjoh, H., Takasu, M., Shishikura, K. et al. (2003). Significant association of the arylalkylamine *N*-acetyltransferase (AA-NAT) gene with delayed sleep phase syndrome. *Neurogenetics* 4 (3): 151–153.

135 Ganguly, S., Gastel, J.A., Weller, J.L. et al. (2001). Role of a pineal cAMP-operated arylalkylamine *N*-acetyltransferase/14–3–3-binding switch in melatonin synthesis. *Proc. Natl. Acad. Sci. U.S.A.* 98 (14): 8083–8088.

136 Dobransky, T., Davis, W.L., Xiao, G.H., and Rylett, R.J. (2000). Expression, purification and characterization of recombinant human choline acetyltransferase: phosphorylation of the enzyme regulates catalytic activity. *Biochem. J* 349 (Pt 1): 141–151.

137 Dobransky, T., Brewer, D., Lajoie, G., and Rylett, R.J. (2003). Phosphorylation of 69-kDa choline acetyltransferase at threonine 456 in response to amyloid-beta peptide 1–42. *J. Biol. Chem.* 278 (8): 5883–5893.

138 Dobransky, T., Davis, W.L., and Rylett, R.J. (2001). Functional characterization of phosphorylation of 69-kDa human choline acetyltransferase at serine 440 by protein kinase C. *J. Biol. Chem.* 276 (25): 22244–22250.

139 Lorenzi, M.V., Trinidad, A.C., Zhang, R., and Strauss, W.L. (1992). Two mRNAs are transcribed from the human gene for choline acetyltransferase. *DNA Cell Biol.* 11 (8): 593–603.

140 Oda, Y., Nakanishi, I., and Deguchi, T. (1992). A complementary DNA for human choline acetyltransferase induces two forms of enzyme with different molecular weights in cultured cells. *Brain Res. Mol. Brain Res.* 16 (3–4): 287–294.

141 Ohno, K., Tsujino, A., Brengman, J.M. et al. (2001). Choline acetyltransferase mutations cause myasthenic syndrome associated with episodic apnea in humans. *Proc. Natl. Acad. Sci. U.S.A.* 98 (4): 2017–2022.

142 Oda, Y. (1999). Choline acetyltransferase: the structure, distribution and pathologic changes in the central nervous system. *Pathol. Int.* 49 (11): 921–937.

143 Carbini, L.A. and Hersh, L.B. (1993). Functional analysis of conserved histidines in choline acetyltransferase by site-directed mutagenesis. *J. Neurochem.* 61 (1): 247–253.

144 Wu, D. and Hersh, L.B. (1995). Identification of an active site arginine in rat choline acetyltransferase by alanine scanning mutagenesis. *J. Biol. Chem.* 270 (49): 29111–29116.

145 Govindasamy, L., Pedersen, B., Lian, W. et al. (2004). Structural insights and functional implications of choline acetyltransferase. *J. Struct. Biol.* 148 (2): 226–235.
146 Cai, Y., Cronin, C.N., Engel, A.G. et al. (2004). Choline acetyltransferase structure reveals distribution of mutations that cause motor disorders. *EMBO J.* 23 (10): 2047–2058.
147 Kim, A.R., Rylett, R.J., and Shilton, B.H. (2006). Substrate binding and catalytic mechanism of human choline acetyltransferase. *Biochemistry* 45 (49): 14621–14631.
148 Yang, L., He, H.Y., and Zhang, X.J. (2002). Increased expression of intranuclear AChE involved in apoptosis of SK-N-SH cells. *Neurosci. Res.* 42 (4): 261–268.
149 Massoulie, J. (2002). The origin of the molecular diversity and functional anchoring of cholinesterases. *NeuroSignals* 11 (3): 130–143.
150 Dudai, Y., Herzberg, M., and Silman, I. (1973). Molecular structures of acetylcholinesterase from electric organ tissue of the electric eel. *Proc. Natl. Acad. Sci. U.S.A.* 70 (9): 2473–2476.
151 Deprez, P. and Inestrosa, N.C. (2000). Molecular modeling of the collagen-like tail of asymmetric acetylcholinesterase. *Protein Eng.Des. Sel.* 13 (1): 27–34.
152 Gennari, K., Brunner, J., and Brodbeck, U. (1987). Tetrameric detergent-soluble acetylcholinesterase from human caudate nucleus: subunit composition and number of active sites. *J. Neurochem.* 49 (1): 12–18.
153 Harel, M., Schalk, I., Ehret-Sabatier, L. et al. (1993). Quaternary ligand binding to aromatic residues in the active-site gorge of acetylcholinesterase. *Proc. Natl. Acad. Sci. USA* 90 (19): 9031–9035.
154 Harel, M., Kleywegt, G.J., Ravelli, R.B. et al. (1995). Crystal structure of an acetylcholinesterase-fasciculin complex: interaction of a three-fingered toxin from snake venom with its target. *Structure* 3 (12): 1355–1366.
155 Bourne, Y., Taylor, P., and Marchot, P. (1995). Acetylcholinesterase inhibition by fasciculin: crystal structure of the complex. *Cell* 83 (3): 503–512.
156 Kryger, G., Harel, M., Giles, K. et al. (2000). Structures of recombinant native and E202Q mutant human acetylcholinesterase complexed with the snake-venom toxin fasciculin-II. *Acta Crystallogr., Sect. D: Biol. Crystallogr.* 56 (Pt 11): 1385–1394.
157 Raves, M.L., Harel, M., Pang, Y.P. et al. (1997). Structure of acetylcholinesterase complexed with the nootropic alkaloid, (−)-huperzine A. *Nat. Struct. Biol.* 4 (1): 57–63.
158 Bourne, Y., Taylor, P., Radic, Z., and Marchot, P. (2003). Structural insights into ligand interactions at the acetylcholinesterase peripheral anionic site. *EMBO J.* 22 (1): 1–12.
159 Katz, F.S., Pecic, S., Tran, T.H. et al. (2015). Discovery of new classes of compounds that reactivate acetylcholinesterase inhibited by organophosphates. *ChemBioChem* 16 (15): 2205–2215.
160 Bourne, Y., Kolb, H.C., Radic, Z. et al. (2004). Freeze-frame inhibitor captures acetylcholinesterase in a unique conformation. *Proc. Natl. Acad. Sci. U.S.A.* 101 (6): 1449–1454.
161 Dvir, H., Harel, M., Bon, S. et al. (2004). The synaptic acetylcholinesterase tetramer assembles around a polyproline II helix. *EMBO J.* 23 (22): 4394–4405.

162 le Du, M.H., Housset, D., Marchot, P. et al. (1996). Structure of fasciculin 2 from green mamba snake venom: evidence for unusual loop flexibility. *Acta Crystallogr., Sect. D: Biol. Crystallogr.* 52 (Pt 1): 87–92.

163 Dvir, H., Silman, I., Harel, M. et al. (2010). Acetylcholinesterase: from 3D structure to function. *Chem. Biol. Interact.* 187 (1-3): 10–22.

164 Ekstrom, F., Akfur, C., Tunemalm, A.K., and Lundberg, S. (2006). Structural changes of phenylalanine 338 and histidine 447 revealed by the crystal structures of tabun-inhibited murine acetylcholinesterase. *Biochemistry* 45 (1): 74–81.

165 Ekstrom, F.J., Astot, C., and Pang, Y.P. (2007). Novel nerve-agent antidote design based on crystallographic and mass spectrometric analyses of tabun-conjugated acetylcholinesterase in complex with antidotes. *Clin. Pharmacol. Ther.* 82 (3): 282–293.

166 Carletti, E., Colletier, J.P., Dupeux, F. et al. (2010). Structural evidence that human acetylcholinesterase inhibited by tabun ages through *O*-dealkylation. *J. Med. Chem.* 53 (10): 4002–4008.

167 Carletti, E., Li, H., Li, B. et al. (2008). Aging of cholinesterases phosphylated by tabun proceeds through *O*-dealkylation. *J. Am. Chem. Soc.* 130 (47): 16011–16020.

168 Allgardsson, A., Berg, L., Akfur, C. et al. (2016). Structure of a prereaction complex between the nerve agent sarin, its biological target acetylcholinesterase, and the antidote HI-6. *Proc. Natl. Acad. Sci. U.S.A.* 113 (20): 5514–5519.

169 Bester, S.M., Guelta, M.A., Cheung, J. et al. (2018). Structural insights of stereospecific inhibition of human acetylcholinesterase by VX and subsequent reactivation by HI-6. *Chem. Res. Toxicol.* **31** (12): 1405–1417.

170 Franklin, M.C., Rudolph, M.J., Ginter, C. et al. (2016). Structures of paraoxon-inhibited human acetylcholinesterase reveal perturbations of the acyl loop and the dimer interface. *Proteins* 84 (9): 1246–1256.

171 Knutsson, S., Engdahl, C., Kumari, R. et al. (2018). Noncovalent inhibitors of mosquito acetylcholinesterase 1 with resistance-breaking potency. *J. Med. Chem.* 61: 10545–10557.

172 Santoni, G., de Sousa, J., de la Mora, E. et al. (2018). Structure-based optimization of nonquaternary reactivators of acetylcholinesterase inhibited by organophosphorus nerve agents. *J. Med. Chem.* 61 (17): 7630–7639.

173 Kryger, G., Silman, I., and Sussman, J.L. (1999). Structure of acetylcholinesterase complexed with E2020 (Aricept): implications for the design of new anti-Alzheimer drugs. *Structure* 7 (3): 297–307.

174 Bar-On, P., Millard, C.B., Harel, M. et al. (2002). Kinetic and structural studies on the interaction of cholinesterases with the anti-Alzheimer drug rivastigmine. *Biochemistry* 41 (11): 3555–3564.

175 Bartolucci, C., Perola, E., Pilger, C. et al. (2001). Three-dimensional structure of a complex of galanthamine (Nivalin) with acetylcholinesterase from Torpedo californica: implications for the design of new anti-Alzheimer drugs. *Proteins* 42 (2): 182–191.

176 Bartolucci, C., Perola, E., Cellai, L. et al. (1999). "Back door" opening implied by the crystal structure of a carbamoylated acetylcholinesterase. *Biochemistry* 38 (18): 5714–5719.

177 Dvir, H., Jiang, H.L., Wong, D.M. et al. (2002). X-ray structures of Torpedo californica acetylcholinesterase complexed with (+)-huperzine A and (−)-huperzine B: structural evidence for an active site rearrangement. *Biochemistry* 41 (35): 10810–10818.
178 Millard, C.B., Kryger, G., Ordentlich, A. et al. (1999). Crystal structures of aged phosphonylated acetylcholinesterase: nerve agent reaction products at the atomic level. *Biochemistry* 38 (22): 7032–7039.
179 Caliandro, R., Pesaresi, A., Cariati, L. et al. (2018). Kinetic and structural studies on the interactions of Torpedo californica acetylcholinesterase with two donepezil-like rigid analogues. *J. Enzyme Inhib. Med. Chem.* 33 (1): 794–803.
180 Cheung, J., Rudolph, M.J., Burshteyn, F. et al. (2012). Structures of human acetylcholinesterase in complex with pharmacologically important ligands. *J. Med. Chem.* 55 (22): 10282–10286.
181 Cheung, J., Gary, E.N., Shiomi, K., and Rosenberry, T.L. (2013). Structures of human acetylcholinesterase bound to dihydrotanshinone I and territrem B show peripheral site flexibility. *ACS Med. Chem. Lett.* 4 (11): 1091–1096.
182 Nachon, F., Carletti, E., Ronco, C. et al. (2013). Crystal structures of human cholinesterases in complex with huprine W and tacrine: elements of specificity for anti-Alzheimer's drugs targeting acetyl- and butyryl-cholinesterase. *Biochem. J* 453 (3): 393–399.
183 Porter, T.G., Spink, D.C., Martin, S.B., and Martin, D.L. (1985). Transaminations catalysed by brain glutamate decarboxylase. *Biochem. J* 231 (3): 705–712.
184 Battaglioli, G., Liu, H., and Martin, D.L. (2003). Kinetic differences between the isoforms of glutamate decarboxylase: implications for the regulation of GABA synthesis. *J. Neurochem.* 86 (4): 879–887.
185 Asada, H., Kawamura, Y., Maruyama, K. et al. (1996). Mice lacking the 65 kDa isoform of glutamic acid decarboxylase (GAD65) maintain normal levels of GAD67 and GABA in their brains but are susceptible to seizures. *Biochem. Biophys. Res. Commun.* 229 (3): 891–895.
186 Namchuk, M., Lindsay, L., Turck, C.W. et al. (1997). Phosphorylation of serine residues 3, 6, 10, and 13 distinguishes membrane anchored from soluble glutamic acid decarboxylase 65 and is restricted to glutamic acid decarboxylase 65alpha. *J. Biol. Chem.* 272 (3): 1548–1557.
187 Kanaani, J., el-Husseini Ael, D., Aguilera-Moreno, A. et al. (2002). A combination of three distinct trafficking signals mediates axonal targeting and presynaptic clustering of GAD65. *J. Cell. Biol.* 158 (7): 1229–1238.
188 Feldblum, S., Erlander, M.G., and Tobin, A.J. (1993). Different distributions of GAD65 and GAD67 mRNAs suggest that the two glutamate decarboxylases play distinctive functional roles. *J. Neurosci. Res.* 34 (6): 689–706.
189 Esclapez, M., Tillakaratne, N.J., Kaufman, D.L. et al. (1994). Comparative localization of two forms of glutamic acid decarboxylase and their mRNAs in rat brain supports the concept of functional differences between the forms. *J. Neurosci.* 14 (3 Pt 2): 1834–1855.
190 Martin, D.L., Martin, S.B., Wu, S.J., and Espina, N. (1991). Regulatory properties of brain glutamate decarboxylase (GAD): the apoenzyme of GAD is present principally as the smaller of two molecular forms of GAD in brain. *J. Neurosci.* 11 (9): 2725–2731.

191 Straub, R.E., Lipska, B.K., Egan, M.F. et al. (2007). Allelic variation in GAD1 (GAD67) is associated with schizophrenia and influences cortical function and gene expression. *Mol. Psychiatry* 12 (9): 854–869.

192 Lynex, C.N., Carr, I.M., Leek, J.P. et al. (2004). Homozygosity for a missense mutation in the 67 kDa isoform of glutamate decarboxylase in a family with autosomal recessive spastic cerebral palsy: parallels with Stiff-Person Syndrome and other movement disorders. *BMC Neurol.* 4 (1): 20.

193 Martin, D.L., Liu, H., Martin, S.B., and Wu, S.J. (2000). Structural features and regulatory properties of the brain glutamate decarboxylases. *Neurochem. Int.* 37 (2-3): 111–119.

194 Kanaani, J., Lissin, D., Kash, S.F., and Baekkeskov, S. (1999). The hydrophilic isoform of glutamate decarboxylase, GAD67, is targeted to membranes and nerve terminals independent of dimerization with the hydrophobic membrane-anchored isoform, GAD65. *J. Biol. Chem.* 274 (52): 37200–37209.

195 Fenalti, G., Law, R.H., Buckle, A.M. et al. (2007). GABA production by glutamic acid decarboxylase is regulated by a dynamic catalytic loop. *Nat. Struct. Mol. Biol.* 14 (4): 280–286.

196 Langendorf, C.G., Tuck, K.L., Key, T.L. et al. (2013). Structural characterization of the mechanism through which human glutamic acid decarboxylase auto-activates. *Biosci. Rep.* 33 (1): 137–144.

197 Eelen, G., Dubois, C., Cantelmo, A.R. et al. (2018). Role of glutamine synthetase in angiogenesis beyond glutamine synthesis. *Nature* 561 (7721): 63–69.

198 Haberle, J., Gorg, B., Rutsch, F. et al. (2005). Congenital glutamine deficiency with glutamine synthetase mutations. *N. Engl. J. Med.* 353 (18): 1926–1933.

199 Rumping, L., Tessadori, F., Pouwels, P.J.W. et al. (2019). GLS hyperactivity causes glutamate excess, infantile cataract and profound developmental delay. *Hum. Mol. Genet.* 28 (1): 96–104.

200 Rose, C.F., Verkhratsky, A., and Parpura, V. (2013). Astrocyte glutamine synthetase: pivotal in health and disease. *Biochem. Soc. Trans.* 41 (6): 1518–1524.

201 Liang, S.L. (2018). The glutamine-glutamate cycle regulates synaptic glutamate release in the ventrolateral ventromedial nucleus of the hypothalamus of perinatal female rats. *J. Neuroendocrinol.* 30 (11): e12642.

202 Suarez, I., Bodega, G., and Fernandez, B. (2002). Glutamine synthetase in brain: effect of ammonia. *Neurochem. Int.* 41 (2–3): 123–142.

203 Chepkova, A.N., Sergeeva, O.A., Gorg, B. et al. (2017). Impaired novelty acquisition and synaptic plasticity in congenital hyperammonemia caused by hepatic glutamine synthetase deficiency. *Sci. Rep.* 7: 40190.

204 Castegna, A. and Menga, A. (2018). Glutamine synthetase: localization dictates outcome. *Genes* 9 (2): 108. https://doi.org/10.3390/genes9020108.

205 Bernstein, H.G., Bannier, J., Meyer-Lotz, G. et al. (2014). Distribution of immunoreactive glutamine synthetase in the adult human and mouse brain. Qualitative and quantitative observations with special emphasis on extra-astroglial protein localization. *J. Chem. Neuroanat.* 61–62: 33–50.

206 Gunnersen, D. and Haley, B. (1992). Detection of glutamine synthetase in the cerebrospinal fluid of Alzheimer diseased patients: a potential diagnostic biochemical marker. *Proc. Natl. Acad. Sci. U.S.A.* 89 (24): 11949–11953.

207 Yamashita, M.M., Almassy, R.J., Janson, C.A. et al. (1989). Refined atomic model of glutamine synthetase at 3.5 A resolution. *J. Biol. Chem.* 264 (30): 17681–17690.

208 Gill, H.S. and Eisenberg, D. (2001). The crystal structure of phosphinothricin in the active site of glutamine synthetase illuminates the mechanism of enzymatic inhibition. *Biochemistry* 40 (7): 1903–1912.

209 Liaw, S.H., Jun, G., and Eisenberg, D. (1994). Interactions of nucleotides with fully unadenylylated glutamine synthetase from *Salmonella typhimurium*. *Biochemistry* 33 (37): 11184–11188.

210 Liaw, S.H., Pan, C., and Eisenbeg, D. (1993). Feedback inhibition of fully unadenylylated glutamine synthetase from *Salmonella typhimurium* by glycine, alanine, and serine. *Proc. Natl. Acad. Sci. U.S.A.* 90 (11): 4996–5000.

211 Gill, H.S., Pfluegl, G.M., and Eisenberg, D. (2002). Multicopy crystallographic refinement of a relaxed glutamine synthetase from Mycobacterium tuberculosis highlights flexible loops in the enzymatic mechanism and its regulation. *Biochemistry* 41 (31): 9863–9872.

212 Krajewski, W.W., Jones, T.A., and Mowbray, S.L. (2005). Structure of Mycobacterium tuberculosis glutamine synthetase in complex with a transition-state mimic provides functional insights. *Proc. Natl. Acad. Sci. U.S.A.* 102 (30): 10499–10504.

213 Murray, D.S., Chinnam, N., Tonthat, N.K. et al. (2013). Structures of the *Bacillus subtilis* glutamine synthetase dodecamer reveal large intersubunit catalytic conformational changes linked to a unique feedback inhibition mechanism. *J. Biol. Chem.* 288 (50): 35801–35811.

214 van Rooyen, J.M., Abratt, V.R., Belrhali, H., and Sewell, T. (2011). Crystal structure of Type III glutamine synthetase: surprising reversal of the inter-ring interface. *Structure* 19 (4): 471–483.

215 Joo, H.K., Park, Y.W., Jang, Y.Y., and Lee, J.Y. (2018). Structural analysis of glutamine synthetase from *Helicobacter pylori*. *Sci. Rep.* 8 (1): 11657.

216 Krajewski, W.W., Collins, R., Holmberg-Schiavone, L. et al. (2008). Crystal structures of mammalian glutamine synthetases illustrate substrate-induced conformational changes and provide opportunities for drug and herbicide design. *J. Mol. Biol.* 375 (1): 217–228.

217 Yoon, C.S., Kim, D.W., Jang, S.H. et al. (2004). Cysteine-321 of human brain GABA transaminase is involved in intersubunit cross-linking. *Mol. Cells* 18 (2): 214–219.

218 Sherif, F., Ahmed, S.S., and Eriksson, L. (1995). Brain aminobutyrate aminotransferase activity in Alzheimer's disease. *Neurodegeneration* 4 (1): 114–115.

219 Sherif, F.M. (1994). GABA-transaminase in brain and blood platelets: basic and clinical aspects. *Prog. Neuro-Psychopharmacol. Biol. Psychiatry* 18 (8): 1219–1233.

220 Morland, C., Froland, A.S., Pettersen, M.N. et al. (2018). Propionate enters GABAergic neurons, inhibits GABA transaminase, causes GABA accumulation and lethargy in a model of propionic acidemia. *Biochem. J* 475 (4): 749–758.
221 Sherif, F.M. and Ahmed, S.S. (1995). Basic aspects of GABA-transaminase in neuropsychiatric disorders. *Clin. Biochem.* 28 (2): 145–154.
222 Sherif, F., Gottfries, C.G., Alafuzoff, I., and Oreland, L. (1992). Brain gamma-aminobutyrate aminotransferase (GABA-T) and monoamine oxidase (MAO) in patients with Alzheimer's disease. *J. Neural. Transm. Park. Dis. Dement Sect.* 4 (3): 227–240.
223 Louro, P., Ramos, L., Robalo, C. et al. (2016). Phenotyping GABA transaminase deficiency: a case description and literature review. *J. Inherit. Metab. Dis.* 39 (5): 743–747.
224 Koenig, M.K., Hodgeman, R., Riviello, J.J. et al. (2017). Phenotype of GABA-transaminase deficiency. *Neurology* 88 (20): 1919–1924.
225 Medina-Kauwe, L.K., Tobin, A.J., De Meirleir, L. et al. (1999). 4-Aminobutyrate aminotransferase (GABA-transaminase) deficiency. *J. Inherit. Metab. Dis.* 22 (4): 414–427.
226 Storici, P., Capitani, G., De Biase, D. et al. (1999). Crystal structure of GABA-aminotransferase, a target for antiepileptic drug therapy. *Biochemistry* 38 (27): 8628–8634.
227 Storici, P., De Biase, D., Bossa, F. et al. (2004). Structures of gamma-aminobutyric acid (GABA) aminotransferase, a pyridoxal 5′-phosphate, and [2Fe-2S] cluster-containing enzyme, complexed with gamma-ethynyl-GABA and with the antiepilepsy drug vigabatrin. *J. Biol. Chem.* 279 (1): 363–373.
228 Liu, W., Peterson, P.E., Carter, R.J. et al. (2004). Crystal structures of unbound and aminooxyacetate-bound *Escherichia coli* gamma-aminobutyrate aminotransferase. *Biochemistry* 43 (34): 10896–10905.
229 Liu, W., Peterson, P.E., Langston, J.A. et al. (2005). Kinetic and crystallographic analysis of active site mutants of Escherichia coli gamma-aminobutyrate aminotransferase. *Biochemistry* 44 (8): 2982–2992.
230 Baugh, L., Phan, I., Begley, D.W. et al. (2015). Increasing the structural coverage of tuberculosis drug targets. *Tuberculosis (Edinb)* 95 (2): 142–148.
231 Lee, H., Doud, E.H., Wu, R. et al. (2015). Mechanism of inactivation of gamma-aminobutyric acid aminotransferase by (1*S*,3*S*)-3-amino-4-difluoromethylene-1-cyclopentanoic acid (CPP-115). *J. Am. Chem. Soc.* 137 (7): 2628–2640.
232 Lee, H., Le, H.V., Wu, R. et al. (2015). Mechanism of Inactivation of GABA aminotransferase by (*E*)- and (*Z*)-(1*S*,3*S*)-3-amino-4-fluoromethylenyl-1-cyclopentanoic acid. *ACS Chem. Biol.* 10 (9): 2087–2098.
233 Juncosa, J.I., Takaya, K., Le, H.V. et al. (2018). Design and mechanism of (*S*)-3-amino-4-(difluoromethylenyl)cyclopent-1-ene-1-carboxylic acid, a highly potent gamma-aminobutyric acid aminotransferase inactivator for the treatment of addiction. *J. Am. Chem. Soc.* 140 (6): 2151–2164.

9

Misfolded Proteins

9.1 Introduction

Proteins are crucial for essentially all the biological processes. When a protein is synthesized, the final polypeptide undergoes a folding process, which gives rise to the tertiary structure. The tertiary structure determines the biological function of the protein; consequently, a correct protein folding is essential for the formation of functional proteins. When a protein folds incorrectly, it tends to aggregate, and the accumulation of such aggregates normally causes important dysfunctions and diseases [1]. In the brain, protein misfolding leads to neurodegenerative diseases [2].

Protein aggregation is a complex process involving several steps leading to ordered fibrils after misfolding and oligomerization [3]. In some cases, aggregation can occur directly from the protein production. The formed aggregates cause neuron toxicity, inducing apoptosis and loss of neurons [4]. In neurodegenerative disorders, misfolding has been suggested for α-synuclein (PD) [5], polyglutamine proteins (HD) [6], and amyloid β (Aβ) (AD) [7].

In AD, aggregations of two types of proteins, Aβ and tau (τ), are produced [8]. Today, there are experimental and clinical evidences supporting the concept that an imbalance between production and clearance of Aβ peptides is a very early, often initiating factor in AD [9, 10]. In the brain, Aβ is found in the extracellular space around neurons, but its exact biological function is unknown. τ is found within neurons and is involved in maintaining the structure of neurons through the stabilization of microtubules. When these proteins become misfolded, they accumulate into clusters. The amyloid plaques are found outside and around neurons in the brain of AD patients; in particular, misfolded Aβ accumulates as plaques in the neuronal endoplasmic reticulum extracellularly. τ clusters are found inside neurons. In PD, the protein called α-synuclein, found in the cytosol of neurons and related to the regulation of neurotransmitters, forms aggregates called Lewy bodies that accumulate in neuronal cell bodies, axons, and synapses of the substantia nigra [11]. In HD, the protein huntingtin accumulates in intranuclear and cytoplasmic bodies [12]. In Lewis body dementia (LBD), Aβ 1-42 has been reported to promote α-synuclein accumulation [13]. All these proteins have a common feature of forming an ordered fibrillar structure called amyloid. The amyloid formation seems to be originated from the exposure of hydrophobic

Chemical Biology of Neurodegeneration: A Molecular Approach, First Edition. Pedro Merino.

surface (normally buried in the core of correctly folded proteins) that increases the London interactions among partially unfolded protein molecules [14].

Similarly to prions, misfolded proteins causing neurodegenerative diseases induce healthy proteins to misfolding, thus forming the typical amyloid plaques [15]. This process called "seeded aggregation" is spread from neuron to neuron, causing a widespread expansion of the disease in the brain regions [16, 17]. However, the precise contribution to the pathology of neurodegenerative disorders, other than neuronal apoptosis [18], is unknown [19]. Although it is still not completely demonstrated that accumulation of misfolded protein aggregates is the underlying cause of those neurodegenerative diseases [20, 21], there are evidences that formation of misfolded protein aggregates causes cellular toxicity and contributes to cellular proteostatic collapse [22]. In this context, molecular chaperones responsible for an efficient folding and degradation or refolding of misfolded proteins [23] are considered potential therapeutic targets against neurodegenerative diseases [24–26]. Molecular chaperones play important roles in neuronal proteostasis [23], and understanding their mechanisms of action is crucial to the development of new treatments [26].

9.1.1 Amyloid β (Aβ)

Amyloid β (Aβ) refers to peptides of c. 37–49 residues formed by the cleavage of a precursor protein called amyloid precursor protein (APP) by β- and γ-secretases [27]. The APP protein is a member of the APP-related proteins, and several isoforms from 695 to 770 residues can be found. In the adult brain, highest expression of APP is found in the frontal lobe of the cortex and in the anterior perisylvian cortex-opercular gyri. Moderate expression is found in the cerebellar cortex, the posterior perisylvian cortex-opercular gyri, and the temporal associated cortex. Weak expression is found in the striate, extrastriate, and motor cortices. APP is generally expressed in cerebrospinal fluid and plasma [28]. Isoform APP695 is the predominant form in neuronal tissue; isoforms APP751 and APP770 are widely expressed in nonneuronal cells. Isoform APP751 is the most abundant form in T-lymphocytes [29]. Isoform APP770 has been chosen as the canonical sequence (Figure 9.1).

APP has been related to iron export [30] and is suggested to be a regulator of neural activity and plasticity [31] and synaptic formation and repair [31]. Recent studies have confirmed that the copper-binding APP has an essential synaptic function [32]. The structure of APP consists of a large extracellular N-terminus, a single transmembrane helical domain, and a short cytoplasmic C-terminus. The APP can present a variety of natural variants at the C-terminus including Asp678Asn, Ala692Gly, Glu693Gly, Ala713Thr, Thr714Ala, Thr714Ile, Val715Met, Ile716Val, Val717Phe, Val717Gly, Val717Ile, Val717Leu [33, 34], and Leu723Phe, most of which related to the mutations involved in AD [35, 36]. APP protein can be N-glycosylated with GlcNAc at Asn542, Asn571 [37]; O-glycosylated with GalNAc at Thr633, Thr651, Thr652, Thr633, Ser66 [38]; with HexNAc at Tyr659 and Tyr681 [39]. APP can also be phosphorylated at different cellular locations. Phosphorylation at serine residues Ser198, Ser206, and Ser442 only takes place at the ectodomain [40].

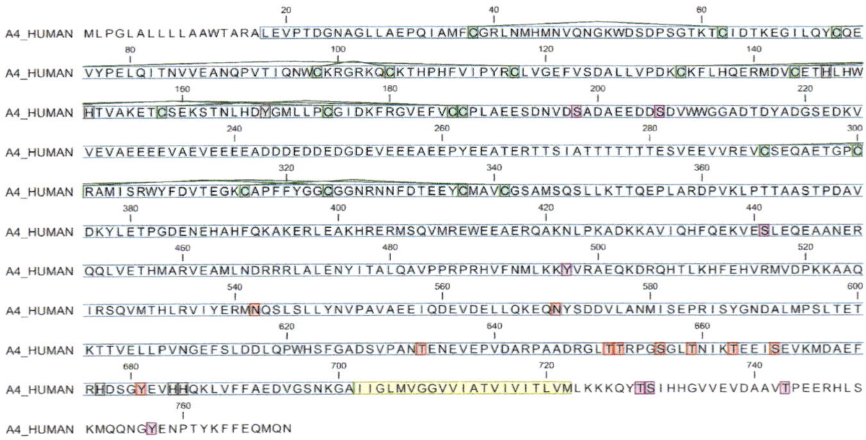

Figure 9.1 Sequence of human amyloid-β precursor protein (APP770). Disulfide bonds (green), metal-binding sites (gray), N-glycosylation sites (red), phosphorylation sites (magenta), extracellular domains (light blue), intracellular domains (white), and transmembrane domains (yellow) are indicated on sequence.

Phosphorylation in the C-terminus at Thr729, Ser730, and Thr743 is neuron specific [41, 42]. Several sites for the cleavage of APP have been identified (Figure 9.2a): caspases cleavage at 197–198, 219–220, 672–673, and 739–740 sites [44]; α-secretase at 687–688 site; β-secretase at 671–672 site; γ-secretase at 711–712, 713–714, and 720–721 sites; θ-secretase at 690–691 site [45].

Some models of C-terminus residues 681–713 of APP have been reported (PDB ID: 2BOM) [46]. Similarly, residues 672–713 (PDB IDs: 1UO7, 1UO8, 1UOA, 1UOI) have been studied by MD simulations [47]. A total of 43 structures corresponding mostly to the cytoplasmic domain have been resolved by nuclear magnetic resonance (NMR) spectroscopy (Table 9.1).

A total of 86 crystal structures have been resolved at different resolutions and correspond to different fragments of APP (Table 9.2). Actually, these fragments corresponding to Aβ peptides are responsible for the formation of regular fibrils found in the brain of AD patients. Description of all these structures is beyond the scope of this chapter, and the reader is directed to the Royal Society of Chemistry (RSC) PDB site to obtain the original references of the corresponding PDB entries.

Aβ cannot be crystallized by common methods, and several studies focus on stabilizing Aβ peptides. A huge number of the above-mentioned structural studies correspond to different fragments of the Aβ peptide. The different structures of Aβ observed in solution and solid state and their implication in aggregation into fibrils have been comprehensively reviewed [43]. The biological and biophysical properties of the Aβ peptide and the biological, biochemical, and neurochemical roles of APP and Aβ in AD have also been discussed in detail [48].

Essentially, the Aβ peptide folds into an α-helical structure that is converted into a β-sheet structure in membrane-like media. Deprotonation of two acidic amino acids in the helix promotes a helix-to-coil conformational transition prior to aggregation of Aβ. Several NMR studies revealed that the C-terminus of the Aβ is critical for the α-helical-to-β-sheet transformation, playing a key role in determining the state of protein aggregation in AD. In the α-helix, the side chains of His13 and Lys16 residing on the same face of the helix are close (Figure 9.2b,c). The peptide collapses into a compact series of loops, strands, and turns without secondary structure. The van der Waals and electrostatic forces are responsible for the conformational stabilization. Approximately 25% of the surface is uninterrupted hydrophobic, and the compact coil structure is metastable. This situation may lead to a global conformational rearrangement and formation of intermolecular β-sheet secondary structure caused by fibrillation (Figure 9.2d,e). Finally, Aβ monomers can form higher order assemblies ranging from low-molecular-weight oligomers, including dimers, trimers, tetramers, and pentamers, to medium–molecular-weight oligomers, including hexamers, nonamers, and dodecamers, to protofibrils and fibrils (Figure 9.2f) [49]. Aβ peptides with a critical concentration of pyroglutamated N-terminus are found to be prone to misfolding into toxic aggregates [50].

Fibrils formed from Aβ40 and Aβ42 peptides are polymorphic, presenting variations in their molecular structure depending on fibril growth conditions. These structural variations correlate with AD clinical subtypes [51]; indeed, the abnormal systemic changes observed in Aβ metabolism affect AD progression [52]. Three structures of fragments 686–713 corresponding to Aβ42 (PDB ID: 5AEF) [53] have been resolved by electron microscopy (Figure 9.3a,b). The observed structure is consistent with earlier structures revealed by NMR spectroscopy and MD simulations (Figure 9.3c,d) [54]. Cryo-EM has also been used for determining the structure of fragments 672–713 (PDB ID: 5OQV) [55] and 695–705 (PDB ID: 5VOS) [56]. By using that technique, a structure showing recognition of APP by human γ-secretase (PDB ID: 6IYC) has been recently reported [57].

Figure 9.2 (a) Human APP proteolytic pathways. Human APP proteolysis in the nonamyloidogenic pathway and amyloidogenic pathway. Nonamyloidogenic processing of APP refers to the sequential processing of APP by membrane-bound α-secretases, which cleave within the Aβ domain to generate the membrane-tethered α-C-terminal fragment CTFα (C83) and the N-terminal fragment sAPPα. CTFα is then cleaved by γ-secretases to generate extracellular P3 and the APP intracellular domain (AICD). Amyloidogenic processing of APP is carried out by the sequential action of membrane-bound β- and γ-secretases. β-Secretase cleaves APP into the membrane-tethered C-terminal fragments β (CTFβ or C99) and N-terminal sAPPβ. CTFβ is subsequently cleaved by γ-secretases into the extracellular Aβ and APP intracellular domain (AICD). (b–f) Structures of Aβ monomer, fibrils and oligomers. (b) The primary amino acid sequence of the 42-amino-acid Aβ isoform Aβ42. (c) The structure of amyloid β peptide (1–28), which forms a predominately α-helical structure that can be converted into a β-sheet structure in membrane-like media (PDB ID: 1AMC, 1AMB). (d) Solution structure of amyloid β peptide (1–40), where in the C-terminal two-thirds of the peptide form an α-helix conformation between residues 15 and 36 with a kink or hinge at 25–27 in aqueous sodium dodecyl sulfate (SDS) micelles with a bend centered at residue 12, while the peptide is unstructured between residues 1 and 14, which are mainly polar and likely solvated by water (PDB IDs: 1BA4, 1BA6). (e) Amyloid β peptide (10–35) forms a collapsed coil structure (PDB ID: 1HZ3). (f) Proposed pathway for the conversion of amyloid β monomers into higher order oligomers, protofibrils, and fibrils. Source: Chen et al. 2017 [43]. Reprinted with permission of Springer Nature.

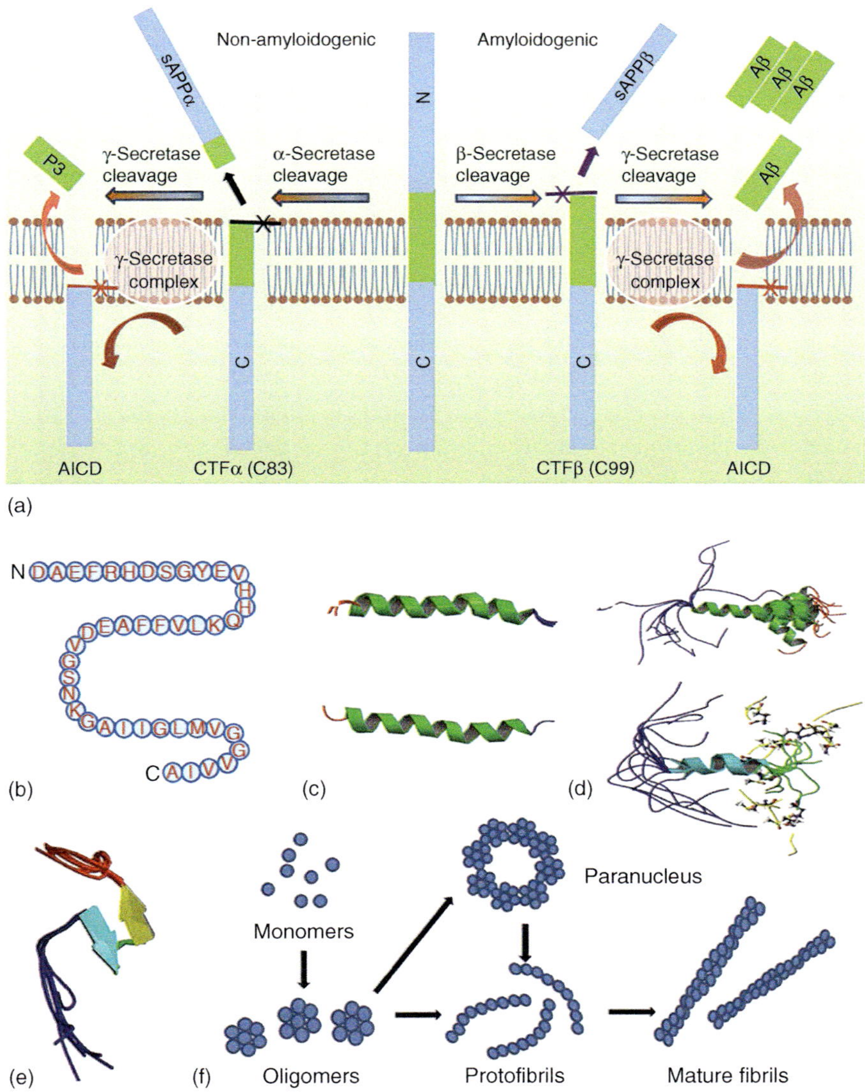

9.1.2 Tau (τ)

Tau (τ) protein is an axonal protein [58], also found in astrocytes [59] and oligodendrocytes [60], that promotes microtubule assembly and stability [61]. The C-terminus binds axonal microtubules, while the N-terminus binds neural plasma membrane components. τ protein is believed to be involved in neuronal polarity [62]. In solution, τ protein is extensively unfolded, and it can form amyloid fibrils, which are present in a variety of neuropathologies known as tauopathies, including AD [63].

The canonical form of protein has 758 residues (Figure 9.4). In the brain, τ protein is a set of six isoforms, whose binding affinity to microtubules can be

Table 9.1 Structures of APP resolved by NMR.

PDB ID	Residues	PDB ID	Residues	PDB ID	Residues
1AMB	672–699	1Z0Q	672–713	2LZ4	699–726
1AMC	672–699	1ZE7	672–687	2M4J	672–711
1AML	672–711	1ZE9	672–687	2M9R	672–711
1BA4	672–711	2BEG	672–713	2M9S	672–711
1BA6	672–711	2BP4	672–687	2MGT	672–687
1BJB	672–699	2LFM	672–711	2MJ1	688–705
1BJC	672–699	2LLM	686–726	2MPZ	686–711
1HZ3	681–706	2LMN	672–711	2MVX	672–711
1IYT	672–713	2LMO	672–711	2MXU	672–713
1OWT	124–189	2LMP	672–711	2NAO	672–713
1QCM	696–706	2LMQ	672–711	2OTK	672–711
1QWP	696–706	2LNQ	672–711	5KK3	672–713
1QXC	696–706	2LOH	686–726	5LFY	672–681
1QYT	696–706	2LP1	671–770		
1TKN	460–569	2LZ3	699–726		

modulated by phosphorylation [64]. It can be phosphorylated by c-AMP (cyclic adenosine monophospate) at Tyr18, Ser29, Ser46, Ser61, Thr69, Thr71, Thr11, Ser214, Ser396, Thr470, Thr486, Thr493, ThrτSer502, Ser508, Tyr514, Ser515, Ser516, Ser519, Thr522, Thr529, Ser531, Thr534, Thr548, Ser552, Ser554, Ser579, Ser602, Ser622, Ser669, Ser673, Tyr711, Ser713, Ser717, Thr720 Ser721, Ser726, Ser733, Ser739, and Thr744 [62, 65–68]. The physiology and pathology of the kinases responsible for τ phosphorylation are closely related to AD [69, 70] as well as dephosphorylation by phosphatases [71]. Some studies suggest that hyperphosphorylation causes detachment of τ protein from microtubules and subsequent aggregation into fibrils [72]. τ protein undergoes *N*-glycosylation at Ly87, Lys383, Lys467, Lys480, Lys491, Ser525, Lys542, Lys551, Ser555, Lys576, Lys597, Lsys598, Lys664, Lys670, and Lys717 [73, 74]. A reduced number of glycosylated sites are related to AD [75]. In addition to phosphorylation and *N*-glycosylation, other post-translational modifications include acetylation at Ala2, Lys480, Lys542, Lys576, Lys607, Lys615, Lys628, Lys648, Lys660, Lys664, Lys686, and Lys702; methylation at Lys480, Lys576, Lys618, and Lys628; and deamidation at Asn484 and Asn596.

Natural variants can include Arg5His, Arg5Leu, Thr17Met, Thr30Ala, Asp285Asn, Val289Ala, Arg370Trp, Tyr441His, Ser447Pro, Lys574Thr, Leu583Val, Gly589Val, Asn596Lys, Asn613His, Val617Ile, Pro618Leu, Pro618Ser, Gly620Val, Ser622Asn, Lys634Met, Ser637Phe, Val654Met, Glu659Val, Ser669Leu, Lys686Ile, Gly706Arg, and Arg723Trp.

The six isoforms have different peptide architectures. The longest isoform contains all four repeats (4R) and the shorter one has only three repeats (3R) lacking repeat 2 (R2). The formation of amyloid aggregates in τ is promoted by two six-residue segments, Val-Gln-Ile-Ile-Asn-Lys at the start of R2 and

Table 9.2 Structures of APP resolved by X-ray crystallography.

PDB ID	Res (Å)	Positions	PDB ID	Res (Å)	Positions	PDB ID	Res (Å)	Positions
1AAP	1.5	287–344	3IFL	1.5	672–678	4MVK	1.5	689–694
1BRC	2.5	287–342	3IFN	1.5	672–711	4MVL	2.3	672–711
1CA0	2.1	289–342	3IFO	2.15	672–678	4NGE	2.7	672–711
1MWP	1.8	28–123	3IFP	2.95	672–678	4OJF	2	672–679
1TAW	1.8	287–344	3JQ5	2.03	672–679	4ONF	2	672–678
1X11	2.5	754–766	3JQL	1.2	687–692	4ONG	2.2	672–711
1ZJD	2.6	289–344	3JTI	1.8	699–706	4PQD	1.33	22–126
2FJZ	1.61	133–189	3KTM	2.7	18–190	4PWQ	1.4	18–190
2FK1	1.6	133–189	3L33	2.48	290–341	4XXD	2.41	683–699
2FK2	1.65	133–189	3L81	1.6	761–767	5AM8	1.9	675–681
2FK3	2.4	133–189	3MOQ	2.05	689–712	5AMB	1.55	706–713
2FKL	2.5	124–189	3MXC	2	754–762	5BUO	2.31	370–710
2FMA	0.85	133–189	3MXY	2.3	754–762	5C67	1.83	294–344
2G47	2.1	672–711	3NYJ	3.2	365–567	5CSZ	1.8	672–682
2IPU	1.65	672–679	3NYL	2.8	365–570	5HOW	2.29	688–705
2R0W	2.5	672–679	3OVJ	1.8	687–692	5HOX	1.9	688–707
2WK3	2.59	672–713	3OW9	1.8	687–692	5HOY	2.29	688–707
2Y29	2.3	687–692	3PZZ	1.29	700–705	5LV0	2.7	706–711
2Y2A	1.91	687–692	3Q2X	1.45	698–703	5MY4	2.21	674–683
2Y3J	1.99	701–706	3SV1	3.3	754–767	5MYO	1.59	674–683
2Y3K	1.9	706–713	3U0T	2.5	701–711	5MYX	1.49	674–689
2Y3L	2.1	706–713	3UMH	2	370–575	5ONP	1.34	700–704
3AYU	2	586–595	3UMI	2.4	370–575	5ONQ	1.17	700–704
3BAE	1.59	672–699	3UMK	2.6	370–575	5TXD	1.45	698–703
3BKJ	1.59	672–687	4HIX	2.2	672–699	5VZY	2.32	682–696
3DXC	2.1	739–770	4JFN	1.75	23–185	5W3P	1.92	672–687
3DXD	2.2	739–770	4M1C	3.5	672–711	6CO3	2.38	672–682
3DXE	2	739–770	4MDR	1.85	758–767	6GFI	2.3	294–346
3GCI	2.04	707–713	4MVI	1.7	672–711			

Val-Gln-Ile-Val-Tyr-Lys at the start of repeat 3 (R3) [76]. The atomic structure of an amyloid fibril formed by Val-Gln-Ile-Val-Tyr-Lys from R3 was resolved, compared to Aβ [77], and used to design inhibitors of aggregation of fragments containing the segment Val-Gln-Ile-Val-Tyr-Lys [78]. However, such inhibitors were ineffective at inhibiting the aggregation of the full-length protein containing the Val-Gln-Ile-Ile-Asn-Lys segment of R2. Further studies using cryo-EM demonstrated that the Val-Gln-Ile-Ile-Asn-Lys segment is the most powerful driver of τ aggregation [79].

Five structures of different fragments of τ have been resolved by NMR (PDB ID: 1I8H, residues 542–554; PDB ID: 2MZ7, residues 584–629; PDB ID: 5N5A,

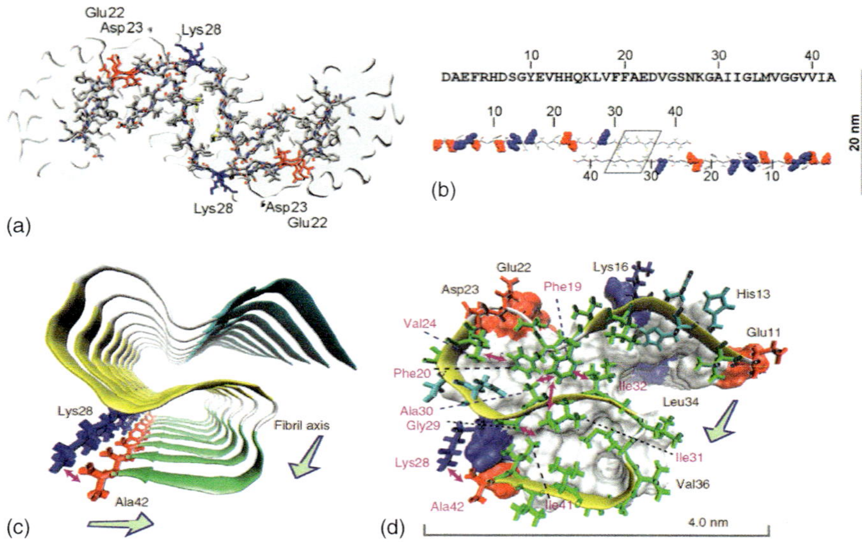

Figure 9.3 Structural models of the Aβ(1–42) fibril. (a) Density superimposed with the family of models producing the best fit to our data. (b) Aβ(1–42) sequence and schematic of the packing of two Aβ(1–42) peptides with the zipper-like region framed. Positively charged amino acids are shown in blue and negatively charged ones in red. Source: (a, b) Reprinted with permission from Schmidt et al. [53]. Copyright 2015, National Academy of Sciences. (c) Structural details of the Aβ(1–42) fibril revealed by the ssNMR analysis. Disordered residues 1–10 are omitted for clarity. Green arrows throughout indicate the fibril axis. (a) View from the fibril axis, showing three β-strand regions (cyan, residues 12–18; yellow, 24–33; green, 36–40) connected by short coil or turn (silver) regions. The unique salt bridge between Lys28 (blue) and Ala42 (red) is shown. (d) Side-chain contacts for a single Aβ chain with a van der Waals surface and polarity diagram for the rest of the Aβ chains. Residues are colored as green, hydrophobic; cyan, polar; red, acidic; and blue, basic. Observed long-range side-chain intramolecular contacts (purple arrows) and intermolecular contacts (blue arrows) are shown. Source: (c, d) Xiao et al. 2015 [54]. Reprinted with permission of Springer Nature.

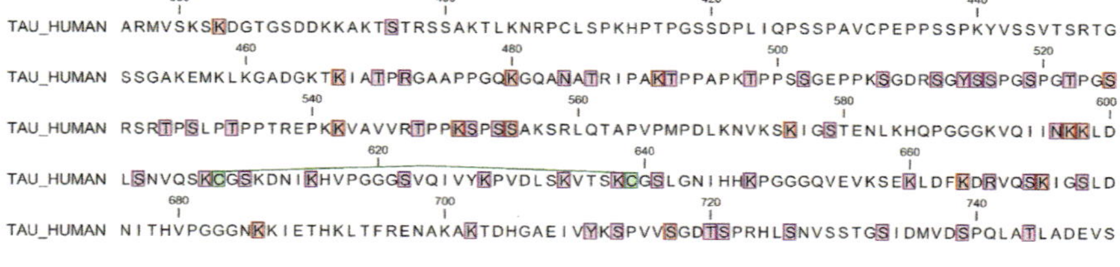

Figure 9.4 Sequence of human τ protein. Disulfide bonds (green), N-glycosylation sites (red), and phosphorylation (magenta) sites are indicated on sequence.

Table 9.3 Structures of τ resolved by X-ray crystallography.

PDB ID	Res (Å)	Positions	PDB ID	Res (Å)	Positions	PDB ID	Res (Å)	Positions
2ON9	1.51	623–628	4Y32	1.7	528–534	5MP3	2.75	609–638
3OVL	1.81	623–628	4Y3B	1.8	528–534	5MP5	2.31	615–628
4E0M	1.75	622–634	4Y5I	1.4	528–534	5ZV3	2.09	52–71
4E0N	1.65	622–634	5DMG	2.5	733–747	6BB4	2.1	703–725
4E0O	1.82	622–634	5E2V	1.64	511–528	6FBW	1.45	528–533
4FL5	1.9	527–536	5E2W	1.5	511–528	6FI5	1.7	529–533
4GLR	1.9	541–557	5HF3	1.8	528–534	6GK7	2.95	625–635
4NP8	1.51	623–628	5MO3	1.69	615–628	6GK8	2.85	52–71
4TQE	1.6	532–547	5MP1	3.1	615–628	6H06	2.63	721–746

residues 571–607; PDB ID: 5N5B, residues 609–636 and PDB ID: 5NVB, residues 571–585). A total of 27 crystal structures have been resolved at different resolutions and correspond to different fragments of τ (Table 9.3). In addition, 11 structures have been resolved by electron microscopy (Table 9.4).

NMR studies in solution confirmed that while τ is disordered in solution, it folds into a stable structure upon binding to microtubules [80]. While the protein is associated with microtubules, a unique conformation is present. On the other hand, dissociation of microtubules (caused by hyperphosphorylation as mentioned earlier) results in a highly flexible structure that adopts a β-sheet structure and promotes the formation of neurotoxic fibrils [81].

Cryo-EM is used to visualize paired helical filaments (PHFs) and straight filaments (SFs) from the brain of AD patients (Figure 9.5) [82]. Filament cores are made of two identical protofilaments comprising residues 306–378 of τ protein, which adopt a combined cross-β/β-helix structure prior to aggregation. Both PHFs and SFs present the same secondary structure consisting of eight β-sheets (β 1–8) running along the length of the protofilaments and adopting a C-shaped architecture. Each C consists of a β-helix region, where three β-sheets are arranged in a triangular form, and two regions with a cross-β architecture, where pairs of β-sheets pack antiparallel to each other (Figure 9.5d), although other out-of-register β-sheets have been found in amyloid fibrils [81]. Differences between PHFs and SFs are found in the interfaces, which present different lateral contacts (Figure 9.5e,f) [82].

By combining cryo-EM with immune-EM, it has been determined that the protein adopts a common fold in patients with AD [83]. Moreover, it has been

Table 9.4 Structures of τ resolved by cryo-EM.

PDB ID	Res (Å)	Positions	PDB ID	Res (Å)	Positions	PDB ID	Res (Å)	Positions
5K7N	1.1	623–628	5V5B	1.5	591–600	6GX5	3.2	602–695
5O3L	3.4	623–695	5V5C	1.25	592–597	6HRE	3.2	1–758
5O3O	3.5	623–695	6CVJ	3.2	514–717	6HRF	3.3	1–758
5O3T	3.4	623–695	6CVN	3.9	514–717			

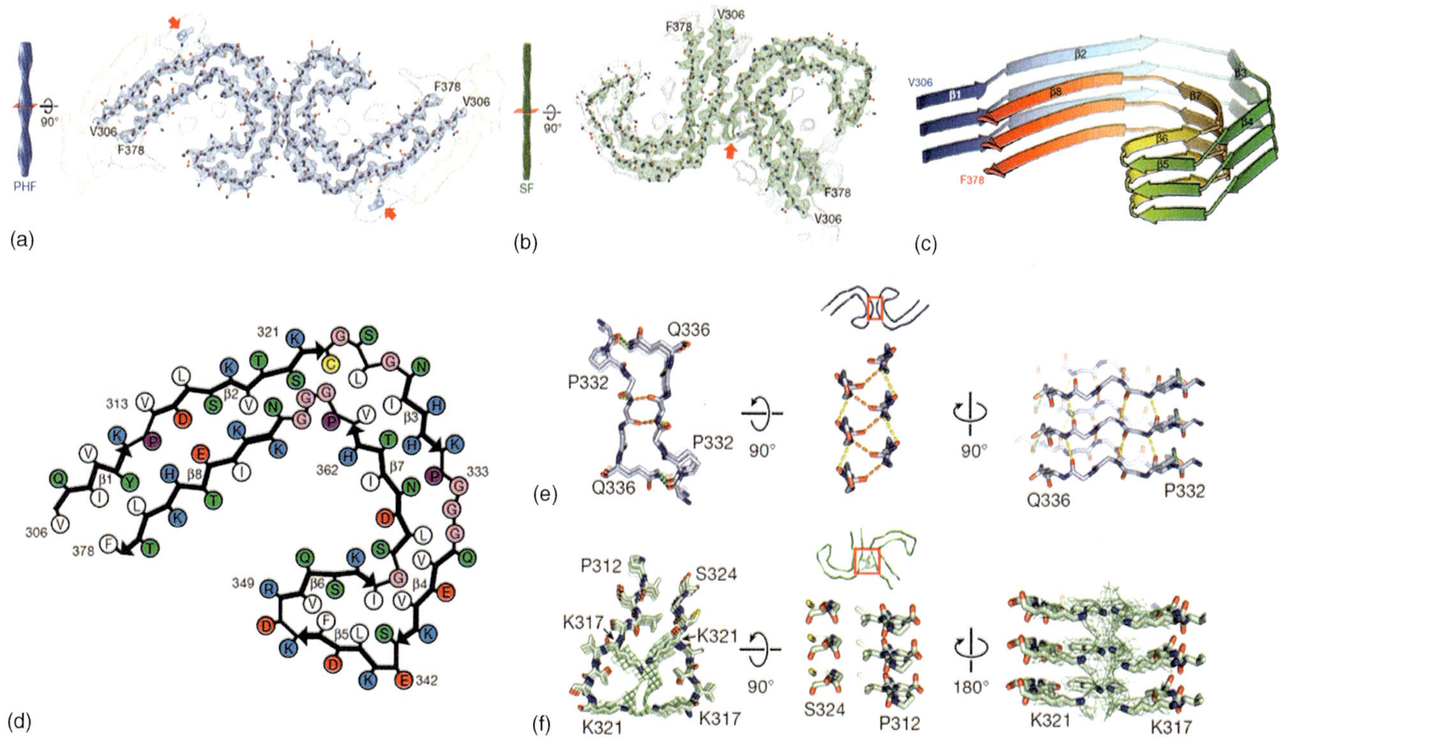

Figure 9.5 (a) Cross sections of the paired helical filaments (PHFs) and cryo-EM structures of straight filaments (SFs). Cryo-EM density and atomic models of PHFs (a) and SFs (b). Overviews of the helical reconstructions (left) show the orientation of the cross-sectional densities (right). Sharpened, high-resolution maps are shown in blue (PHFs) and green (SFs). Red arrows indicate additional densities in contact with K317 and K321. Unsharpened, 4.5 Å low-pass filtered density is shown in gray. Unsharpened density highlighted with an orange background is reminiscent of a less-ordered β-sheet and could accommodate additional 16 amino acids, which would correspond to a mixture of residues 259–274 (R1) from 3R τ and residues 290–305 (R2) from 4R τ. (c) Rendered view of the secondary structure elements in three successive rings. (d) Schematic view of the protofilament core. (e) Packing between residues 332PGGGQ336 of the two protofilaments in PHFs. Interprotofilament hydrogen bonds are shown in orange and green. Intraprotofilament hydrogen bonds are shown in yellow. (f) Packing between residues 317–324 and residues 312–321 of the protofilaments in SFs. Additional density between the side chains of lysines 317 and 321 is shown as a green mesh. Source: Fitzpatrick et al. 2017 [82]. Reproduced with permission of Elsevier.

observed that the protein presents distinct folds in different neurodegenerative diseases [84]. Cryo-EM combined with computational approaches has also been used to visualize different τ constructs on microtubules and to generate atomic models of interactions between τ and tubulin [85].

9.1.3 α-Synuclein

α-Synuclein is a highly conserved presynaptic protein. It is primarily localized in the cytoplasm and its normal function is not fully understood [86] although it is known that it inhibits neurotransmitter release when overexpressed [87]. However, the knockout has little effect on synaptic transmission [88] although it affects learning ability [89]. α-Synuclein is believed to be involved in a variety of cellular functions, including vesicular dopamine release [90] and synaptic plasticity [91].

α-Synuclein exhibits an unstructured shape in solution; abnormal aggregation of misfolded forms of the protein leads to insoluble fibrillar aggregates in characteristic inclusions known as "Lewy bodies" [92], which are found in the brain of PD patients [93]. There is a solid evidence of the toxic effect of α-synuclein in PD. [94, 95] As an example, the requirement of α-synuclein for neurodegeneration is established in mouse models of PD. [96, 97] α-Synuclein also plays a significant role in maintaining cellular iron homeostasis. Consequently, aggregation of α-synuclein and iron dyshomeostasis in PD may be causally related [98].

α-Synuclein is composed of 140 amino acids (Figure 9.6). The *N*-terminus of α-synuclein contains seven 11 residue repeats that form an amphipathic α-helix. Posttranslational modifications of α-synuclein include *N*-acetylation at the *N*-terminus (Met1), which is critical for forming α-helical oligomers, that is, proper folding leads to the native oligomeric structure [99]. α-Synuclein can be phosphorylated at Ser87, Tyr125 [100], and Ser129 [101]. Phosphorylation at Ser129 is the characteristic of synucleinopathy lesions including LBD [102].

Natural variants include Ala30Pro [103], Glu46Lys [104, 105], His50Gln [106–108], and Ala53Thr [109, 110]. With the exception of Ala30Pro, all the other variants increase oligomerization. Two isoforms that are missing residues 103–130 and 42–54 have been found.

The precise way in which α-synuclein is involved in PD pathogenesis is unknown [111, 112], and several alternatives have been proposed [113]. In fact, there is growing evidence for the toxicity of α-synuclein fibrils via multiple mechanisms including impairment of mitochondrial function, disturbance of intracellular protein homeostasis, chronic endoplasmic reticulum stress, formation of pore-like structures with perforation of membranes including the plasma membrane of neurons, and glutamate receptor dysfunction [114]. Although it is known that in PD, the insoluble filaments of α-synuclein are hyperphosphorylated [115, 116] and ubiquitinated [117], the molecular mechanisms of these processes are

```
                        20                       40                       60
SYUA_HUMAN  MDVFMKGLSKAKEGVVAAAEKTKQGVAEAAGKTKEGVLYVGSKTKEGVVHGVATVAEKTKEQVTNVGGAVVTGVT
                 80                     100                      120                140
SYUA_HUMAN  AVAQKTVEGAGSIAAATGFVKKDQLGKNEEGAPQEGILEDMPVDPDNEAYEMPSEEGYQDYEPEA
```

Figure 9.6 Sequence of human α-synuclein. Phosphorylation sites (magenta) and metal-binding sites (gray) are indicated on sequence.

unknown [118]. However, a 12-amino-acid stretch (Val71-Thr-Gly-Val-Thr-Ala-Val-Ala-Gln-Lys-Thr-Val82, termed NACore) in the middle of the hydrophobic domain of human α-synuclein was identified as necessary and sufficient for its fibrillization [119]. Similarly, a nine-residue peptide motif, in the same domain, Val66-Gly-Gly-Ala-Val-Val-Thr-Gly-Val74 named GAV motif, was also found to be responsible for aggregation of α-synuclein [120]. The biochemical characterization of α-synuclein filaments extracted from the brain of PD patients revealed that amino acid residues 31–109 of α-synuclein constitute the core unit of the filaments [121]. Notably, it has been demonstrated that deletion of residues 71–82 abolishes neurotoxicity and aggregation of α-synuclein in a Drosophila model of PD [122], and the isolated fragment from the rest of α-synuclein readily forms amyloid fibrils and is highly cytotoxic [123].

Six structures of different fragments of α-synuclein have been resolved by NMR (PDB ID: 1XQ8, residues 1–140; PDB ID: 2JN5, residues 1–19; PDB ID: 2KKW, residues 1–140; PDB ID: 2M55, residues 1–19; PDB ID: 2N0A, residues 1–140 and PDB ID: 4BXL, residues 35–36). Ten crystal structures have been resolved at different resolutions (Table 9.5). In addition, eight structures have been resolved by electron microscopy (Table 9.6).

With microelectron diffraction, the above-mentioned toxic core of synuclein has been resolved at 1.4 Å resolution (PDB ID: 4ZNN) [124]. A high-resolution structure of an α-synuclein fibril has been determined by solid-state NMR spectroscopy and validated by EM and X-ray fiber diffraction (PDB ID: 2N0A) [125]. These studies revealed a Greek-key β-sheet motif in the hydrophobic core of a single-fibril filament, where salt bridges between Glu46-Lys80, a glutamine ladder (Gln79), and hydrophobic packing of aromatic residues (Phe94) contribute to the stability of the in-register β-sheet. Further cryo-EM studies of full-length α-synuclein revealed two polymorphs (called rod and twister), which showed

Table 9.5 Structures of α-synuclein resolved by X-ray crystallography.

PDB ID	Res (Å)	Positions	PDB ID	Res (Å)	Positions	PDB ID	Res (Å)	Positions
2X6M	1.62	132–140	3Q28	1.6	58–79	4R0W	1.5	70–76
3Q25	1.9	1–19	3Q29	2.3	1–19	4RIK	1.85	69–77
3Q26	1.54	10–42	5CRW	1.6	31–41			
3Q27	1.3	32–57	4R0U	1.38	72–78			

Table 9.6 Structures of α-synuclein resolved by cryo-EM.

PDB ID	Res (Å)	Positions	PDB ID	Res (Å)	Positions	PDB ID	Res (Å)	Positions
4RIL	1.43	68–78	6CU7	3.5	1–140	6H6B	3.4	1–121
4ZNN	1.41	47–56	6CU8	3.6	1–140			
6A6B	3.07	37–99	6FLT	3.42	1–121			

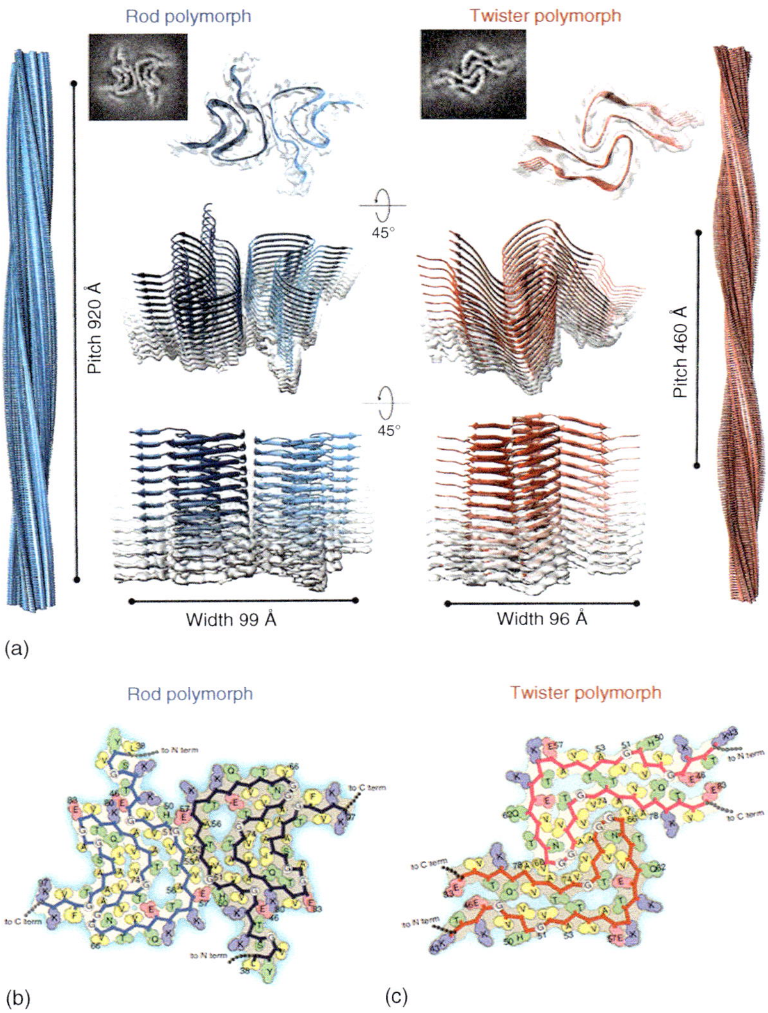

Figure 9.7 (a) The cryo-EM structures of the rod (left) and twister (right) polymorphs of the full-length α-synuclein fibrils shown as density slices (top inset), as semitransparent surfaces overlaid with their atomic models viewed from two different angles (lower panels). The rod (blue) and twister (red) polymorphs contain two protofilaments composed of stacked β-sheets and packed by an approximate 21 screw axis of symmetry. Shown on the left and right sides are the 3D model of the rod and twister fibril polymorphs, respectively, with their distinctively different helical pitches. (b, c) Residue interactions of two asymmetric units in two opposing protofilaments elucidate packing between and within these two protofilaments in the rod (b) and twister (c) polymorphs (viewed down the fibril axis). Residues are colored by hydrophobicity (yellow: hydrophobic; green: polar; red: negative charge; blue: positive charge). (d, e) The two protofilaments in the rod (d) and twister (e) polymorphs contact by different residues (space-filled) and have a distinct fibril core of tightly packed steric zippers of preNAC (blue) and NACore (red), as previously observed in those peptide fibril structures. The cryo-EM density maps are shown as gray mesh surfaces. Intraprotofilament hydrogen bonds are shown in black dashed lines, and interprotofilament hydrogen bonds are in magenta. Source: Li et al. 2018 [126]. Reproduced with permission of Springer Nature.

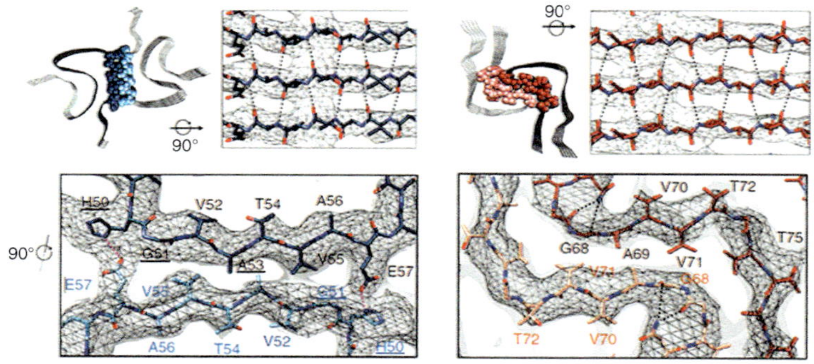

(d) preNAC zipper in rod polymorph (e) NACore zipper in twister polymorph

Figure 9.7 *(Continued)*

a pair of β-sheet protofilaments sharing a conserved kernel consisting of a bent β-arch motif (Figure 9.7a) [126]. A large number of branched amino acid residues (Thr, Val) is involved in the main hydrophobic core of the bent β-arch (Figure 9.7b,c). For both structures, the presence of glycines (Gly67, Gly84), stabilizing hydrogen bonds (Asn65 and Gly68, Gln79 and Gly86), and solvent-exposed charged residues (Glu57, Lys58) are found in major turns in the backbones. Interestingly, the preNAC steric zipper in the rod structure is associated with six PD familial mutation sites (Glu46Lys, His50Gln, Gly51Asp, Ala53Glu, Ala53Thr, and Ala53Val). Residues involved in the mutation sites form the interface between the two protofilaments and contribute to the stability of fibrils [127]. In fact, those mutations have the potential to disrupt the preNAC zipper of the fibril core in the rod structure but with little effect on the stability of the twister structure (Figure 9.7d,e) [128].

References

1 Ramirez-Alvarado, M., Kelly, J. W., Dobson, C. M. (2010). Protein misfolding diseases: Current and emerging principles and therapies. Wiley, Hoboken, p. 1045.
2 Ovádi, J. and Orosz, F. (2009). *Focus on Structural Biology*, vol. 7, 277. Heidelberg: Springer.
3 Harris, J.R. (2012). *Subcellular Biochemistry* (ed. J.R. Harris), 648. Heidelberg: Springer.
4 Bartolini, M. and Andrisano, V. (2010). Strategies for the inhibition of protein aggregation in human diseases. *ChemBioChem* 11 (8): 1018–1035.
5 Uversky, V.N. and Eliezer, D. (2009). Biophysics of Parkinson's disease: structure and aggregation of alpha-synuclein. *Curr. Protein Pept. Sci.* 10 (5): 483–499.
6 Nagai, Y. and Popiel, H.A. (2008). Conformational changes and aggregation of expanded polyglutamine proteins as therapeutic targets of the polyglutamine diseases: exposed beta-sheet hypothesis. *Curr. Pharm. Des.* 14 (30): 3267–3279.
7 Glabe, C.C. (2005). Amyloid accumulation and pathogenesis of Alzheimer's disease: significance of monomeric, oligomeric and fibrillar Abeta. *Subcell. Biochem.* 38: 167–177.

8 Selkoe, D.J. (2001). Alzheimer's disease: genes, proteins, and therapy. *Physiol. Rev.* 81 (2): 741–766.
9 Sadigh-Eteghad, S., Sabermarouf, B., Majdi, A. et al. (2015). Amyloid-beta: a crucial factor in Alzheimer's disease. *Med. Princ. Pract.* 24 (1): 1–10.
10 Selkoe, D.J. and Hardy, J. (2016). The amyloid hypothesis of Alzheimer's disease at 25 years. *EMBO Mol. Med.* 8 (6): 595–608.
11 Goedert, M. (2001). Alpha-synuclein and neurodegenerative diseases. *Nat. Rev. Neurosci.* 2 (7): 492–501.
12 Zoghbi, H.Y. and Orr, H.T. (2000). Glutamine repeats and neurodegeneration. *Annu. Rev. Neurosci.* 23: 217–247.
13 Hashimoto, M., Rockenstein, E., Crews, L., and Masliah, E. (2003). Role of protein aggregation in mitochondrial dysfunction and neurodegeneration in Alzheimer's and Parkinson's diseases. *Neuromolecular Med.* 4 (1–2): 21–36.
14 Muchowski, P.J. (2002). Protein misfolding, amyloid formation, and neurodegeneration: a critical role for molecular chaperones? *Neuron* 35 (1): 9–12.
15 Soto, C. and Pritzkow, S. (2018). Protein misfolding, aggregation, and conformational strains in neurodegenerative diseases. *Nat. Neurosci.* 21 (10): 1332–1340.
16 Brettschneider, J., Del Tredici, K., Lee, V.M.Y., and Trojanowski, J.Q. (2015). Spreading of pathology in neurodegenerative diseases: a focus on human studies. *Nat. Rev. Neurosci.* 16 (2): 109–120.
17 Jucker, M. and Walker, L.C. (2018). Propagation and spread of pathogenic protein assemblies in neurodegenerative diseases. *Nat. Neurosci.* 21 (10): 1341–1349.
18 Soto, C. (2003). Unfolding the role of protein misfolding in neurodegenerative diseases. *Nat. Rev. Neurosci.* 4: 49.
19 Soto, C. and Estrada, L.D. (2008). Protein misfolding and neurodegeneration. *Arch. Neurol.* 65 (2): 184–189.
20 Lansbury, P.T. and Lashuel, H.A. (2006). A century-old debate on protein aggregation and neurodegeneration enters the clinic. *Nature* 443 (7113): 774–779.
21 Winklhofer, K.F., Tatzelt, J., and Haass, C. (2008). The two faces of protein misfolding: gain- and loss-of-function in neurodegenerative diseases. *EMBO J.* 27 (2): 336–349.
22 Sweeney, P., Park, H., Baumann, M. et al. (2017). Protein misfolding in neurodegenerative diseases: implications and strategies. *Transl. Neurodegener.*6: 6. https://doi.org/10.1186/s40035-017-0077-5.
23 Smith, H.L., Li, W., and Cheetham, M.E. (2015). Molecular chaperones and neuronal proteostasis. *Semin. Cell Dev. Biol.* 40: 142–152.
24 Muchowski, P.J. and Wacker, J.L. (2005). Modulation of neurodegeneration by molecular chaperones. *Nat. Rev. Neurosci.* 6 (1): 11–22.
25 Chaudhuri, T.K. and Paul, S. (2006). Protein-misfolding diseases and chaperone-based therapeutic approaches. *FEBS J.* 273 (7): 1331–1349.
26 Lindberg, I., Shorter, J., Wiseman, R.L. et al. (2015). Chaperones in neurodegeneration. *J. Neurosci.* 35 (41): 13853–13859.
27 Nunan, J. and Small, D.H. (2000). Regulation of APP cleavage by alpha-, beta- and gamma-secretases. *FEBS Lett.* 483 (1): 6–10.

28 Seubert, P., Vigo-Pelfrey, C., Esch, F. et al. (1992). Isolation and quantification of soluble Alzheimer's beta-peptide from biological fluids. *Nature* 359 (6393): 325–327.

29 Tang, K., Wang, C., Shen, C. et al. (2003). Identification of a novel alternative splicing isoform of human amyloid precursor protein gene, APP639. *Eur. J. Neurosci.* 18 (1): 102–108.

30 Duce, J.A., Tsatsanis, A., Cater, M.A. et al. (2010). Iron-export ferroxidase activity of beta-amyloid precursor protein is inhibited by zinc in Alzheimer's disease. *Cell* 142 (6): 857–867.

31 Priller, C., Bauer, T., Mitteregger, G. et al. (2006). Synapse formation and function is modulated by the amyloid precursor protein. *J. Neurosci.* 26 (27): 7212–7221.

32 Baumkotter, F., Schmidt, N., Vargas, C. et al. (2014). Amyloid precursor protein dimerization and synaptogenic function depend on copper binding to the growth factor-like domain. *J. Neurosci.* 34 (33): 11159–11172.

33 Chartier-Harlin, M.C., Crawford, F., Houlden, H. et al. (1991). Early-onset Alzheimer's disease caused by mutations at codon 717 of the beta-amyloid precursor protein gene. *Nature* 353 (6347): 844–846.

34 Maruyama, K., Tomita, T., Shinozaki, K. et al. (1996). Familial Alzheimer's disease-linked mutations at Val717 of amyloid precursor protein are specific for the increased secretion of A beta 42(43). *Biochem. Biophys. Res. Commun.* 227 (3): 730–735.

35 Denman, R.B., Rosenzcwaig, R., and Miller, D.L. (1993). A system for studying the effect(s) of familial Alzheimer disease mutations on the processing of the beta-amyloid peptide precursor. *Biochem. Biophys. Res. Commun.* 192 (1): 96–103.

36 Wakutani, Y., Watanabe, K., Adachi, Y. et al. (2004). Novel amyloid precursor protein gene missense mutation (D678N) in probable familial Alzheimer's disease. *J. Neurol. Neurosurg. Psychiatry* 75 (7): 1039–1042.

37 Liu, T., Qian, W.J., Gritsenko, M.A. et al. (2005). Human plasma *N*-glycoproteome analysis by immunoaffinity subtraction, hydrazide chemistry, and mass spectrometry. *J. Proteome Res.* 4 (6): 2070–2080.

38 Halim, A., Brinkmalm, G., Ruetschi, U. et al. (2011). Site-specific characterization of threonine, serine, and tyrosine glycosylations of amyloid precursor protein/amyloid beta-peptides in human cerebrospinal fluid. *Proc. Natl. Acad. Sci. U.S.A.* 108 (29): 11848–11853.

39 Brinkmalm, G., Portelius, E., Ohrfelt, A. et al. (2012). An online nano-LC-ESI-FTICR-MS method for comprehensive characterization of endogenous fragments from amyloid beta and amyloid precursor protein in human and cat cerebrospinal fluid. *J. Mass Spectrom.* 47 (5): 591–603.

40 Walter, J., Capell, A., Hung, A.Y. et al. (1997). Ectodomain phosphorylation of beta-amyloid precursor protein at two distinct cellular locations. *J. Biol. Chem.* 272 (3): 1896–1903.

41 Standen, C.L., Brownlees, J., Grierson, A.J. et al. (2001). Phosphorylation of thr(668) in the cytoplasmic domain of the Alzheimer's disease amyloid precursor protein by stress-activated protein kinase 1b (Jun N-terminal kinase-3). *J. Neurochem.* 76 (1): 316–320.

42 Tarr, P.E., Roncarati, R., Pelicci, G. et al. (2002). Tyrosine phosphorylation of the beta-amyloid precursor protein cytoplasmic tail promotes interaction with Shc. *J. Biol. Chem.* 277 (19): 16798–16804.
43 Chen, G.F., Xu, T.H., Yan, Y. et al. (2017). Amyloid beta: structure, biology and structure-based therapeutic development. *Acta Pharmacol. Sin.* 38 (9): 1205–1235.
44 Gervais, F.G., Xu, D., Robertson, G.S. et al. (1999). Involvement of caspases in proteolytic cleavage of Alzheimer's amyloid-beta precursor protein and amyloidogenic A beta peptide formation. *Cell* 97 (3): 395–406.
45 Weidemann, A., Eggert, S., Reinhard, F.B. et al. (2002). A novel epsilon-cleavage within the transmembrane domain of the Alzheimer amyloid precursor protein demonstrates homology with Notch processing. *Biochemistry* 41 (8): 2825–2835.
46 Mathura, V.S., Paris, D., Ait-Ghezala, G. et al. (2005). Model of Alzheimer's disease amyloid-beta peptide based on a RNA binding protein. *Biochem. Biophys. Res. Commun.* 332 (2): 585–592.
47 Brown, A.M. and Bevan, D.R. (2016). Molecular dynamics simulations of amyloid beta-peptide (1–42): tetramer formation and membrane interactions. *Biophys. J.* 111 (5): 937–949.
48 Hamley, I.W. (2012). The amyloid beta peptide: a chemist's perspective. role in Alzheimer's and fibrillization. *Chem. Rev.* 112 (10): 5147–5192.
49 Mirza, Z., Pillai, V.G., and Kamal, M.A. (2014). Protein interactions between the C-terminus of Abeta-peptide and phospholipase A2 – a structure biology based approach to identify novel Alzheimer's therapeutics. *CNS Neurol. Disord. Drug Targets* 13 (7): 1224–1231.
50 Galante, D., Ruggeri, F.S., Dietler, G. et al. (2016). A critical concentration of N-terminal pyroglutamylated amyloid beta drives the misfolding of Ab1-42 into more toxic aggregates. *Int. J. Biochem. Cell Biol.* 79: 261–270.
51 Qiang, W., Yau, W.M., Lu, J.X. et al. (2017). Structural variation in amyloid-β fibrils from Alzheimer's disease clinical subtypes. *Nature* 541 (7636): 217–221.
52 Wang, J., Gu, B.J., Masters, C.L., and Wang, Y.-J. (2017). A systemic view of Alzheimer disease – insights from amyloid-β metabolism beyond the brain. *Nat. Rev. Neurol.* 13: 612.
53 Schmidt, M., Rohou, A., Lasker, K. et al. (2015). Peptide dimer structure in an Abeta(1–42) fibril visualized with cryo-EM. *Proc. Natl. Acad. Sci. U.S.A.* 112 (38): 11858–11863.
54 Xiao, Y., Ma, B., McElheny, D. et al. (2015). Abeta(1–42) fibril structure illuminates self-recognition and replication of amyloid in Alzheimer's disease. *Nat. Struct. Mol. Biol.* 22 (6): 499–505.
55 Gremer, L., Scholzel, D., Schenk, C. et al. (2017). Fibril structure of amyloid-beta(1–42) by cryo-electron microscopy. *Science* 358 (6359): 116–119.
56 Krotee, P., Griner, S.L., Sawaya, M.R. et al. (2018). Common fibrillar spines of amyloid-beta and human islet amyloid polypeptide revealed by microelectron diffraction and structure-based inhibitors. *J. Biol. Chem.* 293 (8): 2888–2902.
57 Zhou, R., Yang, G., Guo, X. et al. (2019). Recognition of the amyloid precursor protein by human gamma-secretase. *Science* 363 (6428): eaaw0930. https://doi.org/10.1126/science.aaw0930.

58 Binder, L.I., Frankfurter, A., and Rebhun, L.I. (1985). The distribution of tau in the mammalian central nervous system. *J. Cell Biol.* 101 (4): 1371–1378.
59 Papasozomenos, S.C. and Binder, L.I. (1987). Phosphorylation determines two distinct species of Tau in the central nervous system. *Cell Motil. Cytoskel.* 8 (3): 210–226.
60 Migheli, A., Butler, M., Brown, K., and Shelanski, M.L. (1988). Light and electron microscope localization of the microtubule-associated tau protein in rat brain. *J. Neurosci.* 8 (6): 1846–1851.
61 Gong, C.X. and Iqbal, K. (2008). Hyperphosphorylation of microtubule-associated protein tau: a promising therapeutic target for Alzheimer disease. *Curr. Med. Chem.* 15 (23): 2321–2328.
62 Yoshida, H. and Goedert, M. (2012). Phosphorylation of microtubule-associated protein tau by AMPK-related kinases. *J. Neurochem.* 120 (1): 165–176.
63 Goedert, M., Eisenberg, D.S., and Crowther, R.A. (2017). Propagation of tau aggregates and neurodegeneration. *Annu. Rev. Neurosci.* 40: 189–210.
64 Flaherty, D.B., Soria, J.P., Tomasiewicz, H.G., and Wood, J.G. (2000). Phosphorylation of human tau protein by microtubule-associated kinases: GSK3beta and cdk5 are key participants. *J. Neurosci. Res.* 62 (3): 463–472.
65 Litersky, J.M., Johnson, G.V., Jakes, R. et al. (1996). Tau protein is phosphorylated by cyclic AMP-dependent protein kinase and calcium/calmodulin-dependent protein kinase II within its microtubule-binding domains at Ser-262 and Ser-356. *Biochem. J.* 316 (Pt. 2): 655–660.
66 Illenberger, S., Zheng-Fischhofer, Q., Preuss, U. et al. (1998). The endogenous and cell cycle-dependent phosphorylation of tau protein in living cells: implications for Alzheimer's disease. *Mol. Biol. Cell* 9 (6): 1495–1512.
67 Lee, G., Thangavel, R., Sharma, V.M. et al. (2004). Phosphorylation of tau by fyn: implications for Alzheimer's disease. *J. Neurosci.* 24 (9): 2304–2312.
68 Sato, S., Cerny, R.L., Buescher, J.L., and Ikezu, T. (2006). Tau-tubulin kinase 1 (TTBK1), a neuron-specific tau kinase candidate, is involved in tau phosphorylation and aggregation. *J. Neurochem.* 98 (5): 1573–1584.
69 Imahori, K. and Uchida, T. (1997). Physiology and pathology of tau protein kinases in relation to Alzheimer's disease. *J. Biochem.* 121 (2): 179–188.
70 Ferrer, I., Gomez-Isla, T., Puig, B. et al. (2005). Current advances on different kinases involved in tau phosphorylation, and implications in Alzheimer's disease and tauopathies. *Curr. Alzheimer Res.* 2 (1): 3–18.
71 Liu, F., Iqbal, K., Grundke-Iqbal, I. et al. (2005). Dephosphorylation of tau by protein phosphatase 5: impairment in Alzheimer's disease. *J. Biol. Chem.* 280 (3): 1790–1796.
72 Wang, J.Z., Xia, Y.Y., Grundke-Iqbal, I., and Iqbal, K. (2013). Abnormal hyperphosphorylation of tau: sites, regulation, and molecular mechanism of neurofibrillary degeneration. *J. Alzheimers Dis.* 33 (Suppl 1): S123–S139.
73 Nacharaju, P., Ko, L., and Yen, S.H. (1997). Characterization of in vitro glycation sites of tau. *J. Neurochem.* 69 (4): 1709–1719.
74 Smet-Nocca, C., Broncel, M., Wieruszeski, J.M. et al. (2011). Identification of O-GlcNAc sites within peptides of the Tau protein and their impact on phosphorylation. *Mol. Biosyst.* 7 (5): 1420–1429.

75 Liu, F., Shi, J., Tanimukai, H. et al. (2009). Reduced O-GlcNAcylation links lower brain glucose metabolism and tau pathology in Alzheimer's disease. *Brain* 132 (Pt 7: 1820–1832.

76 von Bergen, M., Friedhoff, P., Biernat, J. et al. (2000). Assembly of tau protein into Alzheimer paired helical filaments depends on a local sequence motif ((306) VQIVYK(311)) forming beta structure. *Proc. Natl. Acad. Sci. U.S.A.* 97 (10): 5129–5134.

77 Sawaya, M.R., Sambashivan, S., Nelson, R. et al. (2007). Atomic structures of amyloid cross-beta spines reveal varied steric zippers. *Nature* 447 (7143): 453–457.

78 Sievers, S.A., Karanicolas, J., Chang, H.W. et al. (2011). Structure-based design of non-natural amino-acid inhibitors of amyloid fibril formation. *Nature* 475 (7654): 96–100.

79 Seidler, P.M., Boyer, D.R., Rodriguez, J.A. et al. (2018). Structure-based inhibitors of tau aggregation. *Nat. Chem.* 10 (2): 170–176.

80 Kadavath, H., Jaremko, M., Jaremko, L. et al. (2015). Folding of the tau protein on microtubules. *Angew. Chem. Int. Ed.* 54 (35): 10347–10351.

81 Liu, C., Zhao, M., Jiang, L. et al. (2012). Out-of-register beta-sheets suggest a pathway to toxic amyloid aggregates. *Proc. Natl. Acad. Sci. U.S.A.* 109 (51): 20913–20918.

82 Fitzpatrick, A.W.P., Falcon, B., He, S. et al. (2017). Cryo-EM structures of tau filaments from Alzheimer's disease. *Nature* 547 (7662): 185–190.

83 Falcon, B., Zhang, W., Schweighauser, M. et al. (2018). Tau filaments from multiple cases of sporadic and inherited Alzheimer's disease adopt a common fold. *Acta Neuropathol.* 136 (5): 699–708.

84 Falcon, B., Zhang, W., Murzin, A.G. et al. (2018). Structures of filaments from Pick's disease reveal a novel tau protein fold. *Nature* 561 (7721): 137–140.

85 Kellogg, E.H., Hejab, N.M.A., Poepsel, S. et al. (2018). Near-atomic model of microtubule-tau interactions. *Science* 360 (6394): 1242–1246.

86 Bendor, J.T., Logan, T.P., and Edwards, R.H. (2013). The function of α-synuclein. *Neuron* 79 (6): 1044–1066.

87 Nemani, V.M., Lu, W., Berge, V. et al. (2010). Increased expression of alpha-synuclein reduces neurotransmitter release by inhibiting synaptic vesicle reclustering after endocytosis. *Neuron* 65 (1): 66–79.

88 Drolet, R.E., Behrouz, B., Lookingland, K.J., and Goudreau, J.L. (2004). Mice lacking alpha-synuclein have an attenuated loss of striatal dopamine following prolonged chronic MPTP administration. *NeuroToxicology* 25 (5): 761–769.

89 Kokhan, V.S., Afanasyeva, M.A., and Van'kin, G.I. (2012). alpha-Synuclein knockout mice have cognitive impairments. *Behav. Brain Res.* 231 (1): 226–230.

90 Butler, B., Sambo, D., and Khoshbouei, H. (2017). Alpha-synuclein modulates dopamine neurotransmission. *J. Chem. Neuroanat.* 83–84: 41–49.

91 Cheng, F., Vivacqua, G., and Yu, S. (2011). The role of alpha-synuclein in neurotransmission and synaptic plasticity. *J. Chem. Neuroanat.* 42 (4): 242–248.

92 Miklya, I., Pencz, N., Hafenscher, F., and Goltl, P. (2014). The role of alpha-synuclein in Parkinson's disease. *Neuropsychopharmacol. Hung* 16 (2): 77–84.

93 Olanow, C.W. and Brundin, P. (2013). Parkinson's disease and alpha synuclein: is Parkinson's disease a prion-like disorder? *Mov. Disord.* 28 (1): 31–40.

94 Lee, V.M. and Trojanowski, J.Q. (2006). Mechanisms of Parkinson's disease linked to pathological alpha-synuclein: new targets for drug discovery. *Neuron* 52 (1): 33–38.
95 Venda, L.L., Cragg, S.J., Buchman, V.L., and Wade-Martins, R. (2010). α-Synuclein and dopamine at the crossroads of Parkinson's disease. *Trends Neurosci.* 33 (12): 559–568.
96 Norris, E.H., Giasson, B.I., and Lee, V.M. (2004). Alpha-synuclein: normal function and role in neurodegenerative diseases. *Curr. Top. Dev. Biol.* 60: 17–54.
97 Mor, D.E., Tsika, E., Mazzulli, J.R. et al. (2017). Dopamine induces soluble α-synuclein oligomers and nigrostriatal degeneration. *Nat. Neurosci.* 20 (11): 1560–1568.
98 Singh, N., Haldar, S., Tripathi, A.K. et al. (2014). Iron in neurodegenerative disorders of protein misfolding: a case of prion disorders and Parkinson's disease. *Antioxid. Redox Signaling* 21 (3): 471–484.
99 Trexler, A.J. and Rhoades, E. (2012). N-Terminal acetylation is critical for forming alpha-helical oligomer of alpha-synuclein. *Protein Sci.* 21 (5): 601–605.
100 Nakamura, T., Yamashita, H., Takahashi, T., and Nakamura, S. (2001). Activated Fyn phosphorylates alpha-synuclein at tyrosine residue 125. *Biochem. Biophys. Res. Commun.* 280 (4): 1085–1092.
101 Okochi, M., Walter, J., Koyama, A. et al. (2000). Constitutive phosphorylation of the Parkinson's disease associated alpha-synuclein. *J. Biol. Chem.* 275 (1): 390–397.
102 Fujiwara, H., Hasegawa, M., Dohmae, N. et al. (2002). Alpha-Synuclein is phosphorylated in synucleinopathy lesions. *Nat. Cell Biol.* 4 (2): 160–164.
103 Kruger, R., Kuhn, W., Muller, T. et al. (1998). Ala30Pro mutation in the gene encoding alpha-synuclein in Parkinson's disease. *Nat. Genet.* 18 (2): 106–108.
104 Choi, W., Zibaee, S., Jakes, R. et al. (2004). Mutation E46K increases phospholipid binding and assembly into filaments of human alpha-synuclein. *FEBS Lett.* 576 (3): 363–368.
105 Zarranz, J.J., Alegre, J., Gomez-Esteban, J.C. et al. (2004). The new mutation, E46K, of alpha-synuclein causes Parkinson and Lewy body dementia. *Ann. Neurol.* 55 (2): 164–173.
106 Appel-Cresswell, S., Vilarino-Guell, C., Encarnacion, M. et al. (2013). Alpha-synuclein p.H50Q, a novel pathogenic mutation for Parkinson's disease. *Mov Disord.* 28 (6): 811–813.
107 Proukakis, C., Dudzik, C.G., Brier, T. et al. (2013). A novel alpha-synuclein missense mutation in Parkinson disease. *Neurology* 80 (11): 1062–1064.
108 Khalaf, O., Fauvet, B., Oueslati, A. et al. (2014). The H50Q mutation enhances alpha-synuclein aggregation, secretion, and toxicity. *J. Biol. Chem.* 289 (32): 21856–21876.
109 Polymeropoulos, M.H., Lavedan, C., Leroy, E. et al. (1997). Mutation in the alpha-synuclein gene identified in families with Parkinson's disease. *Science* 276 (5321): 2045–2047.
110 Tsigelny, I.F., Sharikov, Y., Kouznetsova, V.L. et al. (2015). Molecular determinants of alpha-synuclein mutants' oligomerization and membrane interactions. *ACS Chem. Neurosci.* 6 (3): 403–416.

111 Stefanis, L. (2012). α-Synuclein in Parkinson's disease. *Cold Spring Harbor Perspect. Med.* 2 (2): a009399–a009399.

112 Ozansoy, M. and Basak, A.N. (2013). The central theme of Parkinson's disease: alpha-synuclein. *Mol. Neurobiol.* 47 (2): 460–465.

113 Benskey, M.J., Perez, R.G., and Manfredsson, F.P. (2016). The contribution of alpha synuclein to neuronal survival and function – Implications for Parkinson's disease. *J. Neurochem.* 137 (3): 331–359.

114 Kalia, L.V., Kalia, S.K., McLean, P.J. et al. (2013). alpha-Synuclein oligomers and clinical implications for Parkinson disease. *Ann. Neurol.* 73 (2): 155–169.

115 Kahle, P.J., Neumann, M., Ozmen, L. et al. (2002). Hyperphosphorylation and insolubility of alpha-synuclein in transgenic mouse oligodendrocytes. *EMBO Rep.* 3 (6): 583–588.

116 Wang, Y., Shi, M., Chung, K.A. et al. (2012). Phosphorylated α-synuclein in Parkinson's disease. *Sci. Transl. Med.* 4 (121): 121ra20.

117 Hasegawa, M., Fujiwara, H., Nonaka, T. et al. (2002). Phosphorylated alpha-synuclein is ubiquitinated in alpha-synucleinopathy lesions. *J. Biol. Chem.* 277 (50): 49071–49076.

118 Kalia, L.V. and Kalia, S.K. (2015). Alpha-synuclein and lewy pathology in Parkinson's disease. *Curr. Opin. Neurol.* 28 (4): 375–381.

119 Giasson, B.I., Murray, I.V., Trojanowski, J.Q., and Lee, V.M. (2001). A hydrophobic stretch of 12 amino acid residues in the middle of alpha-synuclein is essential for filament assembly. *J. Biol. Chem.* 276 (4): 2380–2386.

120 Du, H.N., Tang, L., Luo, X.Y. et al. (2003). A peptide motif consisting of glycine, alanine, and valine is required for the fibrillization and cytotoxicity of human alpha-synuclein. *Biochemistry* 42 (29): 8870–8878.

121 Miake, H., Mizusawa, H., Iwatsubo, T., and Hasegawa, M. (2002). Biochemical characterization of the core structure of alpha-synuclein filaments. *J. Biol. Chem.* 277 (21): 19213–19219.

122 Periquet, M., Fulga, T., Myllykangas, L. et al. (2007). Aggregated alpha-synuclein mediates dopaminergic neurotoxicity in vivo. *J. Neurosci.* 27 (12): 3338–3346.

123 El-Agnaf, O.M., Jakes, R., Curran, M.D. et al. (1998). Aggregates from mutant and wild-type alpha-synuclein proteins and NAC peptide induce apoptotic cell death in human neuroblastoma cells by formation of beta-sheet and amyloid-like filaments. *FEBS Lett.* 440 (1–2): 71–75.

124 Rodriguez, J.A., Ivanova, M.I., Sawaya, M.R. et al. (2015). Structure of the toxic core of alpha-synuclein from invisible crystals. *Nature* 525 (7570): 486–490.

125 Tuttle, M.D., Comellas, G., Nieuwkoop, A.J. et al. (2016). Solid-state NMR structure of a pathogenic fibril of full-length human alpha-synuclein. *Nat. Struct. Mol. Biol.* 23 (5): 409–415.

126 Li, B., Ge, P., Murray, K.A. et al. (2018). Cryo-EM of full-length alpha-synuclein reveals fibril polymorphs with a common structural kernel. *Nat. Commun.* 9 (1): 3609.

127 Guerrero-Ferreira, R., Taylor, N.M., Mona, D. et al. (2018). Cryo-EM structure of alpha-synuclein fibrils. *Elife* 7: e36402.

128 Li, Y., Zhao, C., Luo, F. et al. (2018). Amyloid fibril structure of alpha-synuclein determined by cryo-electron microscopy. *Cell Res.* 28 (9): 897–903.

10

Therapeutic Agents for Neurodegenerative Disorders

10.1 Alzheimer's Disease (AD)

Despite the dominant view that aggregation of amyloid (amyloid-cascade hypothesis) is the cause of Alzheimer's disease (AD), there is an urgent need of shedding light on its pathogenesis and pathophysiology in order to develop target-oriented therapeutic strategies. Novel strategies for AD treatment target amyloid-β clearance, τ protein deposits, and apolipoprotein-E function among other approaches involving neuroprotection and neurostimulation [1]. Unfortunately, the lack of clarity about the origin of the disease considerably limits mechanism-based approaches, and symptomatic treatments are the only existing options [2]. Current treatment of AD focuses on delaying the symptoms of the disease [3] and helping the patients maintain mental functions and manage behavioral symptoms [4, 5]. The Food and Drugs Administration (FDA) has approved two types of medications: cholinesterase inhibitors (donepezil, rivastigmine, galantamine) and memantine (a *N*-methyl-D-aspartate [NMDA] receptors antagonist) to treat the cognitive symptoms (memory loss, confusion, and problems with thinking and reasoning) of Alzheimer's disease (Figure 10.1) [6].

Since there are no drugs specifically approved by the FDA to treat behavioral and psychiatric dementia symptoms, nondrug approaches are usually applied. For patients with severe symptoms, different *off-label* drugs can be prescribed, including antidepressants (citalopram, fluoxetine, paroxetine, sertraline, trazodone), anxiolytics (lorazepam and oxazepam), and antipsychotics (aripiprazole, clozapine, haloperidol, olanzapine, quetiapine, risperidone, ziprasidone). Sleep changes are often an additional problem for AD patients. To treat sleep changes, in addition to the aforementioned antidepressants and antipsychotic drugs, tricyclic antidepressants such as nortriptyline and trazodone, and "sleeping agents" such as zolpidem or zaleplon are prescribed [7]. In this section, only those drugs specifically designed for treating cognitive symptoms caused by AD are discussed.

10.1.1 Acetylcholinesterase (AChE) Inhibitors

Acetylcholinesterase (AChE) inhibitors increase levels of acetylcholine, which is depleted in the brain of AD patients, although they can also cause other

Chemical Biology of Neurodegeneration: A Molecular Approach, First Edition. Pedro Merino.

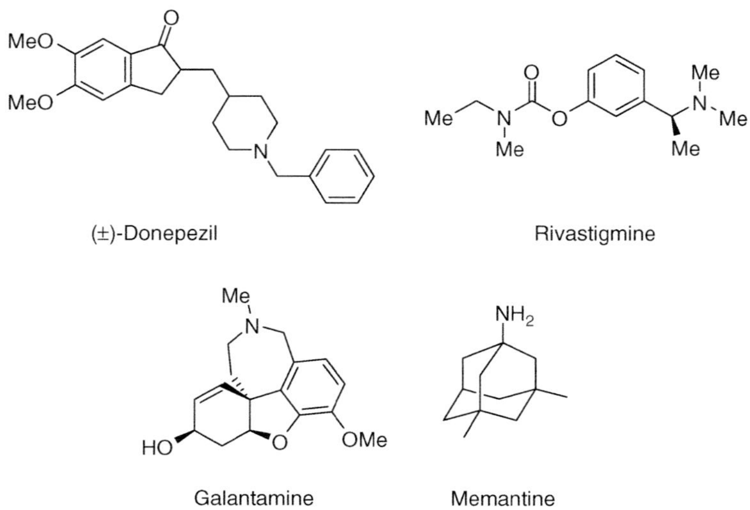

Figure 10.1 FDA-approved drugs for AD treatment.

neuropsychiatric symptoms such as agitation or depression. The three approved cholinesterase inhibitors for the treatment of AD are donepezil (Aricept®), to treat all stages of AD [8], and rivastigmine (Exelon®) and galantamine (Razadyne®), to treat mild to moderate AD.

Donepezil is a reversible AChE inhibitor that increases cortical acetylcholine and promotes central cholinergic neurotransmission [9], presenting several advantages such as nonhepatotoxicity, good blood–brain barrier permeability, and a single daily dose with moderate side effects [10]. It is specific of AChE (AChE:BuChE binding ratio of 1265 : 1) [11] and binds at two different anionic sites on AChE [12, 13]. The catalytic triad in AChE has a narrow 20 Å gorge with a close peripheral anionic site in which the aromatic side chains of residues interact with cationic ligands. The catalytic active site is at the bottom of the gorge. Donepezil has been crystallized with *Tetronarce californica* AChE (PDB ID: 1EVE) [14] and human AChE (PDB ID: 4EY7) (Figure 10.2) [15]. Although the racemic mixture is administered, AChE interacts with the (*R*)-enantiomer. Several analogues have been tested by variations at the benzyl moiety and the piperidine ring lowers the affinity considerably. Donepezil binds differently to human AChE than to *Tc*AChE (see Chapter 8) involving residues beyond the active-site gorge. Overall, interactions in *h*AChE and *Tc*AChE are similar, but a key difference exists: the aromatic groups at the two ends of donepezil are in the same positions giving rise to hydrophobic interactions (the benzyl ring with Trp86-Trp84 in *Tc*AChE- in the active site and the indanone ring with Trp286-Trp279 in *Tc*AChE-), whereas the piperidine ring is 142° rotated about the C10—C11 bond). In fact, the gorge in human AChE created by Tyr337 and Ty341 generates a conformation that allows C15 and C16 of the piperidine ring orientate toward the hydrophobic portions of Tyr337, Phe338, and His447. These interactions

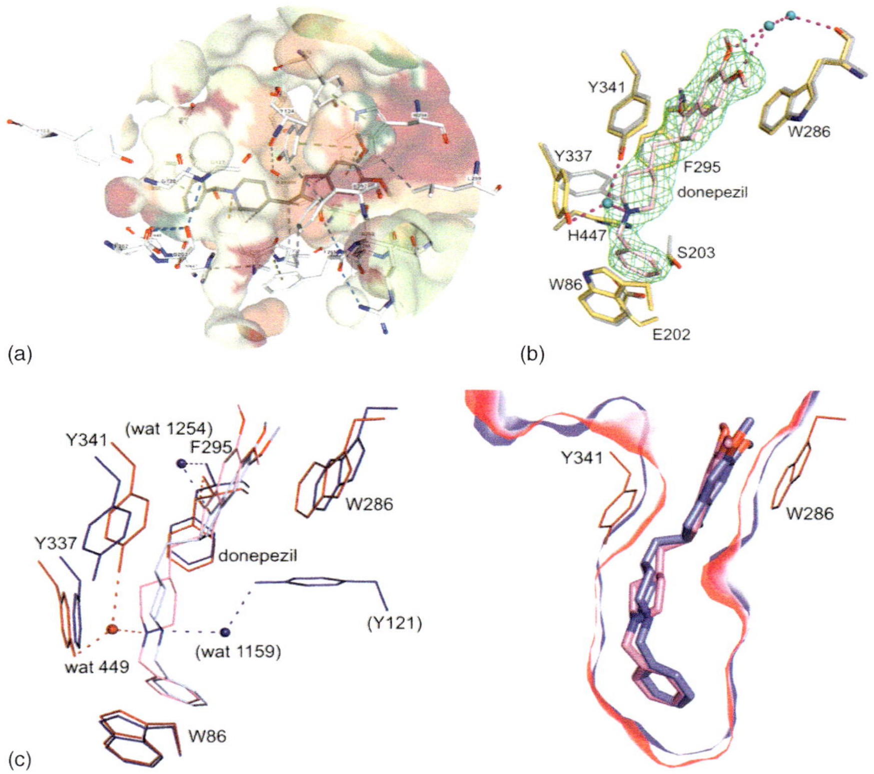

Figure 10.2 (a) Interactions in ligand pockets of *h*AChE for donepezil. Source: Image from the RCSB PDB (www.rcsb.org) of PDB ID 4EY7 and Ref. [15]. H-bonds in blue, halogen bonds in turquoise, hydrophobic contacts in gray, π-interactions in orange/green, metal interactions in purple. Ligand pocket colored by hydrophobicity. (b) Active-site bound (*R*)-donepezil. Carbons are colored pink in the ligands and light yellow in residues of the complex. Residues of the apo state are shown superimposed and colored semitransparent gray. All oxygen and nitrogen atoms are colored red and blue, respectively. Waters are shown as turquoise spheres, and hydrogen bonds are drawn as purple dashes. An Fo – Fc simulated-annealing omit map, contoured at 4σ and colored green, is shown around the ligand. (b) A region near the active site of the human acetylcholinesterase:donepezil complex is shown, with residues from *Torpedo* acetylcholinesterase superimposed. Ligand and protein atoms are shown as sticks, with waters shown as spheres. Protein residues, waters, and hydrogen bond dash representations are colored red in the *h*AChE and blue in the *t*AChE structure. Donepezil nitrogen and oxygen atoms are colored blue and red, respectively, while carbon atoms are colored pink in the human-bound conformer and light blue in the *Torpedo*-bound conformer. Residue numbering refers to human acetylcholinesterase, with equivalent numbers for tAChE in parentheses. (c) The active-site gorge of the superimposed human and *Torpedo* AChE:donepezil complexes is shown; the red and blue "ribbons" represent a 0.5 Å wide slice of the solvent-accessible surface of the human and *Torpedo* structures, respectively. The bound donepezil molecules for both structures are depicted as in (c). Residues W286 and Y341 of the human structure are shown in stick form; the equivalent *t*AChE residues are omitted for clarity. Source: (b, c) Image from the RCSB PDB (www.rcsb.org) of PDB ID 4EY7 and Reprinted with permission from Cheung et al. 2012 [15]. Copyright 2012, American Chemical Society.

reflect a dual-binding mode of action for donepezil, the *N*-benzylpiperidine ring interacting in the catalytic active site, and the indanone moiety binding to the peripheral anionic site [16]. Taking into consideration the interactions of the piperidine ring, several donepezil analogues in which the piperidine ring has been replaced by a pyridinium ring have been evaluated, which have shown promising results [17, 18].

Rivastigmine is a pseudo-reversible AChE inhibitor [12] that has shown effectivity for treating cognitive and behavioral symptoms in AD [19–21]; it also contributes to improve or prevent psychotic and nonpsychotic symptoms associated with AD [22] and may have increased selectivity in the hippocampus and neocortex, two areas typically affected by AD.

Rivastigmine inhibits both AChe and BuChE by covalent binding to active sites, but the precise mechanism of action has not been fully determined. It has been suggested that dual AChE–BuChE inhibitory activity may afford additional therapeutic potential in subcortical dementias, improving cognition and behavioral symptoms [23, 24]. Comparative studies with donepezil revealed that, in the absence of AChE, rivastigmine enhances the levels of extracellular acetylcholine by inhibiting BuChE [25]. Rivastigmine has been crystallized with *T. californica* AChE (PDB ID: 1GQR) [26]. The crystal structure revealed that the carbamyl moiety links covalently to Ser200 at the active site, while the leaving phenol group (NAP, (−)-*S*-3-[1-(dimethylamino)ethyl] phenol) is retained in the anionic site. The crystal structures of *h*BuAChE carbamylated (PDB ID: 6EZ2) and AChE complexed with NAP (PDB ID: 1GQS) have been resolved. A comparison between the native enzyme and that linked to rivastigmine showed a significant change in the orientation of His440, increasing the distance with Asp327 thereby causing disruption of the catalytic triad (Figure 10.3).

Galantamine has a unique dual mode of action [27]. It is a competitive inhibitor of AChE as well as an allosteric modulator of nicotinic acetylcholine receptors (nAChRs), leading to increased central cholinergic neurotransmission [28]. Galantamine response has been analyzed by single-photon emission computed tomography [29], and it is suggested to be related to aggression and dysfunction of the prefrontal cortex [30]. Its effects in reducing caregiver burden have been demonstrated [31]. Galantamine has been crystallized with *T. californica* AChE (PDB ID: 1W76 [32] and PDBID: 1DX6 [33]) and human AChE (PDB ID: 4EY6) [15]. The interaction sites along the active-site gorge of *t*AChE have been investigated with *N*-piperidinopropyl-galantamine (PDB ID: 3I6M) and *N*-saccharinohexyl-galantamine (PDB ID: 3I6Z) [34]. The binding of galantamine to *h*AChE (Figure 10.4) is similar to that observed in *t*AChE in which the inhibitor binds at the base of the active-site gorge through interactions with both the acyl-binding pocket and the principal quaternary ammonium-binding site [36]. A difference found in *h*AChE is the formation of an additional hydrogen bond between the tertiary amine of galantamine and Tyr337 ,which is oriented in a different way than the corresponding Phe330 in *t*AChE.

The affinity of galantamine for nAChRs has been investigated through the X-ray analysis of the complex formed with the soluble acetylcholine-binding

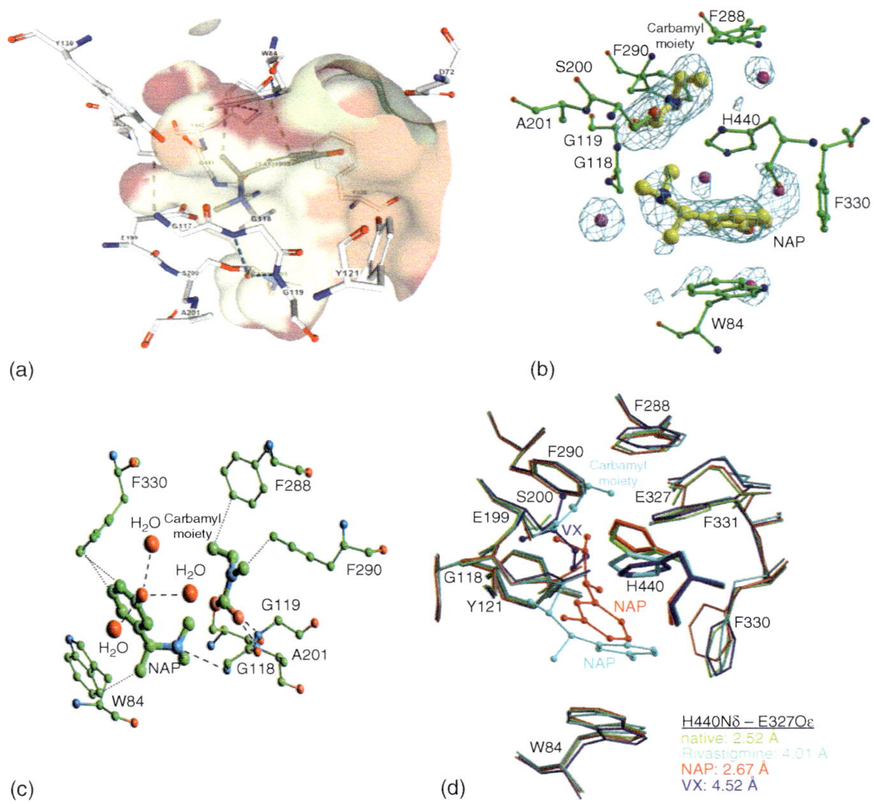

Figure 10.3 (a) Interactions in ligand pockets of *h*AChE for rivastigmine. Source: Image from the RCSB PDB (www.rcsb.org) of PDB ID 1GQR; ref. [26]. H-bonds in blue, halogen bonds in turquoise, hydrophobic contacts in gray, π-interactions in orange/green, metal interactions in purple. Ligand pocket colored by hydrophobicity. (b) Electron density map of the *t*AChE-rivastigmine conjugate. The map shows that both parts of rivastigmine, the carbamyl moiety and NAP, are retained in the active-site gorge, but detached from each other. Rivastigmine is rendered as a ball–and-stick model, with carbon atoms colored yellow, oxygen atoms colored red, and nitrogen atoms colored blue. Selected key protein residues in the vicinity of the conjugate are also rendered in ball-and-stick format, with carbon atoms colored green, and labeled appropriately. (c) Active site of *t*AChE after inhibition with rivastigmine. Amino acid residues within the active site of AChE that may interact with the inhibitor are shown. Rivastigmine is depicted with larger spheres and thicker lines for emphasis. The carbamyl portion of rivastigmine is positioned to make two H-bonds (dashed lines) with the amide nitrogens of Ala201 and Gly119, as well as nonbonded contacts (dotted lines) with Phe288 and Phe290 in the acyl pocket. The aromatic leaving group of rivastigmine (NAP) remains in the active site of crystalline AChE. NAP is within H-bonding distance of three water molecules (large red spheres), as well as of the amide nitrogen of Gly118. Nonbonding contacts and Π–Π interactions with Trp84 and Phe330 are depicted as dotted lines. (d) Comparison of the orientation of His440 in native *t*AChE, the *t*AChE-rivastigmine conjugate, the *t*AChE-NAP complex, and the *t*AChE-VX conjugate. An overlay of the active-site gorge in the four structures is presented. His440 is displayed using thicker lines for emphasis, and the inhibitors are shown as ball-and-stick models. Native *t*AChE is colored green. The *t*AChE-NAP complex is colored red. The *t*AChE-rivastigmine conjugate is colored cyan. The *t*AChE-VX conjugate is colored blue. Source: (b–d) Reprinted with permission from Bar-On et al. 2002 [26]. Copyright 2002, American Chemical Society.

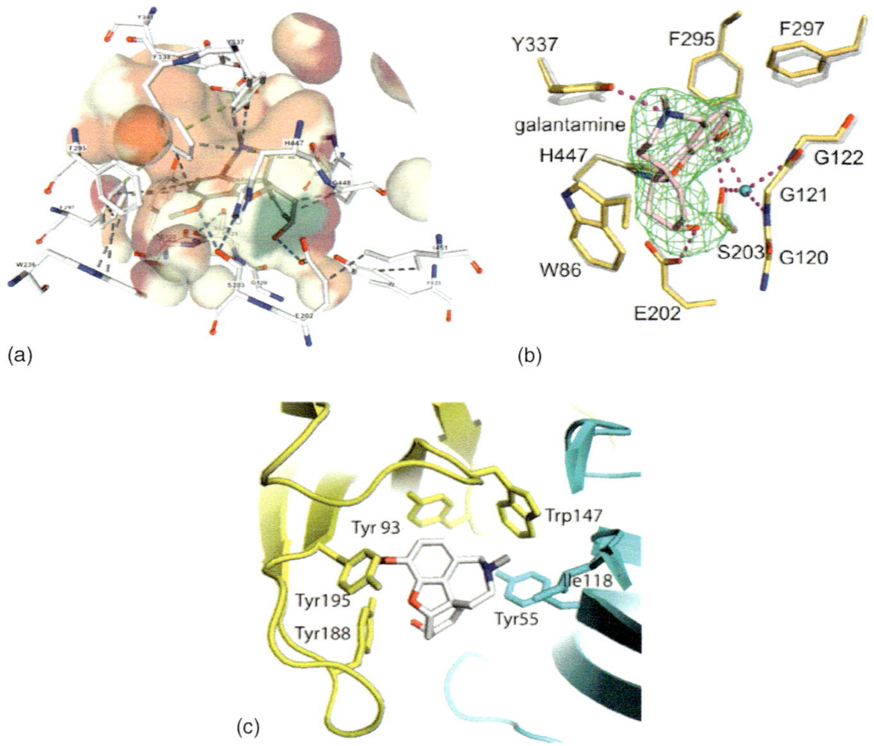

Figure 10.4 (a) Interactions in ligand pockets of *h*AChE for galantamine. Source: Image from the RCSB PDB (www.rcsb.org) of PDB ID 4EY6 and Ref. [15]. H-bonds in blue, halogen bonds in turquoise, hydrophobic contacts in gray, π-interactions in orange/green, metal interactions in purple. Ligand pocket colored by hydrophobicity. (b) Close-up views of the *h*AChE active site with bound galantamine. Residues of the apo state are shown superimposed and colored semitransparent gray. All oxygen and nitrogen atoms are colored red and blue, respectively. Waters are shown as turquoise spheres, and hydrogen bonds are drawn as purple dashes. An $F_o - F_c$ simulated-annealing omit map, contoured at 4σ and colored green, is shown around the ligand. Source: Reprinted with permission from Cheung et al. 2012 [15]. Copyright 2012, American Chemical Society. (c) X-ray crystal structure of galantamine (PDB ID: 2PH9). The subunit interface is shown between two subunits colored yellow and cyan. Source: Hansen and Taylor 2007 [35]. Reprinted with permission of Elsevier.

protein (AChBP) (PDB ID: 2PH9) [35]. AChBP is a soluble homopentamer homologous to the extracellular domain of nAChRs and contains all ligand-binding determinants for competitive nicotinic agonists and antagonists [37]. Consequently, it can be used for studying interactions between galantamine and nAChRs. Galantamine bound AChBP in two different conformations. Binding takes place in four of five subunits in equal proportions, and in all binding pockets containing galantamine loop c packed against the ligand. On the other hand,the vicinal Cys, which shows important interactions with several competitive agonist and antagonists, were positioned away from the ligand. The amine nitrogen of galantamine is placed between Trp147, Tyr55, and Tyr93 but the binding to AChBP is mostly governed by the conserved tryptophan residues at the core of the aromatic pocket and, to a lesser extent, by Tyr195 on loop c (Figure 10.4c) [35].

Bifunctional derivatives of galantamine designed to interact with both the active site and the peripheral anionic site have been studied by X-ray crystallography [32].

10.1.2 Memantine (Namenda®)

Memantine is used in patients with moderate to severe AD [38]. It regulates the activity of glutamate by partially blocking NMDA receptors. It is a noncompetitive *N*-methyl-D-aspartate receptor (NMDAR) antagonist and can be used for treating AD in combination with galantamine [39, 40]. Memantine blocks ion conduction at pentameric ligand-gated ion channels by plugging the channel pore. The X-ray structure of a pentameric ligand-gated ion channel from *Erwinia chrysanthemi* (ELIC) in complex with memantine (PDB ID: 4TWD) and its bromo derivative (PDB ID: 4TWF) showed that the ligand is localized at the extracellular entryway of the channel pore [41]. A comparison with the wild-type ELIC in the apo form (PDB ID: 2VL0) [42] revealed a considerable widening of the channel pore to accommodate the inhibitor at the binding site in the extracellular entryway of the pore (Figure 10.5). These studies suggested that memantine induces a short-lived state different from that induced by an agonist.

Recent studies carried out with crystal structures of GluN1-GluN2B NMDA receptor in which the amino terminal domains were removed (PDB ID: 5UN1) have allowed mapping the binding site of MK-801, an NMDA receptor blocker [43]. These studies together with long-timescale molecular dynamics (MD) simulations showed the mechanism of action of memantine. In summary, memantine physically blocks ion permeation by binding within the vestibule of the ion channel, promoting closure of the ion channel gate and lodging between the M3-helix-bundle crossing and the M2-pore loops.

10.2 Parkinson's Disease (PD)

Parkinson's disease is typically characterized by neurodegeneration of nigrostriatal dopaminergic neurons of the substantia nigra pars compacta, leading to a severe disruption of motor function. In patients with Parkinson's disease (PD), the efficacy of treatment depends on its ability to restore dopamine neurotransmission in the denervated areas of the forebrain, above all in the striatum.

Current treatment of PD focuses on controlling the symptoms by (i) replacing the dopamine lost in PD using L-DOPA (dihydroxyphenylalanine), (ii) stimulating dopamine receptors by the action of dopaminergic agonists, which also have neuroprotective effects, and (iii) inhibiting enzymes catechol-*O*-methyltransferase (COMT) and monoamine oxidase (MAO)-B that catabolize dopamine and its precursor L-DOPA (Figure 10.6).

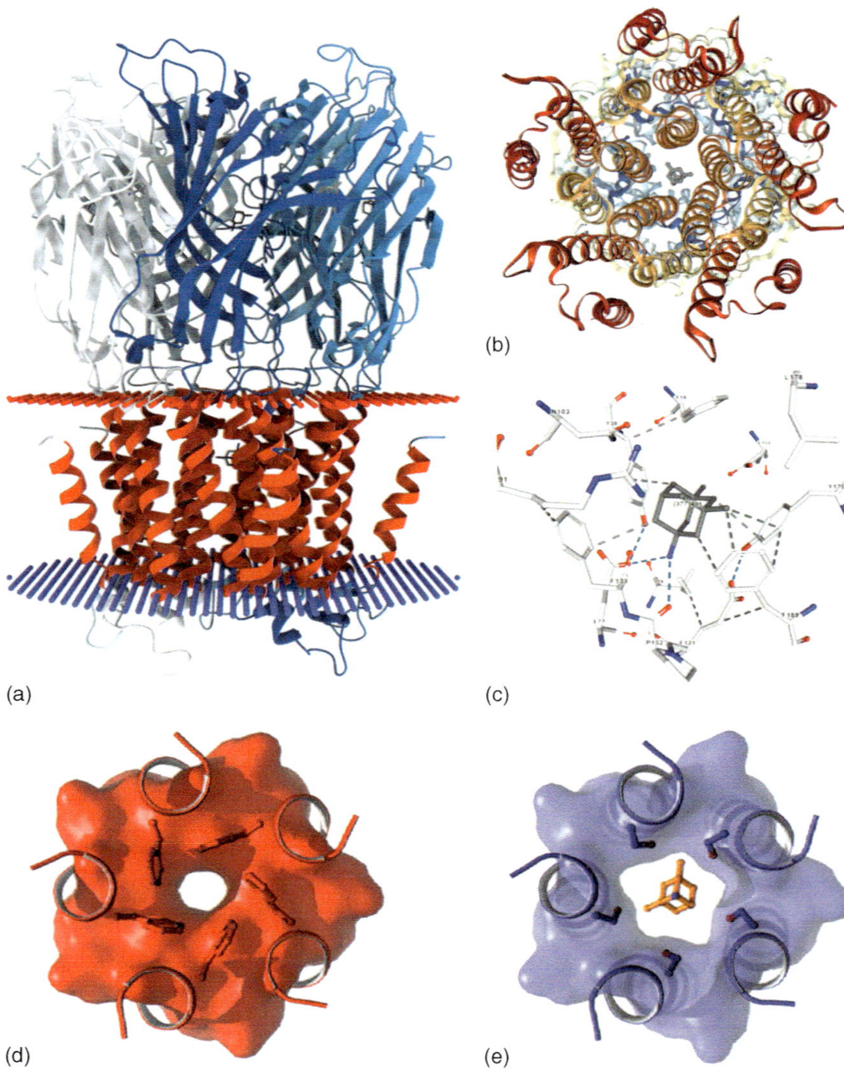

Figure 10.5 X-ray structure of a pentameric ligand-gated ion channel from *Erwinia chrysanthemi* (ELIC) in complex with memantine. (a) General structure showing transmembrane regions. Memantine is shown as sticks in gray. (b) Top cartoon view of ELIC in complex with memantine. (c) Interactions of memantine in ligand pocket. Source: (a–c) Images from the RCSB PDB (www.rcsb.org) of PDB ID 4TWD and Ref. [41]. H-bonds in blue, halogen bonds in turquoise, hydrophobic contacts in gray, π-interactions in orange/green, metal interactions in purple. (d, e) Surface representation of the extracellular entryway of the ELIC channel pore for wild-type ELIC apo (PDB: 2VLO) [42] (red transparent surface, (d)) and ELIC F16′S (blue transparent surface (e)). 16′ residues are shown as sticks. Memantine is shown in orange sticks. Source: Ulens et al. 2014 [41]. Reprinted with permission of Elsevier.

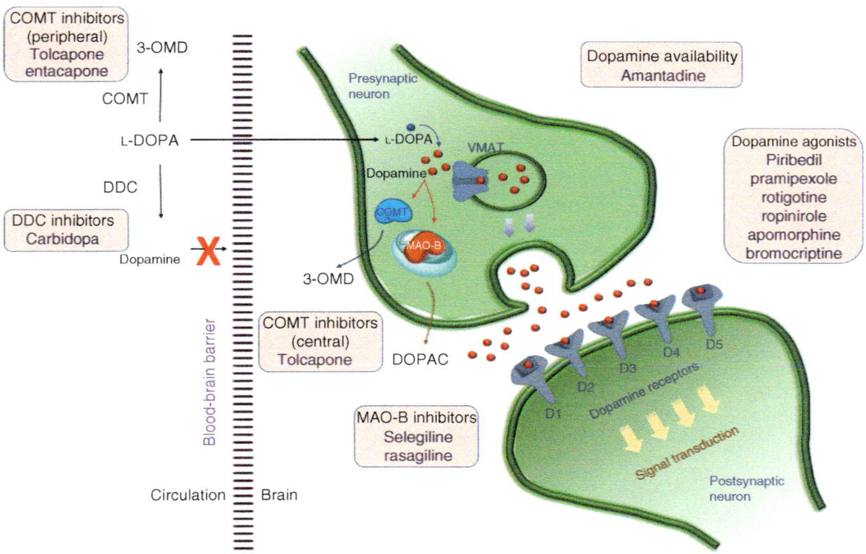

Figure 10.6 Sites of action of PD drugs.

10.2.1 L-DOPA

L-DOPA (dihydroxyphenylalanine) is the metabolic precursor of dopamine by action of DOPA decarboxylase (see Chapter 8), and it is well established as a treatment for the motor deficits of PD and parkinsonism. Unlike dopamine, L-DOPA crosses the brain–blood barrier [44], and it is immediately taken up by dopaminergic neurons and converted into dopamine. Since L-DOPA is an aromatic amino acid, it is also decarboxylated outside the central nervous system (CNS) by peripheral aromatic L-amino acid decarboxylase. L-DOPA may also be methoxylated by COMT to give 3-*O*-methyldopa, which is not converted into dopamine, due to which L-DOPA is usually administered with agents that inhibit peripheral decarboxylases but do not cross the brain–blood barrier, such as carbidopa, and with a COMT inhibitor. Carbidopa has been crystallized in complex with DOPA decarboxylase (PDB ID: 1JS3) [45]. Inhibition of DOPA decarboxylase by carbidopa is exerted by forming a hydrazone bond with the pyridoxal-5'-phosphate (PLP) cofactor through its hydrazine moiety. The resulting complex mimics the external aldimine enzyme–substrate intermediate required for decarboxylation. The catechol ring of carbidopa is inserted in the active-site cleft entering behind the cofactor ring plane (Figure 10.7). The 4′ catechol hydroxyl group of carbidopa forms a hydrogen bond with the hydroxyl group of Thr82. The phosphate group of the cofactor is also involved in substrate binding, by forming a H-bond interaction with the 3′ catechol hydroxyl group of carbidopa. The carboxylate group of carbidopa forms an additional H-bond interaction with His192. It is believed that His192 could assist a catalytic active water molecule (or, alternatively the side chain of Tyr332).

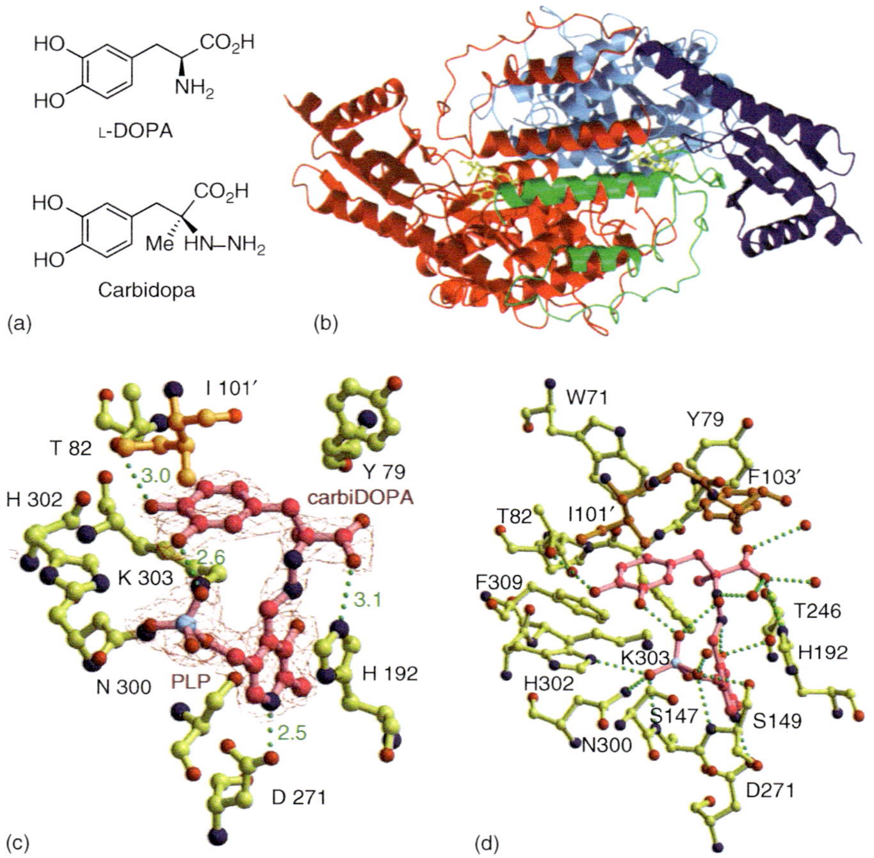

Figure 10.7 (a) Chemical structures of L-DOPA and carbidopa. (b) Ribbon diagram of the polypeptide backbone of DOPA decarboxylase. The view is directly down the twofold symmetry axis. One monomer is completely red, whereas the other one is green (*N*-terminal domain), cyan (large domain), and blue (small domain). The cofactors (PLP) and the inhibitors (carbidopa) are shown in ball-and-stick representation in yellow. (c) Electron density of carbidopa. The difference electron density ($|F_o| - |F_c|$ map with the inhibitor excluded from the phase calculation) in red, contoured at 4σ, is superimposed onto the inhibitor model. Nitrogen, phosphate, and oxygen atoms are marked blue, cyan, and red, respectively. Carbon atoms are colored in yellow for the enzyme, in magenta for the PLP–carbidopa complex, and in orange for the residues of the other monomer. Hydrogen bonds are indicated in green dotted lines. (d) Detailed view of the hydrogen bond interactions, including all structural water molecules in the active site. Color code as in (c). Source: Reprinted with permission from Burkhard et al. 2001 [45]. Copyright 2001, Springer Nature.

Dopamine levels in PD can decrease by 60–80%, and exogenous supplementation of dopamine contributes to compensate the depleted endogenous dopamine. However, only a small percentage of orally administered L-DOPA crosses the brain–blood barrier, and after years of treatment, L-DOPA frequently causes undesired abnormal involuntary movements (L-DOPA-induced dyskinesias, LID) [46] that are poorly understood [47] and with very limited treatments available [48, 49]. The use of mGlu5 negative allosteric modulators can serve attenuated LIDs [50]. Despite the assumption that L-DOPA is transformed in residual striatal dopaminergic neurons into dopamine, the mechanism of action of L-DOPA is not simple and some aspects remain unclear. Thus, the transformation of L-DOPA into dopamine seems to have a more important role for serotonergic than dopaminergic projections. In fact, L-DOPA-induced release of dopamine is widespread and related to serotonergic innervation. [51] L-DOPA produces a substantial increase in nonstriatal regions such as cortex, nucleus accumbens, and the subthalamic nucleus. L-DOPA has also a marked influence upon cholinergic, GABAergic, and glutamatergic neurons. All these characteristics indicate L-DOPA may exert a complex pattern of neurochemical effects of much greater scope than mere transformation into dopamine in dopaminergic neurons [52].

It is also important to point out that dopamine does not readily diffuse into the synapse upon synthesis. Rather, the concentration of dopamine is a function of neuroplasticity and it depends on transporters and regulation (see Chapter 1). Consequently, although dopamine should be available in PD patients, an acute excess of the neurotransmitter can render dopamine to be catabolized by the action of enzymes such as MAO and COMT. For these reasons, administration of L-DOPA is usually complemented with inhibitors of enzymes processing dopamine.

10.2.2 Dopaminergic Agonists

Dopaminergic agonists directly stimulate synaptic dopamine receptors. They have a longer half-life than L-DOPA, providing a more stable clinical effect. Dopaminergic agonists are employed both as monotherapy and as adjuncts to L-DOPA in advanced PD although several factors are needed to be considered for selecting an appropriate treatment [53]. All dopaminergic agonists recognize D_2 receptors, but different affinities with other classes of receptors including serotonergic receptors are observed [54, 55]. In PD striatal D_2 receptors are very sensitive; consequently, partial agonism is sufficient for relief of motor dysfunction. α_2-Adrenergic receptor antagonism enhances adrenergic, dopaminergic, and cholinergic transmission to improve motor function, cognition, mood, and the integrity of dopaminergic neurons. The main prescribed dopaminergic agonists are piribedil, pramipexole, ropinirole, rotigotine, apomorphine, bromocriptine, and amantadine (Figure 10.8).

Piribedil is a nonergoline dopamine agonist indicated for treatment of all stages of PD, either as monotherapy or in combination with L-DOPA. Piribedil is only available in Europe since it is not approved by US FDA. Piribedil has preferred binding affinity for dopamine D_2 and D_3 receptors [56]. Also, it is an antagonist of α_2-adrenergic receptors but show modest affinity for α_1-adrenergic receptors [57]. The characteristic mechanistic profile of piribedil in comparison with other antiparkinsonian agents has been analyzed in a very comprehensive way, demonstrating a very singular profile consisting of (i) partial agonism of D_2 and D_3 receptors, (ii) antagonism of α_2-adrenergic receptors, and (iii) minimal interactions with serotonergic, cholinergic, and histamine

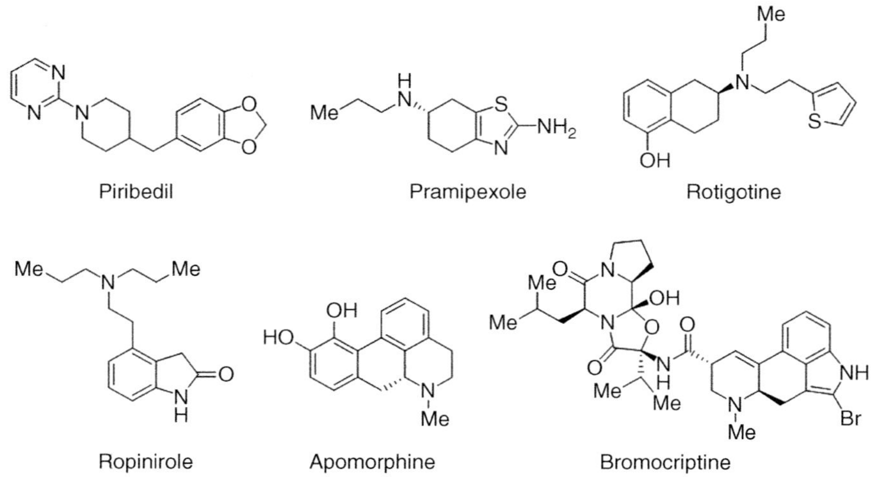

Figure 10.8 Dopaminergic agonists used for the treatment of PD.

receptors [58]. These properties make piribedil a well-established medication for treating early and advanced PD [59]. Recent studies confirmed neuroprotective effects of piribedil on learning and memory [60]. On the other hand, some adverse effects, e.g. hypotension and bradycardia, have been recently reported [61].

Pramipexole is a nonergoline dopamine agonist [62] indicated as monotherapy in early disease and adjunctive therapy with L-DOPA in late disease. The precise mechanism of action of pramipexole is unknown. Pramipexole is an agonist of the dopamine D_2 subfamily receptors showing five- to sevenfold greater affinity for the D_3 receptors with lower affinities for the D_2 and D_4 receptor [63]. Thus, the therapeutic effect of pramipexole is reflected in the caudate [64]. Pramipexole has only minimal α_2-adrenergic receptor activity and virtually no other receptor agonism or antagonism [65]. However, it has been reported that sustained administration of pramipexole contributes to increased serotonin neurotransmission [66]. Further studies confirmed the contribution to serotonergic metabolism but did not support any role of brain-derived neurotrophic factor (BDNF) [67]. On the other hand, recent studies showed that pramipexole promotes structural plasticity in dopaminergic neurons via BDNF and mammalian target of rapamycin (mTOR) signaling, suggesting a key interaction between BDNF and D_3 receptor signaling pathways [68]. It has also been demonstrated that prolonged treatment with pramipexole induces a reduction in dopamine reuptake by dopaminergic neurons due to a drug-induced physical interaction between D_3 receptors and dopamine transporters [69].

Ropinirole is a nonergoline dopamine agonist [70] indicated as monotherapy and adjunctive therapy with L-DOPA [71]. Ropinirole exhibits high specificity for dopamine D_3 receptors with lower but significant affinity for D_2 receptors in the striatum. Minimal or negligible affinity for D_1-like receptors, very low affinity for α_2-adrenoreceptors and 5-HT_2 receptors has been observed and no affinity for $5HT_1$ receptors or GABA receptors [72]. Ropinirole activates the glutathione system, suggesting a mode of action for its neuroprotective effect [73]. As pramipexole, ropinirole also promotes structural plasticity in dopaminergic neurons [68].

Rotigotine is a nonergoline dopamine agonist indicated as monotherapy in early disease and adjunctive therapy with L-DOPA in advanced disease [74], showing a good tolerability among elderly patients [75]. Transdermal application of rotigotine has demonstrated to be an attractive treatment improving morning motor functioning and reduced sleep disturbances, nighttime motor symptoms, depressive symptoms, pain and functioning [74], as well as neuropsychiatric symptoms [60] and blood pressure in PD [76]. The binding affinity is specific for dopamine D_{1-5} receptors with higher affinity for D_3 receptors. Rotigotine is also an antagonist of α_{2B}-adrenergic receptors and a partial agonist of serotonin 5-HT_{1A} receptors [77]. The precise mode of action at a molecular level is unknown, even though it has been suggested that it stimulates receptors within the caudate putamen. It has been found that rotigotine increases BDNF protein in the cortex and hippocampus of rats [78].

Apomorphine ((*R*)-enantiomer) is a nonergoline dopamine agonist [79] indicated for the treatment of "off" episodes in PD [80, 81]. Apomorphine is nonselective, showing binding affinities for both dopamine D_1 and D_2 subfamilies of receptors [82]. The highest affinities were measured for D_3 and D_4 receptors (K_i values in nM are D_1: 484; D_2: 52; D_3: 26; D_4: 4.4; D_5: 189) [54, 55]. Apomorphine also possesses high binding affinity for serotonergic (5-HT_{1A}, 5-$HT_{2A\text{-}C}$) and adrenergic receptors ($\alpha_{1B,D}$, $\alpha_{2A\text{-}C}$). Apomorphine activates dopamine receptors in the nigrostriatal pathway, the limbic system, the hypothalamus, and the pituitary gland although the exact mode of action is unknown. The binding interactions of dopamine and apomorphine at their high- and low-affinity sites of D_2 receptors have been studied and compared using model structures composed of adrenergic crystal structures as templates. The interactions of both ligands have been fully determined for active/inactive states for monomeric and dimeric structures (Figure 10.9) [83]. In agreement with previous biological investigations, the allosteric receptor–receptor interaction for dopamine D_2 receptor dimer is correlated with the accessibility of adjacent residues of transmembrane region 4. The model also predicts correctly the negative cooperativity effect on apomorphine binding.

The model reveals a higher binding affinity of apomorphine at the active state of D_2 receptors showing a more favorable binding of apomorphine than dopamine at both D_{2H} and D_{2L} conformers.

Bromocriptine is the first dopamine agonist successfully employed for the treatment of PD [84]. It is a nonselective ergoline derivative [85, 86] with potent agonist activity of D_2 receptors with special affinity for D_2 and D_3 receptors (K_i values for D_1: 682 nM; D_5: 496 nM; D_2: 2.96 nM; D_3: 5.42 nM; D_4: 328 nM) [87] and various serotonin (K_i values for 5-HT_{1A}: 12.9 nM; 5-HT_{1B}: 355 nM; 5-HT_{1D}: 10.7 nM; 5-HT_{2A}: 107 nM; 5-HT_{2B}: 56.2 nM; 5-HT_{2C}: 741 nM; 5-HT_6: 33 nM) [55, 88] and adrenergic receptors (K_i values for: α_{1A}: 4.17 nM; α_{1B}: 1.38 nM; α_{1D}: 1.12 nM; α_{2A}: 11.0 nM; α_{2B}: 34.7 nM; α_{2C}: 28.2 nM; β_1: 589 nM; β_2: 741 nM) [89]. Bromocriptine also reverts the glutamate transporter GLT1 inhibiting the release of glutamate [90].

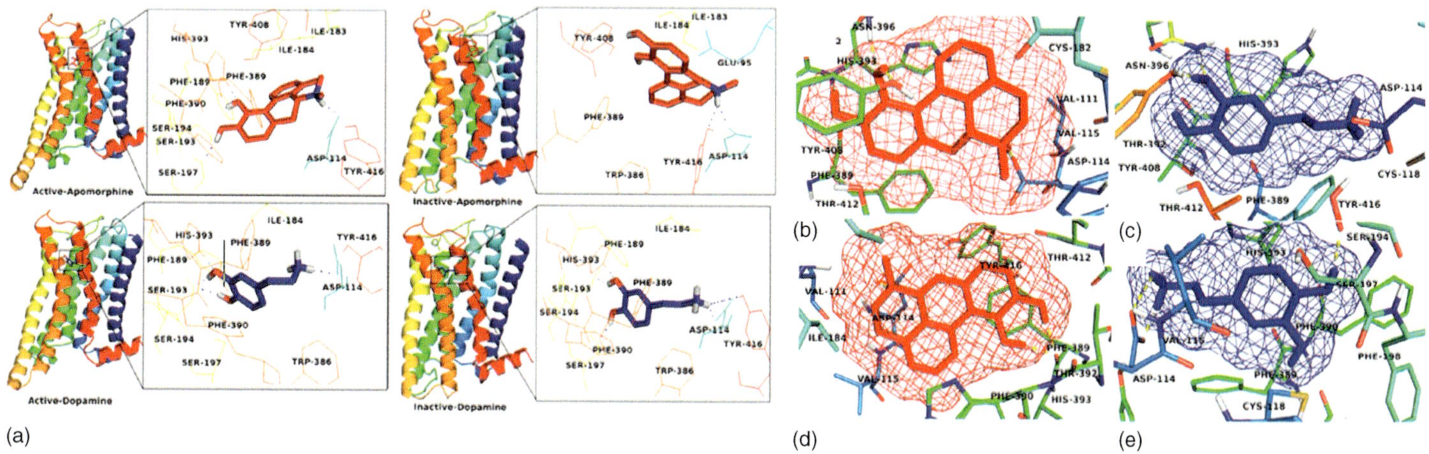

Figure 10.9 (a) Top docking poses from IFD simulations of D_2 receptor monomers were selected for this profile. Each TM is highlighted by different colors (TM1–8). TM cavities of individual complexes were focused and their key amino acid residues. Hydrogen bonding interactions are represented by blue dashes. (b–e) 3D docking poses of apomorphine and dopamine into the active site of active (b, c) and inactive (d, e) D_2 receptor dimers. Electron density of each ligand is illustrated via lines surrounded the atoms. The critical residues that are making polar and nonpolar interactions with compounds are represented. (f–m) 2D ligand interaction diagrams for docking poses of apomorphine and dopamine into the active D_{2R} dimer (f, g); inactive D_2 receptor dimer (h, i), active D_2 receptor monomer (j, k) and inactive D_2 receptor monomer (l, m) are diagrammed. Key amino acids within 4 Å of docked compounds and their binding interactions are identified. Source: Reprinted with permission from Durdagi et al. 2016 [83]. Copyright 2016, American Chemical Society.

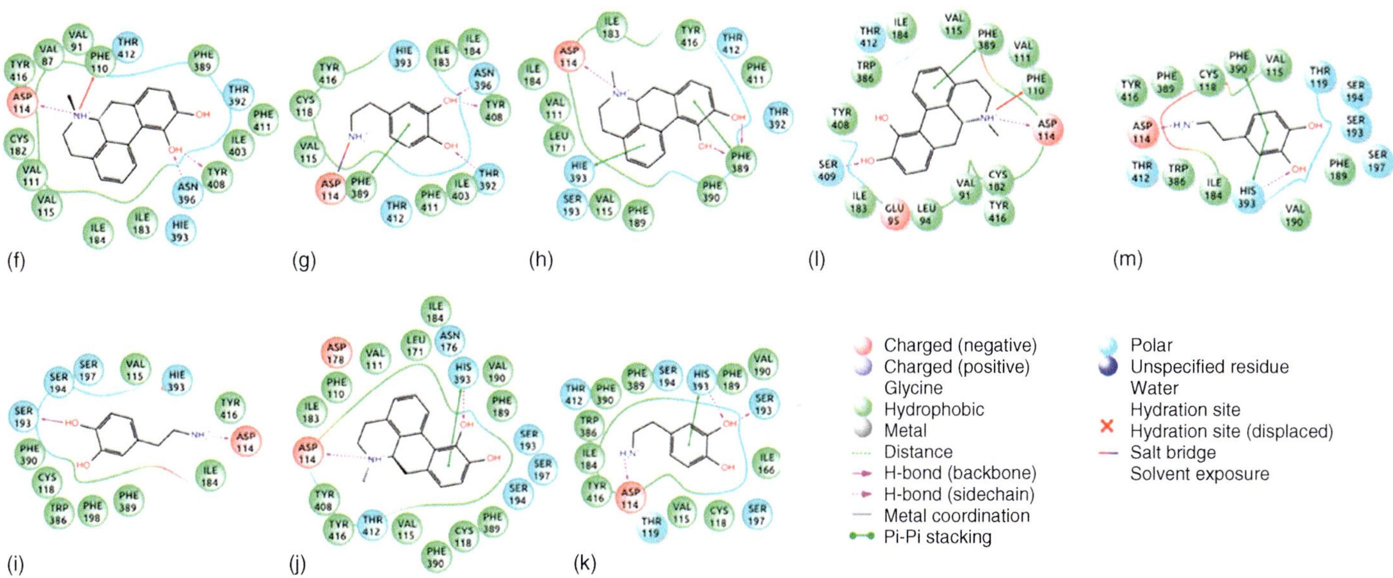

Figure 10.9 *(Continued)*

Bromocriptine has the potential for drug–drug interactions mediated by the cytochrome *P*450 enzyme system, and it is associated with a lower risk of dyskinesia, but fibrosis is a class effect typical of the ergot-derived dopamine agonists [91]. A model for the interaction between cytochrome *P*450 and bromocriptine has been obtained by X-ray crystallography (PDB ID: 3UA1) [92]. The mechanism of action of bromocriptine has been studied in rats, and it is suggested that bromocriptine can inhibit secretion of catecholamines by stimulation of cholinergic receptors, as well as by membrane depolarization [93].

10.2.3 *MAO-B Inhibitors*

MAOs are flavin adenine dinucleotide (FAD)-containing enzymes associated with the mitochondrial outer membrane that catabolize dopamine [94]. Consequently, their inhibition should contribute to an increase in the half-life of the neurotransmitter. The major side effect with the use of first-generation non-selective MAO inhibitors as neuropsychiatric drugs was the so-called "cheese effect," a hypertensive response to excess dietary tyramine, a biogenic sympathomimetic amine. The cheese effect is a consequence of inhibition of MAO-A, the enzyme responsible for metabolism of noradrenaline and serotonin, located in peripheral adrenergic neurons [95]. The identification of two forms of MAO [96, 97] and the development of selective irreversible inhibitors resolved some of these problems. The selective inhibition of brain MAO-B results in elevation of synaptosomal dopamine concentrations [98], and it is currently a field of interest in the treatment of PD [99]. Consequently, MAO-B inhibitors are considered valuable drugs in the treatment of PD [100]. Although there exists some controversy regarding future development of that class of compounds [101], solid evidences that the treatment with MAO-B inhibitors is advantageous as it enables sparing of dopamine agonist and levodopa dosing have been reported [102]. The three MAO-B inhibitors used for treating PD are selegiline, rasagiline, and safinamide (Figure 10.10) [103].

Selegiline (L-deprenyl) is the first selective MAO-B inhibitor [104] developed to treat PD [105, 106]. It blocks microsomal metabolism of dopamine and enhances the dopaminergic activity in the substantia nigra as well as the efficacy of L-DOPA [107]. Several studies suggested that it also enhances the level of cAMP [108] and prevents the undesired action of neurotoxin 1-methyl-4-phenyl-1,2,3,6-tetrahydropyridine (MPTP), decreasing the oxidative stress [109]. In a particular case of selegiline, the molecular mechanism of inhibition of MAO-B is not fully understood, but it has been found that selegiline binds to MAO-B in the

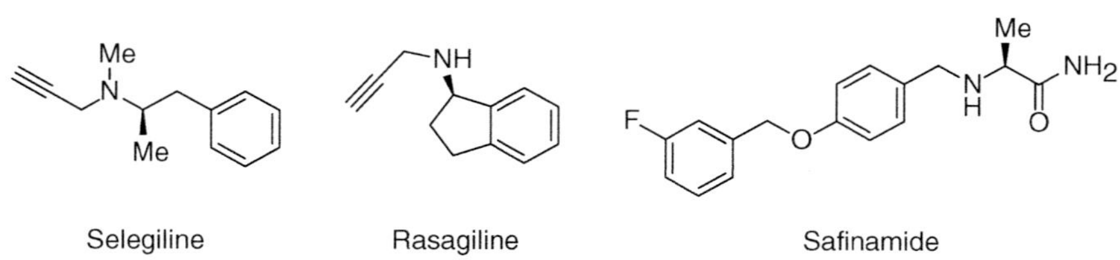

Figure 10.10 Selective MAO-B inhibitors used for treating PD.

basal ganglia [110], within the nigrostriatal pathways in the CNS, including the thalamus, the striatum, the cortex, and the brainstem [111]. Selegiline may also increase dopaminergic activity through mechanisms other than inhibition of MAO-B. The metabolism of selegiline is complex and the metabolites also take part in the complex pharmacological activity of the drug. Selegiline is rapidly metabolized by the microsomal enzymes to amphetamine, methamphetamine, and desmethyldeprenyl [112] although more than 40 metabolites have been detected or suggested [113]. During the past years, rasagiline has been introduced to replace selegiline in clinical application for the treatment of PD even though the recent development of selegiline nanoparticles has rendered again this drug as a promising approach for treating PD [114, 115]. Indeed, selegiline is more effective than rasagiline as MAO-B inhibitor [116].

Rasagiline is an irreversible MAO-B inhibitor indicated for the treatment of PD [117, 118] as initial monotherapy and as adjunct therapy to levodopa [119, 120]. In contrast to selegiline, the prototype of MAO-B inhibitors, rasagiline having five times the potency of selegiline [98], is not metabolized to potentially toxic amphetamine metabolites [121]. The crystal structure of MAO-B in complex with rasagiline (PDB ID: 1S2Q) revealed that irreversible inhibition is due to the covalent bonds formed between flavin and the propargyl chain (Figure 10.11a) [122]. The iminopropene unit formed upon binding adopts an extended conformation, demonstrating that the covalent bonds are fully conjugated (Figure 10.11b). The interactions between rasagiline and MAO-B also involve a number of van der Waals contacts with several amino acid residues (Figure 10.11c). The indan ring is oriented perpendicular to the flavin so that its projection onto the cofactor plane falls into the central pyrazine ring. Moreover, C4 of the indan ring is located at the pore that connects the substrate cavity to the entrance cavity, paving the way to prepare derivatives by introducing substituents at that position [123].

In contrast to selegiline and rasagiline, safinamide is a reversible selective inhibitor of MAO-B [124] that also inhibits glutamate release [125] and dopamine and serotonin reuptake [126]. Safinamide binds to the sigma receptors as well [127], and additionally, it blocks sodium and calcium channels [125, 128]. Safinamide is used for the treatment of mid-to-late stage PD [129, 130], providing enhanced symptom control of motor function [131]. The addition of safinamide to dopamine agonists may also be an effective treatment strategy in early PD patients, prolong the use of dopamine agonists, and/or delay the introduction of L-DOPA [132]. The mechanism of action of safinamide is double: as a MAO-B inhibitor it reduces dopamine reuptake at synaptic junctions increasing the time dopamine is available as a neurotransmitter. The nondopaminergic mechanism of action of safinamide consists of blocking site 2 of voltage-gated N-type sodium channels and calcium channels during their inactivated state inhibiting glutamate release. Glutamate, dopamine, and GABA receptors themselves are not affected by safinamide [133]. Some studies revealed neuroprotective effects for safinamide, but they are not conclusive [134].

The X-ray structure of MAO-B in complex with safinamide has been resolved (PDB ID: 2V5Z) [135]. The MAO-B active site consists of two cavities, the substrate cavity in front of the flavin and the entrance cavity located underneath the

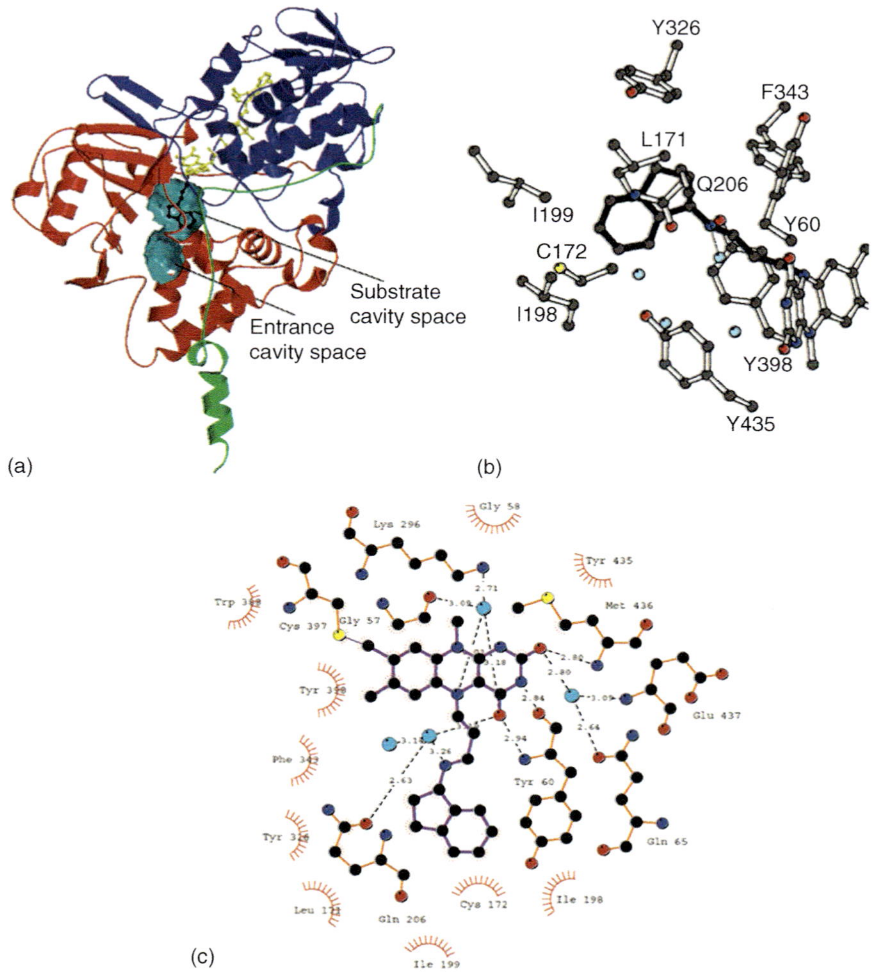

Figure 10.11 (a) Schematic representation of the MAO-B subunit in complex with rasagiline. The FAD-binding domain is in blue, the substrate-binding domain is in red, and the membrane-binding C-terminal region is in green. The FAD cofactor and rasagiline are shown as yellow and black ball-and-stick figures, respectively. The protein cavities are outlined by a cyan semitransparent surface. (b) Details of binding mode of rasagiline. Carbon atoms are in black, oxygen atoms are in red, nitrogen atoms are in blue, and sulfur atoms are in yellow. Water molecules are shown as cyan spheres. H-bonds to inhibitor atoms are outlined by dotted lines. (c) LIGPLOT schematic drawing showing the interactions of rasagiline with MAO-B. Atom colors are as in (b). Dashed lines indicate all potential H-bonds. "Radiating" spheres indicate hydrophobic contacts between carbon atoms of the inhibitor and the neighboring residues. Source: Reprinted with permission from Binda et al. 2004 [122]. Copyright 2004, American Chemical Society.

protein surface and closed by the loop formed by residues 99–112 (Figure 10.12a). Safinamide binds noncovalently to MAO-B in an extended conformation occupying both cavities (Figure 10.12b) in a similar way to other inhibitors such as 7-(3-chlorobenzyloxy)-4-carboxaldehyde-coumarin. While the 3-fluorobenzyloxy

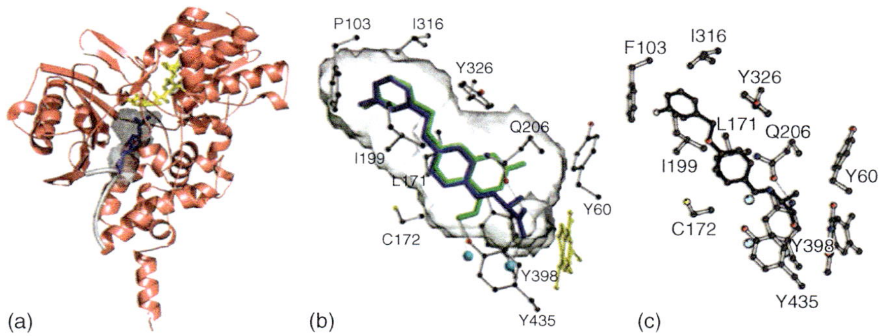

Figure 10.12 (a) Ribbon diagram of MAO-B overall structure. The FAD cofactor and safinamide are highlighted in yellow and blue stick representation, respectively. The semitransparent surface represents the active-site cavity, whose inner part (dark gray) forms the substrate binding site, whereas the outward-facing space (light gray) corresponds to the entrance cavity. Loop formed by residues 99–112, which admits ligands into the active site, is highlighted as gray thick ribbon. (b) Safinamide (blue) and 7-(3-chlorobenzyloxy)-4-carboxaldehyde-coumarin (green) bound in the MAO-B cavity. The structure is rotated approximately 90° with respect to (a). The active-site residues and the flavin are drawn in gray and yellow, respectively. Hydrogen bonds established by the two inhibitors with protein residues and water molecules (cyan spheres) are shown as dashed lines. (c) Detail on safinamide bound to MAO-B. Color code and orientation are as in (b). Active-site residues and the inhibitor are drawn in gray and black, respectively. Water molecules are shown as cyan spheres. Dashed lines show hydrogen bonds between the inhibitor and protein residues. Source: Reprinted with permission from Binda et al. 2007 [135]. Copyright 2004, American Chemical Society.

moiety is in the entrance cavity space, the other aromatic ring occupies the substrate cavity with the primary amide group oriented toward the flavin cofactor.

10.2.4 COMT Inhibitors

COMT metabolizes dopamine to produce 3-*O*-methyldopa (3-OMD), an inactive metabolite that, however, can compete with L-DOPA for translocation into the brain by facilitated transporters [136]. For this reason, COMT inhibitors are commonly used as an adjunct to L-DOPA in PD [137]. The use of COMT inhibitors contributes to lengthen the presence of L-DOPA in the plasma by preventing its peripheral metabolism [138] and thus increasing the bioavailability in the brain [139]. The three COMT inhibitors used for treating PD are entacapone, tolcapone, and opicapone (Figure 10.13).

Entacapone and tolcapone are selective and reversible inhibitors of COMT that prevents the conversion of L-DOPA to 3-OMD and other nonbeneficial metabolites. Both are used as an adjunct treatment to L-DOPA /carbidopa to increase its duration of action. Some studies with entacapone demonstrated its inhibitory central activity of COMT. Similarly, though the precise mechanism of action of tolcapone in PD is unknown, it is suggested that contributes, by means of peripheral effects, to enhance the plasma pharmacokinetics of levodopa [140]. Like entacapone, tolcapone also prevents the conversion of L-DOPA to 3-OMD

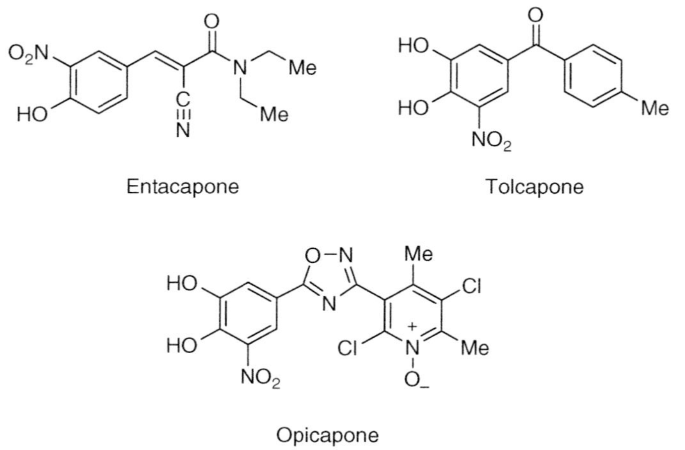

Figure 10.13 Selective COMT inhibitors used for treating PD.

by inhibiting COMT. Both compounds uncouple oxidative phosphorylation and inhibit mitochondrial enzyme complexes, causing mitochondrial toxicity (higher for tolcapone), which could be a possible mechanism for tolcapone-associated hepatotoxicity [141].

The X-ray structure of rat COMT in complex with *S*-adenosyl methionine (SAM), magnesium, and tolcapone (PDB ID: 3S68) has been resolved [142]. A humanized rat COMT in complex with sinefungin, magnesium, and tolcapone has also been crystallized (PDB ID: 4PYL) but a part of tolcapone is disordered [143]. The quaternary complex between rat COMT, magnesium, SAM, and tolcapone showed the SAM methyl group close to the hydroxyl group at the ortho position to the nitro group of tolcapone (Figure 10.14a) but methylation does not take place due to the lack of nucleophilicity of the nitro group. The catechol moiety of tolcapone forms two dative bonds to magnesium ion.

Opicapone is the last approved COMT inhibitor for the treatment of PD [144]. It is used as an adjunctive of L-DOPA and L-DOPA decarboxylase (DDC) inhibitors [145] in adult PD patients [146] and end-of-dose motor fluctuations [147]. Opicapone is a good alternative to the moderate efficacy of entacapone and the risk of liver toxicity for tolcapone [148].

10.2.5 Amantadine

Amantadine is an antiviral agent that also presents anti-PD activity. It may be prescribed alone in early-stage PD and also in combination with carbidopa–levodopa therapy [149] during the later stages of PD to prevent dyskinesias [150]. In fact, amantadine is used for the treatment of L-DOPA-induced dyskinesia [151, 152]. The mode of action of amantadine in PD consists of stimulating the release of dopamine from striatal dopaminergic nerve terminals and inhibiting its presynaptic reuptake [153]. Postsynaptically, amantadine acts directly on the dopamine D_2 receptor [154]. The amantadine-induced hypersensitivity of dopamine

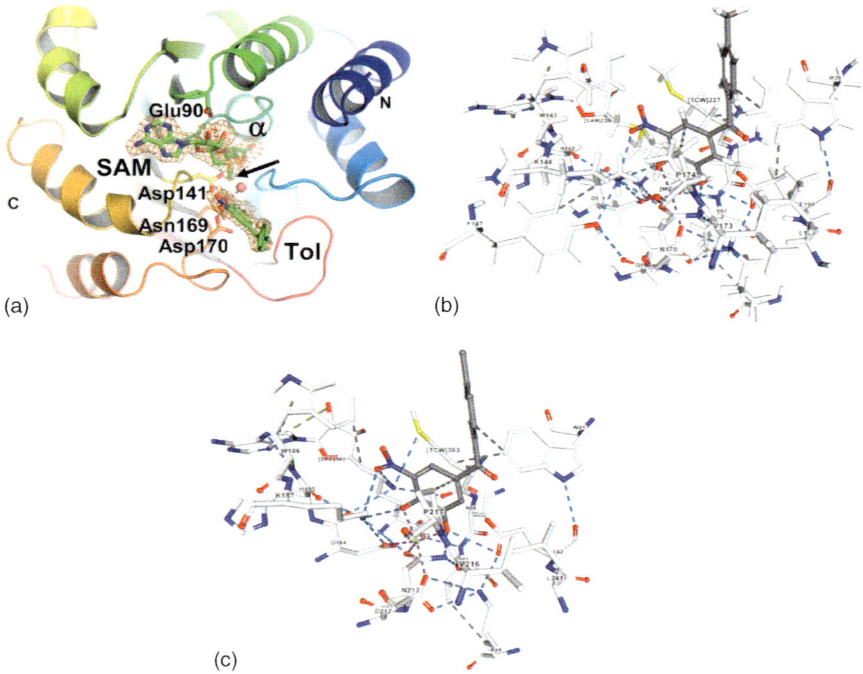

Figure 10.14 (a) Ribbon diagram of the COMT–SAM–Mg^{2+}–tolcapone complex (PDB ID: 3S68), spectrally colored from the N-terminus to the C-terminus. SAM and tolcapone are drawn as stick models with C atoms colored green. Three aspartate side chains (yellow and orange) and a water molecule coordinating the Mg^{2+} are shown as stick models and as a sphere, respectively. The arrow indicates the methyl group of SAM that is usually transferred to the substrate, marking the exit vector for connection with the catechol in bisubstrate inhibitors. Source: Reprinted with permission from Ref. [142]. (b) Details on interactions of tolcapone in ratCOMT. Source: Image from the RCSB PDB (rcsb.org) of PDB ID 3S68 and Ref. [142]. (c) Details on interactions of tolcapone in humanized ratCOMT. Source: Image from the RCSB PDB (rcsb.org) of PDB ID 4PYL and Ref. [143].

receptors has been demonstrated in rats, although the effects were only transient [155]. Amantadine has also anticholinergic activity through inhibition of NMDA receptor-mediated stimulation of acetylcholine [156]. Recent studies showed that amantadine enhances nigrostriatal and mesolimbic dopamine function in the rat brain in relation to motor and exploratory activity [157].

10.3 Lewy Body Disease (LBD)

Lewy body disease (LBD) is one of the most common causes of dementia in the elderly. LBD happens when abnormal structures, called Lewy bodies, build up in areas of the brain. The disease may cause a wide range of symptoms, including changes in alertness and attention, hallucinations, problems with movement and posture, muscle stiffness, confusion, and loss of memory. LBD can be hard to diagnose, because Parkinson's disease and Alzheimer's disease cause similar symptoms. Scientists think that LBD might be related to these diseases or that they sometimes happen together. There is no cure. Treatment focuses on drugs to help symptoms, and there are no approved drugs for specifically treating LBD. Cognitive symptoms are treated with cholinesterase inhibitors [158, 159] developed for AD like donepezil [160, 161]. Some studies revealed that dementia associated with LBD can be treated with rivastigmine [162]. Movement symptoms are treated with L-DOPA used for treating PD. Other symptoms such as visual hallucinations, rapid eye movement sleep behavior disorder (RBD), or neuroleptic sensitivity are treated with specific drugs for treating the corresponding symptoms although in some cases can be problematic (e.g. traditional antipsychotic drugs).

10.4 Huntington's Disease (HD)

Huntington's disease (HD) is an inherited disease that causes certain nerve cells in the brain to waste away. People are born with a defective gene, but symptoms usually don't appear until middle age. Early symptoms of HD may include uncontrolled movements, clumsiness, and balance problems. Later, HD can take away the ability to walk, talk, and swallow. Some people stop recognizing family members. Others are aware of their environment and are able to express emotions. There is no cure. Drugs can help manage some of the symptoms, but cannot slow down or cure the disease. There are no approved drugs for specifically treating Huntington's disease.

10.5 Friedreich's Ataxia (FA)

Friedreich's ataxia (FA) is an inherited disease that damages the nervous system. The damage affects spinal cord and the nerves that control muscle movement in arms and legs. Symptoms usually begin at early ages (5–15). The main symptom is ataxia, which means trouble coordinating movements. Specific symptoms include difficulty walking, muscle weakness, speech problems, involuntary eye movements, scoliosis, and heart palpitations. People with Friedreich's ataxia usually need a wheelchair 15 to 20 years after the first symptoms appear. In severe cases, people become incapacitated. There is no cure. Symptoms can be treated with drugs, braces, surgery, and physical therapy, but there are no approved drugs for specifically treating Friedreich's ataxia.

10.6 Amyotrophic Lateral Sclerosis (ALS)

Amyotrophic lateral sclerosis (ALS) is a nervous system disease that attacks neurons in the brain and spinal cord. In the first stages, mild muscle problems are noticed causing trouble walking, writing, and speech problems. During advanced stages, patients lose strength and cannot move and when muscles in the chest fail, it is not possible to breathe. A breathing machine can help, but most people with ALS die from respiratory failure. Causes of ALS are unknown and there is no cure. Palliative treatments can relieve symptoms, sometimes, prolong survival, but there are no approved drugs for specifically treating ALS.

10.7 Spinal Muscular Atrophy (SMA)

Spinal muscular atrophy (SMA) is a genetic disease that attacks motor neurons in the spinal cord, affecting walking, crawling, breathing, swallowing, and head and neck control. There are many types of SMA. Some of them are fatal. Some people have a normal life expectancy. It depends on the type and how it affects breathing. There is no cure. Treatments help with symptoms and prevent complications. They may include machines to help with breathing, nutritional support, physical therapy, and medicines. There are no approved drugs for specifically treating SMA.

References

1 Cao, J., Hou, J., Ping, J., and Cai, D. (2018). Advances in developing novel therapeutic strategies for Alzheimer's disease. *Mol Neurodegener* 13 (1): 64.
2 Beier, M.T. (2007). Treatment strategies for the behavioral symptoms of Alzheimer's disease: focus on early pharmacologic intervention. *Pharmacotherapy* 27 (3): 399–411.
3 Tariot, P.N. (2006). Contemporary issues in the treatment of Alzheimer's disease: tangible benefits of current therapies. *J Clin Psychiatry* 67 (Suppl 3): 15–22. quiz 23.
4 Standridge, J.B. (2004). Pharmacotherapeutic approaches to the treatment of Alzheimer's disease. *Clin Ther* 26 (5): 615–630.
5 Cummings, J.L., Mackell, J., and Kaufer, D. (2008). Behavioral effects of current Alzheimer's disease treatments: a descriptive review. *Alzheimers Dement* 4 (1): 49–60.
6 Grossberg, G.T. (2005). Rationalizing therapeutic approaches in Alzheimer's disease. *CNS Spectr* 10 (11 Suppl 18): 17–21.
7 Urrestarazu, E. and Iriarte, J. (2016). Clinical management of sleep disturbances in Alzheimer's disease: current and emerging strategies. *Nat Sci Sleep* 8: 21–33.
8 Lee, J.H., Jeong, S.K., Kim, B.C. et al. (2015). Donepezil across the spectrum of Alzheimer's disease: dose optimization and clinical relevance. *Acta Neurol Scand* 131 (5): 259–267.
9 Seltzer, B. (2005). Donepezil: a review. *Expert Opin Drug Metab Toxicol* 1 (3): 527–536.
10 Wilkinson, D. and Murray, J. (2001). Galantamine: a randomized, double-blind, dose comparison in patients with Alzheimer's disease. *International Journal of Geriatric Psychiatry* 16 (9): 852–857.

11 Ogura, H., Kosasa, T., Kuriya, Y., and Yamanishi, Y. (2000). Comparison of inhibitory activities of donepezil and other cholinesterase inhibitors on acetylcholinesterase and butyrylcholinesterase in vitro. *Methods Find Exp Clin Pharmacol* 22 (8): 609–613.
12 Imbimbo, B.P. (2001). Pharmacodynamic-tolerability relationships of cholinesterase inhibitors for Alzheimer's disease. *CNS Drugs* 15 (5): 375–390.
13 Jann, M.W., Shirley, K.L., and Small, G.W. (2002). Clinical pharmacokinetics and pharmacodynamics of cholinesterase inhibitors. *Clin Pharmacokinet* 41 (10): 719–739.
14 Kryger, G., Silman, I., and Sussman, J.L. (1999). Structure of acetylcholinesterase complexed with E2020 (Aricept): implications for the design of new anti-Alzheimer drugs. *Structure* 7 (3): 297–307.
15 Cheung, J., Rudolph, M.J., Burshteyn, F. et al. (2012). Structures of human acetylcholinesterase in complex with pharmacologically important ligands. *J Med Chem* 55 (22): 10282–10286.
16 Vitorović-Todorović, M.D., Koukoulitsa, C., Juranić, I.O. et al. (2014). Structural modifications of 4-aryl-4-oxo-2-aminylbutanamides and their acetyl- and butyrylcholinesterase inhibitory activity. Investigation of AChE-ligand interactions by docking calculations and molecular dynamics simulations. *European Journal of Medicinal Chemistry* 81: 158–175.
17 Lan, J.S., Zhang, T., Liu, Y. et al. (2017). Design, synthesis and biological activity of novel donepezil derivatives bearing N-benzyl pyridinium moiety as potent and dual binding site acetylcholinesterase inhibitors. *Eur J Med Chem* 133: 184–196.
18 Peauger, L., Azzouz, R., Gembus, V. et al. (2017). Donepezil-Based Central Acetylcholinesterase Inhibitors by Means of a "Bio-Oxidizable" Prodrug Strategy: Design, Synthesis, and in Vitro Biological Evaluation. *J. Med. Chem.* 60 (13): 5909–5926.
19 Grossberg, G.T. (2005). Effect of rivastigmine in the treatment of behavioral disturbances associated with dementia: review of neuropsychiatric impairment in Alzheimer's disease. *Curr Med Res Opin* 21 (10): 1631–1639.
20 Farlow, M.R. (2003). Update on rivastigmine. *Neurologist* 9 (5): 230–234.
21 Müller, T. (2007). Rivastigmine in the treatment of patients with Alzheimer's disease. *Neuropsychiatr. Dis. Treat.* 3 (2): 211–218.
22 Finkel, S.I. (2004). Effects of rivastigmine on behavioral and psychological symptoms of dementia in Alzheimer's disease. *Clin Ther* 26 (7): 980–990.
23 Bartorelli, L., Giraldi, C., Saccardo, M. et al. (2005). Effects of switching from an AChE inhibitor to a dual AChE-BuChE inhibitor in patients with Alzheimer's disease. *Curr Med Res Opin* 21 (11): 1809–1818.
24 Kandiah, N., Pai, M.C., Senanarong, V. et al. (2017). Rivastigmine: the advantages of dual inhibition of acetylcholinesterase and butyrylcholinesterase and its role in subcortical vascular dementia and Parkinson's disease dementia. *Clin Interv Aging* 12: 697–707.
25 Naik, R.S., Hartmann, J., Kiewert, C. et al. (2009). Effects of rivastigmine and donepezil on brain acetylcholine levels in acetylcholinesterase-deficient mice. *J Pharm Pharm Sci* 12 (1): 79–85.

26 Bar-On, P., Millard, C.B., Harel, M. et al. (2002). Kinetic and structural studies on the interaction of cholinesterases with the anti-Alzheimer drug rivastigmine. *Biochemistry* 41 (11): 3555–3564.

27 Lilienfeld, S. (2002). Galantamine--a novel cholinergic drug with a unique dual mode of action for the treatment of patients with Alzheimer's disease. *CNS Drug Rev* 8 (2): 159–176.

28 Razay, G. and Wilcock, G.K. (2008). Galantamine in Alzheimer's disease. *Expert Rev Neurother* 8 (1): 9–17.

29 Oka, M., Nakaaki, S., Negi, A. et al. (2016). Predicting the neural effect of switching from donepezil to galantamine based on single-photon emission computed tomography findings in patients with Alzheimer's disease. *Psychogeriatrics* 16 (2): 121–134.

30 Nakayama, S., Suda, A., Nakanishi, A. et al. (2017). Galantamine Response Associates with Agitation and the Prefrontal Cortex in Patients with Alzheimer's Disease. *J Alzheimers Dis* 57 (1): 267–273.

31 Kaufer, D.I., Borson, S., Kershaw, P., and Sadik, K. (2005). Reduction of caregiver burden in Alzheimer's disease by treatment with galantamine. *CNS Spectr* 10 (6): 481–488.

32 Greenblatt, H.M., Guillou, C., Guenard, D. et al. (2004). The complex of a bivalent derivative of galanthamine with torpedo acetylcholinesterase displays drastic deformation of the active-site gorge: implications for structure-based drug design. *J Am Chem Soc* 126 (47): 15405–15411.

33 Greenblatt, H.M., Kryger, G., Lewis, T. et al. (1999). Structure of acetylcholinesterase complexed with (−)-galanthamine at 2.3 A resolution. *FEBS Lett* 463 (3): 321–326.

34 Bartolucci, C., Haller, L.A., Jordis, U. et al. (2010). Probing *Torpedo californica* acetylcholinesterase catalytic gorge with two novel bis-functional galanthamine derivatives. *J Med Chem* 53 (2): 745–751.

35 Hansen, S.B. and Taylor, P. (2007). Galanthamine and non-competitive inhibitor binding to ACh-binding protein: evidence for a binding site on non-alpha-subunit interfaces of heteromeric neuronal nicotinic receptors. *J Mol Biol* 369 (4): 895–901.

36 Bartolucci, C., Perola, E., Pilger, C. et al. (2001). Three-dimensional structure of a complex of galanthamine (Nivalin) with acetylcholinesterase from *Torpedo californica*: implications for the design of new anti-Alzheimer drugs. *Proteins* 42 (2): 182–191.

37 Smit, A.B., Syed, N.I., Schaap, D. et al. (2001). A glia-derived acetylcholine-binding protein that modulates synaptic transmission. *Nature* 411 (6835): 261–268.

38 Grossberg, G.T., Pejovic, V., Miller, M.L., and Graham, S.M. (2009). Memantine therapy of behavioral symptoms in community-dwelling patients with moderate to severe Alzheimer's disease. *Dement Geriatr Cogn Disord* 27 (2): 164–172.

39 Grossberg, G. T., Edwards, K. R., Zhao, Q. (2006). Rationale for combination therapy with galantamine and memantine in Alzheimer's disease. *J Clin Pharmacol,* 46 (7 Suppl 1), 17s-26s

40 Porsteinsson, A.P., Grossberg, G.T., Mintzer, J., and Olin, J.T. (2008). Memantine treatment in patients with mild to moderate Alzheimer's disease already

receiving a cholinesterase inhibitor: a randomized, double-blind, placebo-controlled trial. *Curr Alzheimer Res* 5 (1): 83–89.
41 Ulens, C., Spurny, R., Thompson, A.J. et al. (2014). The prokaryote ligand-gated ion channel ELIC captured in a pore blocker-bound conformation by the Alzheimer's disease drug memantine. *Structure* 22 (10): 1399–1407.
42 Hilf, R.J. and Dutzler, R. (2008). X-ray structure of a prokaryotic pentameric ligand-gated ion channel. *Nature* 452 (7185): 375–379.
43 Song, X., Jensen, M.O., Jogini, V. et al. (2018). Mechanism of NMDA receptor channel block by MK-801 and memantine. *Nature* 556 (7702): 515–519.
44 Kageyama, T., Nakamura, M., Matsuo, A. et al. (2000). The 4F2hc/LAT1 complex transports L-DOPA across the blood-brain barrier. *Brain Res* 879 (1–2): 115–121.
45 Burkhard, P., Dominici, P., Borri-Voltattorni, C. et al. (2001). Structural insight into Parkinson's disease treatment from drug-inhibited DOPA decarboxylase. *Nat Struct Biol* 8 (11): 963–967.
46 Huot, P., Johnston, T.H., Koprich, J.B. et al. (2013). The pharmacology of L-DOPA-induced dyskinesia in Parkinson's disease. *Pharmacol. Rev.* 65 (1): 171–222.
47 Niccolini, F., Rocchi, L., and Politis, M. (2015). Molecular imaging of levodopa-induced dyskinesias. *Cell. Mol. Life Sci.* 72 (11): 2107–2117.
48 Bargiotas, P. and Konitsiotis, S. (2013). Levodopa-induced dyskinesias in Parkinson's disease: emerging treatments. *Neuropsychiatr. Dis. Treat.* 9: 1605–1617.
49 Al Dakheel, A., Beaulieu-Boire, I., and Fox, S.H. (2014). Emerging drugs for levodopa-induced dyskinesia. *Expert Opin. Emerging Drugs* 19 (3): 415–429.
50 Morin, N., Morissette, M., Gregoire, L., and Di Paolo, T. (2016). mGlu5, Dopamine D2 and Adenosine A2A Receptors in L-DOPA-induced Dyskinesias. *Curr. Neuropharmacol.* 14 (5): 481–493.
51 Navailles, S., Lagiere, M., Contini, A., and De Deurwaerdere, P. (2013). Multisite Intracerebral Microdialysis to Study the Mechanism of L-DOPA Induced Dopamine and Serotonin Release in the Parkinsonian Brain. *ACS Chem. Neurosci.* 4 (5): 680–692.
52 De Deurwaerdere, P., Di Giovanni, G., Millan, M. J. (2017). Expanding the repertoire of L-DOPA's actions: A comprehensive review of its functional neurochemistry. *Prog. Neurobiol.* 151 (4): 57–100.
53 Latt, M.D., Lewis, S., Zekry, O., and Fung, V.S.C. (2019). Factors to Consider in the Selection of Dopamine Agonists for Older Persons with Parkinson's Disease. *Drugs Aging* 36 (3): 189–202.
54 Newman-Tancredi, A., Cussac, D., Audinot, V. et al. (2002). Differential actions of antiparkinson agents at multiple classes of monoaminergic receptor. II. Agonist and antagonist properties at subtypes of dopamine D(2)-like receptor and alpha(1)/alpha(2)-adrenoceptor. *J Pharmacol Exp Ther* 303 (2): 805–814.
55 Newman-Tancredi, A., Cussac, D., Quentric, Y. et al. (2002). Differential actions of antiparkinson agents at multiple classes of monoaminergic receptor. III. Agonist and antagonist properties at serotonin, 5-HT(1) and 5-HT(2), receptor subtypes. *J Pharmacol Exp Ther* 303 (2): 815–822.

56 Millan, M.J., Peglion, J.L., Vian, J. et al. (1995). Functional correlates of dopamine D3 receptor activation in the rat in vivo and their modulation by the selective antagonist, (+)-S 14297: 1. Activation of postsynaptic D3 receptors mediates hypothermia, whereas blockade of D2 receptors elicits prolactin secretion and catalepsy. *J Pharmacol Exp Ther* 275 (2): 885–898.
57 Millan, M.J., Cussac, D., Milligan, G. et al. (2001). Antiparkinsonian Agent Piribedil Displays Antagonist *Properties at Native, Rat, and Cloned, Human* α_2 *– Adrenoceptors: Cellular and Functional Characterization.* 297 (3): 876–887.
58 Millan, M.J. (2010). From the cell to the clinic: A comparative review of the partial D2/D3 receptor agonist and α2-adrenoceptor antagonist, piribedil, in the treatment of Parkinson's disease. *Pharmacol. Ther.* 128 (2): 229–273.
59 Perez-Lloret, S. and Rascol, O. (2016). Piribedil for the Treatment of Motor and Non-motor Symptoms of Parkinson Disease. *CNS Drugs* 30 (8): 703–717.
60 Wang, H.T., Wang, L., He, Y., and Yu, G. (2018). Rotigotine transdermal patch for the treatment of neuropsychiatric symptoms in Parkinson's disease: A meta-analysis of randomized placebo-controlled trials. *J Neurol Sci* 393: 31–38.
61 Zhang, P., Li, Y., Nie, K. et al. (2018). Hypotension and bradycardia, a serious adverse effect of piribedil, a case report and literature review. *BMC Neurol* 18 (1): 221.
62 Bennett, J.P. Jr. and Piercey, M.F. (1999). Pramipexole--a new dopamine agonist for the treatment of Parkinson's disease. *J Neurol Sci* 163 (1): 25–31.
63 Piercey, M.F. (1998). Pharmacology of pramipexole, a dopamine D3-preferring agonist useful in treating Parkinson's disease. *Clin. Neuropharmacol.* 21 (3): 141–151.
64 Piercey, M.F., Hoffmann, W.E., Smith, M.W., and Hyslop, D.K. (1996). Inhibition of dopamine neuron firing by pramipexole, a dopamine D3 receptor-preferring agonist: comparison to other dopamine receptor agonists. *Eur J Pharmacol* 312 (1): 35–44.
65 Hubble, J.P. (2000). Pre-clinical studies of pramipexole: clinical relevance. *Eur J Neurol* 7 (Suppl 1): 15–20.
66 Chernoloz, O., El Mansari, M., and Blier, P. (2009). Sustained administration of pramipexole modifies the spontaneous firing of dopamine, norepinephrine, and serotonin neurons in the rat brain. *Neuropsychopharmacology* 34 (3): 651–661.
67 Schulte-Herbrueggen, O., Vogt, M.A., Hoertnagl, H. et al. (2012). Pramipexole is active in depression tests and modulates monoaminergic transmission, but not brain levels of BDNF in mice. *Eur. J. Pharmacol.* 677 (1–3): 77–86.
68 Collo, G., Cavalleri, L., Bono, F. et al. (2018). Ropinirole and Pramipexole Promote Structural Plasticity in Human iPSC-Derived Dopaminergic Neurons via BDNF and mTOR Signaling. *Neural Plast* 2018: https://doi.org/10.1155/2018/4196961.
69 Castro-Hernández, J., Afonso-Oramas, D., Cruz-Muros, I. et al. (2015). Prolonged treatment with pramipexole promotes physical interaction of striatal dopamine D3 autoreceptors with dopamine transporters to reduce dopamine uptake. *Neurobiol. Dis.* 74: 325–335.

70 Rascol, O., Lees, A.J., Senard, J.-M. et al. (1996). A placebo-controlled study of ropinirole, a new D2 agonist, in the treatment of motor fluctuations of L-DOPA-treated Parkinsonian patients. *Adv. Neurol.* 69 (Parkinson's Disease): 531–534.

71 Jost, W.H. and Angersbach, D. (2005). Ropinirole, a non-ergoline dopamine agonist. *CNS Drug Rev* 11 (3): 253–272.

72 Shill, H.A. and Stacy, M. (2009). Update on ropinirole in the treatment of Parkinson's disease. *Neuropsychiatr. Dis. Treat.* 5: 33–36.

73 Tanaka, K.-I., Miyazaki, I., Fujita, N. et al. (2001). Molecular mechanism in activation of glutathione system by ropinirole, a selective dopamine D2 agonist. *Neurochem. Res.* 26 (1): 31–36.

74 Sanford, M. and Scott, L.J. (2011). Rotigotine transdermal patch: a review of its use in the treatment of Parkinson's disease. *CNS Drugs* 25 (8): 699–719.

75 Nomoto, M., Iwaki, H., Kondo, H., and Sakurai, M. (2018). Efficacy and safety of rotigotine in elderly patients with Parkinson's disease in comparison with the non-elderly: a post hoc analysis of randomized, double-blind, placebo-controlled trials. *J Neurol* 265 (2): 253–265.

76 Oka, H., Nakahara, A., and Umehara, T. (2018). Rotigotine Improves Abnormal Circadian Rhythm of Blood Pressure in Parkinson's Disease. *Eur Neurol* 79 (5–6): 281–286.

77 Scheller, D., Ullmer, C., Berkels, R. et al. (2009). The in vitro receptor profile of rotigotine: a new agent for the treatment of Parkinson's disease. *Naunyn Schmiedebergs Arch Pharmacol* 379 (1): 73–86.

78 Adachi, N., Yoshimura, A., Chiba, S. et al. (2018). Rotigotine, a dopamine receptor agonist, increased BDNF protein levels in the rat cortex and hippocampus. *Neurosci Lett* 662: 44–50.

79 Subramony, J.A. (2006). Apomorphine in Dopaminergic Therapy. *Mol. Pharm.* 3 (4): 380–385.

80 Menon, R. and Stacy, M. (2007). Apomorphine in the treatment of Parkinson's disease. *Expert Opin. Pharmacother.* 8 (12): 1941–1950.

81 Titova, N. and Chaudhuri, K.R. (2016). Apomorphine therapy in Parkinson's and future directions. *Parkinsonism Relat Disord* 33: **Suppl 1** S56-s60.

82 Neef, C. and van Laar, T. (1999). Pharmacokinetic-pharmacodynamic relationships of apomorphine in patients with Parkinson's disease. *Clin Pharmacokinet* 37 (3): 257–271.

83 Durdagi, S., Salmas, R.E., Stein, M. et al. (2016). Binding Interactions of Dopamine and Apomorphine in D2High and D2Low States of Human Dopamine D2 Receptor Using Computational and Experimental Techniques. *ACS Chem. Neurosci.* 7 (2): 185–195.

84 Michael Besser, G., Pfeiffer, R.F., and Thorner, M.O. (2018). ANNIVERSARY REVIEW: 50 years since the discovery of bromocriptine. *Eur J Endocrinol* 179 (2): R69–r75.

85 Roehrich, H., Dackis, C.A., and Gold, M.S. (1987). Bromocriptine. *Med Res Rev* 7 (2): 243–269.

86 Lieberman, A.N. and Goldstein, M. (1985). Bromocriptine in Parkinson disease. *Pharmacol Rev* 37 (2): 217–227.

87 de Leeuw van Weenen, J.E., Parlevliet, E.T., Maechler, P. et al. (2010). The dopamine receptor D2 agonist bromocriptine inhibits glucose-stimulated insulin secretion by direct activation of the alpha2-adrenergic receptors in beta cells. *Biochem Pharmacol* 79 (12): 1827–1836.

88 Maj, J., Gancarczyk, L., and Rawlow, A. (1977). The influence of bromocriptine on serotonin neurons. *J Neural Transm* 41 (4): 253–264.

89 Lew, J.Y., Hata, F., Ohashi, T., and Goldstein, M. (1977). The interactions of bromocriptine and lergotrile with dopamine and alpha-adrenergic receptors. *J Neural Transm* 41 (2–3): 109–121.

90 Shirasaki, Y., Sugimura, M., and Sato, T. (2010). Bromocriptine, an ergot alkaloid, inhibits excitatory amino acid release mediated by glutamate transporter reversal. *Eur J Pharmacol* 643 (1): 48–57.

91 Kvernmo, T., Hartter, S., and Burger, E. (2006). A review of the receptor-binding and pharmacokinetic properties of dopamine agonists. *Clin Ther* 28 (8): 1065–1078.

92 Sevrioukova, I.F. and Poulos, T.L. (2012). Structural and mechanistic insights into the interaction of cytochrome P4503A4 with bromoergocryptine, a type I ligand. *J Biol Chem* 287 (5): 3510–3517.

93 Lim, D.-Y., Lee, Y.-G., and Kim, I.-H. (2002). Inhibitory mechanism of bromocriptine on catecholamine release evoked by cholinergic stimulation and membrane depolarization from the rat adrenal medulla. *Arch. Pharmacal Res.* 25 (4): 511–521.

94 Tipton, K.F., Boyce, S., O'Sullivan, J. et al. (2004). Monoamine oxidases: certainties and uncertainties. *Curr Med Chem* 11 (15): 1965–1982.

95 Youdim, M.B. and Weinstock, M. (2004). Therapeutic applications of selective and non-selective inhibitors of monoamine oxidase A and B that do not cause significant tyramine potentiation. *NeuroToxicology* 25 (1–2): 243–250.

96 Bach, A.W., Lan, N.C., Johnson, D.L. et al. (1988). cDNA cloning of human liver monoamine oxidase A and B: molecular basis of differences in enzymatic properties. *Proc Natl Acad Sci USA* 85.

97 Abell, C.W. and Kwan, S.W. (2001). Molecular characterization of monoamine oxidases A and B. *Prog Nucleic Acid Res Mol Biol* 65: 129–156.

98 Fernandez, H.H. and Chen, J.J. (2007). Monoamine oxidase inhibitors: current and emerging agents for Parkinson disease. *Clin Neuropharmacol* 30 (3): 150–168.

99 Carradori, S. and Silvestri, R. (2015). New Frontiers in Selective Human MAO-B Inhibitors. *J Med Chem* 58 (17): 6717–6732.

100 Dezsi, L. and Vecsei, L. (2017). Monoamine Oxidase B Inhibitors in Parkinson's Disease. *CNS Neurol. Disord.: Drug Targets* 16 (4): 425–439.

101 Yiannakis, A. and Konitsiotis, S. (2016). The improvement of MAO-B inhibitors in Parkinson's disease is clinically irrelevant: a review. *Int. J. Clin, Neurosci, Mental Health* 3 (3): S05.

102 Riederer, P. and Muller, T. (2018). Monoamine oxidase-B inhibitors in the treatment of Parkinson's disease: clinical-pharmacological aspects. *J Neural Transm* 125 (11): 1751–1757.

103 Alborghetti, M. and Nicoletti, F. (2018). Different generations of type-B monoamine oxidase inhibitors in Parkinson's disease: from bench to bedside. *Curr Neuropharmacol* Ahead of print. DOI: 10.2174/1570159X16666180830100754.

104 Heinonen, E.H. and Lammintausta, R. (1991). A review of the pharmacology of selegiline. *Acta Neurol Scand Suppl* (136): 44–59.

105 Birkmayer, W., Riederer, P., and Youdim, M.B. (1982). (−)Deprenyl in the treatment of Parkinson's disease. *Clin Neuropharmacol* 5 (2): 195–230.

106 Miklya, I. (2016). The significance of selegiline/(−)-deprenyl after 50 years in research and therapy (1965-2015). *Mol. Psychiatry* 21 (11): 1499–1503.

107 Wessel, K. and Szelenyi, I. (1992). Selegiline--an overview of its role in the treatment of Parkinson's disease. *Clin Investig* 70 (5): 459–462.

108 Ebadi, M., Brown-Borg, H., Ren, J. et al. (2006). Therapeutic efficacy of selegiline in neurodegenerative disorders and neurological diseases. *Curr. Drug Targets* 7 (11): 1513–1529.

109 Ebadi, M., Sharma, S., Shavali, S., and El Refaey, H. (2002). Neuroprotective actions of selegiline. *J. Neurosci. Res.* 67 (3): 285–289.

110 Youdim, M.B. (1986). Pharmacology of MAO B inhibitors: mode of action of (−)deprenyl in Parkinson's disease. *J Neural Transm Suppl* 22: 91–105.

111 Gerlach, M., Youdim, M.B., and Riederer, P. (1996). Pharmacology of selegiline. *Neurology* 47 (6 Suppl 3): S137–S145.

112 Magyar, K. (2011). The pharmacology of selegiline. *Int Rev Neurobiol* 100: 65–84.

113 Kalasz, H., Magyar, K., Szoke, E. et al. (2014). Metabolism of selegiline [(−)-deprenyl]. *Curr Med Chem* 21 (13): 1522–1530.

114 Kumar, S., Dang, S., Nigam, K. et al. (2018). Selegiline Nanoformulation in Attenuation of Oxidative Stress and Upregulation of Dopamine in the Brain for the Treatment of Parkinson's Disease. *Rejuvenation Res* 21 (5): 464–476.

115 Sridhar, V., Gaud, R., Bajaj, A., and Wairkar, S. (2018). Pharmacokinetics and pharmacodynamics of intranasally administered selegiline nanoparticles with improved brain delivery in Parkinson's disease. *Nanomedicine* 14 (8): 2609–2618.

116 Binde, C.D., Tvete, I.F., Gasemyr, J. et al. (2018). A multiple treatment comparison meta-analysis of monoamine oxidase type B inhibitors for Parkinson's disease. *Br J Clin Pharmacol* 84 (9): 1917–1927.

117 Guay, D.R. (2006). Rasagiline (TVP-1012): a new selective monoamine oxidase inhibitor for Parkinson's disease. *Am J Geriatr Pharmacother* 4 (4): 330–346.

118 Chen, J.J., Swope, D.M., and Dashtipour, K. (2007). Comprehensive review of rasagiline, a second-generation monoamine oxidase inhibitor, for the treatment of Parkinson's disease. *Clin Ther* 29 (9): 1825–1849.

119 Rabey, J.M., Sagi, I., Huberman, M. et al. (2000). Rasagiline mesylate, a new MAO-B inhibitor for the treatment of Parkinson's disease: a double-blind study as adjunctive therapy to levodopa. *Clin Neuropharmacol* 23 (6): 324–330.

120 Chen, J.J. and Wilkinson, J.R. (2012). The monoamine oxidase type B inhibitor rasagiline in the treatment of Parkinson disease: is tyramine a challenge? *J Clin Pharmacol* 52 (5): 620–628.

121 Lecht, S., Haroutiunian, S., Hoffman, A., and Lazarovici, P. (2007). Rasagiline - a novel MAO B inhibitor in Parkinson's disease therapy. *Therapeutics and clinical risk management* 3 (3): 467–474.

122 Binda, C., Hubalek, F., Li, M. et al. (2004). Crystal Structures of Monoamine Oxidase B in Complex with Four Inhibitors of the N-Propargylaminoindan Class. *J. Med. Chem.* 47 (7): 1767–1774.

123 Binda, C., Hubalek, F., Li, M. et al. (2005). Binding of rasagiline-related inhibitors to human monoamine oxidases: a kinetic and crystallographic analysis. *J Med Chem* 48 (26): 8148–8154.

124 Deeks, E.D. (2015). Safinamide: First Global Approval. *Drugs* 75 (6): 705–711.

125 Caccia, C., Maj, R., Calabresi, M. et al. (2006). Safinamide: from molecular targets to a new anti-Parkinson drug. *Neurology* 67 (7 Suppl 2): S18–S23.

126 Fabbri, M., Rosa, M.M., Abreu, D., and Ferreira, J.J. (2015). Clinical pharmacology review of safinamide for the treatment of Parkinson's disease. *Neurodegener Dis Manag* 5 (6): 481–496.

127 Salvati, P., Maj, R., Caccia, C. et al. (1999). Biochemical and electrophysiological studies on the mechanism of action of PNU-151774E, a novel antiepileptic compound. *J Pharmacol Exp Ther* 288 (3): 1151–1159.

128 Pevarello, P., Bonsignori, A., Caccia, C. et al. (1999). Sodium channel activity and sigma binding of 2-aminopropanamide anticonvulsants. *Bioorg Med Chem Lett* 9 (17): 2521–2524.

129 Kandadai, R.M., Jabeen, S.A., Kanikannan, M.A., and Borgohain, R. (2014). Safinamide for the treatment of Parkinson's disease. *Expert Rev Clin Pharmacol* 7 (6): 747–759.

130 Muller, T. (2017). Pharmacokinetic drug evaluation of safinamide mesylate for the treatment of mid-to-late stage Parkinson's disease. *Expert Opin Drug Metab Toxicol* 13 (6): 693–699.

131 Blair, H.A. and Dhillon, S. (2017). Safinamide: A Review in Parkinson's Disease. *CNS Drugs* 31 (2): 169–176.

132 Stocchi, F. and Torti, M. (2016). Adjuvant therapies for Parkinson's disease: critical evaluation of safinamide. *Drug Des Devel Ther* 10: 609–618.

133 deSouza, R.M. and Schapira, A. (2017). Safinamide for the treatment of Parkinson's disease. *Expert Opin Pharmacother* 18 (9): 937–943.

134 Teixeira, F.G., Gago, M.F., Marques, P. et al. (2018). Safinamide: a new hope for Parkinson's disease? *Drug Discov Today* 23 (3): 736–744.

135 Binda, C., Wang, J., Pisani, L. et al. (2007). Structures of Human Monoamine Oxidase B Complexes with Selective Noncovalent Inhibitors: Safinamide and Coumarin Analogs. *J. Med. Chem.* 50 (23): 5848–5852.

136 Mannisto, P.T. and Kaakkola, S. (1999). Catechol-O-methyltransferase (COMT): biochemistry, molecular biology, pharmacology, and clinical efficacy of the new selective COMT inhibitors. *Pharmacol Rev* 51 (4): 593–628.

137 Bonifacio, M.J., Palma, P.N., Almeida, L., and Soares-da-Silva, P. (2007). Catechol-O-methyltransferase and its inhibitors in Parkinson's disease. *CNS Drug Rev* 13 (3): 352–379.

138 Muller, T. (2015). Catechol-O-methyltransferase inhibitors in Parkinson's disease. *Drugs* 75 (2): 157–174.

139 Nissinen, E. and Mannisto, P.T. (2010). Biochemistry and pharmacology of catechol-O-methyltransferase inhibitors. *Int Rev Neurobiol* 95: 73–118.

140 Truong, D.D. (2009). Tolcapone: review of its pharmacology and use as adjunctive therapy in patients with Parkinson's disease. *Clin. Interventions Aging* 4 109–4 113.

141 Grunig, D., Felser, A., Bouitbir, J., and Krahenbuhl, S. (2017). The catechol-O-methyltransferase inhibitors tolcapone and entacapone uncouple and inhibit the mitochondrial respiratory chain in HepaRG cells. *Toxicol In Vitro* 42: 337–347.

142 Ellermann, M., Lerner, C., Burgy, G. et al. (2012). Catechol-O-methyltransferase in complex with substituted 3′-deoxyribose bisubstrate inhibitors. *Acta Crystallogr D Biol Crystallogr* 68 (Pt 3: 253–260.

143 Ehler, A., Benz, J., Schlatter, D., and Rudolph, M.G. (2014). Mapping the conformational space accessible to catechol-O-methyltransferase. *Acta Crystallogr D Biol Crystallogr* 70 (Pt 8: 2163–2174.

144 Fabbri, M., Ferreira, J.J., Lees, A. et al. (2018). Opicapone for the treatment of Parkinson's disease: A review of a new licensed medicine. *Mov Disord* 33 (10): 1528–1539.

145 Fabbri, M., Rosa, M.M., and Ferreira, J.J. (2016). Clinical pharmacology review of opicapone for the treatment of Parkinson's disease. *Neurodegener Dis Manag* 6 (5): 349–362.

146 Lees, A.J., Ferreira, J., Rascol, O. et al. (2017). Opicapone as Adjunct to Levodopa Therapy in Patients With Parkinson Disease and Motor Fluctuations: A Randomized Clinical Trial. *JAMA Neurol* 74 (2): 197–206.

147 Lees, A.J., Ferreira, J., Rascol, O. et al. (2017). Opicapone for the management of end-of-dose motor fluctuations in patients with Parkinson's disease treated with L-DOPA. *Expert Rev Neurother* 17 (7): 649–659.

148 Scott, L.J. (2016). Opicapone: A Review in Parkinson's Disease. *Drugs* 76 (13): 1293–1300.

149 Rinne, U.K., Sonninen, V., and Siirtola, T. (1972). Treatment of Parkinson's disease with amantadine and L-Dopa. *Eur Neurol* 7 (4): 228–240.

150 Verhagen Metman, L., Del Dotto, P., van den Munckhof, P. et al. (1998). Amantadine as treatment for dyskinesias and motor fluctuations in Parkinson's disease. *Neurology* 50 (5): 1323–1326.

151 Elkurd, M.T., Bahroo, L.B., and Pahwa, R. (2018). The role of extended-release amantadine for the treatment of dyskinesia in Parkinson's disease patients. *Neurodegener Dis Manag* 8 (2): 73–80.

152 Perez-Lloret, S. and Rascol, O. (2018). Efficacy and safety of amantadine for the treatment of L-DOPA-induced dyskinesia. *J Neural Transm (Vienna)* 125 (8): 1237–1250.

153 Von Voigtlander, P.F. and Moore, K.E. (1971). Dopamine: release from the brain in vivo by amantadine. *Science* 174 (4007): 408–410.

154 Moresco, R.M., Volonte, M.A., Messa, C. et al. (2002). New perspectives on neurochemical effects of amantadine in the brain of parkinsonian patients: a PET - [(11)C]raclopride study. *J Neural Transm* 109 (10): 1265–1274.

155 Peeters, M., Page, G., Maloteaux, J.M., and Hermans, E. (2002). Hypersensitivity of dopamine transmission in the rat striatum after treatment with the NMDA receptor antagonist amantadine. *Brain Res* 949 (1–2): 32–41.

156 Stoof, J.C., Booij, J., Drukarch, B., and Wolters, E.C. (1992). The anti-parkinsonian drug amantadine inhibits the N-methyl-D-aspartic acid-evoked release of acetylcholine from rat neostriatum in a non-competitive way. *Eur J Pharmacol* 213 (3): 439–443.

157 Nikolaus, S., Wittsack, H.J., Beu, M. et al. (2019). Amantadine enhances nigrostriatal and mesolimbic dopamine function in the rat brain in relation to motor and exploratory activity. *Pharmacol Biochem Behav* .

158 Wild, R., Pettit, T., Burns, A. (2003). Cholinesterase inhibitors for dementia with Lewy bodies. *Cochrane Database Syst Rev,* (3), Cd003672

159 Rolinski, M., Fox, C., Maidment, I., McShane, R. (2012). Cholinesterase inhibitors for dementia with Lewy bodies, Parkinson's disease dementia and cognitive impairment in Parkinson's disease. *Cochrane Database Syst Rev,* (3), Cd006504

160 Szigeti, K. and Hafeez, M.U. (2015). Exploring the role of donepezil in dementia with Lewy bodies. *Drugs Today (Barc)* 51 (10): 579–590.

161 Cummings, J., Lai, T.J., Hemrungrojn, S. et al. (2016). Role of Donepezil in the Management of Neuropsychiatric Symptoms in Alzheimer's Disease and Dementia with Lewy Bodies. *CNS Neurosci Ther* 22 (3): 159–166.

162 Gabelli, C. (2003). Rivastigmine: an update on therapeutic efficacy in Alzheimer's disease and other conditions. *Curr Med Res Opin* 19 (2): 69–82.

Index

Chemical Biology of Neurodegeneration: A Molecular Approach, First Edition. Pedro Merino.

dihydroxyphenylalanine (DOPA) 3, 230, 249, 331, 333